String-Math 2022

Proceedings of Symposia in
PURE MATHEMATICS

Volume 107

String-Math 2022

String-Math 2022
July 11–15, 2022
University of Warsaw, Warsaw, Poland

Ron Donagi
Adrian Langer
Piotr Sułkowski
Katrin Wendland
Editors

AMERICAN MATHEMATICAL SOCIETY
Providence, Rhode Island

2020 *Mathematics Subject Classification.* Primary 81-06, 14-06, 57-06, 17-06, 00B25.

Library of Congress Cataloging-in-Publication Data

Names: String-Math (Conference) (2012 : Warsaw, Poland), author. | Donagi, Ron, editor. | Langer, Adrian, 1973- editor. | Sułkowski, Piotr (Physicist), editor. | Wendland, Katrin, 1970- editor.

Title: String-Math 2022 / Ron Donagi, Adrian Langer, Piotr Sułkowski, Katrin Wendland, editors.

Description: Providence, Rhode Island : American Mathematical Society, [2024] | Series: Proceedings of symposia in pure mathematics, 0082-0717 ; volume 107 | "July 11-15, 2022, University of Warsaw, Warsaw, Poland." | Includes bibliographical references.

Identifiers: LCCN 2023053391 | ISBN 9781470472405 (paperback) | ISBN 9781470476755 (ebook)

Subjects: LCSH: Geometry, Algebraic–Congresses. | Quantum theory–Mathematics–Congresses. | AMS: Quantum theory – Proceedings, conferences, collections, etc.. | Algebraic geometry – Proceedings, conferences, collections, etc.. | Manifolds and cell complexes – Proceedings, conferences, collections, etc.. | Nonassociative rings and algebras – Proceedings, conferences, collections, etc.. | General – Conference proceedings and collections of papers – Proceedings of conferences of miscellaneous specific interest.

Classification: LCC QA564 .S77 2024 | DDC 516.3/5–dc23/eng/20240116

LC record available at https://lccn.loc.gov/2023053391

♾ The paper used in this book is acid-free and falls within the guidelines
established to ensure permanence and durability.
Visit the AMS home page at `https://www.ams.org/`

10 9 8 7 6 5 4 3 2 1 29 28 27 26 25 24

Contents

Preface

This is a proceedings volume from the String-Math conference which took place at the University of Warsaw in 2022. In the past years the interactions between mathematics and high energy theoretical physics have been rapidly developing. Among others, they resulted in the rise of a new subfield, which is often referred to as physical mathematics. Quantum field theory and string theory play an important role in these developments, in particular as a source of often unexpected conjectures at the frontier of mathematics. Without a physical insight, many of these conjectures would not exist. The series of String-Math conferences is the natural and most prestigious venue, where such developments can be presented and discussed altogether by both physicists and mathematicians.

The String-Math conference in 2022 focused on several research areas actively developing these days. They included generalized (categorical) symmetries in quantum field theory and their relation to topological phases of matter; formal aspects of quantum field theory, in particular twisted holography; various developments in supersymmetric gauge theories, BPS counting and Donaldson-Thomas invariants. Other topics discussed included new advances in Gromov-Witten theory, curve counting, and Calabi-Yau manifolds. Another broad topic concerned algebraic aspects of conformal field theory, vertex operator algebras, and quantum groups. Furthermore, several other recent developments were presented during the conference, such as understanding the role of operator algebras in the presence of gravity, derivation of gauge-string duality, complexity of black holes, or mathematical aspects of the amplituhedron. We expect that all these topics will be actively developed also in the forthcoming future. This proceedings volume contains articles summarizing 14 conference lectures, devoted to some of the above topics.

The conference in Warsaw was the 12th conference in the String-Math series. These conferences take place annually, and previously were hosted by: University of Pennsylvania, University of Bonn, Simons Center for Geometry and Physics (Stony Brook), University of Alberta, Tsinghua University, College de France, University of Hamburg, Tohoku University, Uppsala Univeristy, Universities of Stellenbosch and Cape Town, IMPA (Rio de Janeiro); the next two meetings are taking place in Melbourne (2023) and Trieste (2024). Over the last decade the series of String-Math conferences has developed into a central event on the interface between mathematics and physics related to string theory, quantum field theory and related topics. The conference in Warsaw provided a special opportunity for community members to convene in person following two prior online conferences held during the Covid pandemic.

Proceedings of Symposia in Pure Mathematics
Volume **107**, 2024
https://doi.org/10.1090/pspum/107/01942

Quiver DT invariants and log Gromov–Witten theory of toric varieties

Hülya Argüz

ABSTRACT. We review how log Gromov–Witten invariants of toric varieties can be used to express quiver Donaldson–Thomas invariants in terms of the simpler attractor Donaldson–Thomas invariants. This is an exposition of joint work with Pierrick Bousseau [Compos. Math., 158(2022), 2206–2249], [Argüz and Bousseau, *Quivers, flow trees, and log curves*, arXiv:2302.02068].

CONTENTS

0. Introduction

Donaldson-Thomas (DT) invariants of a quiver with potential can be expressed in terms of simpler attractor DT invariants by a universal formula. The coefficients in this formula are calculated combinatorially using attractor flow trees. In this paper, we give an introduction to the main result of [**8**] showing that these coefficients are genus 0 log Gromov–Witten invariants of d-dimensional toric varieties, where d is the number of vertices of the quiver.

1. Quiver DT invariants

1.1. Quiver representations and moduli spaces. A **quiver** is a finite oriented graph Q. We denote by

Q_0: the finite set of vertices of Q,
Q_1: the finite set of arrows (oriented edges) of Q,
$s : Q_1 \to Q_0$ the source map which maps an arrow to its source, and;
$t : Q_1 \to Q_0$ the target map, which maps an arrow to its target.

We illustrate a quiver in Figure 1.1.

DEFINITION 1.1. A **representation** of a quiver Q is an assignment of

(i) a finite-dimensional vector space V_i over $\mathbb{C}$ for each vertex $i \in Q_0$, and
(ii) a $\mathbb{C}$-linear transformation $f_\alpha \in \mathrm{Hom}(V_{s(\alpha)}, V_{t(\alpha)})$ for each arrow $\alpha \in Q_1$.

The research of Hülya Argüz is partially supported by the NSF grant DMS-2302116.

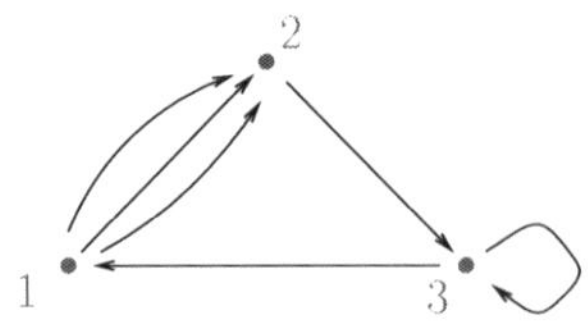

FIGURE 1.1. A quiver

Representations of Q naturally form an abelian category: a morphism between (V_i, f_α) and (V_i', f_α') is the data of linear maps $V_i \to V_i'$ intertwining the maps f_α and f_α'. In what follows, we fix

$$N = \mathbb{Z}^{Q_0} = \bigoplus_{i \in Q_0} \mathbb{Z} e_i \,,$$

and $N^\oplus = \mathbb{N}^{Q_0} \subset N$. The **dimension** of a quiver representation (V_i, f_α) is defined as the vector

$$\gamma = (\gamma_i)_{i \in Q_0} \in N^\oplus \subset N \,,$$

where $\gamma_i := \dim V_i$ – see Figure 1.2 for an illustration.

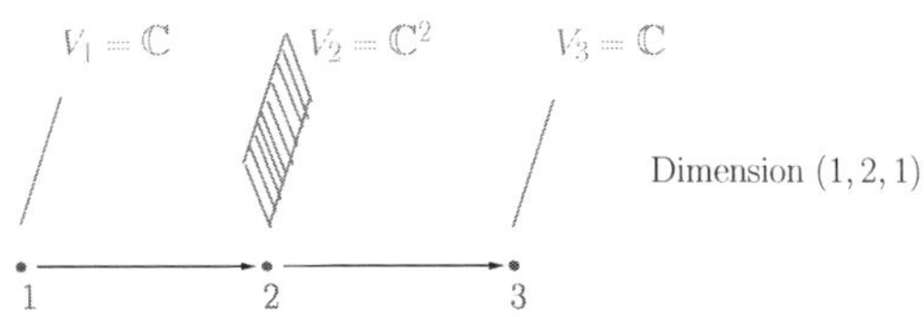

FIGURE 1.2. A quiver with dimension vector $\gamma = (1, 2, 1)$

There is a notion of stability due to King in the context of quiver representations, defined as follows [**32**]. Denote $M := \mathrm{Hom}(N, \mathbb{Z})$ the dual lattice to N and set

$$M_\mathbb{R} = \mathrm{Hom}(N, \mathbb{R}) = M \otimes \mathbb{R}.$$

DEFINITION 1.2 (King's stability). A **stability parameter** for $\gamma \in N$ is a point $\theta \in \gamma^\perp := \{\theta \in M_\mathbb{R} \,, \theta(\gamma) = 0\} \subset M_\mathbb{R}$. A quiver representation V of dimension γ is **θ-stable** (resp. **θ-semistable**) if for all non-zero strict subrepresentation V' of V we have $\theta(\dim(V')) < 0$ (resp. $\theta(\dim(V')) \leq 0$).

Given a dimension vector $\gamma \in N^\oplus$ and a stability parameter $\theta \in M_\mathbb{R}$, one obtains by geometric representation theory [**32**, Proposition 4.2] a moduli space $\mathcal{M}_\gamma^\theta$ of S-equivalence classes of θ-semistable quiver representations of Q of dimension γ. It is a quasi-projective variety over $\mathbb{C}$, which is actually projective if Q is acyclic.

When working with a non-acyclic quiver Q, we consider the additional data of a **potential** W, that is, a formal linear combination of oriented cycles of Q. A choice of potential W defines a **trace function** $\mathrm{Tr}(W)_\gamma^\theta : \mathcal{M}_\gamma^\theta \to \mathbb{C}$ on each of the moduli spaces $\mathcal{M}_\gamma^\theta$: if c is an oriented cycle of arrows $\alpha_1, \ldots, \alpha_n$ of Q, $\mathrm{Tr}(c)_\gamma^\theta : \mathcal{M}_\gamma^\theta \to \mathbb{C}$ is defined by

$$V = (V_i, f_\alpha) \longmapsto \mathrm{Tr}(f_{\alpha_n} \circ \ldots \circ f_{\alpha_1})$$

and we set

$$\mathrm{Tr}(W)^{\theta}_{\gamma} = \sum_{c} \lambda_c \mathrm{Tr}(c)^{\theta}_{\gamma}$$

if W is the linear combination $\sum_c \lambda_c c$ of oriented cycles.

1.2. DT invariants. Let (Q, W) be a quiver with potential, $\gamma \in N^{\oplus}$ a dimension vector, and θ a generic stability parameter for γ. The Donaldson-Thomas (DT) invariant $\Omega^{+,\theta}_{\gamma}$ is an integer which is a virtual count of the critical points of the trace function $\mathrm{Tr}(W)^{\theta}_{\gamma}$ on the moduli space $\mathcal{M}^{\theta}_{\gamma}$ of θ-semistable representations of dimension γ. We provide a more concrete description of this count in what follows.

In the situation when the θ-stable locus in $\mathcal{M}^{\theta}_{\gamma}$ is empty, we have $\Omega^{+,\theta}_{\gamma} = 0$, and else, $\Omega^{+,\theta}_{\gamma}$ is the Euler characteristic of $\mathcal{M}^{\theta}_{\gamma}$ valued in the perverse sheaf obtained by applying the vanishing-cycle functor for $\mathrm{Tr}(W)^{\theta}_{\gamma}$ to the intersection cohomology sheaf [**17, 18, 38**]:

$$\Omega^{+,\theta}_{\gamma} := e(M^{\theta}_{\gamma}, \phi_{\mathrm{Tr}(W)^{\theta}_{\gamma}}(IC[-\dim \mathcal{M}^{\theta}_{\gamma}])) \, .$$

For example, if γ is primitive and $W = 0$, then $\Omega^{+,\theta}_{\gamma}$ is simply the topological Euler characteristic[1] of $\mathcal{M}^{\theta}_{\gamma}$. The DT invariants $\Omega^{+,\theta}_{\gamma}$ can be repackaged into the **rational DT invariants**

$$\overline{\Omega}^{+,\theta}_{\gamma} := \sum_{\substack{\gamma' \in N \\ \gamma = k\gamma', \, k \in \mathbb{Z}_{\geq 1}}} \frac{(-1)^{k-1}}{k^2} \Omega^{+,\theta}_{\gamma'} \, ,$$

which can also be defined using the motivic Hall algebra [**30, 33, 44, 45**].

The quiver DT invariants $\Omega^{\theta,+}_{\gamma}$ have a piecewise constant dependence on $\theta \in \gamma^{\perp}$: there is a natural wall and chamber structure on $M_{\mathbb{R}}$ as illustrated in Figure 1.3, and while the quiver DT invariants remain constant as long as θ remains inside a chamber, they change by a universal **wall-crossing formula** [**30, 33**] once we cross a wall and pass to another chamber – see 1.3.

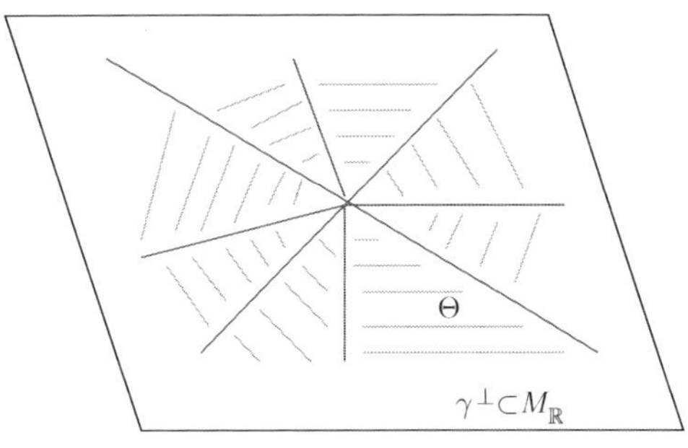

FIGURE 1.3. A wall and chamber structure

EXAMPLE 1.3. Let Q be the n-Kronecker quiver, consisting of two vertices connected by n arrows as illustrated in Figure 1.4. Fix the dimension vector $\gamma := \dim(V) = (1, 1) \in N$ and a choose a stability parameter $\theta = (\theta_1, -\theta_1) \in \gamma^{\perp} \subset M_{\mathbb{R}}$.

[1] Here, we are considering the "unsigned" DT invariants $\Omega^{+,\theta}_{\gamma}$, and not the signed DT invariants Ω^{θ}_{γ}, which are equal to $(-1)^{\dim \mathcal{M}^{\theta}_{\gamma}} e(\mathcal{M}^{\theta}_{\gamma})$ when γ is primitive and $W = 0$.

The data of n linear maps $\mathbb{C} \to \mathbb{C}$ is given by a tuple of n complex numbers $(\xi_1, \ldots, \xi_n)$, and a global rescaling of $(\xi_1, \ldots, \xi_n)$ by a non-zero constant produces isomorphic representations. For $\theta_1 > 0$, θ-semistability requires $(\xi_1, \ldots, \xi_n) \neq (0, \ldots, 0)$, and in this case, the moduli space of θ-semistable quiver respresentations $\mathcal{M}_\gamma^\theta$ is $\mathbb{CP}^{n-1}$, and so

$$\Omega_\gamma^{\theta,+} = e(\mathbb{CP}^{n-1}) = n.$$

On the other hand, for $\theta_1 < 0$ we have $\mathcal{M}_\gamma^\theta = \emptyset$, and so $\Omega_\gamma^{\theta,+} = 0$.

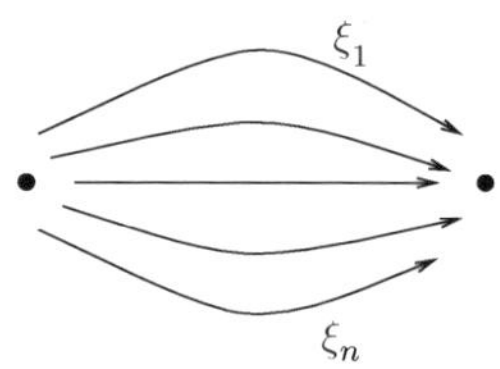

FIGURE 1.4. The n-Kronecker quiver

Quiver DT invariants are extensively studied both in the physics as well as in the mathematics literature. In physics they appear in the context of supersymmetric quantum mechanics, and of BPS states in $\mathcal{N} = 2$ four-dimensional quantum field theories and string theory compactifications [**5, 10, 37**], whereas in mathematics they are closely related to geometric DT theory, which concerns counts of coherent sheaves or special Lagrangians in Calabi–Yau threefolds [**30, 33, 49**] – see Table 1.

EXAMPLE 1.4. The 3-Kronecker quiver appears in the description of the BPS spectrum of $\mathcal{N} = 2$, four-dimensional $SU(3)$ super Yang-Mills theory [**22**]. In this context, explicit values of quiver DT invariants are calculated –see [**22**, A.2] and Figure 1.5. Here, for every $\gamma = (\gamma_1, \gamma_2) \in N^\oplus$, the stability parameter $\theta = (\theta_1, \theta_2) \in \gamma^\perp$, and so with $n_1\theta_1 + n_2\theta_2 = 0$, satisfies $\theta_1 > 0$ (and so $\theta_2 < 0$). If $\theta_1 < 0$, the DT invariant is zero, as in Example 1.3.

1.3. The flow tree formula. Quiver DT invariants can be expressed in terms of a simpler set of quiver DT invariants, known as attractor DT invariants defined as follows. Let $\langle -, - \rangle$ be the skew-symmetrized Euler form on

$$N = \mathbb{Z}^{Q_0} = \bigoplus_{i \in Q_0} \mathbb{Z} e_i.$$

It is the skew-symmetric form on N defined by $\langle e_i, e_j \rangle = a_{ij} - a_{ji}$, where a_{ij} is the number of arrows from i to j in Q. Given $\gamma \in N$, $\langle \gamma, - \rangle$ is a point in the hyperplane $\gamma^\perp$ of stability parameters for γ, referred to as the **attractor point**.

DEFINITION 1.5. [**4, 34, 41**] Fix a dimension vector $\gamma \in N^\oplus$. The **attractor DT invariant** for γ is defined by $\Omega_\gamma^\star := \Omega_\gamma^\theta$, where θ is a small generic perturbation in $\gamma^\perp$ of the attractor point $\langle \gamma, - \rangle$ [2].

The attractor DT invariants turn out to be particularly simple in many situations.

[2] One can show that the attractor DT invariant is independent of the choice of perturbation, see [**41**, Thm 3.7].

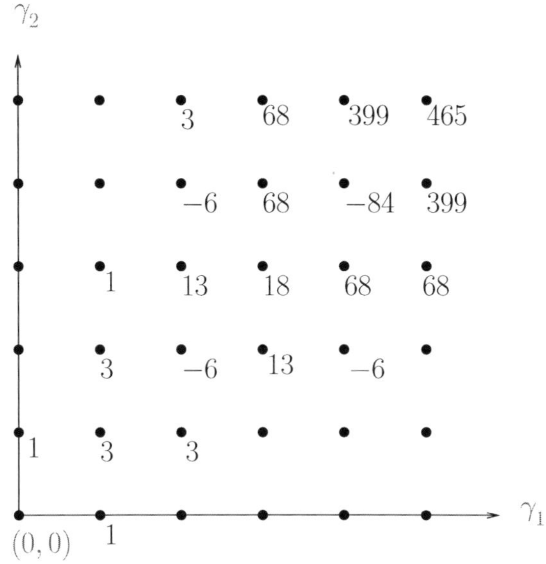

FIGURE 1.5. Values of Ω^θ_γ for the 3-Kronecker quiver [**22**, A.2]

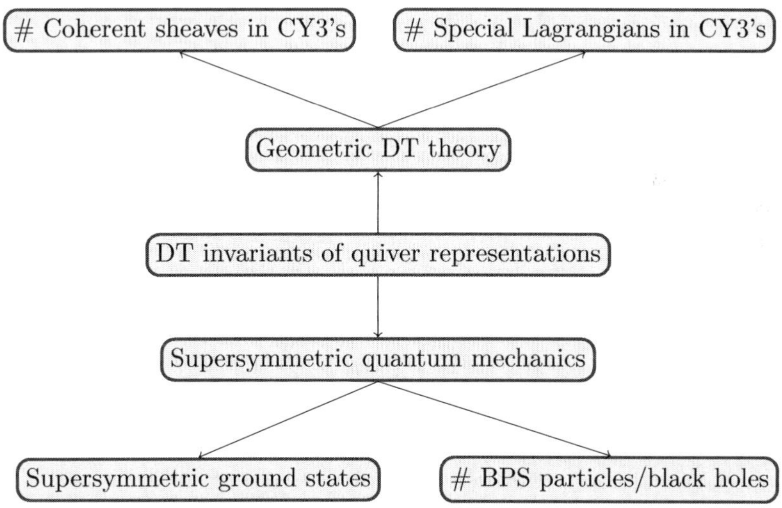

TABLE 1. DT invariants of quivers in maths and physics

EXAMPLE 1.6. It is shown by Bridgeland [?bridgeland] that if Q is acyclic then

$$\Omega^\star_\gamma = \begin{cases} 1 & \text{if } \gamma = e_i \text{ for some } i \in Q_0 \\ 0 & \text{otherwise} \end{cases}$$

This theorem has been generalized by Lang Mou to some situations when Q is not acyclic, but admits a so called red to green sequence [**39**].

EXAMPLE 1.7. Given a toric Calabi-Yau 3-fold, one can construct a quiver with a potential (Q, W) such that

$$D^b Rep(Q, W) \cong D^b Coh(X),$$

where $D^b Rep(Q, W)$ is the bounded derived category of representations of (Q, W) and $D^b Coh(X)$ is the bounded derived category of coherent sheaves on X [**40**].

Beaujard–Manschot–Pioline [**11**] and Mozgovoy–Pioline [**41**] conjectured in this situation that one gets $\Omega^\star_\gamma = 0$ unless either $\gamma = e_i$ for some $i \in Q_0$ (in which case one has $\Omega^\star_\gamma = 1$) or γ is in the kernel of $\langle -, - \rangle$. A more precise conjecture including the values of all non-trivial attractor inariants has been formulated by Descombes [**19**]. This conjecture is proven for local $\mathbb{P}^2$ by Bousseau–Descombes–Le Floch–Pioline [**14**].

More generally, attractor DT invariants are expected to take a simple form in physics set-ups without dynamical gravity, such as string compactifications on non-compact Calabi-Yau 3-folds. For theories with dynamical gravity, such as string compactifications on compact Calabi-Yau 3-folds, one typically expects attractor invariants to be sufficiently non-trivial to account for the entropy of black holes [**37**].

Fix a dimension vector $\gamma \in N^\oplus$, and a generic stability parameter $\theta \in \gamma^\perp$. By an iterative application of the wall-crossing formula [**34**], quiver DT invariants can be obtained in terms of attractor DT invariants by a formula of the form

$$\overline{\Omega}^\theta_\gamma = \sum_{\gamma=\gamma_1+\cdots+\gamma_r} \frac{1}{|\mathrm{Aut}((\gamma_i)_i)|} F^\theta_r(\gamma_1, \ldots, \gamma_r) \prod_{i=1}^r \overline{\Omega}^\star_{\gamma_i}, \tag{1.1}$$

where the sum is over all possible decompositions of γ into $\gamma_1 + \cdots + \gamma_r$ such that $\gamma_i \in N^\oplus$ for $1 \leq i \leq r$ and $r \in \mathbb{N}$, the term $|\mathrm{Aut}((\gamma_i)_i)|$ is the order of the group of permutation symmetries of the decomposition $\gamma = \gamma_1 + \cdots + \gamma_r$, and the coefficients $F^\theta_r(\gamma_1, \ldots, \gamma_r)$ are sums of contributions from attractor trees with leaves decorated by $\gamma_1, \ldots, \gamma_r$ and with root at θ:

$$F^\theta_r(\gamma_1, \ldots, \gamma_r) = \sum_{T \in \mathcal{T}^\theta_{\gamma_1,\ldots,\gamma_r}} F^\theta_{r,T}(\gamma_1, \ldots, \gamma_r).$$

These trees, as illustrated in Figure 1.6, are oriented trees with one root vertex corresponding to the point $\theta \in M_\mathbb{R}$ and edges have direction vectors determined by the decompositions of the dimension vector γ.

By construction, attractor trees satisfy the "tropical balancing condition" – that is, weighted directions around edges add up to zero. Hence, they can be viewed as tropical trees. As illustrated in Figure 1.6, these trees have in general vertices of valency bigger than 3, and calculating their contributions in this case is technically challenging, as it requires the application of the wall-crossing formula.

In joint work with Bousseau [**6**], we proved the **flow tree formula** conjectured by Alexandrov-Pioline [**4**] computing the coefficients $F^\theta_r(\gamma_1, \ldots, \gamma_t)$ in terms of binary trees as illustrated in Figure 1.7, counted with simple tropical multiplicities attached to the trivalent vertices. These binary trees are obtained as generic trivalent perturbations of the non-trivalent attractor trees.

In the following section we explain how to interpret the coefficients $F^\theta_r(\gamma_1, \ldots, \gamma_r)$ in (1.1) geometrically in terms of counts of rational curves in toric varieties, by relating contributions of binary trees to such curve counts.

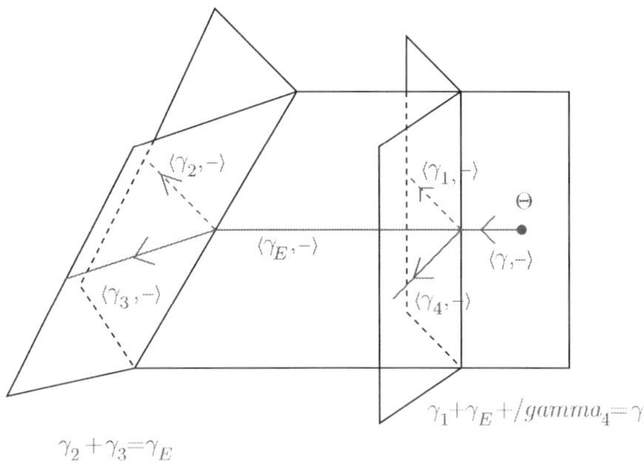

FIGURE 1.6. An attractor tree with higher valent vertices

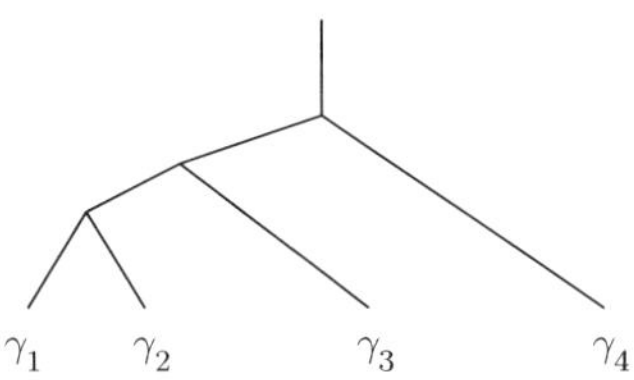

FIGURE 1.7. A binary tree with leaves labelled by $\gamma_1, \ldots, \gamma_n$, for $n = 4$

2. Log Gromov–Witten invariants of toric varieties

The main result of [**6**] relates DT invariants of quivers to log Gromov–Witten invariants given by counts of log curves toric varieties. After briefly explaining the toric varieties we consider and the enumerative problem of curves, we provide further details of log geometry and log Gromov–Witten theory. In the final section we state the relationship between counts of log curves in toric varieties and quiver DT invariants, and provide a brief summary of the proof.

2.1. From quivers to enumerative geometry in toric varieties. Fix a quiver Q, a dimension vector $\gamma \in N^{\oplus}$ and a decomposition

$$\gamma = \gamma_1 + \ldots + \gamma_r.$$

Recall that we denote by $\langle -, - \rangle$ the skew-symmetrized Euler form on N. We assume that $\langle \gamma_i, - \rangle \neq 0$ for all $1 \leq i \leq r$, and $\langle \gamma, - \rangle \neq 0$.

Let Σ be a fan in $M_{\mathbb{R}}$ of a smooth projective toric variety X_Σ containing the rays $\mathbb{R}_{\geq 0}\langle \gamma_i, - \rangle$ for all $1 \leq i \leq r$. We consider hypersurfaces $H_i \subset D_i$ defined by equations

$$(z^{\frac{\gamma_i}{|\gamma_i|}} + c_i)|_{D_i} = 0\,,$$

where $|\gamma_i|$ is the divisibility of γ_i in N, and c_i is a general constant.

The count of log curves which we show are related to quiver DT invariants in [**6**] are given by appropriately defined counts of marked genus 0 stable maps

$$f : (C, \{p_1, \ldots, p_{r+1}\}) \to X_\Sigma$$

satisfying

(i) $f(p_i) \in H_i$ for all $1 \leq i \leq r$

(ii) The contact order of f with D_i at p_i is the divisibility of $\langle \gamma_i, - \rangle$ in M.

See Figure 2.1 for an illustration.

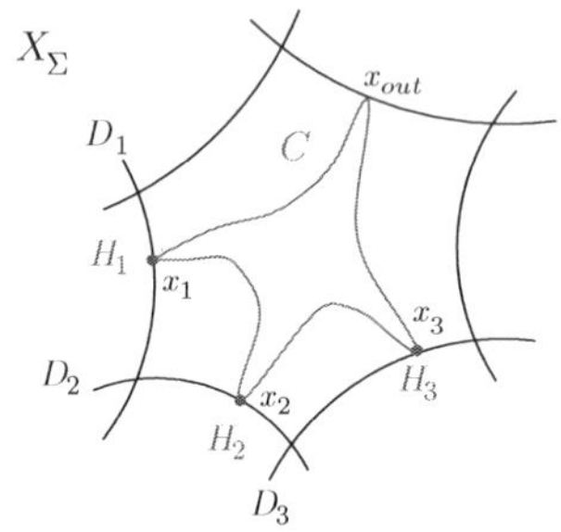

FIGURE 2.1. A log curve in a toric variety with prescribed tangencies along the toric boundary

To count such curves with prescribed tangency conditions, we use log Gromov–Witten theory developed by Abramovich–Chen [**1**] and Gross–Siebert [**28**]. In the following sections we briefly review log geometry and log Gromov–Witten theory before stating a precise relationship between counts of log curves and quiver DT invariants.

2.2. Log schemes. We start reviewing a couple of definitions from log geometry to fix our notation. For familiarity with log geometry we refer to [**31**, **43**]. All the monoids considered below are assumed to be commutative.

DEFINITION 2.1. Let X be a scheme. A **prelog structure** on X is a sheaf of monoids $\mathcal{M}$ on X together with a homomorphism of monoids $\alpha\colon \mathcal{M} \longrightarrow \mathcal{O}_X$ where we consider the structure sheaf $\mathcal{O}_X$ as a monoid with respect to multiplication. A pre log structure on X is called a **log structure** if α induces an isomorphism

$$\alpha\big|_{\alpha^{-1}(\mathcal{O}_X^\times)} : \alpha^{-1}(\mathcal{O}_X^\times) \longrightarrow \mathcal{O}_X^\times.$$

We call a scheme X endowed with a log structure a **log scheme**, and denote the log structure on X by $(\mathcal{M}_X, \alpha_X)$, or sometimes by omitting the structure homomorphism from the notation, just by $\mathcal{M}_X$. We denote a scheme X, endowed with a log structure by $(X, \mathcal{M}_X)$.

All log schemes below are assumed to be fine and saturated (fs) as in [**43**, I,§1.3]. The combinatorial information of the log structure is encoded via tropical geometry using the ghost sheaf as defined below.

DEFINITION 2.2. Let X be a log scheme. The **ghost sheaf** of X is the sheaf on X defined by

$$\overline{\mathcal{M}}_X\colon = \mathcal{M}_X/\alpha^{-1}(\mathcal{O}_X^\times).$$

The **tropicalization** $\Sigma(X)$ of X is an abstract polyhedral cone complex: $\Sigma(X)$ is the collection of cones $\sigma_{\bar{x}} := \mathrm{Hom}(\overline{\mathcal{M}}_{X,\bar{x}}, \mathbb{R}_{\geq 0})$ for every geometric point $\bar{x} \to X$, glued together using the natural generization and specialization maps, see [**2**, §2.1] for details.

EXAMPLE 2.3. Let $D \subset X$ be a divisor. Let $j\colon X \setminus D \hookrightarrow X$ denote the inclusion map. The *divisorial log structure* on X is the pair $(\mathcal{M}_{(X,D)}, \alpha_X)$, where $\mathcal{M}_{(X,D)}$ is the sheaf of regular functions on X, that restrict to units on $X \setminus D$. That is,

$$\mathcal{M}_{(X,D)} = j_*(\mathcal{O}^\times_{X \setminus D}) \cap \mathcal{O}_X$$

The structure homomorphism α_X is given by the inclusion $\alpha_X : \mathcal{M}_{(X,D)} \hookrightarrow \mathcal{O}_X$.

REMARK 2.4. The log structure we consider on the toric variety X_Σ constructed from the data of a quiver, as discussed in 2.1, is the divisorial log structure defined by the toric boundary divisor D_Σ in X_Σ, that is, the natural anti-canonical divisor formed by the union of divisors fixed under the torus action on X_Σ.

Note that, for any toric variety X_Σ endowed with the divisorial log structure defined by the toric boundary divisor, the tropicalization of the resulting log scheme is naturally identified with the toric fan Σ associated to X.

Analogous to assigning a sheaf to a presheaf, we can assign a log structure to a prelog structure, using a fibered coproduct, as follows.

DEFINITION 2.5. Let $\alpha\colon \mathcal{P} \longrightarrow \mathcal{O}_X$ be a prelog structure on X. We define the *log structure associated to the prelog structure* $(\mathcal{P}, \alpha)$ on X as follows. Set

$$\mathcal{P}^a = \mathcal{P} \oplus \mathcal{O}^\times_X \Big/ \{(p, \alpha(p)^{-1}) \,|\, p \in \alpha^{-1}(\mathcal{O}^\times_X)\}$$

and define the structure homomorphism $\alpha^a\colon \mathcal{P}^a \longrightarrow \mathcal{O}_X$ by

$$\alpha^a(p,h) = h \cdot \alpha(p)$$

One can easily check that $(\mathcal{P}^a, \alpha^a)$ is a log structure on X.

Let Y be a log scheme, and let $f\colon X \to Y$ be a scheme theoretic morphism. Then, the log structure on Y, given by $\alpha_Y : \mathcal{M}_Y \longrightarrow \mathcal{O}_Y$, induces a log structure on X defined as follows. First define a prelog structure, by considering the composition

$$f^{-1}\mathcal{M}_Y \to f^{-1}\mathcal{O}_Y \to \mathcal{O}_X\,.$$

Then, we endow X with the log structure associated to this prelog structure, as in Definition 2.5. We refer to this log structure as the *pull back log structure* or the *induced log structure* on X, and denote it by $\mathcal{M}_X = f^*\mathcal{M}_Y$.

EXAMPLE 2.6. For every scheme X, $\mathcal{M}_X = \mathcal{O}^\times_X$ define a log structure on X, called the *trivial log structure*. In particular, taking $X = \operatorname{Spec}\mathbb{C}$, we call $(\operatorname{Spec}\mathbb{C}, \mathbb{C}^\times)$ the *trivial log point.*

EXAMPLE 2.7. Let $X = \operatorname{Spec}\mathbb{C}$. Define $\mathcal{M}_X = \mathbb{C}^\times \oplus \mathbb{N}$, and $\alpha_X\colon \mathcal{M}_X \to \mathbb{C}$ as follows.

$$\alpha_X(x,n) = \begin{cases} x & \text{if } n = 0 \\ 0 & \text{if } n \neq 0 \end{cases}$$

The corresponding log scheme $O_0 = (\operatorname{Spec}\mathbb{C}, \mathbb{C}^\times \oplus \mathbb{N})$ is called the *standard log point.* One can check that the log structure on the standard log point is the same as the pull-back of the divisorial log structure on $\mathbb{A}^1$ with the divisor $D = \{0\} \subset \mathbb{A}^1$. One also obtains a more general log point $(\operatorname{Spec}\mathbb{C}, \mathbb{C}^\times \oplus Q)$ by replacing $\mathbb{N}$ by a more general monoid Q such that $Q^\times = \{0\}$. It is the pullback to the torus fixed point of the divisorial log structure on the affine toric variety $\operatorname{Spec}\mathbb{C}[Q]$. Its tropicalization is the cone $\operatorname{Hom}(Q, \mathbb{R}_{\geq 0})$.

DEFINITION 2.8. A **morphism of log schemes** $f\colon (X,\mathcal{M}_X)\to(Y,\mathcal{M}_Y)$ is a morphism of schemes $f\colon X\to Y$ along with a homomorphism of sheaves of monoids $f^\sharp\colon f^{-1}\mathcal{M}_Y\to\mathcal{M}_X$ such that the diagram

$$\begin{array}{ccc} f^{-1}\mathcal{M}_Y & \xrightarrow{f^\sharp} & \mathcal{M}_X \\ {\scriptstyle\alpha_Y}\big\downarrow & & \big\downarrow{\scriptstyle\alpha_X} \\ f^{-1}\mathcal{O}_Y & \xrightarrow{f^*} & \mathcal{O}_X. \end{array}$$

is commutative. Here, f^* is the usual pull-back of regular functions defined by the morphism f. Given a morphism of log spaces $f:(X,\mathcal{M}_X)\to(Y,\mathcal{M}_Y)$, we denote by $\underline{f}\colon X\to Y$ the underlying morphism of schemes. By abuse of notation, the underlying morphism on topological spaces is also denoted by $\underline{f}$.

We refer to [**24**, Defn. 3.23] for the notion of *log smooth morphism*. We recall that a toric variety is log smooth over the trivial log point. If $\pi\colon\mathcal{X}\to\mathbb{A}^1$ is a toric degeneration of toric varieties, then π can be naturally viewed as a log smooth morphism. In restriction to $t=0$, we get a log smooth morphism $X_0\to O_0$, that is, X_0 is log smooth over the standard log point O_0. We also refer to [**24**, Remark 3.25] for the notion of **integral** morphism of log schemes.

2.3. Stable log maps. We recall that a ℓ-marked stable map with target a scheme X is a map $f\colon C\to X$, where C is a proper nodal curve with ℓ marked smooth points $\mathbf{x}=(x_1,\ldots,x_\ell)$, such that the group of automorphisms of C fixing $\mathbf{x}$ and commuting with f is finite.

The notion of stable map has a natural generalization to the setting of logarithmic geometry [**1**, **28**].

DEFINITION 2.9. Let $X\to B$ be a log morphism between log schemes. A **ℓ-marked stable log map with target** $X\to B$ is a commutative diagram of log schemes

$$\begin{array}{ccc} C & \xrightarrow{f} & X \\ {\scriptstyle\pi}\big\downarrow & & \big\downarrow \\ W & \longrightarrow & B\,, \end{array}$$

together with a tuple of sections $\mathbf{x}=(x_1,\ldots,x_\ell)$ of π, where π is a proper log smooth and integral morphism of log schemes, such that, for every geometric point w of W, the restriction of $\underline{f}$ to w with the marked points $\underline{\mathbf{x}}(w)$ is an ordinary stable map, and such that, if $U\subset C$ is the non-critical locus of $\underline{\pi}$, we have $\overline{\mathcal{M}}_C|_U\simeq\underline{\pi}^*\overline{\mathcal{M}}_W\oplus\bigoplus_{j=1}^{\ell}(x_j)_*\mathbb{N}_W$.

When the scheme underlying W in Definition 2.9 is a point, then W is a log point $(\operatorname{Spec}\mathbb{C},\mathbb{C}^*\oplus Q)$ as in Example 2.7. The general enumerative theory of stable log maps, called log Gromov–Witten theory and developed in [**1**, **28**], is based on the notion of **basic stable log map**. For a basic stable log map over a log point $(\operatorname{Spec}\mathbb{C},\mathbb{C}^*\oplus Q)$, the monoid Q is the so-called **basic monoid**. The basic monoid is uniquely determined by the combinatorial type of the stable log map and has a natural tropical geometric interpretation: the dual cone $\operatorname{Hom}(Q,\mathbb{R}_{\geq 0})$ is the base of a universal family of tropical curves with given combinatorial type. We provide a more precise description of the combinatorial type of a stable log map, and of basic monoids below.

The **combinatorial type** τ of a stable log map $f\colon C/W \to X/S$ consists of:

(i) The dual intersection graph $G = G_C$ of C, with set of vertices $V(G)$, set of edges $E(G)$, and set of legs $L(G)$.

(ii) The map $\sigma : V(G)\cup E(G)\cup L(G) \to \Sigma(X)$ mapping $x \in C$ to $\left(\overline{\mathcal{M}}_{X,f(x)}\right)^\vee_{\mathbb{R}}$.

(iii) The contact data $u_p \in \overline{\mathcal{M}}^\vee_{X,f(p)} = \mathrm{Hom}(\overline{\mathcal{M}}_{X,f(p)}, \mathbb{N})$ and

$$u_q \in \mathrm{Hom}(\overline{\mathcal{M}}_{X,f(q)}, \mathbb{Z})$$

at marked points p and nodes q of C.

DEFINITION 2.10. Given a combinatorial type τ of a stable log map $f\colon C/W \to X/S$, we define the associated **basic monoid** Q by first defining its dual

$$Q^\vee_\tau = \left\{ ((V_\eta)_\eta, (e_q)_q) \in \bigoplus_\eta \overline{\mathcal{M}}^\vee_{X,f(\eta)} \oplus \bigoplus_q \mathbb{N} \,\middle|\, \forall q : V_{\eta_2} - V_{\eta_1} = e_q u_q \right\}$$

where the sum is over generic points η of C and nodes q of C. We then set

$$Q_\tau := \mathrm{Hom}(Q^\vee_\tau, \mathbb{N}).$$

Note that, by definition Q_τ indeed only depends on the combinatorial type of $f\colon C/W \to X/S$. Geometrically, one can interpret $Q^\vee_{\tau,\mathbb{R}} := \mathrm{Hom}(Q_\tau, \mathbb{R}_{\geq 0})$ as the moduli cone of tropical curves of fixed combinatorial type.

Given a stable log map $f\colon C/W \to X/S$, one can show that there is a canonical map $Q \to \overline{\mathcal{M}}_W$, where Q is the basic monoid defined by the combinatorial type of f.

DEFINITION 2.11. A stable log map $f\colon C/W \to X/S$ is said to be **basic** if the natural map of monoids $Q \to \overline{M}_W$ is an isomorphism.

For every $g \in \mathbb{Z}_{\geq 0}$, $\beta \in H_2(X,\mathbb{Z})$ and $u = (u_1, \dots, u_k)$ with $u_i \in |\Sigma(X)|$, we denote by $\mathcal{M}_{g,u}(X/S, \beta)$ the moduli space of genus g basic stable log maps to X/S of class β and with k marked points of contact data

$$u = u_1, \dots, u_k.$$

THEOREM 2.12 (Abramovich–Chen [**1**], Gross–Siebert [**28**], 2011). *If X/S is projective, then the moduli space $\mathcal{M}_{g,u}(X/S,\beta)$ is a proper Deligne-Mumford stack. Furthermore, If X/S is log smooth, then the moduli space $\mathcal{M}_{g,u}(X/S,\beta)$ admits a natural virtual fundamental class $[\mathcal{M}_{g,u}(X/S,\beta)]^{virt}$.*

2.4. Flow trees to tropical and log curves. Recall from §1.3 that quiver DT invariants can be calculated using a universal formula from simpler attractor DT invariants. Further, the coefficients $F^\theta_{r,T}(\gamma_1, \dots, \gamma_r)$ are indexed by attractor tropical trees T in $M_{\mathbb{R}}$. In this section we explain how to interpret these coefficients in terms of counts of log curves.

In [**42**], Nishinou-Siebert used toric degenerations and log deformation theory to prove a correspondence theorem between counts of genus 0 log curves in n-dimensional toric varieties and counts of tropical curves in $\mathbb{R}^n$. The counts of genus zero log curves considered in [**42**] are shown to correspond to genus zero log Gromov–Witten invariants by Mandel–Ruddat [**36**]. This tropical-log correspondence theorem is established in the context when the basic monoid is trivial. In our paper with Bousseau [**8**], we generalize this tropical-log correspondence to the situation of non-trivial basic monoids. In other words, we work with "families" of tropical trees corresponding to log curves over basic monoids of rank equal to the dimension of the family.

Fix a quiver Q, a dimension vector $\gamma \in N^{\oplus}$ and a decomposition $\gamma = \gamma_1 + \cdots + \gamma_r$, such that $\langle \gamma_i, - \rangle \neq 0$ for all $1 \leq i \leq r$ and $\langle \gamma, - \rangle \neq 0$. We denote by d the number of vertices of Q, so that $N \simeq \mathbb{Z}^d$ and $M_{\mathbb{R}} \simeq \mathbb{R}^d$. Let X_Σ be a toric variety as in §2.1, with toric divisors D_i containing hypersurfaces H_i. Writing $\mathbf{H} = (H_1, \ldots, H_r)$, let $\mathcal{M}^{\log}_{\mathbf{H}}(X_\Sigma)$ be the moduli space of genus 0 basic stable log maps $f : (C, \{p_1, \ldots, p_{r+1}\}) \to X_\Sigma$ satisfying

(i) $f(p_i) \in H_i$ for all $1 \leq i \leq r$
(ii) The contact order of f with D_i at p_i is $\langle \gamma_i, - \rangle \in M$.
(iii) The contact order of f with the toric boundary divisor of X_Σ at p_{r+1} is $-\langle \gamma, - \rangle$.

For general $\mathbf{H}$, one shows in [**8**, Theorem 2.8] that the moduli space $\mathcal{M}^{\log}_{\mathbf{H}}(X_\Sigma)$ is $(d-2)$-dimensional and naturally stratified by the combinatorial types of stable log curves. For every combinatorial type ρ, there is a stratum $\mathcal{M}^{\log}_{\rho,\mathbf{H}}(X_\Sigma)$ consisting of stable log maps of type ρ, and which is of codimension $\operatorname{rk} Q^{\text{gp}}_\rho$ in $\mathcal{M}^{\log}_{\mathbf{H}}(X_\Sigma)$, where Q_ρ is the basic monoid of ρ. In particular, types ρ with $\operatorname{rk} Q^{\text{gp}}_\rho = (d-2)$ correspond to 0-dimensional strata of $\mathcal{M}^{\log}_{\mathbf{H}}(X_\Sigma)$, and we denote by $N^{\text{toric}}_{\rho,H}(X_\Sigma)$ the log Gromov–Witten counts of the stable log maps of type ρ. The tropicalizations of these stable log maps are $(d-2)$-dimensional families of tropical curves parametrized by the cones $\operatorname{Hom}(Q_\rho, \mathbb{R}_{\geq 0})$.

We construct such $(d-2)$-dimensional families ρ_T of tropical curves in $M_{\mathbb{R}}$ from an attractor tree T, contributing in the formula (1.1) for quiver DT invariants. To do this, we extend the root of T to infinity to obtain a tropical curve with leaves constrained to lie in the hyperplanes $\gamma_i^\perp$. We then consider a deformation of this tropical curve while preserving the combinatorial type and the constraints on the leaves—see Figure 2.2.

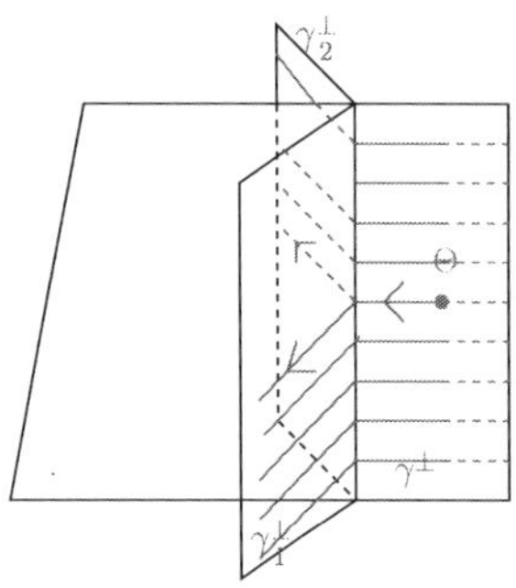

FIGURE 2.2. A family of tropical curves associated to an attractor tree

We denote by $N^{\text{toric}}_{\rho_T,\mathbf{H}}(X_\Sigma)$ the corresponding log Gromov–Witten invariant counting the finitely many stable log maps in $\mathcal{M}^{\log}_{\mathbf{H}}(X_\Sigma)$ of type ρ_T– see [**8**, Defn. 1.20]. Our main result is [**8**, Theorem B]:

THEOREM 2.13. *The coefficients $F^\theta_{r,T}(\gamma_1, \ldots, \gamma_r)$ expressing the contribution to $F^\theta_r(\gamma_1, \ldots, \gamma_r)$ of an attractor tree T in 1.1 satisfy*

$$F^\theta_{r,T}(\gamma_1, \ldots, \gamma_r) = N^{\text{toric}}_{\rho_T,\mathbf{H}}(X_\Sigma).$$

We provide a brief summary of the main steps of the proof here. The first step is to construct a toric degeneration $\mathcal{X} \to \mathbb{A}^1$ of X_Σ and of the constraints $\mathbf{H}$. We then use a degeneration formula expressing the invariants $N^{\text{toric}}_{\rho_T,H}(X_\Sigma)$ of the general fibers X_Σ as a sum of invariants $N^{\text{toric}}_{\rho_S}(\mathcal{X}_0)$ of the special fiber $\mathcal{X}_0$, where S are binary trees in $M_\mathbb{R}$ deforming T. We then prove an explicit formula

$$N^{\text{toric}}_{\rho_S}(\mathcal{X}_0) = \prod_v |\langle \gamma_{v'}, \gamma_{v''} \rangle|$$

computing $N^{\text{toric}}_{\rho_S}(\mathcal{X}_0)$ as a product over local contributions of the trivalent vertices v of S, where v' and v'' denote the children of v, that is, each of the vertices v' and v'' are connected to v by a single edge which is not contained in the unique path lying between v and the root. This formula is proved using the theory of punctured log maps due to Abramovich–Chen–Gross–Siebert [**3**] to count log maps by gluing punctured log maps. Finally, by the flow tree formula proved in [**6**] and briefly reviewed at the end of §1.3, one also calculates,

$$F^\theta_{r,T}(\gamma_1, \ldots, \gamma_r) = \sum_S \prod_v |\langle \gamma_{v'}, \gamma_{v''} \rangle| \,.$$

We then conclude that $F^\theta_{r,T}(\gamma_1, \ldots, \gamma_r) = N^{\text{toric}}_{\rho_T,H}(X_\Sigma)$.

Another way to construct a geometrical object out of a quiver is based on the theory of cluster algebras [**21**] and cluster varieties [**20**]. Starting with a quiver Q with d vertices, one can construct a d-dimensional cluster variety using cluster transformations in a way prescribed by the combinatorics of Q.

In a paper in preparation with Bousseau, we provide a correspondence between quiver DT invariants, in the situation when attractor DT invariants are trivial, and counts of log curves in cluster varieties. The proof will use a comparison between the stability scattering diagram of Bridgeland [**16**] for quiver DT invariants and the canonical scattering diagram of Gross-Siebert [**29**] for log Gromov–Witten invariants of log Calabi-Yau varieties, using as intermediate steps the cluster scattering diagram of Gross–Hacking–Keel–Kontsevich [**26**] and the comparison results obtained in [**7, 9**]. This will generalize the Kronecker-Gromov–Witten correspondence previously shown in the $d = 2$ case [**12, 23, 27, 46, 47**]. For variants of the Kronecker-Gromov–Witten correspondence in dimension two see also [**13, 15, 48**]. Such a correspondence between quiver DT invariants and log curve counts in cluster varieties is compatible with Theorem 2.13 using that cluster varieties are obtained by non-toric blow-ups of toric varieties [**25**], and so admit degenerations into a toric variety as in [**9, 27, 35**]. In particular, log curve counts in cluster varieties are naturally related to log curve counts in toric varieties appearing in Theorem 2.13.

Acknowledgments

Hülya Argüz thanks the organisers of the String Math 2022 conference, held at the University of Warsaw, Poland for the wonderful event.

References

[1] Dan Abramovich and Qile Chen, *Stable logarithmic maps to Deligne-Faltings pairs II*, Asian J. Math. **18** (2014), no. 3, 465–488, DOI 10.4310/AJM.2014.v18.n3.a5. MR3257836

[2] Dan Abramovich, Qile Chen, Mark Gross, and Bernd Siebert, *Decomposition of degenerate Gromov-Witten invariants*, Compos. Math. **156** (2020), no. 10, 2020–2075, DOI 10.1112/s0010437x20007393. MR4177284

[3] Dan Abramovich, Qile Chen, Mark Gross, and Bernd Siebert, *Punctured logarithmic maps*, Preprint, `arXiv:2009.07720`, 2020.

[4] Sergei Alexandrov and Boris Pioline, *Attractor flow trees, BPS indices and quivers*, Adv. Theor. Math. Phys. **23** (2019), no. 3, 627–699, DOI 10.4310/ATMP.2019.v23.n3.a2. MR4049073

[5] Murad Alim, Sergio Cecotti, Clay Córdova, Sam Espahbodi, Ashwin Rastogi, and Cumrun Vafa, *$\mathcal{N}=2$ quantum field theories and their BPS quivers*, Adv. Theor. Math. Phys. **18** (2014), no. 1, 27–127. MR3268234

[6] Hülya Argüz and Pierrick Bousseau, *The flow tree formula for Donaldson-Thomas invariants of quivers with potentials*, Compos. Math. **158** (2022), no. 12, 2206–2249, DOI 10.1112/s0010437x22007801. MR4524360

[7] Hülya Argüz and Pierrick Bousseau, *Fock-Goncharov dual cluster varieties and Gross-Siebert mirrors*, J. Reine Angew. Math. **802** (2023), 125–171, DOI 10.1515/crelle-2023-0043. MR4635344

[8] Hülya Argüz and Pierrick Bousseau, *Quivers, flow trees, and log curves*, Preprint, `arXiv:2302.02068`, 2023.

[9] Hülya Argüz and Mark Gross, *The higher-dimensional tropical vertex*, Geom. Topol. **26** (2022), no. 5, 2135–2235, DOI 10.2140/gt.2022.26.2135. MR4520304

[10] Paul S. Aspinwall, Tom Bridgeland, Alastair Craw, Michael R. Douglas, Mark Gross, Anton Kapustin, Gregory W. Moore, Graeme Segal, Balázs Szendrői, and P. M. H. Wilson, *Dirichlet branes and mirror symmetry*, Clay Mathematics Monographs, vol. 4, American Mathematical Society, Providence, RI; Clay Mathematics Institute, Cambridge, MA, 2009. MR2567952

[11] Guillaume Beaujard, Jan Manschot, and Boris Pioline, *Vafa-Witten invariants from exceptional collections*, Comm. Math. Phys. **385** (2021), no. 1, 101–226, DOI 10.1007/s00220-021-04074-2. MR4275783

[12] Pierrick Bousseau, *The quantum tropical vertex*, Geom. Topol. **24** (2020), no. 3, 1297–1379, DOI 10.2140/gt.2020.24.1297. MR4157555

[13] Pierrick Bousseau, *On an example of quiver Donaldson-Thomas/relative Gromov-Witten correspondence*, Int. Math. Res. Not. IMRN **15** (2021), 11845–11888, DOI 10.1093/imrn/rnz368. MR4294135

[14] Pierrick Bousseau, Pierre Descombes, Bruno Le Floch, and Boris Pioline, *BPS Dendroscopy on Local P2*, Preprint, `arXiv:2210.10712`, 2022.

[15] Pierrick Bousseau and Longting Wu, *All-genus WDVV recursion, quivers, and BPS invariants*, Preprint, `arXiv:2303.00503`, 2023.

[16] Tom Bridgeland, *Scattering diagrams, Hall algebras and stability conditions*, Algebr. Geom. **4** (2017), no. 5, 523–561, DOI 10.14231/2017-027. MR3710055

[17] Ben Davison and Sven Meinhardt, *Donaldson-Thomas theory for categories of homological dimension one with potential*, Preprint, `arXiv:1512.08898`, 2015.

[18] Ben Davison and Sven Meinhardt, *Cohomological Donaldson-Thomas theory of a quiver with potential and quantum enveloping algebras*, Invent. Math. **221** (2020), no. 3, 777–871, DOI 10.1007/s00222-020-00961-y. MR4132957

[19] Pierre Descombes, *Cohomological DT invariants from localization*, J. Lond. Math. Soc. (2) **106** (2022), no. 4, 2959–3007. MR4524190

[20] Vladimir V. Fock and Alexander B. Goncharov, *Cluster ensembles, quantization and the dilogarithm* (English, with English and French summaries), Ann. Sci. Éc. Norm. Supér. (4) **42** (2009), no. 6, 865–930, DOI 10.24033/asens.2112. MR2567745

[21] Sergey Fomin and Andrei Zelevinsky, *Cluster algebras. I. Foundations*, J. Amer. Math. Soc. **15** (2002), no. 2, 497–529, DOI 10.1090/S0894-0347-01-00385-X. MR1887642

[22] Davide Gaiotto, Gregory W. Moore, and Andrew Neitzke, *Wall-crossing, Hitchin systems, and the WKB approximation*, Adv. Math. **234** (2013), 239–403, DOI 10.1016/j.aim.2012.09.027. MR3003931

[23] Sheel Ganatra and Daniel Pomerleano, *Symplectic cohomology rings of affine varieties in the topological limit*, Geom. Funct. Anal. **30** (2020), no. 2, 334–456, DOI 10.1007/s00039-020-00529-1. MR4108612

[24] Mark Gross, *Tropical geometry and mirror symmetry*, CBMS Regional Conference Series in Mathematics, vol. 114, Published for the Conference Board of the Mathematical Sciences, Washington, DC; by the American Mathematical Society, Providence, RI, 2011, DOI 10.1090/cbms/114. MR2722115

[25] Mark Gross, Paul Hacking, and Sean Keel, *Birational geometry of cluster algebras*, Algebr. Geom. **2** (2015), no. 2, 137–175, DOI 10.14231/AG-2015-007. MR3350154

[26] Mark Gross, Paul Hacking, Sean Keel, and Maxim Kontsevich, *Canonical bases for cluster algebras*, J. Amer. Math. Soc. **31** (2018), no. 2, 497–608, DOI 10.1090/jams/890. MR3758151

[27] Mark Gross, Rahul Pandharipande, and Bernd Siebert, *The tropical vertex*, Duke Math. J. **153** (2010), no. 2, 297–362, DOI 10.1215/00127094-2010-025. MR2667135

[28] Mark Gross and Bernd Siebert, *Logarithmic Gromov-Witten invariants*, J. Amer. Math. Soc. **26** (2013), no. 2, 451–510, DOI 10.1090/S0894-0347-2012-00757-7. MR3011419

[29] Mark Gross and Bernd Siebert, *The canonical wall structure and intrinsic mirror symmetry*, Invent. Math. **229** (2022), no. 3, 1101–1202, DOI 10.1007/s00222-022-01126-9. MR4462625

[30] Dominic Joyce and Yinan Song, *A theory of generalized Donaldson-Thomas invariants*, Mem. Amer. Math. Soc. **217** (2012), no. 1020, iv+199, DOI 10.1090/S0065-9266-2011-00630-1. MR2951762

[31] Kazuya Kato, *Logarithmic structures of Fontaine-Illusie*, Algebraic analysis, geometry, and number theory (Baltimore, MD, 1988), Johns Hopkins Univ. Press, Baltimore, MD, 1989, pp. 191–224. MR1463703

[32] A. D. King, *Moduli of representations of finite-dimensional algebras*, Quart. J. Math. Oxford Ser. (2) **45** (1994), no. 180, 515–530, DOI 10.1093/qmath/45.4.515. MR1315461

[33] Maxim Kontsevich and Yan Soibelman, *Stability structures, motivic Donaldson-Thomas invariants and cluster transformations*, Preprint, `arXiv:0811.2435`, 2008.

[34] Maxim Kontsevich and Yan Soibelman, *Wall-crossing structures in Donaldson-Thomas invariants, integrable systems and mirror symmetry*, Homological mirror symmetry and tropical geometry, Lect. Notes Unione Mat. Ital., vol. 15, Springer, Cham, 2014, pp. 197–308, DOI 10.1007/978-3-319-06514-4_6. MR3330788

[35] Travis Mandel, *Theta bases and log Gromov-Witten invariants of cluster varieties*, Trans. Amer. Math. Soc. **374** (2021), no. 8, 5433–5471, DOI 10.1090/tran/8398. MR4293777

[36] Travis Mandel and Helge Ruddat, *Descendant log Gromov-Witten invariants for toric varieties and tropical curves*, Trans. Amer. Math. Soc. **373** (2020), no. 2, 1109–1152, DOI 10.1090/tran/7936. MR4068259

[37] Jan Manschot, Boris Pioline, and Ashoke Sen, *From black holes to quivers*, J. High Energy Phys. **11** (2012), 023, front matter + 53, DOI 10.1007/JHEP11(2012)023. MR3036499

[38] Sven Meinhardt and Markus Reineke, *Donaldson-Thomas invariants versus intersection cohomology of quiver moduli*, J. Reine Angew. Math. **754** (2019), 143–178, DOI 10.1515/crelle-2017-0010. MR4000572

[39] Lang Mou, *Scattering diagrams of quivers with potentials and mutations*, Preprint, `arXiv:1910.13714`, 2019.

[40] Sergey Mozgovoy, *Crepant resolutions and brane tilings I: Toric realization*, Preprint, `arXiv:0908.3475`, 2009.

[41] Sergey Mozgovoy and Boris Pioline, *Attractor invariants, brane tilings and crystals*, Preprint, `arXiv:2012.14358`, 2020.

[42] Takeo Nishinou and Bernd Siebert, *Toric degenerations of toric varieties and tropical curves*, Duke Math. J. **135** (2006), no. 1, 1–51, DOI 10.1215/S0012-7094-06-13511-1. MR2259922

[43] Arthur Ogus, *Lectures on logarithmic algebraic geometry*, Cambridge Studies in Advanced Mathematics, vol. 178, Cambridge University Press, Cambridge, 2018, DOI 10.1017/9781316941614. MR3838359

[44] Markus Reineke, *Poisson automorphisms and quiver moduli*, J. Inst. Math. Jussieu **9** (2010), no. 3, 653–667, DOI 10.1017/S1474748009000176. MR2650811

[45] Markus Reineke, *Cohomology of quiver moduli, functional equations, and integrality of Donaldson-Thomas type invariants*, Compos. Math. **147** (2011), no. 3, 943–964, DOI 10.1112/S0010437X1000521X. MR2801406

[46] Markus Reineke, Jacopo Stoppa, and Thorsten Weist, *MPS degeneration formula for quiver moduli and refined GW/Kronecker correspondence*, Geom. Topol. **16** (2012), no. 4, 2097–2134, DOI 10.2140/gt.2012.16.2097. MR3033514

[47] Markus Reineke and Thorsten Weist, *Refined GW/Kronecker correspondence*, Math. Ann. **355** (2013), no. 1, 17–56, DOI 10.1007/s00208-012-0778-0. MR3004575

[48] Markus Reineke and Thorsten Weist, *Moduli spaces of point configurations and plane curve counts*, Int. Math. Res. Not. IMRN **13** (2021), 10339–10372, DOI 10.1093/imrn/rnz118. MR4283580

[49] R. P. Thomas, *A holomorphic Casson invariant for Calabi-Yau 3-folds, and bundles on $K3$ fibrations*, J. Differential Geom. **54** (2000), no. 2, 367–438. MR1818182

Department of Mathematics, University of Georgia, Athens, Georgia 30605
Email address: Hulya.Arguz@uga.edu

Proceedings of Symposia in Pure Mathematics
Volume **107**, 2024
https://doi.org/10.1090/pspum/107/01943

Root bundles: Applications to F-theory Standard Models

Martin Bies

Abstract. The study of vector-like spectra in 4-dimensional F-theory compactifications involves root bundles, which are important for understanding the Quadrillion F-theory Standard Models (F-theory QSMs) and their potential implications in physics. Recent studies focused on a superset of physical root bundles whose cohomologies encode the vector-like spectra for certain matter representations. It was found that more than 99.995% of the roots in this superset for the family $B_3(\Delta_4^\circ)$ of $\mathcal{O}(10^{11})$ different F-theory QSM geometries had no vector-like exotics, indicating that this scenario is highly likely.

To study the vector-like spectra, the matter curves in the F-theory QSMs were analyzed. It was found that each of them can be deformed to nodal curve that is identical across all spaces in $B_3(\Delta^\circ)$. Therefore, from studying a few nodal curves, one can probe the vector-like spectra of a large fraction of F-theory QSMs. To this end, the cohomologies of all limit roots were determined, with line bundle cohomology on rational nodal curves playing a major role. A computer algorithm was used to enumerate all limit roots and analyze the global sections of all tree-like limit roots. For the remaining circuit-like limit roots, the global sections were manually determined. These results were organized into tables, which represent – to the best knowledge of the author – the first arithmetic steps towards Brill-Noether theory of limit roots.

Contents

1. Introduction

String Theory is a top contender for a theory that unifies quantum field theory and gravity, because it skillfully combines the dynamics of gauge fields with gravity. This means that String Theory must be able to explain all of the various aspects of the physical world that we observe, including the behavior of particles at low energy

This work was supported by the SFB-TRR 195 *Symbolic Tools in Mathematics and their Application* of the German Research Foundation (DFG). The author is also grateful for finanical support by the Forschunginitiative des Landes Rheinland-Pfalz through *SymbTools – Symbolic Tools in Mathematics and their Application*.

levels. Most importantly, there is a strong desire to see an explicit representation of the Standard Model of particle physics within String Theory.

A lot of work has been put into achieving this goal using perturbative String Theory, including the $E_8 \times E_8$ heterotic string [**5–7, 25, 26, 30, 33, 61**] and intersecting brane models in type II [**2, 3, 13, 24, 49, 50, 66**], as described in [**23**] and sources therein. These String Theory compactifications were among the first to produce the Standard Model gauge sector, complete with chiral and, in [**25, 26**], even vector-like spectrum. However, these perturbative models typically suffer from exotic chiral or vector-like matter. The first models which surpassed this hurdle and managed to provide a globally consistent construction of the Minimal Supersymmetric Standard Model (MSSM) using String Theory are [**25, 26**]. The works [**27, 57**] provide more details on the subtle global conditions for slope-stability of vector bundles.

To accurately describe the gauge dynamics on 7-branes, including their back-reactions on the compactification geometry B_n at all string coupling orders, a non-perturbative extension of type IIB String Theory is required. This can be achieved elegantly through F-theory, as explained in [**84, 85, 94**]. The back-reactions in question are encoded in the geometry of an elliptically fibered Calabi-Yau space, denoted by $\pi \colon Y_{n+1} \twoheadrightarrow B_n$. The properties of this space can be investigated using established tools of algebraic geometry, which allows for the study and satisfaction of the global consistency conditions of the physics in $10 - 2n$ non-compact dimensions. For a current overview of the recent developments towards connecting F-theory and the Standard Model, interested readers are directed to [**45**].

The chiral fermionic spectrum is a significant feature of particle physics that must be replicated in a realistic 4d $\mathcal{N} = 1$ F-theory compactification, specifically when $n = 3$. In F-theory, the chiral spectrum is determined by a background gauge flux that can be described by the internal C_3 profile in the dual M-theory geometry. The chiral spectrum only relies on $G_4 = dC_3 \in H^{(2,2)}(Y_4)$. Several toolkits have been designed to create and count the primary vertical subspace of G_4 configurations and identify physical quantities determined by such fluxes [**29, 43, 46, 62, 69, 71, 75, 79, 80**]. These tools have been used to develop chiral F-theory models that are globally consistent [**46, 47, 70, 74, 80**]. The largest currently-known group of explicit string vacua that realize the Standard Model gauge group with their exact chiral spectrum and gauge coupling unification is the *Quadrillion F-theory Standard Models* (QSMs) [**44**].

In a 4-dimensional compactification of F-theory with $\mathcal{N} = 1$ supersymmetry, there exist chiral (super)fields in a particular representation $\mathbf{R}$, as well as their charge conjugate counterparts in the representation $\overline{\mathbf{R}}$. The difference between the number of fields in these two representations determines the chiral spectrum. The counting of these fields is done using the vector-like spectrum, which presents additional challenges since the zero modes depend on the potential's flat directions C_3 and not just on $G_4 = dC_3$. An F-theory "gauge field" is used to specify the vector-like spectrum and is represented by an element of the Deligne cohomology. The program described in [**14, 21, 22**] relies on the fact that some Deligne cohomology subsets can be expressed as Chow classes. This enables the extraction of line bundles $L_{\mathbf{R}}$ defined on curves $C_{\mathbf{R}}$ in B_3, which can be interpreted as the localization of gauge flux on matter curves $C_{\mathbf{R}}$ in the dual IIB picture. This in turn renders vector-like pairs on these curves massive. The counting of the zero modes is carried out using the sheaf cohomology groups of the line bundle $L_{\mathbf{R}}$. Specifically, there are $h^0(C_{\mathbf{R}}, L_{\mathbf{R}})$ massless chiral superfields in the representation $\mathbf{R}$ on $C_{\mathbf{R}}$, while there are $h^1(C_{\mathbf{R}}, L_{\mathbf{R}})$ massless chiral superfields in the representation $\overline{\mathbf{R}}$ on the same curve $C_{\mathbf{R}}$.

The aforementioned process is theoretically sound, but in practice, technical difficulties arise due to the intricate relationship between line bundle cohomologies and the complex structure of Y_4. Even with state-of-the-art algorithms [**52, 53, 58, 93**] (see also [**14, 22**]) and supercomputers, such as `Plesken` at *Siegen University*, the necessary computations cannot be performed in realistic compactification geometries. As a result, previous works such as [**14, 21, 22**] focused on models with computationally simple geometries, although they had unrealistically large numbers of chiral fermions.

Motivated by the progress made in using machine learning techniques for applications in String Theory [**1, 35, 63**] (see also [**92**]), recent studies have utilized machine learning techniques in an attempt to overcome this limitation. For instance, the work in [**17**] investigated the complex structure dependency of line bundle cohomologies in a systematic way. A large dataset [**16**] of line bundle cohomologies for different complex structure moduli was generated with the algorithms in [**90**]. This dataset was analyzed with with data science techniques. These insights were put on a sound theoretical foundation by use of Brill-Noether theory [**32**].[1] These insights led to a quantitative study of jumps in charged matter vector pairs as a function of the complex structure moduli of the matter curve $C_{\mathbf{R}}$.

The quantization condition for a G_4-flux – $G_4 + \frac{1}{2}c_2(Y_4) \in H^4_{\mathbb{Z}}(Y_4)$ – is often difficult to verify computationally. The authors of [**44**] examined *necessary* conditions for a specific G_4-flux to satisfy the quantization conditions. We will proceed under the assumption that the quantization condition is satisfied for the F-theory QSMs. It is important to mention that the proposed G_4-flux candidate for the F-theory QSMs cancels the D3-tadpole and renders the $U(1)$-gauge boson massless.

As a result, there is a need to investigate the vector-like spectra present in QSMs, which are localized on five matter curves and were first examined in [**18**]. The study showed that on three of these curves, the line bundle $L\mathbf{R}$ must be a fractional power $P\mathbf{R}$ of the canonical bundle, while on the remaining two curves, the line bundles are altered by contributions from the Yukawa points. These fractional powers of line bundles are known as root bundles, which can be seen as generalizations of spin bundles. The term "P" refers to root bundles throughout the article, inspired by the Greek word "Ρίζα", meaning "root". Similar to spin bundles, root bundles are not unique. The mathematics behind root bundles implies that we should consider different root bundles as being induced from non-equivalent gauge potentials for a given G_4-flux.

Typically, we cannot assume that all allowable root bundles on the matter curve $C_{\mathbf{R}}$ can be induced by F-theory gauge potentials in Deligne cohomology. This is similar to how only certain spin bundles on the matter curves are expected to be consistent with the F-theory geometry Y_4. Rather, an F-theory gauge potential generates a set $P_{\mathbf{R}}$ that includes one root bundle for each matter curve $C_{\mathbf{R}}$. By repeating this process for all physically relevant F-theory gauge potentials, we should anticipate that only a subset of all root bundles on $C_{\mathbf{R}}$ will generally be reached. Conversely, if we are given a set of root bundles $P_{\mathbf{R}}$ (one for each matter curve),

[1] For a contemporary presentation of Brill-Noether theory, we refer the interested reader to [**54**]. An earlier usage of Brill-Noether theory in the context of F-theory can be found in [**95**].

then this set may not originate from an F-theory gauge potential. It is possible for one of the root bundles to not be induced at all or not be induced in combination with one of the other roots.

To determine which root bundles are physical (i.e., induced from an F-theory gauge potential) is crucial for the physics, but it is also a very challenging question. Therefore, [**18**] chose to conduct a systematic study of all root and spin bundles on each matter curve instead of investigating this question. This allowed the identification of a root bundle and a spin bundle on all matter curves except for the Higgs curve in a particular QSM geometry [**44**], such that their tensor product is a line bundle with the cohomologies desired for vector-like spectra. Those bundles are found by deforming the smooth, physical matter curve $C_{\mathbf{R}}$ into a nodal curve $C^{\bullet}_{\mathbf{R}}$ and describing root bundles on the latter in a diagrammatic way using *limit roots* [**34**]. Counting the global sections of limit root bundles on the full blow-up of $C^{\bullet}_{\mathbf{R}}$ can be done using the techniques developed in [**18**].

In [**20**], the study was expanded by using a computer algorithm to count all root bundles with three global sections on the full blow-up of $C^{\bullet}_{\mathbf{R}}$ for various QSM geometries. This required examining families of toric spaces $B_3(\Delta^\circ)$ associated with polytopes Δ° in the Kreuzer-Skarke list [**77**] (see also [**64**]). These toric spaces are different desingularizations of toric K3-surfaces [**11, 28, 40, 87, 91**]. It was found that the nodal curve $C^{\bullet}_{\mathbf{R}}$, and thus the number of limit root bundles, only depends on the polytope Δ°. With a computer scan, results were obtained that apply to a large number of QSM geometries. These techniques were refined in [**19**], so that not only full blow-up limit roots but also certain partial blow-up limit roots could be handled. This computer scan indicates that absence of vector-like exotics in the $(\mathbf{3}, \mathbf{2})_{1/6}$ representation is a very likely scenario within the F-theory QSMs. It must be emphasized that this finding has a statistical meaning, as the top-down origin of the limit roots in F-theory gauge potentials has not been clarified yet.

Outline. This note summarizes and presents the findings of three research works: [**18–20**]. It starts by reviewing how vector-like spectra in F-theory are related to Deligne cohomology and Chow groups in section 2. Then, section 3 describes the study of root bundles in the context of F-theory, followed by an explanation of nodal deformations and limit roots in section 4.1. In section 4.2, it is shown that for F-theory QSMs, the nodal curve is invariant across different geometries, allowing for the approximation of vector-like spectra for a class of F-theory QSMs from a single study. These techniques are then applied to 33 families of F-theory QSMs. A computer scan is used to gauge the statistical likelihood of absence of vector-like exotics, and the cases left by the computer scan are treated manually, as exemplified for the QSM family $B_3(\Delta_4^\circ)$ in section 4.5. Finally, a summary and outlook are given in section 5. The studies in [**18–20**] do not provide a clear picture of which root bundles are realized by F-theory gauge potentials, but statistically speaking, for the selected 33 QSM families, it is likely that many roots are realized in this top-down fashion.

2. Deligne cohomology and vector-like spectra in F-theory

Type IIB string theory has an effective super-gravity description at the massless level. In flat Minkowski space with 10 dimensions, the effective field theory has supersymmetry $\mathcal{N} = (2, 0)$. Its bosonic fields include the dilaton ϕ and the

Ramond-Ramond 0-form field C_0. The value of the dilaton ϕ_0 far away from D_p-branes determines the string coupling constant g_s as $g_s = e^{\phi_0}$ [**59, 60, 88, 89**]. It is convenient to combine the dilaton and the Ramond-Ramond field into a single complex-valued field called the axio-dilaton:

$$\tau\colon \mathcal{E}_4 \times B_3 \to \mathbb{C},\ x^\mu \mapsto C_0(x^\mu) + ie^{-\phi(x^\mu)}. \tag{2.1}$$

In this equation, $\mathcal{E}_4$ refers to the internal 4-dimensional real space while B_3 refers to the internal 3-dimensional complex space. The axio-dilaton remains constant on $\mathcal{E}_4$. However, due to the backreactions of D_7-branes, it varies across B_3. Additionally, unless the theory is trivial, the axio-dilaton τ must become infinite at specific loci of B_3, thereby indicating strong-coupling effects in g_s.

The type IIB supergravity action has an $\mathrm{SL}(2, \mathbb{Z})$-symmetry that can be used to interpret the axio-dilaton τ as the complex structure modulus of an elliptic curve $\mathbb{C}_{1,\tau(x^\mu)}$ for each $x^\mu \in B_3$. More information on this can be found in [**96**] and related works. An elliptic fibration $\pi\colon Y_4 \twoheadrightarrow B_3$ with $\pi^{-1}(x^\mu) = \mathbb{C}_{1,\tau(x^\mu)}$ serves as a geometric tool that records the value of τ over B_3. Since τ diverges over certain loci of B_3 (unless the type IIB supergravity theory is trivial), this elliptic fibration must be singular. This makes the study of singular elliptic fibrations a common topic in F-theory.

To study F-theory, one could try to work with singular fibrations directly. Among others, [**8, 9, 36, 38, 39**] follow this line of thought. However, in this article, the assumption is made that Y_4 has a smooth, crepant resolution which is a flat elliptic fibration $\widehat{\pi}\colon \widehat{Y_4} \twoheadrightarrow B_3$. Such fibrations may or may not have sections. Studies on F-theory without sections include [**4, 31, 42, 56, 72, 73, 79, 81–83, 97**], whereas this article assumes that $\widehat{\pi}$ has a section.

The irreducible components of the codimension-1 locations in B_3 where the elliptic curve $\mathbb{C}_{1,\tau(x^\mu)}$ becomes singular are known as gauge divisors. The singularity type over each gauge divisor contains information about an affine Dynkin diagram and thereby relates each gauge divisor to a Lie group. This Lie group is referred to as the gauge group of the corresponding gauge divisor. More details can be found in [**96**] and references therein.

The irreducible components of the codimension-2 locus of B_3 over which the elliptic curve $\mathbb{C}_{1,\tau(x^\mu)}$ is singular, are termed matter curves. The singularity type of each matter curve determines a representation of each gauge group associated to a gauge divisor in which the matter curve is contained. Therefore, matter curves are typically denoted as $C_{\mathbf{R}}$ with $\mathbf{R}$ the representation in question. In quantum field theories, representations are the mathematical counterpart for matter particles. Therefore, it is common to say that matter in representation $\mathbf{R}$ is localized on $C_{\mathbf{R}}$.

Yukawa points refer to the irreducible components of the codimension-3 locus of B_3 over which the elliptic curve $\mathbb{C}_{1,\tau(x^\mu)}$ singular. These points are located at the intersections of matter curves. In terms of physics, these points represent interactions among the matter fields that are present on the matter curves which intersect at that point. In F-theory Standard models, these intersections produce the Yukawa interactions in the Standard model.

The G_4-flux $G_4 \in H^{(2,2)}(\widehat{Y}_4)$ needs to satisfy the quantization condition [**98**]

$$G_4 + \frac{1}{2} c_2(T_{\widehat{Y}_4}) \in H^{(2,2)}_{\mathbb{Z}}(\widehat{Y}_4) = H^{(2,2)}(\widehat{Y}_4) \cap H^4(\widehat{Y}_4, \mathbb{Z}). \tag{2.2}$$

If $\widehat{Y}_4$ is a smooth Calabi-Yau 4-fold that has a globally defined Weierstrass model, then $G_4 \in H^{(2,2)}\mathbb{Z}(\widehat{Y}4)$ due to the evenness of the class $c_2(T\widehat{Y}_4)$ [**37**]. It can be challenging to verify this condition due to computational limitations. The authors of the *Quadrillion F-theory Standard Models* [**44**] evaluated necessary conditions and confirmed that the chosen G_4-flux candidate satisfies both the D3-tadpole cancellation and the masslessness of the $U(1)$-gauge boson.

There is a surjection from the Deligne cohomology group

$$\widehat{c} \colon H^4_D(\widehat{Y}_4, \mathbb{Z}(2)) \twoheadrightarrow H^{(2,2)}_{\mathbb{Z}}(\widehat{Y}_4), \tag{2.3}$$

and a G_4-flux can be lifted to an F-theory gauge potential A in the Deligne cohomology group $H^4_D(\widehat{Y}_4, \mathbb{Z}(2))$. While this group is difficult to work with in explicit computations, a subset of it can be parametrized by the map

$$\widehat{\gamma} \colon \mathrm{CH}^2(\widehat{Y}_4, \mathbb{Z}) \to H^4_D(\widehat{Y}_4, \mathbb{Z}(2)). \tag{2.4}$$

Therefore, we focus on investigating $\mathcal{A} \in \mathrm{CH}^2(\widehat{Y}_4, \mathbb{Z})$ that satisfies $(\widehat{c} \circ \widehat{\gamma})(\mathcal{A}) = G_4$.

We can link each $\mathcal{A} \in \mathrm{CH}^2(\widehat{Y}_4, \mathbb{Z})$ to a divisor on the matter curve $C_{\mathbf{R}}$ through the cylinder map $D_{\mathbf{R}} : \mathrm{CH}^2(\widehat{Y}_4, \mathbb{Z}) \to \mathrm{CH}^1(C_{\mathbf{R}}, \mathbb{Z}) \cong \mathrm{Pic}(C_{\mathbf{R}})$. For this, we use the matter surface $S_{\mathbf{R}} \subseteq \widehat{Y}4$ – actually a 2-cycle and not a surface – which is a linear combination of $\mathbb{P}^1$-fibrations over $C_{\mathbf{R}}$, with the coefficients determined by the weight vector $\mathbf{w}$ of the representation $\mathbf{R}$ (see [**96**] and references therein). In terms of $\iota_{S_{\mathbf{R}}} : S_{\mathbf{R}} \hookrightarrow \widehat{Y}_4$ and $\pi_{S_{\mathbf{R}}} : S_{\mathbf{R}} \twoheadrightarrow C_{\mathbf{R}}$, the cylinder map is given by

$$D_{\mathbf{R}}(\mathcal{A}) = \pi_{S_{\mathbf{R}}*}\left(\iota^*_{S_{\mathbf{R}}}(\mathcal{A})\right) \in \mathrm{Pic}(C_{\mathbf{R}}). \tag{2.5}$$

Note that $\iota^*_{S_{\mathbf{R}}}(\mathcal{A}) \in \mathrm{CH}^2(S_{\mathbf{R}}, \mathbb{Z})$ describes the restriction of $\mathcal{A}$ to the matter surface $S_{\mathbf{R}}$. The pushforward along $\pi_{S_{\mathbf{R}}}$ then projects this algebraic cycle to a divisor class $D_{\mathbf{R}}(\mathcal{A}) \in \mathrm{CH}^1(C_{\mathbf{R}}, \mathbb{Z}) \cong \mathrm{Pic}(C_{\mathbf{R}})$.[2]

The divisor $D_{\mathbf{R}}(\mathcal{A})$ gives rise to the line bundle [**14, 21, 22**]

$$P_{\mathbf{R}}(\mathcal{A}) = \mathcal{O}_{C_{\mathbf{R}}}(D_{\mathbf{R}}(\mathcal{A})) \otimes_{\mathcal{O}_{C_{\mathbf{R}}}} \mathcal{O}^{\mathrm{spin}}_{C_{\mathbf{R}}}, \tag{2.6}$$

where $\mathcal{O}^{\mathrm{spin}}_{C_{\mathbf{R}}}$ is a spin bundle on $C_{\mathbf{R}}$ compatible with the global structure of the F-theory compactification on $\widehat{Y}_4$. We will elaborate on $\mathcal{O}^{\mathrm{spin}}_{C_{\mathbf{R}}}$ momentarily. The sheaf cohomologies of $P_{\mathbf{R}}(\mathcal{A})$ encode the vector-like spectrum on $C_{\mathbf{R}}$:

$$h^0(C_{\mathbf{R}}, P_{\mathbf{R}}(\mathcal{A})) \leftrightarrow \text{chiral zero modes in representation } \mathbf{R}, \tag{2.7}$$

$$h^1(C_{\mathbf{R}}, P_{\mathbf{R}}(\mathcal{A})) \leftrightarrow \text{chiral zero modes in charge conjugate rep. } \overline{\mathbf{R}}. \tag{2.8}$$

Note that the Riemann-Roch theorem implies

$$\chi(P_{\mathbf{R}}(\mathcal{A})) = h^0(C_{\mathbf{R}}, P_{\mathbf{R}}(\mathcal{A})) - h^1(C_{\mathbf{R}}, P_{\mathbf{R}}(\mathcal{A})) = \deg(D_{\mathbf{R}}(\mathcal{A})) = \int_{S_{\mathbf{R}}} G_4. \tag{2.9}$$

For simplicity, let us assume $\chi \geq 0$. Then, in the case where there are no unusual vector-like pairs present, the value of $h^0(C_{\mathbf{R}}, P_{\mathbf{R}}(\mathcal{A}))$ is equal to $\chi(P_{\mathbf{R}}(\mathcal{A}))$. On the other hand, if there is exactly one vector-like pair, then $h^0(C_{\mathbf{R}}, P_{\mathbf{R}}(\mathcal{A}))$ is equal to $\chi(P_{\mathbf{R}}(\mathcal{A})) + 1$. This second scenario is particularly relevant in the context of

[2]Just as in [**21**], we adopt a notation where superscripts indicate the codimension of algebraic cycles within the containing variety, while subscripts represent the actual dimension of the algebraic cycles. For example, we have the equivalence $\mathrm{CH}^1(C_{\mathbf{R}}, \mathbb{Z}) \equiv \mathrm{CH}_0(C_{\mathbf{R}}, \mathbb{Z}) \cong \mathrm{Pic}(C_{\mathbf{R}})$. It is important to note that the use of subscripts and superscripts in the Chow groups is purely symbolic, and no conversions between homology and cohomology classes, such as Poincaré duality, are involved.

constructing an F-theory MSSM, where having precisely one vector-like pair on the Higgs curve is desirable.

We will now discuss the spin bundle $\mathcal{O}_{C_{\mathbf{R}}}^{\text{spin}}$. A smooth genus g curve has 2^{2g} spin bundles, so it is necessary to specify which one is being used for $\mathcal{O}_{C_{\mathbf{R}}}^{\text{spin}}$. The cancellation of the Freed-Witten anomaly requires spin^c-structures on $D7$-branes in perturbative IIB compactifications, as shown in [**55**]. This requirement extends to the demand for spin^c-structures on gauge surfaces $S \subset B_3$ in F-theory GUTs, as explained in [**12**]. Then a choice of spin^c-structure on $N_{C_{\mathbf{R}}|S}$ induces a unique spin^c-structure on $C_{\mathbf{R}}$ as shown in [**78**].

In general, it is difficult to determine the spin structure on the matter curve $C_{\mathbf{R}}$ needed to analyze the zero mode spectrum. The procedure outlined in [**12**] assumes that the matter curve $C_{\mathbf{R}}$ belongs to the canonical class of a GUT gauge divisor, but this is not the case for the matter curves in F-theory QSMs [**44**] or the works in [**14, 21, 22**]. In the latter cases, a non-trivial G_4-flux was chosen to cancel the ambiguity introduced by the choice of one of the 2^{2g} spin bundles on the matter curve $C_{\mathbf{R}}$. To identify the spin bundles in F-theory compactifications such as [**14, 21, 22, 44**], it seems necessary to repeat the analysis of [**12**] for these more general cases, while also investigating the role of 2-torsion F-theory gauge potentials $A_2 \in J^2(\widehat{Y}_4)$ in the intermediate Jacobian of $\widehat{Y}4$ and their effect on the spin structure of the matter curve $C_{\mathbf{R}}$. However, this is a technically challenging study that has not yet been attempted.

Instead, this note will provide a necessary condition for both the spin bundle and the line bundles induced by the G_4-flux, and study all possible combinations of these line bundles to gain a statistical understanding of the likelihood of finding the absence of vector-like exotics in certain representations of the Quadrillion F-theory Standard Models [**44**].

3. Necessary root bundle constraints in F-theory

Let us return to the topic of F-theory gauge potentials and examine them more closely. These potentials are elements of the Deligne cohomology group $H_D^4(\widehat{Y}_4, \mathbb{Z}(2))$. In practical terms, a gauge potential is often defined as $\widehat{\gamma}(\mathcal{A})$ for a "potential" $\mathcal{A} \in \mathrm{CH}^2(\widehat{Y}_4, \mathbb{Z})$. From the geometry, we can derive a class $\widehat{\gamma}(\mathcal{A}') \in H_D^4(\widehat{Y}_4, \mathbb{Z}(2))$ and an integer $\xi \in \mathbb{Z}> 0$ that subject $\mathcal{A}$ to two conditions:

$$\gamma(\mathcal{A}) = G_4, \qquad \xi \cdot \widehat{\gamma}(\mathcal{A}) \sim \widehat{\gamma}(\mathcal{A}'). \tag{3.1}$$

The statement $\gamma(\mathcal{A}) = G_4$ implies that $\widehat{\gamma}(\mathcal{A})$ is a valid F-theory gauge potential for the specific G_4-flux. While the G_4-flux has a significant effect on the chiral spectrum, the absence of chiral exotics is determined by the second condition in some compactifications. This applies, for example, to the F-theory QSMs [**44**].

Crucially, the two conditions in equation eq. (3.1) do not necessarily specify a unique gauge potential $\mathcal{A}$. It is challenging to make assertions about solutions in the Chow group, but in the Deligne cohomology, it is known that if the collection of all ξ-th roots of $\widehat{\gamma}(\mathcal{A}')$ is not empty, it is a coset of the group of all ξ-th roots of 0. Consequently, the number of solutions is $\xi^{2\cdot \dim_{\mathbb{C}}\left(J^2(\widehat{Y}_4)\right)}$. These solutions yield the same chiral spectrum, but they may differ in their vector-like spectrum. This additional flexibility appears to be the primary tool for generating a desirable vector-like spectrum like that of the MSSM.

In many practical scenarios, we can construct $\mathcal{A}' \in \mathrm{CH}^2(\widehat{Y}_4, \mathbb{Z})$ explicitly. As a result, we have a sufficient level of arithmetic control over $\widehat{\gamma}(\mathcal{A}')$, and it is reasonable to inquire about the relationship between the induced divisors $D_{\mathbf{R}}(\mathcal{A}')$ and $D_{\mathbf{R}}(\mathcal{A})$. They satisfy **[18]**

$$\xi \cdot D_{\mathbf{R}}(\mathcal{A}) \sim D_{\mathbf{R}}(\mathcal{A}') \in \mathrm{Pic}(C_{\mathbf{R}}). \tag{3.2}$$

To clarify, when we consider the F-theory gauge potential $A = \widehat{\gamma}(\mathcal{A})$, it induces a specific divisor $D_{\mathbf{R}}(\mathcal{A})$ such that its ξ-th multiple is linearly equivalent to the divisor induced by the F-theory gauge potential $A' = \widehat{\gamma}(\mathcal{A}')$. Consequently, $D_{\mathbf{R}}(\mathcal{A})$ is a ξ-th root of $D_{\mathbf{R}}(\mathcal{A}')$.

Usually, ξ-th roots of $D_{\mathbf{R}}(\mathcal{A}')$ do not exist, and even when they do, they are not unique. For the case where $\xi = 2$ and $D_{\mathbf{R}}(\mathcal{A}') = K_{\mathbf{R}}$, the existence of second roots of the canonical bundle is well-known, and these are the spin structures on $C_{\mathbf{R}}$. If $C_{\mathbf{R}}$ is a smooth curve of genus g, then it has 2^{2g} spin structures **[10, 86]**. This result can be generalized to $\xi > 2$ and $D_{\mathbf{R}}(\mathcal{A}') \neq K_{\mathbf{R}}$. Specifically, ξ-th roots of $D_{\mathbf{R}}(\mathcal{A}')$ exist if and only if ξ divides the degree of $D_{\mathbf{R}}(\mathcal{A}')$ and then, there are ξ^{2g} ξ-th roots of $D_{\mathbf{R}}(\mathcal{A}')$ on a smooth genus g curve **[18]**.

In general, it cannot be assumed that there is a one-to-one correspondence between the ξ-th roots of $D_{\mathbf{R}}(\mathcal{A}') \in \mathrm{Pic}(C_{\mathbf{R}})$ and the ξ-th roots of $A' = \widehat{\gamma}(\mathcal{A}') \in H^4_D(\widehat{Y}_4, \mathbb{Z}(2))$. Only a subset of the ξ-th roots of $D_{\mathbf{R}}(\mathcal{A}')$ can be realized from F-theory gauge potentials in $H^4_D(\widehat{Y}_4, \mathbb{Z}(2))$. Thus, the root bundle constraint in eq. (3.2) is necessary but not sufficient to determine whether the divisor $D_{\mathbf{R}}(\mathcal{A})$ arises from an F-theory gauge potential. This is analogous to the challenge of identifying physical spin bundles. The author is unaware of a solution to this problem and will not attempt to provide one in this note. Rather, the goal is to systematically examine all ξ-th roots of $D_{\mathbf{R}}(\mathcal{A}')$ and all spin bundles on $C_{\mathbf{R}}$ in order to identify combinations of root and spin bundles that result in a line bundle whose cohomologies satisfy the physical demand for the presence or absence of vector-like pairs. This study is a first step towards understanding the likelihood of finding solutions to 4-dimensional F-theory compactifications without vector-like exotics.

As an example, look at the F-theory QSMs **[44]**. These geometries are defined in terms of toric 3-folds with $\overline{K}^3_{B_3} \in \{6, 10, 18, 30\}$. The matter curves are defined by sections $s_i \in H^0\left(B_3, \overline{K}_{B_3}\right)$. An explicit study finds the following root bundle constraints **[18]**:

(3.3)

curve	root bundle constraint
$C_{(\mathbf{3},\mathbf{2})_{1/6}} = V(s_3, s_9)$	$P_{(\mathbf{3},\mathbf{2})_{1/6}}^{\otimes 2\overline{K}^3_{B_3}} = K_{(\mathbf{3},\mathbf{2})_{1/6}}^{\otimes\left(6+\overline{K}^3_{B_3}\right)}$
$C_{(\mathbf{1},\mathbf{2})_{-1/2}} = V(s_3, P_H)$	$P_{(\mathbf{1},\mathbf{2})_{-1/2}}^{\otimes 2\overline{K}^3_{B_3}} = K_{(\mathbf{1},\mathbf{2})_{-1/2}}^{\otimes\left(4+\overline{K}^3_{B_3}\right)} \otimes \mathcal{O}_{(\mathbf{1},\mathbf{2})_{-1/2}}(-30Y_1)$
$C_{(\overline{\mathbf{3}},\mathbf{1})_{-2/3}} = V(s_5, s_9)$	$P_{(\overline{\mathbf{3}},\mathbf{1})_{-2/3}}^{\otimes 2\overline{K}^3_{B_3}} = K_{(\overline{\mathbf{3}},\mathbf{1})_{-2/3}}^{\otimes\left(6+\overline{K}^3_{B_3}\right)}$
$C_{(\overline{\mathbf{3}},\mathbf{1})_{1/3}} = V(s_9, P_R)$	$P_{(\overline{\mathbf{3}},\mathbf{1})_{1/3}}^{\otimes 2\overline{K}^3_{B_3}} = K_{(\overline{\mathbf{3}},\mathbf{1})_{1/3}}^{\otimes\left(4+\overline{K}^3_{B_3}\right)} \otimes \mathcal{O}_{(\overline{\mathbf{3}},\mathbf{1})_{1/3}}(-30Y_3)$
$C_{(\mathbf{1},\mathbf{1})_1} = V(s_1, s_5)$	$P_{(\mathbf{1},\mathbf{1})_1}^{\otimes 2\overline{K}^3_{B_3}} = K_{(\mathbf{1},\mathbf{1})_1}^{\otimes\left(6+\overline{K}^3_{B_3}\right)}$

In this table, we make use of two polynomials: $P_H = s_2 s_5^2 + s_1(s_1 s_9 - s_5 s_6)$ and $P_R = s_3 s_5^2 + s_6(s_1 s_6 - s_2 s_5)$. It is important to note that the line bundles on both the Higgs curve $C_{(\mathbf{1},\mathbf{2})_{-1/2}}$ and the curve $C(\overline{\mathbf{3}},\mathbf{1})_{1/3}$ are dependent on the Yukawa points $Y_1 = V(s_3, s_5, s_9)$ and $Y_3 = V(s_3, s_6, s_9)$, respectively. Furthermore, it is important to keep in mind that if two divisors D and E are linearly equivalent, i.e., $D \sim E$, then $n \cdot D \sim n \cdot E$ for any integer n. However, the converse is not true, and that is why we do not cancel common factors.

The root bundle constraints in eq. (3.3) can be connected to the original work in [**44**]. In that work, a G_4-flux was divided by $\overline{K}_{B_3}^3$ to eliminate chiral exotics. Equivalently, this division ensured that the line bundles on all matter curves possess an Euler characteristic/chiral index of exactly 3. In moving from the degree of a line bundle to the actual line bundle, one must wonder what the proper interpretation of "dividing a line bundle by $\overline{K}_{B_3}^3$" should be. Indeed, there is a natural mathematical analogue as we explained above, and this is to take roots of line bundles. Note also that these roots exist by virtue of an integral chiral index, which in turn ensures that the degree of the line bundle on the RHS in eq. (3.3) is divisible by the exponents on the LHS of eq. (3.3). In comparison, the integrability of the G_4-flux itself matters only so much to the root bundles in that this guarantees the existence of an F-theory gauge potential $\mathcal{A} \in \mathrm{CH}^2(\widehat{Y}_4, \mathbb{Z})$ with $\widehat{c}(\mathcal{A}) = G_4$, where $\widehat{c}$ is the map discussed around eq. (2.3). Finally, note that we are taking $2\overline{K}_{B_3}^3$-th roots in eq. (3.3). This is necessary to account for the spin bundle. The interested reader can find more details in [**18**].

4. Vector-like spectra of F-theory QSMs from limit roots

To construct root bundles that meet the constraints in eq. (3.3) and have the phenomenologically desired cohomologies, we begin by deforming the curve $C_{\mathbf{R}}$ to a nodal curve $C_{\mathbf{R}}^{\bullet}$. Deforming to a nodal curve has two benefits. Firstly, constructing root bundles on smooth curves is difficult, while constructing them on nodal curves via so-called limit roots has been thoroughly detailed and involves combinatorial tasks [**34**]. Secondly, for F-theory QSMs, the nodal curve is invariant across a range of different geometries, so we can approximate the vector-like spectra for many F-theory QSMs with a single computation. To fully appreciate these advantages, we will provide a summary of limit root bundles in section 4.1 and explain the invariance of the nodal matter curve in section 4.2.

Certainly, we eventually want to determine the cohomologies of root bundles on smooth, irreducible matter curves, so we study how those cohomologies relate to the cohomologies of limit roots. The pushforward of limit roots $P_{\mathbf{R}}^{\circ}$ through the blow-up map $\pi\colon C_{\mathbf{R}}^{\circ} \to C_{\mathbf{R}}^{\bullet}$ keeps the number of global sections the same, that is, $h^0(C_{\mathbf{R}}^{\circ}, P_{\mathbf{R}}^{\circ}) = h^0(C_{\mathbf{R}}^{\bullet}, P_{\mathbf{R}}^{\bullet})$. In principle, we can trace these roots $P_{\mathbf{R}}^{\bullet}$ along the deformation $C_{\mathbf{R}}^{\bullet} \to C_{\mathbf{R}}$ to find roots $P_{\mathbf{R}}$ on the original curve $C_{\mathbf{R}}$. However, this deformation can alter the number of sections. For the deformation $C_{\mathbf{R}}^{\bullet} \to C_{\mathbf{R}}$, where a nodal curve is transformed into a smooth, irreducible curve, it is known that the number of sections is an upper semi-continuous function. Thus, the number of sections either stays the same or decreases:

$$h^0(C_{\mathbf{R}}, P_{\mathbf{R}}) \leq h^0(C_{\mathbf{R}}^{\bullet}, P_{\mathbf{R}}^{\bullet}) = \chi(P_{\mathbf{R}}) + \delta, \qquad \delta \in \mathbb{Z}_{\geq 0}. \tag{4.1}$$

In order to find roots for which the number of sections remains constant upon deformation to the original curve, we can look for instances where equality holds. One such instance is the generic case $\delta = 0$, in which $h^0(C_{\mathbf{R}}^{\bullet}, P_{\mathbf{R}}^{\bullet}) = \chi(P_{\mathbf{R}})$. This is because the number of sections is already minimal on $C_{\mathbf{R}}^{\bullet}$ and therefore must remain constant throughout the deformation. As a result, the number of limit roots on the partial blow-ups $C^{\circ}_{\mathbf{R}}$ of the nodal curve $C^{\bullet}_{\mathbf{R}}$ that have the generic number of global sections provides a lower bound on the number of roots $P_{\mathbf{R}}$ without vector-like exotics.

For 33 F-theory QSMs families, which we select in section 4.3, the predictive powers of these counts are strongest. This is due to the limitation that it is not yet clear which of the mathematical root bundles stem from F-theory gauge potentials. In section 4.4, we then outline techniques to count the number of limit roots with prescribed number of global sections. This allows us to gauge – statistically speaking – how likely absence of vector-like exotics is realized within the F-theory QSMs. The computer implementation used for to obtain these counts cannot (yet) determine the number of global sections for all limit roots. Rather, if we want to see the complete picture, a few remaining cases must be treated by hand. For the QSM familiy $B_3(\Delta^{\circ}_4)$, this is outlined in section 4.5. We find that more than 99.995% of the limit roots admit exactly three global sections, just as desired for string phenomenology.

4.1. Limit roots. To recap, on a smooth curve C, n-th roots P of a line bundle L exist when $\deg(L)$ is divisible by n. However, this is not the case for reducible, nodal curves $C^{\bullet}$. In this situation, a root $P^{\bullet}$ of a line bundle $L^{\bullet}$ on such curves should restrict to a root on the irreducible components, but even if n divides the degree of $L^{\bullet}$, it may not divide $\deg(L|_Z)$ for some irreducible component Z of $C^{\bullet}$. This issue is elegantly avoided by utilizing limit n-th root bundles P° on (partial) blow-ups C° of $C^{\bullet}$, as originally introduced in [**34**]. These limit n-th root bundles P° are determined by weighted graphs, which can be exploited to make the task of section-counting manageable. In the following section, we will briefly outline this combinatorial approach to limit roots. Those interested in further information on limit roots may consult [**67, 68**].

To represent a connected nodal curve $C^{\bullet}$ with arithmetic genus g, we create a *dual graph* $\Pi_{C^{\bullet}}$. Each vertex of $\Pi_{C^{\bullet}}$ corresponds to an irreducible component $C^{\bullet}_i$ of $C^{\bullet}$. Each half-edge extending from a vertex $C^{\bullet}_i$ is a node on $C^{\bullet}_i$. If a node is present on both $C^{\bullet}_i$ and $C^{\bullet}_j$, the half-edges connected to $C^{\bullet}_i$ and $C^{\bullet}_j$ come together to form an edge.

EXAMPLE 4.1. We can represent the dual graph $\Pi_{E^{\bullet}}$ of the nodal curve $E^{\bullet}$ with two rational components C_1 and C_2 intersecting at two nodes as follows:

(4.2)

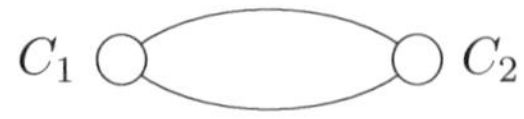

Suppose we have a nodal curve $C^{\bullet}$ and a blow-up $\pi : C^{\circ} \to C^{\bullet}$. If n_i is a node on $C^{\bullet}$, then the exceptional components over n_i are denoted by $\mathcal{E}_i$, each of which is isomorphic to $\mathbb{P}^1$. Let C^N be the complement of the exceptional components in C°. Then, the map $\pi|_{C^N} : C^N \to C^{\bullet}$ is the normalization of $C^{\bullet}$. The points in $(\pi|_{C^N})^{-1}(n_i)$ are called the exceptional nodes and consist of the intersection of the exceptional components with C^N. A *full blow-up* of $C^{\bullet}$ is a blow-up at all the nodes, while a *partial blow-up* is a blow-up at a proper subset of the nodes.

To begin with, consider a positive integer n and a line bundle $L^{\bullet}$ on the nodal curve $C^{\bullet}$ such that n divides the degree of $L^{\bullet}$. Let $\Delta_{C^{\bullet}}$ be a subset of the nodes of $C^{\bullet}$, which may or may not be the entire set.

DEFINITION 4.2. A *limit n-th root* of $L^{\bullet}$ associated to $\Delta_{C^{\bullet}}$ is a triple $(C^{\circ}, P^{\circ}, \alpha)$ consisting of:

- the (partial) blow-up $\pi : C^\circ \to C^\bullet$,
- a line bundle P° on C°,
- a homomorphism $\alpha : (P^\circ)^n \to \pi^*(L^\bullet)$,

satisfying the following properties:

(1) $\deg(P^\circ|_{\mathcal{E}_i}) = 1$ for every exceptional component $\mathcal{E}_i$,
(2) α is an isomorphism at all points of C° except for the exceptional components,
(3) for every exceptional component $\mathcal{E}_i$ of C°, the sum of the orders of vanishing of α at the exceptional nodes p_i and q_i is equal to n.

The combinatorial information of limit n-th roots over $C^\bullet$ is represented by weighted graphs, which are based on the dual graph of the nodal curve $C^\bullet$. Conversely, these graphs allow for the construction and retrieval of limit n-th roots. Although there is not a one-to-one correspondence between the weighted graphs and limit roots, these graphs provide a useful way to parametrize the limit roots. To begin, let us define the weighted graphs in question. The set of exceptional nodes corresponding to $\Delta_{C^\bullet}$ is denoted by $\widetilde{\Delta}_{C^\bullet}$.

DEFINITION 4.3. The *weighted graph linked to a limit n-th root $(C^\circ, P^\circ, \alpha)$ of $L^\bullet$* refers to the dual graph $\Pi_{C^\bullet}$ with weights assigned using a weight function

$$w : \widetilde{\Delta}_{C^\bullet} \to 1, \ldots, n-1, \tag{4.3}$$

where $w(p_i) = u_i$ and $w(q_i) = v_i$ denote the orders at which α vanishes at p_i and q_i respectively.

These weighted graphs have two inherent characteristics:

(A) The sum of weights assigned to vertices p_i and q_i equals $u_i + v_i$, which in turn is equal to n.
(B) For each irreducible component $C_i^\bullet$ of $C^\bullet$, the sum of weights assigned to the vertex corresponding to $C_i^\bullet$ is congruent to $\deg_{C_i^\bullet} L^\bullet$ modulo n.

EXAMPLE 4.4. We return to the nodal curve $E^\bullet$ and wish to find the limit 2nd roots of $K_{E^\bullet}$. If $C_i^\bullet$ is a component of $E^\bullet$, then set $k_i = \#C_i^\bullet \cap (\overline{H^\bullet \setminus C_i^\bullet})$. Therefore, $\deg(K_{E^\bullet}|_{C_i^\bullet}) = 2g(C_i^\bullet) - 2 + k_i = -2 + k_i$, and the multi-degree of $K_{E^\bullet}$ is

$$(\deg(K_{E^\bullet}|_{C_1}), \deg(K_{H^\bullet}|_{C_2}) = (0,0), \tag{4.4}$$

which has total degree is 0. Here are the descriptions of the two weighted graphs connected to the (partial) blow-up limit 2nd roots of $K_{E^\bullet}$:

(4.5)

The labels inside the vertices are the multi-degrees of $K_{E^\bullet}$, while the labels outside the vertices are the weights.

It is important to acknowledge that any weighted graph with an underlying graph of $\Pi_{C^\bullet}$ and a weight function $w : \widetilde{\Delta}_{C^\bullet} \to 1, \ldots, n-1$ that satisfies the conditions (A) and (B) contains information about a limit n-th root $(C^\circ, P^\circ, \alpha)$ of L. Indeed, this specific weighted graph is the same as the weighted graph linked to $(C^\circ, P^\circ, \alpha)$ of $L^\bullet$.

For any nodal curve $C^\bullet$ and line bundle $L^\bullet$ on $C^\bullet$, there exists a total of $n^{b_1(\Pi_{C^\bullet})}$ weighted subgraphs that fulfill conditions (A) and (B), where

$$b_1(\Pi_{C^\bullet}) = \#\text{edges} + \#\text{connected components} - \#\text{vertices},, \tag{4.6}$$

represents the first Betti number of $\Pi_{C^\bullet}$. It is important to note that this includes all of the weighted subgraphs whose edge sets correspond to subsets of $\Delta_{C^\bullet}$.

The relationship between limit n-th roots and weighted graphs that fulfill conditions (A) and (B) is not a one-to-one mapping. In fact, creating a limit root involves selecting a root P^N of $(\pi|_{C^N})^*(L^\bullet)(-\sum(u_ip_i + v_iq_i))$. By doing a precise calculation, it can be determined that there are n^{2g} limit n-th roots [**34**].

EXAMPLE 4.5. We describe the full blow-up limit 2nd roots of $K_{E^\bullet}$:

(1) Blow-up at the set of all nodal singularities, and denote the exceptional component at the i-th node by $\mathcal{E}_i \cong \mathbb{P}^1$. This $\mathbb{P}^1$ touches E_i at the exceptional node p_i and Γ at q_i.
(2) Let E^N be the (full) normalization of $E^\bullet$, and consider the bundle with multi-degree $(-2,-2)$. This bundle admits exactly one root and this root has multi-degree $(-1,-1)$.
(3) Pick the 2nd root P^N, and glue to it a degree one bundle over every $\mathcal{E}_i$. The resulting *limit 2nd root* P° *of* $K_{E^\bullet}$ can be represented as follows:

(4.7)
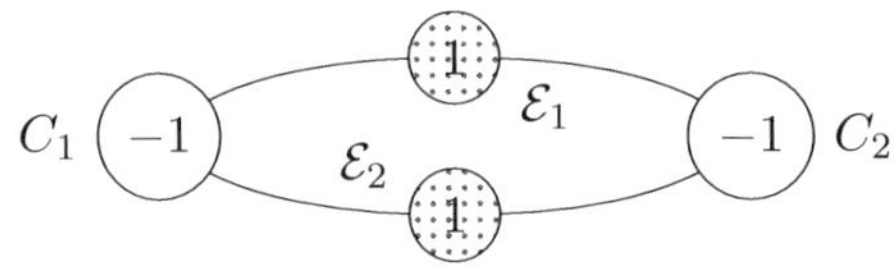

We obtain a distinct limit 2nd root of $K_{E^\bullet}$ by not performing any node blow-ups:

(4.8)
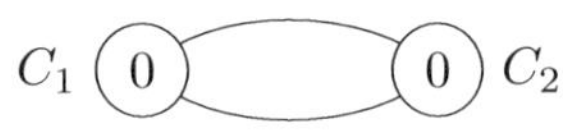

On a smooth curve with genus $g = 1$, there exist four different 2nd roots of K_C, which seems to be in contrast to the case of a nodal curve with genus $g = 1$ for which we only found two distinct weighted subgraphs. To obtain r^{2g} limit rth roots, we note that each weighted graph accounts for multiple limit roots. This multiplicity is known as the *geometric multiplicity* [**34**]. The number of distinct limit roots associated to a weighted subgraph Δ^w is given by $\mu = r^{2g^\nu + b_1(\Gamma_C)}$, where g^ν is the arithmetic genus of the normalization [**19**].

EXAMPLE 4.6. In the case of our example of a nodal $g = 1$ curve, we have $g^\nu = 0$, which means that the geometric multiplicity is $\mu = r^{2g^\nu + b_1(\Gamma_C)} = r^{b_1(\Gamma_C)} = 2$. This implies that the weighted graphs in eq. (4.7) and eq. (4.8) each encode two distinct limit root bundles, and thus we obtain a total of $r^{2g} = 4$ limit spin bundles, as expected.

4.2. FRST-invariant graphs. The F-theory QSM geometries arise from desingularizations of toric K3-surfaces. Such desingularizations were first studied in [**11**] and correspond to 3-dimensional, reflexive lattice polytopes $\Delta \subset M_{\mathbf{R}}$ and their polar duals $\Delta^\circ \subset N_{\mathbf{R}}$ defined by $\langle \Delta, \Delta^\circ \rangle \geq -1$. The complete list of such 3-dimensional polytopes is available in [**77**]. We follow [**20**] and denote the i-th polytope in this list as $\Delta_i^\circ \subset N_{\mathbf{R}}$.

Given a lattice polytope $\Delta \subset M_{\mathbf{R}}$, one can derive its normal fan Σ_Δ by using the facet normals of Δ as ray generators. The vertices of Δ correspond to the maximal cones of Σ_Δ. Although a CY-hypersurface in the toric variety $X_\Delta \equiv X_{\Sigma_\Delta}$ may not be smooth, its singularities can be resolved through a process known as a *maximal projective crepant partial* (MPCP) desingularization, as discussed in [**11**]. The aforementioned resolutions are sometimes referred to as maximal projective subdivisions of the normal fan [**40**]. An MPCP desingularization corresponds to a fine, regular, star triangulation (FRST) $\Sigma(T)$ of the lattice polytope's polar dual, Δ°. In this context, *star* means that every simplex in the triangulation contains the origin, *fine* indicates that every lattice point in Δ° is used as a ray generator[3], and *regular* implies that $X_{\Sigma(T)}$ is a projective variety. Consequently, $\Sigma(T)$ defines a *maximal* projective refinement of the normal fan Σ_Δ. In applications involving toric K3-surfaces, the resulting $X_{\Sigma(T)}$ is smooth [**20**], due to results in [**11, 40, 41**].

The F-theory QSM geometries have a requirement that $\overline{K}^3_{X_{\Sigma(T)}}$ must be one of the values $6, 10, 18, 30$. Out of the 4319 polytopes listed in [**77**], only 708 satisfy this condition [**44**]. These polytopes have multiple FRSTs, with an approximate count provided in [**64**]. The family of toric 3-folds obtained from the various FRSTs of Δ° is denoted by $B_3(\Delta^\circ)$. For each F-theory QSM, a nodal quark-doublet curve can be introduced [**18**]

$$C^\bullet_{(\mathbf{3},\mathbf{2})_{1/6}} = \bigcup_{i \in L} V(x_i, s_9), \tag{4.9}$$

where L is the collection of all lattice points of Δ° and $V(s_9)$ is the K3-surface.

We make the assumption that the element $s_9 \in H^0(X_{\Sigma(T)}, \overline{K}_{X_{\Sigma(T)}})$ is generic. In [**20**], it was argued that the set $V(x_i, s_9)$ is either empty, consists of a single irreducible component, or is a finite collection of smooth $\mathbb{P}^1$s (further background on this topic can be found in [**11, 40, 76, 87, 91**]). The number of intersection points among the irreducible components of $C^\bullet_{(\mathbf{3},\mathbf{2})_{1/6}}$ is equal to their topological intersection number, which in turn depends only on Δ° [**20**]. Therefore, the dual graph of $C^\bullet_{(\mathbf{3},\mathbf{2})_{1/6}}$ is the same for all FRSTs of Δ°, meaning that it is identical for all spaces in $B_3(\Delta^\circ)$. This is an impressive finding! Despite there being an enormous number of F-theory QSM geometries in $B_3(\Delta^\circ_8)$ (around 10^{15}), it is possible to estimate the vector-like spectra of all of them by counting the global sections of limit roots on a single nodal curve.

The interested reader may wonder about the physics of the nodal curves and if triangulation independence could be expected from the physics. Sadly, the physics behind this particular nodal configuration exceeds the author's current knowledge and abilities. Consequently, it cannot be ruled out that a physics-inspired argument would predict the triangulation independence. Let us elaborate more on the challenges that hindering the physics interpretation of the nodal configuration $C^\bullet_{(\mathbf{3},\mathbf{2})_{1/6}}$.

First, the physics interpretation of the nodal limit by itself is challenging. While we deform the F-theory geometry such that the smooth irreducible matter curve $C_{(\mathbf{3},\mathbf{2})_{1/6}}$ turns into the canonical nodal matter curve $C^\bullet_{(\mathbf{3},\mathbf{2})_{1/6}}$, it is likely that this deformation introduces significant singularities outside of $C(\mathbf{3},\mathbf{2})1/6^\bullet$. This complicates the physics interpretation of the global setup. Even when focusing solely on $C(\mathbf{3},\mathbf{2})_{1/6}{}^\bullet$, challenges persist. In the usual F-theory approach, the smooth irreducible components of this nodal curve are treated as independent matter curves.

[3]The described property is also known as being *full*[**51**].

However, this approach faces an obstacle, as the chiral indices on the smooth irreducible components are typically not divisible by $2\overline{K}^3_{X_{\Sigma(T)}}$. In other words, taking local roots on the smooth irreducible components is not possible, and this is precisely why we turned to the techniques described in [**34**]. This procedure handles the limit roots by blowing up a subset of the nodal singularities, replacing the blown-up nodes with a pair of an exceptional $\mathbb{P}^1$ and the line bundle $\mathcal{O}\mathbb{P}^1(1)$ thereon. Consequently, additional sections or zero modes are introduced. These modes may have already existed at the nodal singularity.

Second, we wish to connect the nodal limit to counting zero modes on the smooth irreducible components. For this, we should anticipate interactions among the modes localized on these smooth irreducible components of $C(\mathbf{3},\mathbf{2})_{1/6}{}^{\bullet}$, potentially mediated by modes localized at nodal singularities. Such interactions then lead to Yukawa interactions, which can render some initially massless modes massive during the smoothing process. This effect is particularly crucial for engineering a Higgs boson. In order to understand this effect, a detailed study similar to [**48**] is needed. However, a clear understanding of the physics on $C^{\bullet}_{(\mathbf{3},\mathbf{2})_{1/6}}$ is a prerequisite, and as previously discussed, this presents a challenging task.

4.3. Selection among the F-theory QSMs. To find the number of limit roots $P^{\bullet}_{(\mathbf{3},\mathbf{2})_{1/6}}$ that satisfy the conditions of having three global sections, we need to investigate the $(6+\overline{K}^3_{B_3})$-th power of the canonical bundle on $C^{\bullet}_{(\mathbf{3},\mathbf{2})1/6}$ (cf. eq. (3.3)). There are $(2\overline{K}^3_{B_3})^{2g}$ such roots on a genus g curve. In comparison, there are $(2\overline{K}^3_{B_3})^{2h^{2,1}(\widehat{Y}_4)}$ inequivalent F-theory gauge potentials, but we do not yet know which of these roots are physical. Therefore, we focus on F-theory QSMs that have at least as many F-theory gauge potentials as roots on $C_{(\mathbf{3},\mathbf{2})_{1/6}}$, which means that many of the roots can be expected to come from F-theory gauge potentials. There are 37 F-theory QSM families that satisfy $h^{2,1}(\widehat{Y}_4) \geq g$, but four of these families have $C^{\bullet}_{(\mathbf{3},\mathbf{2})_{1/6}}$ with a component whose genus is greater than one. Since this makes it difficult to compute the number of global sections, we ignore these four QSM families for now. The triangulations of the remaining 33 polytopes provide most of the $\mathcal{O}(10^{15})$ F-theory QSMs. [**44, 64**].

4.4. First steps towards Brill-Noether theory of limit roots. We will demonstrate how to use section counting techniques to analyze limit roots in F-theory QSMs [**44**], by using the example of the fourth polytope Δ_4° in the Kreuzer-Skarke database [**77**]. We examine the dual graph of $C^{\bullet}_{(\mathbf{3},\mathbf{2})_{1/6}}$ for the $\mathcal{O}(10^{11})$ geometries $B_3(\Delta_4^\circ)$ [**64**], which can be represented as follows:

(4.10)
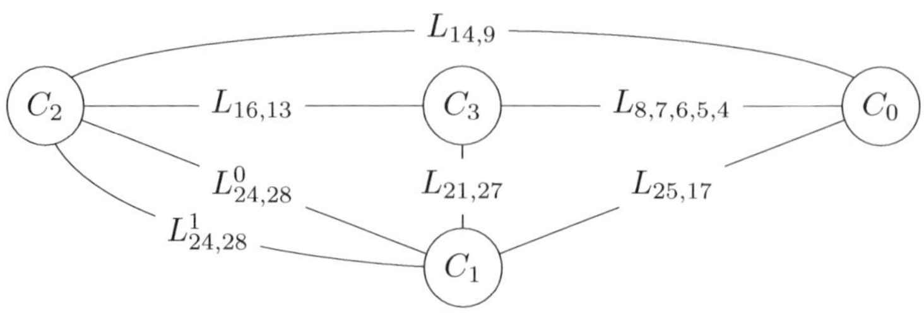

In this diagram, the edges indicate nodal singularities, while the vertices represent the irreducible components of $C^{\bullet}_{(\mathbf{3},\mathbf{2})1/6}$. The four curves C_i are all smooth $\mathbb{P}^1$s.

Additionally, $L_{i_1,\ldots,i_k}$ denotes a link of k smooth, genus-zero curves C_{i_j}, such that each pair of adjacent curves intersects at exactly one node. Therefore, as a dual graph, $L_{i_1,\ldots,i_k}$ corresponds to a link of k vertices, with the vertices ordered such that C_{i_1} intersects the $\mathbb{P}^1$ on the left and C_{i_k} intersects the $\mathbb{P}^1$ on the right. For example, in the case of $L_{21,27}$, C_{21} intersects C_3, while C_{27} intersects C_1.

Remember that for a curve C with dual graph Γ_C, there are $r^{b_1(\Gamma_C)}$ weighted subgraphs that meet the conditions (C1) and (C2) in **[34]**. In the case where C is a link of rational curves, we have $b_1(\Gamma_C) = 0$, and therefore there is only one possible weighted subgraph, which corresponds to the full blow-up. As a result, we can substitute the dual graph in eq. (4.10) with the following graph, which simplifies the process of counting limit root bundles **[19, 20]**:

(4.11)

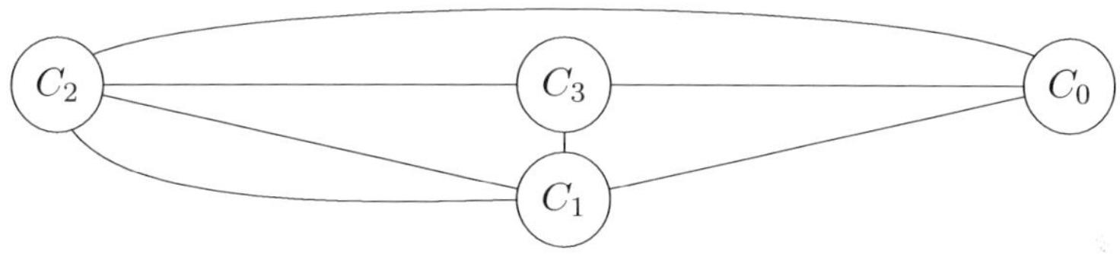

The first counts of limit roots with exactly three global sections were presented in **[20]**. The focus of this study was on full blow-up limit roots, which means that each nodal singularity is substituted by an exceptional $\mathbb{P}^1$. In the case of Δ_4°, the dual graph of this curve $C^\circ_{(\mathbf{3},\mathbf{2})_{1/6}}$ takes the following shape:

(4.12)

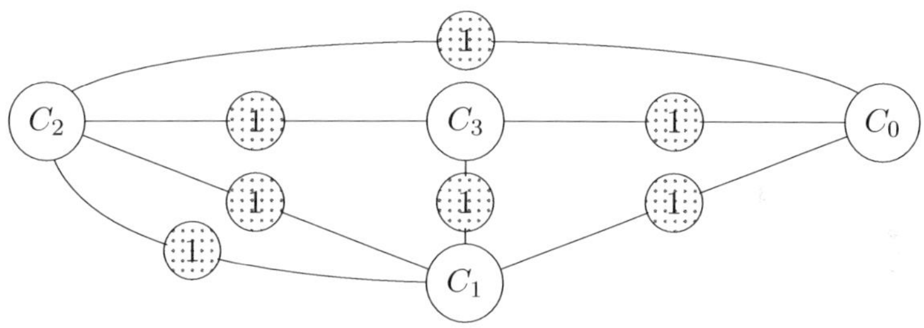

In this diagram, the exceptional $\mathbb{P}^1$s are marked with grey dots in the background. It's worth noting that for each exceptional $\mathbb{P}^1$, $P|_{E_i} \cong \mathcal{O}_{\mathbb{P}^1}(1)$, as established in **[34]**. The (local) sections on the exceptional $\mathbb{P}^1$s are uniquely determined by the need to connect the sections on the components C_i of $C^\bullet_{(\mathbf{3},\mathbf{2})_{1/6}}$, which is why they were referred to as "bridging sections" in **[18]**. Consequently, the number of global sections of $P^\bullet$ is the same as that of the restriction of $P^\bullet$ to the following completely disconnected curve:

(4.13)

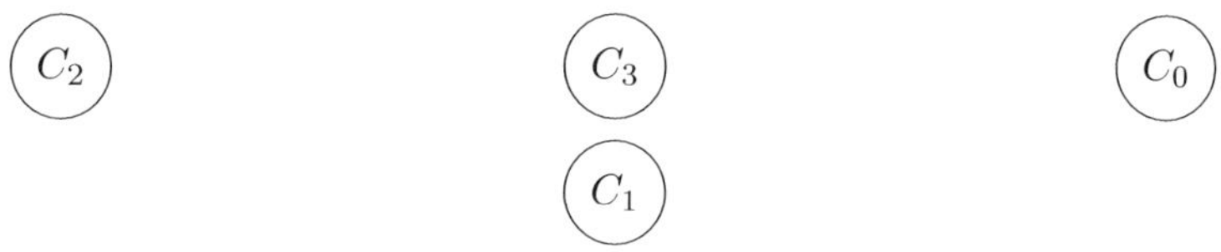

Consequently, it holds

$$h^0(C^\circ_{\mathbf{R}}, P^\circ_{\mathbf{R}}) = \sum_Z h^0(Z, P^\circ_{\mathbf{R}}|_Z), \tag{4.14}$$

where Z is an irreducible component of $C^\bullet_{\mathbf{R}}$. A computational analysis of the 33 chosen F-theory QSMs yielded the following counts of limit roots possessing precisely three global sections [**19**, **20**]:

(4.15)

$\overline{K}^3_{B_3}$	Polytope	$\check{N}_P^{(3)}$	μ	N_{total}	$\frac{\check{N}_P^{(3)} \cdot \mu}{N_{\text{total}}}$ [%]
6	Δ°_8	142560	12^3	12^8	57.3
	Δ°_4	11110	12^4	12^8	53.6
	Δ°_{134}	10010	12^4	12^8	48.7
	Δ°_{128}, Δ°_{130}, Δ°_{136}, Δ°_{236}	8910	12^4	12^8	42.0
10	Δ°_{88}	781.680.888	20^5	20^{12}	61.1
	Δ°_{110}	738.662.983	20^5	20^{12}	57.8
	Δ°_{272}, Δ°_{274}	736.011.640	20^5	20^{12}	57.5
	Δ°_{387}	733.798.300	20^5	20^{12}	57.3
	Δ°_{798}, Δ°_{808}, Δ°_{810}, Δ°_{812}	690.950.608	20^5	20^{12}	54.0
	Δ°_{254}	35.004.914	20^6	20^{12}	54.7
	Δ°_{302}	34.908.682	20^6	20^{12}	54.7
	Δ°_{52}	34.980.351	20^6	20^{12}	54.7
	Δ°_{786}	32.860.461	20^6	20^{12}	51.3
	Δ°_{762}	32.858.151	20^6	20^{12}	51.3
	Δ°_{417}	32.857.596	20^6	20^{12}	51.3
	Δ°_{838}	32.845.047	20^6	20^{12}	51.3
	Δ°_{782}	32.844.379	20^6	20^{12}	51.3
	Δ°_{377}, Δ°_{499}, Δ°_{503}	30.846.440	20^6	20^{12}	48.2
	Δ°_{1348}	30.845.702	20^6	20^{12}	48.2
	Δ°_{882}, Δ°_{856}	30.840.098	20^6	20^{12}	48.2
	Δ°_{1340}	28.954.543	20^6	20^{12}	45.2
	Δ°_{1879}	28.950.852	20^6	20^{12}	45.2
	Δ°_{1384}	27.178.020	20^6	20^{12}	42.5

The table presented here groups together polytopes whose dual graph of $C^\bullet_{(\mathbf{3},\mathbf{2})_{1/6}}$ are the same. Based on this initial outcome, it can be inferred that around 50% of all limit root bundles have exactly three global sections, which is promising news for string phenomenology. It is worth reiterating that these counts are FRST-invariant, meaning they apply to the entire family of F-theory QSM geometries $B_3(\Delta^\circ)$.

The method for counting limit roots presented in [**20**] was expanded in [**19**] to include the counting of global sections of specific partial blow-up limit roots. As an example, we can examine a limit root on a partial blow-up where we do not blow up the nodal singularity identified by the dashed edge:

(4.16)

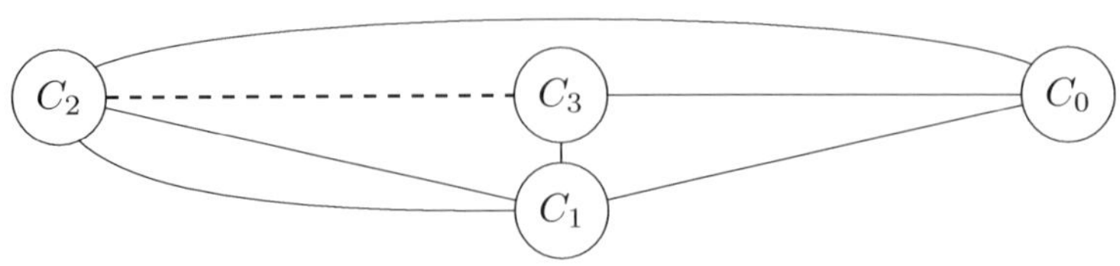

Such limit roots $P^\bullet$ correspond to line bundles on the following nodal curve:

(4.17)
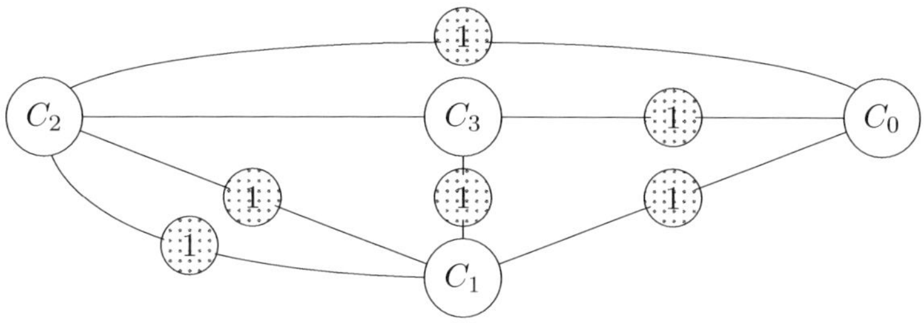

Once more, the sections on the exceptional $\mathbb{P}^1$s play a role in connecting the sections on neighboring components C_i. However, in this case, one node (marked by the dashed edge in eq. (4.16)) has not been blown up, resulting in the number of global sections of $P^\bullet$ matching those of the restriction of $P^\bullet$ to the following nodal curve:

(4.18)
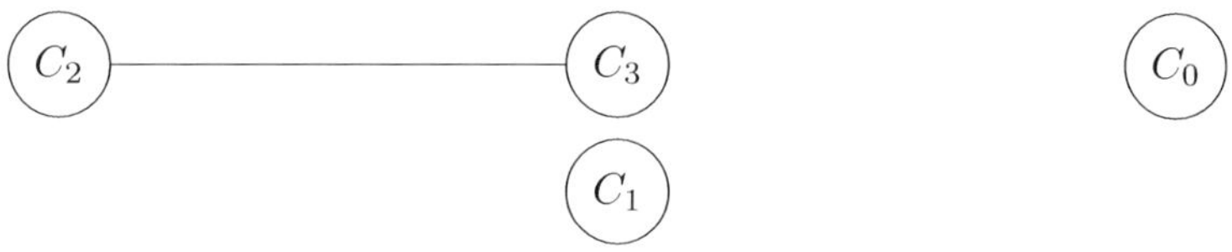

It should be noted that $C_2 \cup C_3$ forms a tree-like nodal curve. In **[19]**, an algorithm was developed for computing the cohomology of line bundles on such curves, which allowed us to determine the number of global sections using the formula:

$$h^0(C^\bullet, P^\bullet) = h^0\left(C_0, P^\bullet|_{C_0}\right) + h^0\left(C_1, P^\bullet|_{C_1}\right) + h^0\left(C_2 \cup C_3, P^\bullet|_{C_2 \cup C_3}\right). \tag{4.19}$$

Our computer implementation determined the number of limit roots for Δ_4° based on the remaining number of nodes, denoted as N:

(4.20)

N	Percentages $h^0 = 3$	Percentages $h^0 \geq 3$	Absolute numbers $h^0 = 4$	Absolute numbers $h^0 \geq 4$
0	53.6		11110	
1	36.7		7601	
2	7.5	0.5	1562	110
3	1.3	0.1	264	11
4		0.3		66
5		0.1		11
7		0.0		1
Σ	99.0	1.0	20537	199

The absolute numbers are expressed as multiples of $\mu = 12^4$. For example, there are 7601×12^4 12-th limits roots of $\overline{K}^{12}_{C^\bullet_{(\mathbf{3},\mathbf{2})_{1/6}}}$ with $h^0 = 3$, resulting from not blowing up precisely $N = 1$ nodes. The algorithm covered all limit roots, as evidenced by $12^4 \cdot (20537 + 199) = 12^8$. Almost 99% of all limit roots have precisely three sections on $C^\bullet_{(\mathbf{3},\mathbf{2})_{1/6}}$, which is a significant improvement over the findings of **[20]**.

This procedure can be repeated for each of the selected 33 QSM polytopes [**19**].

(4.21)

Polytope	$=3$	≥ 3	$=4$	≥ 4	$=5$	≥ 5	$=6$	≥ 6
Δ_8°	76.4	23.6						
Δ_4°	99.0	1.0						
Δ_{134}°	99.8	0.2						
Δ_{128}°, Δ_{130}°, Δ_{136}°, Δ_{236}°	99.9	0.1						
Δ_{88}°	74.9	22.1	2.5	0.5	0.0	0.0		
Δ_{110}°	82.4	14.1	3.1	0.4	0.0			
Δ_{272}°, Δ_{274}°	78.1	18.0	3.4	0.5	0.0	0.0		
Δ_{387}°	73.8	21.9	3.5	0.8	0.0	0.0		
Δ_{798}°, Δ_{808}°, Δ_{810}°, Δ_{812}°	77.0	17.9	4.4	0.7	0.0	0.0		
Δ_{254}°	95.9	0.5	3.5	0.0	0.0	0.0		
Δ_{52}°	95.3	0.7	3.9	0.0	0.0	0.0		
Δ_{302}°	95.9	0.5	3.5	0.0	0.0			
Δ_{786}°	94.8	0.3	4.8	0.0	0.0	0.0		
Δ_{762}°	94.8	0.3	4.9	0.0	0.0	0.0		
Δ_{417}°	94.8	0.3	4.8	0.0	0.0	0.0	0.0	
Δ_{838}°	94.7	0.3	5.0	0.0	0.0	0.0		
Δ_{782}°	94.6	0.3	5.0	0.0	0.0	0.0		
Δ_{377}°, Δ_{499}°, Δ_{503}°	93.4	0.2	6.2	0.0	0.1	0.0		
Δ_{1348}°	93.7	0.0	6.2	0.0	0.1		0.0	
Δ_{882}°, Δ_{856}°	93.4	0.3	6.2	0.0	0.1	0.0	0.0	
Δ_{1340}°	92.3	0.0	7.6	0.0	0.1		0.0	
Δ_{1879}°	92.3	0.0	7.5	0.0	0.1		0.0	
Δ_{1384}°	90.9	0.0	8.9	0.0	0.2		0.0	

Based on these findings, we can infer that the majority of (partial) limit root bundles have exactly three global sections. This is excellent news for string phenomenology, as it implies that it is highly probable that there are no vector-like exotics on the quark-doublet matter curve.

The observant reader may have noticed that our results are considerably more conclusive for some polytopes than for others. For example, we establish a lower bound for h^0 in roughly 23% of the limit roots for Δ_8°, while for Δ_4°, we cannot determine h^0 for only about 1% of the limit roots. The primary factor influencing this discrepancy is the presence of an elliptic curve component in Δ_8°. When the nodal quark-doublet curve contains at least one elliptic curve, we face limitations with partial blow-up root bundles. Firstly, the line bundle cohomology technology on tree-like nodal curves presented in [**19**] is no longer applicable if the elliptic curve has one unblown node. Secondly, if there is a $d = 0$ line bundle on an elliptic curve, and we are searching for r-th roots, then $r^2 - 1$ of these roots are guaranteed to lack global sections. The number of global sections in the one remaining root is determined by whether the original bundle was trivial or not, and distinguishing between the two situations is extremely challenging. Therefore, we can only provide a lower bound in such cases. Indeed, all entries in eq. (4.21) that indicate an uncertainty greater than 1% contain an elliptic curve.

It is intriguing to compare these results to the classical Brill-Noether theory, which deals with a continuous space of line bundles and estimates the dimension of the variety of line bundles with a specific h^0. However, this notion does not apply to root bundles since there are only finitely many of them. To the best of our knowledge, eq. (4.21) (further information in [**19**]) represents the first counts/estimates for such bundles. As a result, we propose that these tables be regarded as an initial response to the question "What is Brill-Noether theory for limit roots?". In particular, we notice that for the QSM configurations with $\overline{K}_{B_3}^3 = 10$, we find limit root bundles that always have more than the minimum number of global sections.

4.5. Probability for absence of vector-like exotics in $\mathbf{B_3(\Delta_4^\circ)}$. To gain a complete understanding, it is important to investigate the limit roots for which the methods mentioned earlier were unable to determine the number of global sections. Therefore, we will now examine the unresolved cases for the polytope Δ_4°. There are only five configurations to be discussed after taking symmetry into account. We will begin with the first configuration:

(4.22)

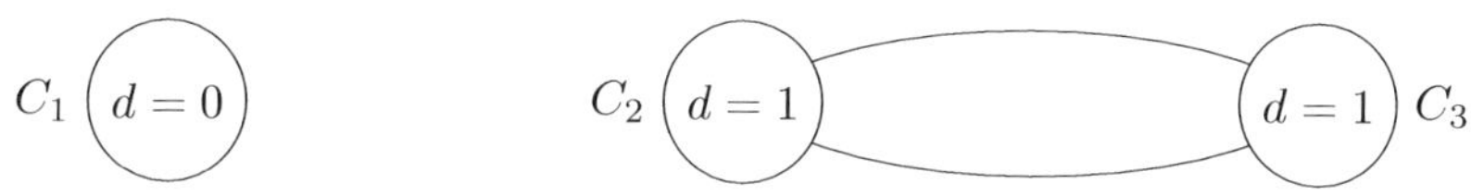

A line bundle on a nodal curve whose irreducible components are all isomorphic to $\mathbb{P}^1$ can be uniquely determined by two pieces of information as stated in [**65**]: the degree of the line bundle on each irreducible component, and the descent data which specifies the gluing conditions at the nodes. The descent data is represented by a number $\lambda \in \mathbb{C}^*$ for each node, and the global sections at each node must be equal up to this factor λ. One approach to investigate the unresolved cases for the polytope Δ_4° is to trace the descent data through the limit root construction of [**34**]. Alternatively, we can study all different descent data that can be put on the nodal curve represented by eq. (4.22), i.e., all line bundles on the curve with the given multidegree. We choose the second approach, as every such line bundle has exactly two global sections on $C_2 \cup C_3$. The tracing of descent data through the limit root prescription of [**34**] will be left for future work. We select coordinates for C_2 and C_3 and parametrize the local sections on C_2, C_3 using $(\alpha_1, \alpha_2, \alpha_3, \alpha_4) \in \mathbb{C}^4$:

(4.23)

Curve	Coordinates	Sections
C_2	$[a : b]$	$\alpha_1 a + \alpha_2 b$
C_3	$[c : d]$	$\alpha_3 c + \alpha_4 d$

By use of a Möbius transformation, we place the nodes at the following positions:

(4.24)

Label	Coordinates in C_2	Coordinates in C_3
n_1	$[a : b] = [1 : 0]$	$[c : d] = [1 : 0]$
n_2	$[a : b] = [0 : 1]$	$[c : d] = [0 : 1]$

Next, let us enforce the gluing conditions:

$$n_1 : \quad (\alpha_1 a + \alpha_2 b)([1:0]) = \lambda_1 \cdot (\alpha_3 c + \alpha_4 d)([1:0]), \qquad \lambda_1 \in \mathbb{C}^*, \tag{4.25}$$

$$n_2 : \quad (\alpha_1 a + \alpha_2 b)([0:1]) = \lambda_2 \cdot (\alpha_3 c + \alpha_4 d)([0:1]), \qquad \lambda_2 \in \mathbb{C}^*. \tag{4.26}$$

Hence, the global sections on $C_2 \cup C_3$ are parametrized by $(\alpha_3, \alpha_4) \in \mathbb{C}^2$ via

(4.27)

Curve	Coordinates	Sections
C_2	$[a : b]$	$\lambda_1 \alpha_3 a + \lambda_2 \alpha_4 b$
C_3	$[c : d]$	$\alpha_3 c + \alpha_4 d$

By also considering C_1, we find that the number of global sections for the configuration in eq. (4.22) is three. It is worth noting that we could have rescaled the sections on C_2 to eliminate one of the λ parameters, resulting in only one remaining parameter, which is equivalent to the first Betti number of the graph [**65**]. We can perform a similar analysis for three other graphs of this type, all of which also have $h^0(C^\bullet, P^\bullet) = 3$ [**19**]. This improves the original Brill-Noether table in eq. (4.20):

(4.28)

	Percentages		Absolute numbers	
N	$h^0 = 3$	$h^0 \geq 3$	$h^0 = 3$	$h^0 \geq 3$
0	53.6		11110	
1	36.7		7601	
2	8.0		1672	
3	1.3		275	
4	0.3		66	
5	0.1		11	
7		0.0		1
Σ	100.0	0.0	20735	1

Recall from eq. (3.3) that we are looking for limit 12th roots of the 12th power of $K_{C^\bullet_{(\mathbf{3},\mathbf{2})_{1/6}}}$. It is evident that $K_{C^\bullet_{(\mathbf{3},\mathbf{2})_{1/6}}}$ is a solution to this root bundle equation and that the number of global sections of the canonical bundle always matches the arithmetic genus, which is four in this case. Since all the limit roots found above had $h^0 = 3$, this final configuration must be $K_{C^\bullet_{(\mathbf{3},\mathbf{2})_{1/6}}}$ and have $h^0 = 4$. Therefore, only more than 99.995% of the limit root bundles on the nodal curves $C^\bullet_{(\overline{\mathbf{3}},\mathbf{2})_{1/6}}$ have exactly three global sections. It is possible to lose sections along $C^\bullet_{(\overline{\mathbf{3}},\mathbf{2})_{1/6}} \to C_{(\overline{\mathbf{3}},\mathbf{2})_{1/6}}$, which means that – statistically speaking – the probability of exotic vector-like quark-doublets being absent in the family $B_3(\Delta_4^\circ)$ is higher than 99.995%.

5. Summary and outlook

Studies on vector-like spectra of 4d F-theory compactifications often feature root bundles. This is particularly true for the Quadrillion F-theory Standard Models [**44**] for which zero modes in the representations $(\mathbf{3},\mathbf{2})_{1/6}$, $(\overline{\mathbf{3}},\mathbf{1})_{-2/3}$, and $(\mathbf{1},\mathbf{1})_1$ are counted by root bundles. A superset of these root bundles has been studied extensively in [**18**–**20**]. In particular, [**19**] focused on the family $B_3(\Delta_4^\circ)$ of $\mathcal{O}(10^{11})$ F-theory QSM. The geometries of these F-theory vacua are defined by toric 3-folds which are defined by full, regular, star triangulations of the 4-th polytope in the Kreuzer-Skarke list of 3-dimensional, reflexive poytopes [**77**]. It was found that more than 99.995% of the root bundles in the superset of studied root bundles have precisely three global sections. This discovery resulted from a thorough investigation of a significant portion of QSM geometries. Although the outcomes are most conclusive for the family $B_3(\Delta_4^\circ)$, the findings for this extensive subset of QSM geometries offer statistical evidence supporting the possibility of the non-existence of vector-like exotics in the representations $(\mathbf{3},\mathbf{2})_{1/6}$, $(\overline{\mathbf{3}},\mathbf{1})_{-2/3}$ and $(\mathbf{1},\mathbf{1})_1$ within the F-theory QSMs.

In QSM geometries, vector-like spectra are calculated by counting the cohomologies of line bundles, which are roots of the (twists of certain powers of) canonical bundle [**18**]. These root bundles $P_{\mathbf{R}}$ on the matter curve $C_{\mathbf{R}}$ are not unique, and it is unclear which ones are induced from F-theory gauge potentials in the Deligne cohomology. Investigating this matter is a complex task beyond our current abilities. Therefore, the studies in [**18**–**20**] analyzed *all* mathematically permissible root bundles, including those that could potentially be induced from the G_4-flux and all spin bundles on the matter curve under examination. This study is limited to a local and bottom-up analysis of one matter curve at a time, neglecting any correlations among the vector-like spectra on different matter curves. Consequently, a (proper) superset of all physically acceptable root bundles was examined.

To count the root bundles on the matter curve $C_{\mathbf{R}}$, it is helpful to deform the curve into a nodal curve. This deformation provides two advantages. Firstly, constructing root bundles on smooth curves is challenging, while constructing them on nodal curves using limit roots has been extensively described and involves combinatorial tasks [**34**]. Secondly, for each family $B_3(\Delta^\circ)$ of F-theory QSMs, there is a canonical nodal curve $C^\bullet_{\mathbf{R}}$ that solely depends on Δ°. Thus, by examining limit roots on this canonical nodal curve $C^\bullet_{\mathbf{R}}$, one can estimate the vector-like spectra for the entire family $B_3(\Delta^\circ)$ with a single calculation.

This method necessitates the capability to <enumerate limit root bundles with specific numbers of global sections, which is precisely within the realm of Brill-Noether theory. For F-theory QSMs, the global sections of most limit roots can be calculated by finding the cohomologies of line bundles on tree-like rational curves, which can be achieved with a computer algorithm described in [**19**]. However, for some limit roots, line bundle cohomology on circuit-like curves is needed, which is currently being researched. This is not a big issue for F-theory QSMs because the nodal curves being studied have relatively small Betti numbers and high root indices, resulting in few contributions from circuits. If the root indices were lower, circuits would become more significant, and for second roots, circuit-like and tree-like graphs would contribute equally, with circuits potentially dominating.

The algorithm used in [**19**] simplifies a pair consisting of a nodal curve $C^\bullet$ and a line bundle L on it by removing components of $C^\bullet$ and adjusting L in such a way that the number of global sections remains the same. This process ends with a final configuration, from which the number of global sections can be easily determined. For partial blow-up limit roots that go beyond the tree-like case, the counting of global sections had to be performed manually. In the case of the $B_3(\Delta_4^\circ)$ family, this involved studying line bundle cohomologies on no more than five nodal, circuit-like curves. This study revealed that on four of these nodal curves, the line bundle had exactly three sections. The remaining case, denoted as a "jumping circuit" in [**19**], was identified as the canonical bundle, which must have as many sections as the arithmetic genus of the nodal case, which turned out to be four. This is why "only" more than 99.995% of limit root bundles on the nodal curves $C^\bullet_{(\mathbf{3},\mathbf{2})_{1/6}}$, $C^\bullet_{(\overline{\mathbf{3}},\mathbf{1})_{-2/3}}$, and $C^\bullet_{(\mathbf{1},\mathbf{1})_1}$ have exactly three global sections.

The ignorances in eq. (4.21) can be attributed to two factors. Firstly, when a node remains on an elliptic curve, the methods used in [**18**–**20**] only yield a lower bound. Secondly, at the time of the publication of [**19**], there was no algorithmic method for computing the line bundle cohomology on circuit-like nodal curves, which meant that even in these cases, only a lower bound was obtained. The employed computer algorithm can be found in [**15**]. Currently, this algorithm is being extended to include functionality to compute line bundle cohomology on circuit-like nodal curves and to provide better lower bounds. Interested readers may be tempted to perform the relevant computations themselves using a personal computer. The results at the top of eq. (4.21) should take only a few minutes to compute while those at the bottom may take more than one day.

The table in eq. (4.21) demonstrates that certain limit roots on nodal matter curves with $\overline{K}^3_{B_3} = 10$ possess more than three global sections. Consequently, it is necessary to establish a set of conditions that maintain the number of sections or reduce it as the matter curve is smoothed out, especially to ensure the presence of a single Higgs pair. One possible avenue to identifying these conditions is to examine Yukawa interactions, which provide information about which fields acquire mass, when they acquire it, and how they do so during deformation. Insights into these conditions could be gained by conducting a similar investigation to that described in [**48**]. Future research on explicit realizations of F-theory MSSMs will likely need to address this issue.

The analysis presented in this paper focuses on the vector-like spectra of F-theory QSMs for particle phenomenology. However, it also represents the – to the best knowledge of the author – *first* computational approach to Brill-Noether theory of limit root bundles. While classical Brill-Noether theory is concerned with the classification of a continuous space of line bundles by their global sections, the number of root bundles in this case is finite and denoted as N_{total}. Thus, the Brill-Noether theory in this context involves finding a partition $N_{\text{total}} = N_0 + N_1 + \ldots$, where N_i represents the number of root bundles with i global sections. The main results presented in eq. (4.21) can be interpreted in this way.

The approach presented in this paper for studying the Brill-Noether theory of limit root bundles may have potential applications in machine learning. By using the algorithm [**15**], an approximation of the Brill-Noether theory of limit roots for a collection of simple graphs can be computed (modulo the mentioned caveats). It would be interesting to investigate if there is a pattern linking the graphs to the corresponding Brill-Noether approximations. This could provide mathematical insights. Moreover, a well-trained algorithm could be used for phenomenological applications, such as estimating the limit root bundles on the Higgs curve in F-theory QSMs, which are currently beyond our computational capabilities.

To conclude, it should be noted that the previous studies [**18**–**20**] did not address the fundamental question of which mathematical roots are induced by F-theory gauge potentials. It is likely that answering this question is complex, and so has not yet been attempted. Nevertheless, even though more than 99.995% of limit root bundles in representation $(\mathbf{3}, \mathbf{2})_{1/6}$ in the family of F-theory QSMs $B_3(\Delta_4^\circ)$ have exactly three global sections, the definitive answer regarding the presence or absence of vector-like exotics can only be obtained by determining which root bundles can be realized by F-theory gauge potentials. The detailed examination of such top-down conditions is planned for future research.

Acknowledgment

The author expresses gratitude towards his colleagues Mirjam Cvetič, Ron Donagi, Muyang Liu and Marielle Ong for their contributions and valuable discussions.

References

[1] Steven Abel, Andrei Constantin, Thomas R. Harvey, and Andre Lukas, *String Model Building, Reinforcement Learning and Genetic Algorithms*, Nankai Symposium on Mathematical Dialogues: In celebration of S.S.Chern's 110th anniversary, 11 2021.

[2] G. Aldazabal, S. Franco, L. E. Ibáñez, R. Rabadán, and A. M. Uranga, *$D = 4$ chiral string compactifications from intersecting branes*, J. Math. Phys. **42** (2001), no. 7, 3103–3126, DOI 10.1063/1.1376157. Strings, branes, and M-theory. MR1840334

[3] G. Aldazabal, S. Franco, Luis E. Ibáñez, R. Rabadan, and A. M. Uranga, *Intersecting brane worlds*, J. High Energy Phys. **02** (2001), 047.

[4] Lara B. Anderson, Iñaki García-Etxebarria, Thomas W. Grimm, and Jan Keitel, *Physics of F-theory compactifications without section*, J. High Energy Phys. **12** (2014), 156, front matter+42, DOI 10.1007/JHEP12(2014)156. MR3303436

[5] Lara B. Anderson, James Gray, Yang-Hui He, and Andre Lukas, *Exploring positive monad bundles and a new heterotic standard model*, J. High Energy Phys. **2** (2010), 054, 49, DOI 10.1007/JHEP02(2010)054. MR2672744

[6] Lara B. Anderson, James Gray, Andre Lukas, and Eran Palti, *Two Hundred Heterotic Standard Models on Smooth Calabi-Yau Threefolds*, Phys. Rev. **D84** (2011), 106005.

[7] Lara B. Anderson, James Gray, Andre Lukas, and Eran Palti, *Heterotic line bundle standard models*, Journal of High Energy Physics **2012** (2012), no. 6.

[8] Lara B. Anderson, Jonathan J. Heckman, and Sheldon Katz, *T-branes and geometry*, J. High Energy Phys. **5** (2014), 080, front matter+67, DOI 10.1007/JHEP05(2014)080. MR3219955

[9] Lara B. Anderson, Jonathan J. Heckman, Sheldon Katz, and Laura P. Schaposnik, *T-branes at the limits of geometry*, J. High Energy Phys. **10** (2017), 058, front matter+55, DOI 10.1007/jhep10(2017)058. MR3731155

[10] Michael F. Atiyah, *Riemann surfaces and spin structures*, Ann. Sci. École Norm. Sup. (4) **4** (1971), 47–62. MR286136

[11] Victor V. Batyrev, *Dual polyhedra and mirror symmetry for Calabi-Yau hypersurfaces in toric varieties*, J. Algebraic Geom. **3** (1994), no. 3, 493–535. MR1269718

[12] Chris Beasley, Jonathan J. Heckman, and Cumrun Vafa, *GUTs and exceptional branes in F-theory. I*, J. High Energy Phys. **1** (2009), 058, 87, DOI 10.1088/1126-6708/2009/01/058. MR2480326

[13] Micha Berkooz, Michael R. Douglas, and Robert G. Leigh, *Branes intersecting at angles*, Nuclear Phys. B **480** (1996), no. 1-2, 265–278, DOI 10.1016/S0550-3213(96)00452-X. MR1421704

[14] Martin Bies, *Cohomologies of coherent sheaves and massless spectra in F-theory*, Ph.D. thesis, Heidelberg U., 2 2018.

[15] Martin Bies, *Root counter*, `https://github.com/Julia-meets-String-Theory/RootCounter`, 2023.

[16] Martin Bies, Mirjam Cvetič, Ron Donagi, Ling Lin, Muyang Liu, and Fabian Ruehle, *Database*, `https://github.com/Learning-line-bundle-cohomology`, 2020.

[17] Martin Bies, Mirjam Cvetič, Ron Donagi, Ling Lin, Muyang Liu, and Fabian Ruehle, *Machine learning and algebraic approaches towards complete matter spectra in 4d F-theory*, J. High Energy Phys. **1** (2021), Paper No. 196, 69, DOI 10.1007/jhep01(2021)196. MR4257645

[18] Martin Bies, Mirjam Cvetič, Ron Donagi, Muyang Liu, and Marielle Ong, *Root bundles and towards exact matter spectra of F-theory MSSMs*, J. High Energy Phys. **9** (2021), Paper No. 076, 63, DOI 10.1007/jhep09(2021)076. MR4327240

[19] Martin Bies, Mirjam Cvetič, Ron Donagi, and Marielle Ong, *Brill-Noether-general limit root bundles: absence of vector-like exotics in F-theory standard models*, J. High Energy Phys. **11** (2022), Paper No. 4, 70, DOI 10.1007/jhep11(2022)004. MR4531615

[20] Martin Bies, Mirjam Cvetič, and Muyang Liu, *Statistics of limit root bundles relevant for exact matter spectra of F-theory MSSMs*, Phys. Rev. D **104** (2021), no. 6, Paper No. L061903, 8, DOI 10.1103/physrevd.104.l061903. MR4335105

[21] Martin Bies, Christoph Mayrhofer, Christian Pehle, and Timo Weigand, *Chow groups, Deligne cohomology and massless matter in F-theory*, (2014).

[22] Martin Bies, Christoph Mayrhofer, and Timo Weigand, *Gauge backgrounds and zero-mode counting in F-theory*, J. High Energy Phys. **11** (2017), 081, front matter+104, DOI 10.1007/jhep11(2017)081. MR3747226

[23] Ralph Blumenhagen, Mirjam Cvetič, Paul Langacker, and Gary Shiu, *Toward realistic intersecting D-brane models*, Annual Review of Nuclear and Particle Science **55** (2005), no. 1, 71–139.

[24] Ralph Blumenhagen, Boris Körs, Dieter Lüst, and Tassilo Ott, *The standard model from stable intersecting brane world orbifolds*, Nuclear Phys. B **616** (2001), no. 1-2, 3–33, DOI 10.1016/S0550-3213(01)00423-0. MR1865632

[25] Vincent Bouchard, Mirjam Cvetič, and Ron Donagi, *Tri-linear couplings in an heterotic minimal supersymmetric standard model*, Nuclear Phys. B **745** (2006), no. 1-2, 62–83, DOI 10.1016/j.nuclphysb.2006.03.032. MR2229502

[26] Vincent Bouchard and Ron Donagi, *An* SU(5) *heterotic standard model*, Phys. Lett. B **633** (2006), no. 6, 783–791, DOI 10.1016/j.physletb.2005.12.042. MR2201999

[27] Vincent Bouchard and Ron Donagi, *On heterotic model constraints*, Journal of High Energy Physics **2008** (2008), no. 08, 060–060.

[28] Andreas P. Braun, Cody Long, Liam McAllister, Michael Stillman, and Benjamin Sung, *The Hodge Numbers of Divisors of Calabi-Yau Threefold Hypersurfaces*, (2017).

[29] Volker Braun, Thomas W. Grimm, and Jan Keitel, *Geometric Engineering in Toric F-Theory and GUTs with U(1) Gauge Factors*, J. High Energy Phys. **12** (2013), 069.

[30] Volker Braun, Yang-Hui He, Burt A. Ovrut, and Tony Pantev, *A heterotic standard model*, Phys. Lett. B **618** (2005), no. 1-4, 252–258, DOI 10.1016/j.physletb.2005.05.007. MR2148973

[31] Volker Braun and David R. Morrison, *F-theory on genus-one fibrations*, J. High Energy Phys. **8** (2014), 132, front matter+45, DOI 10.1007/JHEP08(2014)132. MR3256108

[32] Alexander Brill and Max Noether, *Ueber die algebraischen Functionen und ihre Anwendung in der Geometrie*, Mathematische Annalen **7** (1874), no. 2, 269–310.

[33] P. Candelas, Gary T. Horowitz, Andrew Strominger, and Edward Witten, *Vacuum configurations for superstrings*, Nuclear Phys. B **258** (1985), no. 1, 46–74, DOI 10.1016/0550-3213(85)90602-9. MR800347

[34] Lucia Caporaso, Cinzia Casagrande, and Maurizio Cornalba, *Moduli of roots of line bundles on curves*, Trans. Amer. Math. Soc. **359** (2007), no. 8, 3733–3768, DOI 10.1090/S0002-9947-07-04087-1. MR2302513

[35] Jonathan Carifio, James Halverson, Dmitri Krioukov, and Brent D. Nelson, *Machine learning in the string landscape*, J. High Energy Phys. **9** (2017), 157, front matter+35, DOI 10.1007/jhep09(2017)157. MR3710457

[36] Andrés Collinucci, Simone Giacomelli, Raffaele Savelli, and Roberto Valandro, *T-branes through 3d mirror symmetry*, J. High Energy Phys. **7** (2016), 093, front matter+40, DOI 10.1007/JHEP07(2016)093. MR3547369

[37] Andrés Collinucci and Raffaele Savelli, *On flux quantization in F-theory*, J. High Energy Phys. **2** (2012), 015, front matter + 49, DOI 10.1007/JHEP02(2012)015. MR2996132

[38] Andres Collinucci and Raffaele Savelli, *F-theory on singular spaces*, J. High Energy Phys. **09** (2015), 100.

[39] Andres Collinucci and Raffaele Savelli, *T-branes as branes within branes*, J. High Energy Phys. **09** (2015), 161.

[40] David A. Cox and Sheldon Katz, *Mirror symmetry and algebraic geometry*, Mathematical Surveys and Monographs, vol. 68, American Mathematical Society, Providence, RI, 1999, DOI 10.1090/surv/068. MR1677117

[41] David A. Cox, John B. Little, and Henry K. Schenck, *Toric varieties*, Graduate Studies in Mathematics, vol. 124, American Mathematical Society, Providence, RI, 2011, DOI 10.1090/gsm/124. MR2810322

[42] Mirjam Cvetič, Ron Donagi, Denis Klevers, Hernan Piragua, and Maximilian Poretschkin, *F-theory vacua with* $\mathbb{Z}_3$ *gauge symmetry*, Nuclear Phys. B **898** (2015), 736–750, DOI 10.1016/j.nuclphysb.2015.07.011. MR3377501

[43] Mirjam Cvetič, Antonella Grassi, Denis Klevers, and Hernan Piragua, *Chiral four-dimensional F-theory compactifications with SU(5) and multiple U(1)-factors*, J. High Energy Phys. **04** (2014), 010.

[44] Mirjam Cvetič, James Halverson, Ling Lin, Muyang Liu, and Jiahua Tian, *Quadrillion F-theory compactifications with the exact chiral spectrum of the standard model*, Phys. Rev. Lett. **123** (2019), no. 10, 101601, 6, DOI 10.1103/PhysRevLett.123.101601. MR4008483

[45] Mirjam Cvetič, James Halverson, Gary Shiu, and Washington Taylor, *Snowmass White Paper: String Theory and Particle Physics*, (2022).

[46] Mirjam Cvetič, Denis Klevers, Damián Kaloni Mayorga Peña, Paul-Konstantin Oehlmann, and Jonas Reuter, *Three-family particle physics models from global F-theory compactifications*, J. High Energy Phys. **8** (2015), 087, front matter + 44, DOI 10.1007/JHEP08(2015)087. MR3400743

[47] Mirjam Cvetič, Ling Lin, Muyang Liu, and Paul-Konstantin Oehlmann, *An F-theory realization of the chiral MSSM with $\mathbb{Z}_2$-parity*, J. High Energy Phys. **9** (2018), 089, front matter+32, DOI 10.1007/jhep09(2018)089. MR3868398

[48] Mirjam Cvetič, Ling Lin, Muyang Liu, Hao Y. Zhang, and Gianluca Zoccarato, *Yukawa hierarchies in global F-theory models*, J. High Energy Phys. **1** (2020), 037, 26, DOI 10.1007/jhep01(2020)037. MR4088282

[49] Mirjam Cvetič, Gary Shiu, and Angel M. Uranga, *Chiral four-dimensional $N = 1$ supersymmetric type IIA orientifolds from intersecting D6-branes*, Nuclear Phys. B **615** (2001), no. 1-3, 3–32, DOI 10.1016/S0550-3213(01)00427-8. MR1863196

[50] Mirjam Cvetič, Gary Shiu, and Angel M. Uranga, *Three family supersymmetric standard - like models from intersecting brane worlds*, Phys. Rev. Lett. **87** (2001), 201801.

[51] Jesús A. De Loera, Jörg Rambau, and Francisco Santos, *Triangulations*, Algorithms and Computation in Mathematics, vol. 25, Springer-Verlag, Berlin, 2010. Structures for algorithms and applications, DOI 10.1007/978-3-642-12971-1. MR2743368

[52] The Sage Developers, *SageMath, the Sage Mathematics Software System (Version 9.8)*, 2023, `https://www.sagemath.org`.

[53] Eder, Christian and Decker, Wolfram and Fieker, Claus and Horn, Max and Joswig, Michael (ed.), *The OSCAR book*, 2024.

[54] David Eisenbud, Mark Green, and Joe Harris, *Cayley-Bacharach theorems and conjectures*, Bull. Amer. Math. Soc. (N.S.) **33** (1996), no. 3, 295–324, DOI 10.1090/S0273-0979-96-00666-0. MR1376653

[55] Daniel S. Freed and Edward Witten, *Anomalies in string theory with D-branes*, Asian J. Math. **3** (1999), no. 4, 819–851, DOI 10.4310/AJM.1999.v3.n4.a6. MR1797580

[56] Iñaki García-Etxebarria, Thomas W. Grimm, and Jan Keitel, *Yukawas and discrete symmetries in F-theory compactifications without section*, J. High Energy Phys. **11** (2014), 125, front matter+23, DOI 10.1007/JHEP11(2014)125. MR3290724

[57] Tomás L. Gómez, Sergio Lukic, and Ignacio Sols, *Constraining the Kähler moduli in the heterotic Standard Model*, Comm. Math. Phys. **276** (2007), no. 1, 1–21, DOI 10.1007/s00220-007-0338-8. MR2342286

[58] Daniel R. Grayson and Michael E. Stillman, *Macaulay2, a software system for research in algebraic geometry*, Available at `http://www.math.uiuc.edu/Macaulay2/`, 2023.

[59] Michael B. Green, John H. Schwarz, and Edward Witten, *Superstring theory. Vol. 1: Introduction*, Cambridge Monographs on Mathematical Physics, Cambridge University Press, Cambridge, 1987. MR878143

[60] Michael B. Green, John H. Schwarz, and Edward Witten, *Superstring theory. Vol. 2: Loop amplitudes, anomalies and phenomenology*, Cambridge Monographs on Mathematical Physics, Cambridge University Press, Cambridge, 1987. MR878144

[61] Brian R. Greene, Kelley H. Kirklin, Paul J. Miron, and Graham G. Ross, *A superstring-inspired standard model*, Phys. Lett. B **180** (1986), no. 1-2, 69–76, DOI 10.1016/0370-2693(86)90137-1. MR865089

[62] Thomas W. Grimm and Hirotaka Hayashi, *F-theory fluxes, chirality and Chern-Simons theories*, J. High Energy Phys. **3** (2012), 027, front matter+53, DOI 10.1007/JHEP03(2012)027. MR2980208

[63] James Halverson, Brent Nelson, and Fabian Ruehle, *Branes with brains: exploring string vacua with deep reinforcement learning*, J. High Energy Phys. **6** (2019), 003, 59, DOI 10.1007/jhep06(2019)003. MR3982507

[64] James Halverson and Jiahua Tian, *Cost of seven-brane gauge symmetry in a quadrillion F-theory compactifications*, Phys. Rev. D **95** (2017), no. 2, 026005, 12, DOI 10.1103/physrevd.95.026005. MR3779800

[65] Joe Harris and Ian Morrison, *Moduli of curves*, Graduate Texts in Mathematics, vol. 187, Springer-Verlag, New York, 1998. MR1631825

[66] Luis E. Ibáñez, Fernando Marchesano, and Raúl Rabadán, *Getting just the standard model at intersecting branes*, J. High Energy Phys. **11** (2001), Paper 2, 40. MR1878513

[67] Tyler J. Jarvis, *Geometry of the moduli of higher spin curves*, Internat. J. Math. **11** (2000), no. 5, 637–663, DOI 10.1142/S0129167X00000325. MR1780734

[68] Tyler J. Jarvis, *The Picard group of the moduli of higher spin curves*, New York J. Math. **7** (2001), 23–47. MR1838471

[69] Patrick Jefferson, Washington Taylor, and Andrew P. Turner, *Chiral matter multiplicities and resolution-independent structure in 4D F-theory models*, (2021).

[70] Patrick Jefferson, Washington Taylor, and Andrew P. Turner, *Chiral spectrum of the universal tuned* $(\mathrm{SU}(3) \times \mathrm{SU}(2) \times \mathrm{U}(1))/\mathbb{Z}_6$ *4D F-theory model*, J. High Energy Phys. **2** (2023), Paper No. 254, 62, DOI 10.1007/jhep02(2023)254. MR4557841

[71] Patrick Jefferson and Andrew P. Turner, *Generating functions for intersection products of divisors in resolved F-theory models*, Nuclear Phys. B **991** (2023), Paper No. 116177, 45, DOI 10.1016/j.nuclphysb.2023.116177. MR4583757

[72] Yusuke Kimura, *Discrete gauge groups in F-theory models on genus-one fibered Calabi-Yau 4-folds without section*, J. High Energy Phys. **4** (2017), 168, front matter+26, DOI 10.1007/JHEP04(2017)168. MR3650100

[73] Denis Klevers, Damian Kaloni Mayorga Pena, Paul-Konstantin Oehlmann, Hernan Piragua, and Jonas Reuter, *F-Theory on all Toric Hypersurface Fibrations and its Higgs Branches*, J. High Energy Phys. **01** (2015), 142.

[74] Sven Krause, Christoph Mayrhofer, and Timo Weigand, *G_4-flux, chiral matter and singularity resolution in F-theory compactifications*, Nuclear Phys. B **858** (2012), no. 1, 1–47, DOI 10.1016/j.nuclphysb.2011.12.013. MR2879318

[75] Sven Krause, Christoph Mayrhofer, and Timo Weigand, *Gauge Fluxes in F-theory and Type IIB Orientifolds*, J. High Energy Physics **2012** (2012), no. 8.

[76] Maximilian Kreuzer, *Toric geometry and Calabi-Yau compactifications*, Ukr. J. Phys. **55** (2010), no. 5, 613–625.

[77] Maximilian Kreuzer and Harald Skarke, *Classification of reflexive polyhedra in three dimensions*, Adv. Theor. Math. Phys. **2** (1998), no. 4, 853–871, DOI 10.4310/ATMP.1998.v2.n4.a5. MR1663339

[78] H. Blaine Lawson Jr. and Marie-Louise Michelsohn, *Spin geometry*, Princeton Mathematical Series, vol. 38, Princeton University Press, Princeton, NJ, 1989. MR1031992

[79] Ling Lin, Christoph Mayrhofer, Oskar Till, and Timo Weigand, *Fluxes in F-theory compactifications on genus-one fibrations*, J. High Energy Phys. **1** (2016), 098, front matter+36, DOI 10.1007/JHEP01(2016)098. MR3471482

[80] Ling Lin and Timo Weigand, *G4-flux and standard model vacua in F-theory*, Nuclear Physics B **913** (2016), 209–247.

[81] Christoph Mayrhofer, Eran Palti, Oskar Till, and Timo Weigand, *Discrete Gauge Symmetries by Higgsing in four-dimensional F-Theory Compactifications*, J. High Energy Phys. **12** (2014), 068.

[82] Christoph Mayrhofer, Eran Palti, Oskar Till, and Timo Weigand, *On Discrete Symmetries and Torsion Homology in F-Theory*, J. High Energy Phys. **06** (2015), 029.

[83] David R. Morrison and Washington Taylor, *Sections, multisections, and $U(1)$ fields in F-theory*, J. Singul. **15** (2016), 126–149, DOI 10.5427/jsing.2016.15g. MR3562858

[84] David R. Morrison and Cumrun Vafa, *Compactifications of F-theory on Calabi-Yau threefolds. I*, Nuclear Phys. B **473** (1996), no. 1-2, 74–92, DOI 10.1016/0550-3213(96)00242-8. MR1409284

[85] David R. Morrison and Cumrun Vafa, *Compactifications of F-theory on Calabi-Yau threefolds. II*, Nuclear Phys. B **476** (1996), no. 3, 437–469, DOI 10.1016/0550-3213(96)00369-0. MR1412112

[86] David Mumford, *Theta characteristics of an algebraic curve*, Ann. Sci. École Norm. Sup. (4) **4** (1971), 181–192. MR292836

[87] Eugene Perevalov and Harald Skarke, *Enhanced gauge symmetry in type II and F-theory compactifications: Dynkin diagrams from polyhedra*, Nuclear Phys. B **505** (1997), no. 3, 679–700, DOI 10.1016/S0550-3213(97)00477-X. MR1490785

[88] Joseph Polchinski, *String theory. Vol. I: An introduction to the bosonic string*, Cambridge Monographs on Mathematical Physics, Cambridge University Press, Cambridge, 1998, DOI 10.1017/CBO9780511618123. MR1648555

[89] J. Polchinski, *String Theory: Volume 2, Superstring Theory and Beyond*, Cambridge Monographs on Mathematical Physics, Cambridge University Press, 2001.

[90] The Toric Varieties project authors, *The* `ToricVarieties` *project*, `https://github.com/homalg-project/ToricVarieties_project`, 2019–2023.

[91] Falk Rohsiepe, *Lattice polarized toric K3 surfaces*, (2004).

[92] Fabian Ruehle, *Data science applications to string theory*, Phys. Rep. **839** (2020), 1–117, DOI 10.1016/j.physrep.2019.09.005. MR4066240

[93] The OSCAR Team, *OSCAR – Open Source Computer Algebra Research system, Version 0.12.0-DEV*, `https://www.oscar-system.org`, 2023.

[94] Cumrun Vafa, *Evidence for F-theory*, Nuclear Phys. B **469** (1996), no. 3, 403–415, DOI 10.1016/0550-3213(96)00172-1. MR1403744

[95] Taizan Watari, *Vector-like pairs and Brill–Noether theory*, Physics Letters B **762** (2016), 145–150.

[96] Timo Weigand, *F-theory*, PoS **TASI2017** (2018), 016.

[97] Edward Witten, *Non-perturbative superpotentials in string theory*, Nuclear Phys. B **474** (1996), no. 2, 343–360, DOI 10.1016/0550-3213(96)00283-0. MR1408164

[98] Edward Witten, *On flux quantization in M theory and the effective action*, J. Geom. Phys. **22** (1997), 1–13.

Department of Mathematics, RPTU Kaiserslautern-Landau, Gottlieb-Daimler-Strasse 48, 67663 Kaiserslautern, Germany

Email address: `bies@mathematik.uni-kl.de`

Proceedings of Symposia in Pure Mathematics
Volume **107**, 2024
https://doi.org/10.1090/pspum/107/01944

Holomorphic Floer theory and Donaldson-Thomas invariants

Pierrick Bousseau

Abstract. We present several expected properties of the holomorphic Floer theory of a holomorphic symplectic manifold. In particular, we propose a conjecture relating holomorphic Floer theory of Hitchin integrable systems and Donaldson-Thomas invariants of non-compact Calabi-Yau 3-folds. More generally, we conjecture that the BPS spectrum of a 4-dimensional $\mathcal{N} = 2$ quantum field theory can be recovered from the holomorphic Floer theory of the corresponding Seiberg-Witten integrable system.

Contents

0. Introduction

The goal of this paper is to present an overview of the expected formal aspects of holomorphic Floer theory of holomorphic symplectic manifolds and to propose a conjectural relation between holomorphic Floer theory and Donaldson-Thomas (DT) theory.

0.1. Holomorphic Floer theory. Holomorphic Floer theory is an example of infinite-dimensional complex Morse theory. We first review the more familiar case of Floer theory as an example of infinite-dimensional Morse theory.

0.1.1. *Floer theory as Morse theory.* Floer theory of a symplectic manifold (M, ω) is formally the result of applying Morse theory to the infinite-dimensional

The research of the author was partially supported by the NSF grant DMS-2302117.

manifold $\mathcal{P}$ of paths between pairs of Lagrangian submanifolds $L_1, L_2 \subset M$, endowed with the action functional

$$\begin{aligned} f : \mathcal{P} &\longrightarrow \mathbb{R} \\ \mathfrak{p} &\longmapsto -\int_{D_{\mathfrak{p}}} \omega, \end{aligned}$$

where $D_{\mathfrak{p}}$ is the surface in M formed by a one-parameter family of paths from a reference path $\mathfrak{p}_0$ to the path $\mathfrak{p}$ [**28**]. Critical points of f are constant paths located at the intersection points $L_1 \cap L_2$. Moreover, the choice of a ω-compatible almost complex structure J on M defines a Riemannian metric on $\mathcal{P}$, and the corresponding gradient flow lines of f are J-holomorphic disks in M with boundary on $L_1 \cup L_2$. Hence, the Morse complex of f is the Floer complex $CF(L_1, L_2)$ generated by the intersection points $L_1 \cap L_2$ and whose differential is given by counts of J-holomorphic disks. More generally, counting J-holomorphic disks with boundary on finite unions of Lagrangian submanifolds leads to the definition of the Fukaya category of (M, ω), whose objects are Lagrangian submanifolds and spaces of morphisms are Floer homology groups [**30**].

0.1.2. *Complex Morse theory.* Whereas Morse theory deals with a real-valued smooth function on a Riemannian manifold, complex Morse theory considers a holomorphic function on a Kähler manifold. This includes the Picard-Lefschetz theory of critical points and vanishing cycles of holomorphic functions, and also its categorification given by the Fukaya-Seidel category [**79, 80**].

The original definition of the Fukaya-Seidel category is based on the geometry of Lefschetz thimbles, but we will use a conjecturally equivalent alternative definition due to Haydys [**46**] and Gaiotto-Moore-Witten [**38**][1]. Given a Kähler manifold $\mathcal{P}$ and a holomorphic Morse function

$$W : \mathcal{P} \longrightarrow \mathbb{C},$$

this definition is based on counts of gradient flow lines of the real Morse functions $\mathrm{Re}(e^{-i\theta} W)$ connecting pairs of critical points, and on counts of solutions to a W-perturbed holomorphic curve equation asymptotically approaching gradient flow lines. In physics, a real-valued Morse function on a Riemannian manifold defines a $\mathcal{N} = 2$ supersymmetric quantum mechanics, whose space of vacua is Morse homology [**86**], whereas a holomorphic Morse function on a Kähler manifold defines a Landau-Ginzburg 2-dimensional $\mathcal{N} = (2, 2)$ supersymmetric quantum field theory, whose category of boundary conditions is the Fukaya-Seidel category [**38**].

0.1.3. *Holomorphic Floer theory as complex Morse theory.* The main advantage of the construction of the Fukaya-Seidel category in terms of counts of gradient flow lines and W-perturbed holomorphic curves is that it can be formally applied to infinite dimensional set-ups [**46**]. Known examples reviewed in more details in §1.4 include the complex Chern-Simons theory for complex connections on a 3-manifold [**46, 87**][**60**, §4.2] and the holomorphic Chern-Simons theory for $\overline{\partial}$-connections on Calabi-Yau 3-folds [**24, 25, 46**].

Holomorphic Floer theory is another example, obtained by taking for $\mathcal{P}$ the space of paths between two holomorphic Lagrangian submanifolds L_1 and L_2 in a

[1] Haydys's proposal for a definition of the Fukaya-Seidel category was made rigorous very recently by Donghao Wang [**84**].

holomorphic symplectic manifold (M, Ω) and for W the holomorphic action functional

$$W : \mathcal{P} \longrightarrow \mathbb{C}$$
$$\mathfrak{p} \longmapsto -\int_{D_{\mathfrak{p}}} \Omega$$

obtained by integrating the holomorphic symplectic form Ω. In this case, critical points are intersection points of L_1 and L_2, gradient flow lines are holomorphic curves in M for hyperkähler rotated complex structures and the W-perturbed holomorphic curve equation is a 3d Fueter equation for maps $\mathbb{R}^2 \times [0,1] \to M$, see §2 for details.

Applying formally the constructions of [**38**], one should be able to define:

(1) a vector space $H(p, p')$ for every pair of intersection points $p, p' \in L_1 \cap L_2$, obtained as the cohomology of a complex generated by holomorphic curves and with differential given by counts of Fueter maps.
(2) a A_∞-category $C(L_1, L_2)$ obtained as the "Fukaya-Seidel category of $(\mathcal{P}, W)$."

The collection of all holomorphic Lagrangian submanifolds in (M, Ω) should form a 2-category[2] $\mathrm{Ft}(M, \Omega)$, that we refer to as the *Fueter 2-category* of (M, Ω), with the categories $C(L_1, L_2)$ as Hom-categories:

$$C(L_1, L_2) = \mathrm{Hom}_{\mathrm{Ft}(M,\Omega)}(L_1, L_2).$$

The holomorphic action functional is in general multivalued, and so considering a cover of the space of paths $\mathcal{P}$ is necessary in general for the definition of the vector spaces $H(p, p')$ and of the categories $C(L_1, L_2)$. This point is essential in the formulation of the relation with Donaldson-Thomas invariants in the following section.

As reviewed in more details in §0.3, various aspects of holomorphic Floer theory are also discussed by Doan-Rezchikov [**23**], Khan [**60**] and Kontsevich-Soibelman [**66**]. The main contribution of the present paper is to formulate several expected properties of holomorphic Floer theory, and in particular its conjectural relation with DT theory.

0.2. DT theory and holomorphic Floer theory.

0.2.1. *DT theory and complex Morse theory.* DT invariants are counts of stable objects in a 3-dimensional Calabi–Yau triangulated category, such as for example the derived category of coherent sheaves or the Fukaya category of a Calabi-Yau 3-fold. As a function of the stability, DT invariants jump across real codimension one loci in a way determined by a universal wall-crossing formula [**51, 64**].

Many of the algebraic structures underlying DT invariants and their wall-crossing are formally similar to the ones appearing in complex Morse theory for a holomorphic Morse function on a Kähler manifold. For example, the wall-crossing formula can be viewed as an iso-Stokes condition [**18, 50**], and the DT invariants naturally define Riemann-Hilbert problems and geometric structures formally similar to Frobenius manifold structures [**15, 16**]. In this analogy, DT invariants play the role of counts of gradient flow lines in complex Morse theory.

[2] More precisely a stable $(\infty, 2)$-category.

Whereas complex Morse theory of a holomorphic Morse function with n critical points leads to algebraic structures based on the finite-dimensional Lie algebra $\mathfrak{gl}_n$ of $n \times n$ matrices, the algebraic structures coming from DT theory are based on the infinite-dimensional Lie algebra $\mathfrak{g}_{DT}$ of vector fields preserving a holomorphic symplectic form on a complex torus. Nevertheless, it is natural to ask if this relation between DT theory and complex Morse theory theory is deeper than a mere analogy:

QUESTION 0.1. Is DT theory actually an example of complex Morse theory?

Answering Question 0.1 means finding a Kähler manifold $\mathcal{P}$ with a holomorphic function $W\colon \mathcal{P} \to \mathbb{C}$ such that a formal application of complex Morse theory to $(\mathcal{P}, W)$ reproduces DT theory. In this context, a natural puzzle is to understand the origin of the Lie algebra $\mathfrak{g}_{DT}$: as $\mathfrak{g}_{DT} \neq \mathfrak{gl}_n$, a necessary condition is that W has infinitely many critical points.

Question 0.1 admits a natural reformulation in terms of physics. DT invariants can often be interpreted as counts of BPS states in 4-dimensional $\mathcal{N} = 2$ supersymmetric quantum field theories or string theory compactifications. On the other hand, a holomorphic Morse function on a Kähler manifold defines a Landau-Ginzburg 2-dimensional $\mathcal{N} = (2,2)$ quantum field theory. In this context, the analogy [**34**] is that the wall-crossing formula for the BPS states of 4-dimensional $\mathcal{N} = 2$ theories is formally similar to the Cecotti-Vafa wall-crossing formula for the BPS states in 2-dimensional $\mathcal{N} = (2,2)$ theories [**20**]. Therefore, answering Question 0.1 seems to require the construction of an appropriate 2-dimensional $\mathcal{N} = (2,2)$ theory from a 4-dimensional $\mathcal{N} = 2$ theory.

0.2.2. *DT invariants and holomorphic Floer theory of complex integrable systems.* We will argue that holomorphic Floer theory gives a positive answer to Question 0.1 in the particular case of DT invariants describing the BPS spectrum of a 4-dimensional $\mathcal{N} = 2$ field theory (without dynamical gravity). This includes for example string compactifications on non-compact Calabi-Yau 3-folds, but not string compactifications on compact Calabi–Yau 3-folds.

Let $\mathcal{T}$ be a 4-dimensional $\mathcal{N} = 2$ field theory. The low energy physics of $\mathcal{T}$ is controlled by its Seiberg-Witten complex integrable system

$$\pi\colon M \longrightarrow B.$$

Here B is the Coulomb branch of $\mathcal{T}$ on $\mathbb{R}^4$ and M is the Coulomb branch of $\mathcal{T}$ on $\mathbb{R}^3 \times S^1$. The space M is hyperkähler and the holomorphic symplectic form Ω_I for the complex structure I in which π is holomorphic. The fibers $F_b := \pi^{-1}(b)$ over points $b \in B$ in the complement of the discriminant locus Δ of π are abelian varieties and holomorphic Lagrangian submanifolds of (M, Ω_I). The lattice of charges for states in the vacuum $b \in B \setminus \Delta$ is

$$\Gamma_b := \pi_2(M, F_b)$$

and for every $\gamma \in \Gamma_b$, we have a space BPS_γ^b of BPS states of charge γ [**34, 37, 76–78**].

In many cases, there is an associated 3-dimensional Calabi-Yau (CY3) triangulated category $\mathcal{C}$, the base $B \setminus \Delta$ maps to the space of Bridgeland stability conditions on $\mathcal{C}$ [**14**], the central charge of the stability corresponding to a point $b \in B \setminus \Delta$ is given by

$$Z_\gamma(b) = \int_\gamma \Omega_I,$$

and the spaces of BPS states BPS_γ^b are mathematically realized as a cohomological DT invariants of $\mathcal{C}$ counting b-stable objects of class γ. A concrete example is provided by class $\mathcal{S}$ theories of type A_1 defined by a Riemann surface C (possibly with punctures)[**34**,**37**]. In this case, M is a moduli space of $SL(2,\mathbb{C})$ Higgs bundles on C, B is a space of quadratic differentials on C, and π is the Hitchin map. Moreover, $\mathcal{C}$ is a 3-dimensional Calabi-Yau triangulated category described by a quiver with potential defined in terms of ideal triangulations of C [**17**], and can also be viewed as the Fukaya category of a non-compact Calabi-Yau 3-fold constructed from C [**81**].

Our main claim is that the DT theory of $\mathcal{C}$ for the stability u is equivalent to the holomorphic Floer theory of M for the pair of holomorphic Lagrangian submanifolds given by the torus fiber $F_b := \pi^{-1}(b)$ and a section S of π. Physically, $S \subset M$ is a holomorphic Lagrangian section of π defined by putting $\mathcal{T}$ on a cigar geometry, see §4.2 for details. For example, S is the Hitchin section when M is a moduli space of Higgs bundles. In other words, our proposed answer to Question 0.1 is positive in this setting and one should recover DT theory by applying complex Morse theory to the infinite dimensional Kähler manifold $\mathcal{P}$ of paths between S and F_b, equipped with the holomorphic action functional $W : \mathfrak{p} \mapsto -\int_{D_\mathfrak{p}} \Omega_I$.

As S is a section of π and F_b is a fiber, the intersection $S \cap F_b$ consists of a single point. However, $\mathcal{P}$ is not simply connected: as we will show in §4.2, we have

$$\pi_1(\mathcal{P}) \simeq \pi_2(M, F_b) = \Gamma_b.$$

The holomorphic action functional W becomes a single-valued function on the universal cover $\widetilde{\mathcal{P}}$ of $\mathcal{P}$, and has infinitely many critical points which form a torsor under Γ_b. Fixing a reference critical point 0, we identify the set of critical points with Γ_b. We can now state our main conjecture relating DT theory of the CY3 category $\mathcal{C}$ and the holomorphic Floer theory of the Seiberg-Witten integrable system (M, Ω_I).

CONJECTURE 0.2. *For every $b \in B \setminus \Delta$ and $\gamma \in \Gamma_b$, the cohomological DT invariant BPS_γ^b is isomorphic to the vector space $H(0,\gamma)$ attached by holomorphic Floer theory of the pair of holomorphic Lagrangian submanifolds S and F_b in (M, Ω_I) to the two critical points 0 and γ of the holomorphic action functional:*

$$BPS_\gamma^b \simeq H(0,\gamma).$$

In §4.3, we give a physics argument in favor of Conjecture 0.2. In particular, we will obtain an answer to the physics reformulation of Question 0.1: the relevant 2d $\mathcal{N} = (2,2)$ theory is obtained by compactification of $\mathcal{T}$ on a cigar geometry with appropriate boundary conditions. In §4.4, we present further conjectures involving the A_∞-category

$$C(S, F_b) = \mathrm{Hom}_{\mathrm{Ft}(M,\Omega_I)}(S, F_b)$$

attached by holomorphic Floer theory to the pair of holomorphic Lagrangian submanifolds S and F_b.

0.3. Related works. While thinking about Question 0.1 in 2017, the author stumbled upon the idea to consider the Fukaya-Seidel of the holomorphic action functional and did the exercise to derive the corresponding Fueter equation. Recently, Doan-Rezchikov [**23**] independently derived the 2-categorical picture of holomorphic Floer theory. More importantly, they started the analysis of the compactness properties of the Fueter equation. Any attempt to rigorously discuss the conjectures of our paper should probably start with their work.

In his thesis [**60**, §1.1.1] Khan also considered the holomorphic area functional and derived the Fueter equation as the corresponding ζ-instanton equation. Another important part of this work is a systematical study of the 2d categorical wall-crossing [**61**]] and of the 2d categorical wall-crossing with "twisted masses" [**60**, §3], which applies in particular to multivalued Landau-Ginzburg potential. In particular, formal aspects of the 4d wall-crossing appear in this context. From the point of view of our Conjecture 0.2, this is not surprizing because we claim that the 4d wall-crossing is an example of 2d wall-crossing for the multivalued holomorphic action functional. In particular, applying [**60**, §3] to the holomorphic area functional should help to make more precise the algebraic structures involved in Conjecture 0.2.

The 2-categories attached to holomorphic symplectic varieties conjecturally enter in 3d mirror symmetry. In this context, the B-side is the Rozansky-Witten 2-category, based on the category of matrix factorizations. The A-side should be a gauged version of the Fueter category, based on the Fukaya-Seidel category [**39**,**40**]: concrete examples of the A-side in the literature rely on explicit algebraic models of the 2-category and it is an interesting problem to clarify the relation with the geometric model given by the Fueter 2-category.

Finally, Kontsevich-Soibelman have an ongoing work on holomorphic Floer theory [**66**], not focused on the 2-categorical aspects related to the Fueter equation, but rather on the exploration of the 1-categorical level in many directions that we do not touch upon, such as the relation with deformation quantization or a general theory of resurgence, see §2.7.

0.4. Plan of the paper. After an introduction to finite-dimensional complex Morse theory in §1, we introduce holomorphic Floer theory in §2. Conjectures relating holomorphic Floer theory and the structure of 2d $\mathcal{N} = (2, 2)$ theories are discussed in §3. Finally, the relation between holomorphic Floer theory and DT invariants is the subject of §4. Whereas the conjectures in §2 are stated entirely mathematically, conjectures in §3-§4 are stated in physical terms for maximal generality, even if they have mathematically well-formulated special cases.

1. Complex Morse theory

As announced in §0.1, holomorphic Floer theory should be viewed as an example of infinite-dimensional complex Morse theory. In this section, we first briefly review aspects of finite-dimensional complex Morse theory.

1.1. Solitons and instantons. Let X be a Kähler manifold, with complex structure J and Kähler form ω, such that $c_1(X) = 0$. Let $W: X \to \mathbb{C}$ be a holomorphic function which is Morse, that is with non-degenerate critical points, and which has distinct critical values. In this section, we assume that $\omega = d\lambda$ is exact and that the set of critical points of W is finite. We will have to eventually relax these assumptions for the infinite-dimensional set-up of holomorphic Floer theory, see §2.5.

Following Haydys [**46**] and Gaiotto-Moore-Witten [**38**], complex Morse theory for (X, W) is based on the following two sets of equations, parametrized by a phase $\zeta \in \mathbb{C}$, $|\zeta| = 1$:

(1) for a path

$$u\colon \mathbb{R} \longrightarrow X \quad x \longmapsto u(x), \tag{1.1}$$

the equation

$$\partial_x u + \operatorname{grad}\left(\operatorname{Re}\left(\zeta^{-1} W(u)\right)\right) = 0 \tag{1.2}$$

is the gradient flow line equation for the Morse function $\operatorname{Re}(\zeta^{-1}W)$, or equivalently the Hamiltonian flow equation for $\operatorname{Im}(\zeta^{-1}W)$.

(2) for a plane

$$u\colon \mathbb{R}^2 \longrightarrow X \quad (t, x) \longmapsto u(t, x), \tag{1.3}$$

the equation

$$\partial_t u + J(u)\left(\partial_x u + \operatorname{grad}\left(\operatorname{Re}\left(\zeta^{-1} W(u)\right)\right)\right) = 0 \tag{1.4}$$

is the holomorphic curve equation perturbed by the Hamiltonian $\operatorname{Im}(\zeta^{-1}W)$.

Equations (1.2) and (1.4) are respectively the critical point equation and the gradient flow line equation for the functional

$$u \mapsto \int_{\mathbb{R}} (u^*\lambda - \operatorname{Im}(\zeta^{-1}W)dx)$$

on the space of paths $\operatorname{Maps}(\mathbb{R}, X) = \{u\colon \mathbb{R} \to X\}$. In [**38**, §11.1], (1.2) and (1.4) are respectively called the *ζ-soliton equation* and the *ζ-instanton equation*[3] of the 2-dimensional $\mathcal{N} = (2,2)$ massive Landau-Ginzburg theory defined by (X, W).[4] Following this terminology, we will use *ζ-soliton* for a solution of (1.2) and *ζ-instanton* for a solution of (1.4).

1.2. 2d BPS states. Let $\mathbb{V}$ be the set of critical points of W. Complex Morse theory attaches a vector space $H(p, p')$, referred to as the *space of 2d BPS states*, to every pair $p, p' \in \mathbb{V}$ of distinct critical points. The vector space $H(p, p')$ should be defined as the cohomology of a complex $(C(p, p'), d)$, where

(1) $C(p, p')$ is the vector space generated by paths

$$u_i\colon \mathbb{R} \longrightarrow X \quad x \longmapsto u_i(x),$$

considered up to $\mathbb{R}$-translation, asymptotic to p and p',

$$\lim_{x \to -\infty} u_i(x) = p, \quad \lim_{x \to +\infty} u_i(x) = p',$$

[3] Our convention for ζ agrees with [**52**], differs from [**61**] by a factor of -1, and differs from [**38**] by a factor of $i = \sqrt{-1}$.

[4] In the context of FJRW invariants, (1.4) is sometimes called the Witten equation [**27**].

and which are $\zeta_{p,p'}$-solitons, that is solutions of the gradient flow line equation (1.2) with

$$\zeta := \zeta_{p,p'} := \frac{W(p) - W(p')}{|W(p) - W(p')|}.$$

(2) $d\colon C(p,p') \to C(p,p')$ is the differential defined by

$$du_i = \sum_j n_{ij} u_j,$$

where n_{ij} is a signed count of planes

$$u\colon \mathbb{R}^2 \longrightarrow X$$
$$(t,x) \longmapsto u(t,x),$$

asymptotic to u_i and u_j,

$$\lim_{t\to-\infty} u(t,x) = u_i(x), \quad \lim_{t\to+\infty} u(t,x) = u_j(x),$$

and which are $\zeta_{p,p'}$-instantons, that is solutions of (1.4) with $\zeta = \zeta_{p,p'}$, and considered up to the action of $\mathbb{R}$ induced by translation of the t-variable, see Figure 1.1.

In the terminology of [**38**], $H(p,p')$ is the space of 2d BPS states connecting the vacua p and p' of the 2-dimensional $\mathcal{N} = (2,2)$ massive Landau-Ginzburg theory defined by (X,W).

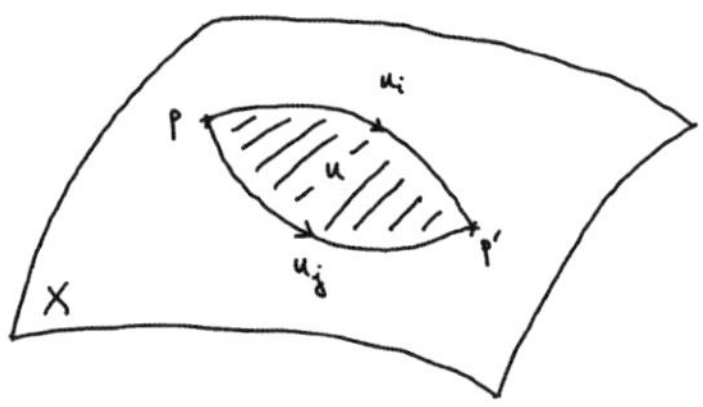

FIGURE 1.1. Contribution to the differential of a $\zeta_{p,p'}$-instantons u asymptotic to gradient flow lines u_i and u_j between critical points p and p'.

1.3. Fukaya-Seidel category. Complex Morse theory also attaches a A_∞-category $FS(X,W)$ to (X,W), which is conjecturally equivalent to the Fukaya-Seidel category of (X,W). We describe in a few words the web-based construction of Gaiotto-Moore-Witten [**38**], reformulated in the dual language of polygons by Kapranov-Kontsevich-Soibelman [**52**] (see also [**54**]). We denote by $\mathbb{C}_w$ the copy of $\mathbb{C}$ where W takes its values. Let $\mathbb{C}_w \cup S^1_\infty$ be the compactification of $\mathbb{C}_w$ obtained by adding a circle of oriented directions at infinity. We fix a phase $\zeta \in \mathbb{C}$, $|\zeta| = 1$, and we denote by $w_\infty \in S^1_\infty$ the corresponding point at infinity. We think of w_∞ as being the limit $R \to +\infty$ of the point $R\zeta_\infty \in \mathbb{C}_w$, with $R > 0$. Finally, we assume that $\zeta \neq \zeta_{p,p'}$ for every $p, p' \in \mathbb{V}$.

For every pair p and p' of distinct critical points, we define the complex

$$R_{p,p'} := \bigoplus_{Q \in \mathcal{Q}_{p,p'}} R_Q, \tag{1.5}$$

where the direct sum is over the set $\mathscr{Q}_{p,p'}$ of all the convex polygons Q in $\mathbb{C}_w \cup S^1_\infty$ with vertices in clockwise order given by $w_\infty, W(p), W(p_1), \dots, W(p_k), W(p')$, where $p_1, \dots, p_k \in \mathbb{V}$ are distinct critical points, also distinct from p and p', see Figure 1.2, and where R_Q is the complex

$$R_Q := C(p, p_1) \otimes C(p_1, p_2) \otimes \cdots \otimes C(p_k, p'). \tag{1.6}$$

Moreover, for every critical point p, we set $R_{p,p} = \mathbb{C}$ with trivial differential.

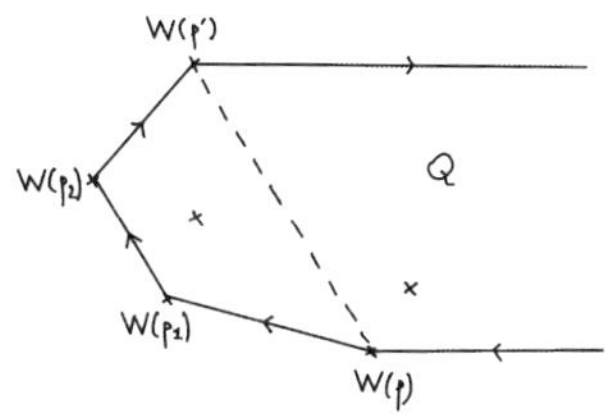

FIGURE 1.2. A polygon $Q \in \mathscr{Q}_{p,p'}$ with $\zeta = 1$.

Let R be the dg-category with objects $T_p(\zeta)$ indexed by the critical points p, Hom complexes given by

$$\mathrm{Hom}_R(T_p(\zeta), T_{p'}(\zeta)) = R_{p,p'},$$

and where the composition maps

$$\begin{aligned} R_{p,p'} \otimes R_{p',p''} &\longrightarrow R_{p,p''} \\ a \otimes b &\longmapsto ab \end{aligned}$$

are defined as follows: for $Q \in \mathscr{Q}_{p,p'}$, $Q' \in \mathscr{Q}_{p',p''}$, $a \in R_Q$ and $b \in R_{Q'}$, $ab \in R_{Q \cup Q'}$ is the natural concatenation of a and b if the polygon $Q \cup Q'$ obtained by gluing Q and Q' along their common edge $[p', w_\infty]$ is convex, and $ab = 0$ otherwise. In what follows, we view the dg-category R as an example of A_∞-category with trivial higher operations $(m_k)_{k \geq 3}$. By construction, $\mathrm{Hom}_R(T_p(\zeta), T'_p(\zeta)) = R_{p,p'} = 0$ if the triangle with ordered vertices $w_\infty, W(p), W(p')$ is oriented counterclockwise, and so R is an example of directed A_∞-category.

To obtain a A_∞-category with non-trivial higher operations, one considers the following counting problems. Let $\mathcal{P}$ be the set of convex polygons P in $\mathbb{C}_w$ with vertices in clockwise order given by $W(p_1), \dots, W(p_k)$, where $p_1, \dots, p_k \in \mathbb{V}$ are distinct critical points. Given such a polygon P, we define a map

$$\mu_P \colon C(p_1, p_2) \otimes \cdots \otimes C(p_{k-1}, p_k) \longrightarrow \mathbb{Z}$$

as follows: given $\zeta_{p_i,p_{i+1}}$-solitons generators α_i of $C(p_i, p_{i+1})$ for $1 \leq i \leq k-1$,

$$\mu_P(\alpha_1, \dots, \alpha_{k-1})$$

is a count of ζ-instantons

$$u \colon \mathbb{C} \longrightarrow X$$

such that the restriction of u to a direction perpendicular to $\mathbb{R}_{\geq 0}(W(p_{i+1}) - W(p_i))$ asymptotically approaches the ζ_{p_{i+1},p_i}-soliton α_i rotated by a phase $\zeta\zeta^{-1}_{p_i,p_{i+1}}$.

The main result of [**38**] is that $\sum_{P \in \mathscr{P}} \mu_P$ naturally defines a Maurer-Cartan element for R, and so one can construct a new A_∞-algebra $\tilde{R}$ by deforming R by this Maurer-Cartan element. Explicitly, higher A_∞-operations of $\tilde{R}$ are defined by deforming the naive concatenation of polygons $Q \in \mathscr{Q}_{p,p'}$ by terms indexed by regular polyhedral decompositions of the concatenated polygons, and where each finite piece P of a polyhedral decomposition contributes a factor μ_P. Passing to the category of twisted complexes of $\tilde{R}$-modules, we obtain a triangulated A_∞-category $FS(X,W)$ which is the model of [**38, 52**] for the Fukaya-Seidel of (X,W).

The categories R and $\tilde{R}$ depend on the choice of the phase ζ, but the triangulated category $FS(X,W)$ is independent of these choices. By construction, for a fixed choice of ζ, we have objects $(T_p(\zeta))_{p\in\mathbb{V}}$ which form an exceptional collection of $FS(X,W)$ when ordered according to the values of $\mathrm{Im}(\zeta^{-1}W(p))$. When ζ crosses a value $\zeta_{p,p'}$, this exceptional collection changes by a mutation. Comparing with the original construction of the Fukaya-Seidel category, $T_p(\zeta)$ corresponds to the Lefschetz thimble emanating from p, which consists of the gradient flow lines starting at p and going to $\mathrm{Re}(\zeta^{-1}W) = -\infty$, and whose image by W is the half-line $W(p) - \mathbb{R}_{\geq 0}\zeta$ in $\mathbb{C}_w$.

Following the terminology of [**54**], the above construction of the Fukaya-Seidel category of [**38, 52**] follows a *rectilinear* approach, based on $\zeta_{p,p'}$-solitons whose images by W are the straight line segments $[W(p), W(p')]$. In [**46**], a *curvilinear* approach is presented instead, based on a system of possibly curvy lines connecting the critical values $W(p)$ to a given base point $-R\zeta$ with $R >> 0$. This approach is in a way more direct, as it avoids the non-trivial web/convex geometry of [**38, 52**], but is also less explicit because it relies on versions of ζ-solitons and ζ-instanton equations where ζ is x or (t,x)-dependent. Intuitively, one goes from the curvilinear approach of [**46**] to the rectilinear approach of [**38, 52**] by starting with a possibly curvy line connecting $W(p)$, $W(p')$ passing through $-R\zeta$, and then trying to straighten it while relaxing the condition to pass through $-R\zeta$: the convex polygonal lines connecting $W(p)$, $W(p')$ appear upon hitting critical values, see Figure 1.3.

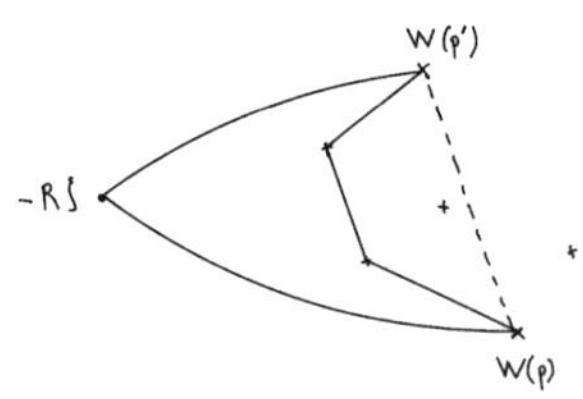

FIGURE 1.3. Comparison of the curvilinear and rectilinear approaches to complex Morse theory.

1.4. Infinite-dimensional examples. Complex Morse theory can be formally applied to pairs (X,W) where X is infinite-dimensional. For example, one can take for X the space of $G_{\mathbb{C}}$-connections on a 3-manifold M, where $G_{\mathbb{C}}$ is a complex reductive group, and for W the complex-valued Chern-Simons functional [**46, 87**][**60**, §4.2]. In this case,

(1) critical points of W are flat $G_{\mathbb{C}}$-connections on M,
(2) the ζ-soliton equation (1.2) is the Kapustin-Witten equation in 4-dimensional gauge theory on $M \times \mathbb{R}$,
(3) the ζ-instanton equation (1.4) is the Haydys-Witten equation in 5-dimensional gauge theory on $M \times \mathbb{R}^2$.

Another example is obtained by taking for X the space of $(0,1)$-connections on a vector bundle V on a Calabi-Yau 3-fold and for W the holomorphic Chern-Simons functional [**24, 25, 46**]. In this case,

(1) critical points of W are holomorphic structures on V,
(2) the ζ-soliton equation (1.2) is the equation for G_2-instantons in 7-dimensional gauge theory on $M \times \mathbb{R}$,
(3) the ζ-instanton equation (1.4) is the equation for $Spin_7$-instantons in 8-dimensional gauge theory on $X \times \mathbb{R}^2$.

As explained in the following section §2, holomorphic Floer theory is another example obtained when X is the space of paths stretched between two holomorphic Lagrangian submanifolds L_1 and L_2 of a holomorphic symplectic manifold (M, Ω), and W is the holomorphic action functional. In this case,

(1) critical points of W are intersection points of L_1 and L_2,
(2) the ζ-soliton equation (1.2) is the equation for holomorphic curves $\mathbb{R} \times [0,1] \to M$ with respect to hyperkähler rotated complex structures (see Lemma 2.1),
(3) the ζ-instanton equation (1.4) is the 3d Fueter equation for maps $\mathbb{R}^2 \times [0,1] \to M$ (see Lemma 2.2).

2. Holomorphic Floer theory

In this section, we introduce holomorphic Floer theory as an example of infinite dimensional complex Morse theory applied to the holomorphic action functional on the space of paths connecting two holomorphic Lagrangian submanifolds in a holomorphic symplectic variety. We only describe a conjectural picture ignoring all analytic difficulties. We refer to the work of Doan-Rezchikov [**23**] for first steps towards solving these analytic problems.

2.1. Holomorphic symplectic geometry. Let M be a holomorphic symplectic manifold, with complex structure I and holomorphic symplectic form Ω_I. We fix a compatible hyperkähler structure on M, that is a Riemannian metric g and complex structures J and K such that:

(1) I, J, K form a quaternionic triple, that is $IJK = -1$.
(2) g is Kähler with respect to I, J, K. We denote by
$$\omega_I := g(I-,-),\ \omega_J := g(J-,-),\ \omega_K := g(K-,-)$$
the corresponding Kähler forms.
(3) $\Omega_I = \omega_J + i\omega_K$.

Given a hyperkähler structure as above, the metric g is Kähler with respect to any complex structure in the *twistor sphere* spanned by I, J, K, that is any complex structure of the form $aI + bJ + cK$ with $a, b, c \in \mathbb{R}$ such that $a^2 + b^2 + c^2 = 1$. For any phase $\zeta \in \mathbb{C}$, $|\zeta| = 1$, we denote

$$J_\zeta := (\operatorname{Re}\zeta)J + (\operatorname{Im}\zeta)K.$$

If I and $-I$ are viewed as respectively the North and South poles of the twistor sphere, then the complex structures of the form J_ζ with $|\zeta| = 1$ are exactly the complex structures on the Equator of the twistor sphere. The Kähler form of the metric g for the complex structure J_ζ is

$$\omega_\zeta := \operatorname{Re}(\zeta^{-1}\Omega_I) = (\operatorname{Re}\zeta)\omega_J + (\operatorname{Im}\zeta)\omega_K. \tag{2.1}$$

2.2. The holomorphic action functional. Let L_1 and L_2 be two connected holomorphic Lagrangian submanifolds of (M, I, Ω_I), that is L_1 and L_2 are I-holomorphic half-dimensional submanifolds of M and $\Omega_I|_{L_1} = \Omega_I|_{L_2} = 0$. Assume that the intersection $L_1 \cap L_2$ consists of finitely many transverse intersection points.

Let $\mathcal{P}$ be the infinite-dimensional manifold of smooth paths stretched between L_1 and L_2:

$$\mathcal{P} := \{\mathfrak{p} : [0,1] \longrightarrow M \,|\, \mathfrak{p}(0) \in L_1, \mathfrak{p}(1) \in L_2\}.$$

It is naturally an infinite-dimensional Kähler manifold, with Riemannian metric

$$g_{\mathcal{P}}(\delta\mathfrak{p}_1, \delta\mathfrak{p}_2) := \int_0^1 g(\delta\mathfrak{p}_1(y), \delta\mathfrak{p}_2(y))dy, \tag{2.2}$$

complex structure

$$(I_{\mathcal{P}}(\mathfrak{p})(\delta\mathfrak{p}))(y) := I(\mathfrak{p}(y))(\delta\mathfrak{p}(y)), \tag{2.3}$$

and Kähler form

$$\omega_{I,\mathcal{P}}(\delta\mathfrak{p}_1, \delta\mathfrak{p}_2) := \int_0^1 \omega(\delta\mathfrak{p}_1(y), \delta\mathfrak{p}_2(y))dy. \tag{2.4}$$

For every intersection point $p \in L_1 \cap L_2$, we still denote by p the corresponding point of $\mathcal{P}$ given by the constant path sitting at p_0. As L_1 and L_2 are connected, there exists a unique connected component $\mathcal{P}_0$ of $\mathcal{P}$ containing all these constant paths sitting at the intersection points of L_1 and L_2. For every $p, p' \in L_1 \cap L_2$, we denote by $\pi_1(p, q)$ the set of homotopy classes of paths in $\mathcal{P}_0$ from p to p'.

We fix an intersection point $p_0 \in L_1 \cap L_2$. For every $\mathfrak{p} \in \mathcal{P}_0$ and w a path in $\mathcal{P}_0$ connecting p_0 to $\mathfrak{p}$, that is a map

$$\begin{aligned} w\colon [0,1] &\longrightarrow \mathcal{P}_0 \\ x &\longmapsto w_x \end{aligned}$$

such that $w_0 = p_0$ and $w_1 = \mathfrak{p}$, one views w as a map

$$\begin{aligned} w\colon [0,1] \times [0,1] &\longrightarrow M \\ (x,y) &\longmapsto w_x(y). \end{aligned}$$

and one defines

$$W(\mathfrak{p}, w) := -\int_{[0,1]\times[0,1]} w^* \Omega_I. \tag{2.5}$$

As Ω_I is closed and L_1, L_2 are I-holomorphic Lagrangian submanifolds ($\Omega_I|_{L_1} = \Omega_I|_{L_2} = 0$), the value $W(\mathfrak{p}, w)$ only depends on the homotopy class $[w] \in \pi_1(p_0, p)$ of w, and so we obtain a well-defined function on the universal cover $\widetilde{\mathcal{P}}_0$ of $\mathcal{P}_0$:

$$\begin{aligned} W\colon \widetilde{\mathcal{P}}_0 &\longrightarrow \mathbb{C} \\ (\mathfrak{p}, [w]) &\longmapsto W(\mathfrak{p}, [w]) := -\int_{[0,1]\times[0,1]} w^* \Omega_I, \end{aligned}$$

called the *holomorphic action functional.* As Ω_I is a I-holomorphic 2-form on M, W is a holomorphic function on the infinite-dimensional Kähler manifold $\widetilde{\mathcal{P}}_0$ endowed with the complex structure $I_{\mathcal{P}}$.

2.3. Infinite-dimensional complex Morse theory. The holomorphic action functional W is a holomorphic function on the infinite-dimensional Kähler manifold $\widetilde{\mathcal{P}}_0$. The critical points of W are of the form (p, γ), where $p \in L_1 \cap L_2$ is an intersection point, and where $\gamma \in \pi_1(p_0, p)$ is a homotopy class of paths from p_0 to p in $\mathcal{P}_0$. These critical points are non-degenerate under the assumption that L_1 and L_2 intersect transversally. Hence, we can apply formally complex Morse theory to $(\widetilde{\mathcal{P}}_0, W)$.

LEMMA 2.1. *For every $\zeta \in \mathbb{C}$ with $|\zeta| = 1$, a map*

$$\begin{aligned} u\colon \mathbb{R} &\longrightarrow \widetilde{\mathcal{P}}_0 \\ x &\longmapsto (y \mapsto u(x,y)) \end{aligned}$$

is a solution of the ζ-soliton equation for $(\widetilde{\mathcal{P}}_0, W)$ if and only if

$$\begin{aligned} u\colon \mathbb{R} \times [0,1] &\longrightarrow M \\ (x,y) &\longmapsto u(x,y) \end{aligned}$$

is a solution of the J_ζ-holomorphic curve equation

$$\partial_x u + J_\zeta(u)\partial_y u = 0. \tag{2.6}$$

PROOF. The ζ-soliton equation (1.2) for $(\widetilde{\mathcal{P}}_0, W)$ is the gradient flow line equation of the functional $\mathrm{Re}(\zeta^{-1}W)$. Using (2.1) and (2.5), one finds that $\mathrm{Re}(\zeta^{-1}W)$ is the action functional for the symplectic form ω_ζ, and so its gradient flow line equation is the J_ζ-holomorphic curve equation in M. This latter result is the well-known starting point of Floer theory [**28**, (1.9-1.10)], but we briefly review its derivation for convenience. Using (2.5), the differential of W at the point $u(x,-) := (y \mapsto u(x,y))$ evaluated on the tangent vector field $\xi = (y \mapsto \xi(y))$ is given by

$$dW_{u(x,-)}(\xi) = -\int_0^1 \omega_\zeta(\xi(y), \partial_y u(x,y)).$$

As $\omega_\zeta(-,-) = -g(-, J_\zeta -)$, it follows from the definition (2.2) of the metric on $\widetilde{\mathcal{P}}_0$ that

$$(\mathrm{grad}_{u(x,-)} W)(y) = J_\zeta(u(x,y))\partial_y u(x,y),$$

and so the gradient flow line equation (1.2) indeed reduces to the J_ζ-holomorphic curve equation (2.6). □

LEMMA 2.2. *For every $\zeta \in \mathbb{C}$ with $|\zeta| = 1$, a map*

$$\begin{aligned} u\colon \mathbb{R}^2 &\longrightarrow \widetilde{\mathcal{P}}_0 \\ (t,x) &\longmapsto (y \mapsto u(t,x,y)) \end{aligned}$$

is a solution of the ζ-instanton equation for $(\widetilde{\mathcal{P}}_0, W)$ if and only if

$$\begin{aligned} u\colon \mathbb{R}^2 \times [0,1] &\longrightarrow M \\ (t,x,y) &\longmapsto u(t,x,y) \end{aligned}$$

is a solution of the 3d ζ-Fueter equation

$$\partial_t u + I(u)(\partial_x u + J_\zeta(u)\partial_y u) = 0. \tag{2.7}$$

PROOF. The result follows using the definition (1.4) of the ζ-instanton equation, definition (2.3) of the complex structure on $\widetilde{\mathcal{P}}_0$, and Lemma 2.1 computing the ζ-soliton equation. □

2.4. Vector spaces from pairs of intersection points. For every pair of intersection points $p, p' \in L_1 \cap L_2$ and homotopy classes of paths $\gamma \in \pi_1(p_0, p)$ and $\gamma' \in \pi_1(p_0, p')$ in $\mathcal{P}_0$, one should be able by formally applying §1.2 to define a vector space

$$H((p,\gamma),(p',\gamma'))$$

of 2d BPS states for $(\widetilde{\mathcal{P}}_0, W)$. Explicitly, $H((p,\gamma),(p',\gamma'))$ is defined as the cohomology of the complex $(C(p,\gamma),(p',\gamma')), d)$, where

(1) $C((p,\gamma),(p',\gamma'))$ is the vector space generated by J_ζ-holomorphic curves

$$\begin{aligned} u_i : \mathbb{R} \times [0,1] &\longrightarrow M \\ (x,y) &\longmapsto u_i(x,y), \end{aligned}$$

considered up to $\mathbb{R}$-translation of x, such that

$$u_i(.,0) \in L_1, u_i(0,.) \in L_2, \lim_{x \to -\infty} u_i(x,.) = p, \lim_{x \to +\infty} u_i(x,.) = p',$$
$$[u_i] = [\gamma' \circ \gamma^{-1}],$$

where u_i is viewed as a path from p to p' in $\widetilde{\mathcal{P}}_0$ and γ^{-1} is the path obtained from γ by reversing the orientation, and

$$\zeta = \zeta_{(p,\gamma),(p',\gamma')} := \frac{\int_{\gamma' \circ \gamma^{-1}} \Omega_I}{|\int_{\gamma' \circ \gamma^{-1}} \Omega_I|}, \tag{2.8}$$

where the class $\gamma' \circ \gamma^{-1} \in \pi_1(p,p')$ is viewed as defining an element in the relative homology group $H_2(M, L_1 \cup L_2, \mathbb{Z})$.

(2) $d : C((p,\gamma),(p',\gamma')) \to C((p,\gamma),(q,\gamma'))$ is the differential defined by

$$du_i = \sum_j n_{ij} u_j, \tag{2.9}$$

where n_{ij} is a signed count of maps

$$\begin{aligned} u\colon \mathbb{R}^2 \times [0,1] &\longrightarrow M \\ (t,x,y) &\longmapsto u_i(t,x,y), \end{aligned}$$

such that

$$u(t,x,0) \in L_1, u(t,x,.) \in L_2, \lim_{x \to -\infty} u(.,x,.) = p, \lim_{x \to +\infty} u(.,x,.) = p',$$

asymptotic to u_i and u_j,

$$\lim_{t \to -\infty} u(t,x,y) = u_i(x,y), \lim_{t \to +\infty} u(t,x) = u_j(x,y),$$

which are solutions of the 3d $\zeta_{p,p'}$-Fueter equation

$$\partial_t u + I(u)(\partial_x u + J_\zeta(u)\partial_y u) = 0,$$

and considered up to $\mathbb{R}$-translation of t.

Remark 2.3. As L_1 and L_2 are holomorphic Lagrangian submanifolds of (M, Ω_I), they are in particular Lagrangian submanifolds of (M, ω_ζ) for every $\zeta \in \mathbb{C}$, $|\zeta| = 1$, and so define natural boundary conditions for J_ζ-holomorphic curves.

Remark 2.4. We are making the simplifying assumption that the sum (2.9) is finite and we are suppressing the discussion of Novikov coefficients which would be required in general.

2.5. Categories from pairs of holomorphic Lagrangian submanifolds. Applying formally §1.3, one should be able to define a A_∞-category

$$FS(\widetilde{\mathcal{P}}_0, W)$$

as the Fukaya-Seidel category of $(\widetilde{\mathcal{P}}_0, W)$. Objects are pairs

$$(p, \gamma)$$

where $p \in L_1 \cap L_2$ and γ is a homotopy class of paths from $\mathfrak{p}_0$ to p in $\widetilde{\mathcal{P}}_0$, spaces of morphisms have bases consisting of formal tensor products of $J_{\zeta_{(p,\gamma),(p',\gamma')}}$-holomorphic curves with boundary on $L_1 \cup L_2$, and A_∞-morphisms are given by counts of solutions of the 3d ζ-Fueter equation asymptotic to sequences of $J_{\zeta_{(p,\gamma),(p',\gamma')}}$-holomorphic curves.

To make sense of this construction, one needs to deal with the issue that the set of critical points (p, γ) of W might be infinite if the fundamental group $\pi_1(\mathcal{P}_0)$ is infinite, whereas we assumed in our review of complex Morse theory in §1 the finiteness of the set of critical points. We address this issue below by considering a categorical version of the Novikov coefficients of Floer theory.

First remark that by construction the chain complexes $C((p,\gamma),(p',\gamma'))$ only depend on the class $\gamma' \circ \gamma^{-1} \in \pi_1(p,p')$. For every $\alpha \in \pi_1(p,p')$, we denote $C(p,p',\alpha)$ for $C((p,\gamma),(p',\gamma'))$ with $\gamma' \circ \gamma^{-1} = \alpha$. For every intersection points $p, p' \in L_1 \cap L_2$, we denote by

$$r : \pi_1(p,p') \to H_2(M, L_0 \cup L_1, \mathbb{Z})$$

the natural projection. By construction, $C(p,p',\alpha) \neq 0$ if and only if there exists J_{ζ_α}-holomorphic curves of class α, where

$$\zeta_\alpha = \frac{\int_{r(\alpha)} \Omega_I}{|\int_{r(\alpha)} \Omega_I|}. \tag{2.10}$$

LEMMA 2.5. *Fix $p, p' \in L_1 \cap L_2$ and $\alpha \in \pi_1(p,p')$. If C is a J_{ζ_α}-holomorphic curve of class α, then the area of C is $|\int_{r(\alpha)} \Omega_I|$.*

PROOF. As C is a J_{ζ_α}-holomorphic curve of class α, its area is $\int_{r(\alpha)} \omega_\zeta$. Using (2.1), this can be rewritten as $\int_{r(\alpha)} \mathrm{Re}(\zeta_\alpha^{-1}\Omega_I) = \mathrm{Re}(\zeta_\alpha^{-1} \int_{r(\alpha)} \Omega_I)$. By (2.10), this is equal to $\mathrm{Re}(|\int_{r(\alpha)} \Omega_I|) = |\int_{r(\alpha)} \Omega_I|$. □

LEMMA 2.6. *Fix a norm $\|-\|$ on $H_2(M, L_1 \cup L_2, \mathbb{Z}) \otimes \mathbb{R}$. There exists a constant $A > 0$ such that for every $p, p' \in L_1 \cap L_2$ and $\alpha \in \pi_1(p,p')$ with $C(p,p',\alpha) \neq 0$, we have*

$$\|r(\alpha)\| \leq A |\int_{r(\alpha)} \Omega_I|.$$

PROOF. Let (h_i) be a basis of harmonic 2-forms on M relative to $L_1 \cap L_2$ [**75**]. It is enough to bound $|\int_{r(\alpha)} h_i|$ for every i. Let A_i be the supremum of $|h_i|$ over all of M, which is finite if M is bounded or appropriately convex at infinity. If $C(p,p',\alpha) \neq 0$, there exists a J_{ζ_α}-holomorphic curve C of class α, and so $|\int_{r(\alpha)} h_i|$ is bounded by A_i times the area of C, which is equal to $|\int_{r(\alpha)} \Omega_I|$ by Lemma 2.5. □

REMARK 2.7. The inequality of Lemma 2.6 is formally similar to the support property in DT theory [**64**] and its proof is parallel to the argument given in [**64**, §1.2, Remark 1] for the validity of the support property in the context of special

Lagrangian submanifolds. Under the correspondence between holomorphic Floer theory and DT theory described in §4, the inequality of Lemma 2.6 will exactly match with the support property in DT theory.

LEMMA 2.8. *Fix $p, p' \in L_1 \cap L_2$ and $\beta \in H_2(M, L_1 \cup L_2, \mathbb{Z})$. Then, there exists finitely many $\alpha \in \pi_1(p, p')$ such that $C(p, p', \alpha) \neq 0$ and $r(\alpha) = \beta$.*

PROOF. The complex structure ζ_α and the area of J_{ζ_α}-holomorphic curves given by Lemma 2.5 only depend on $r(\alpha)$. Hence, we are considering a set of holomorphic curves with respect to a fixed complex structure and with fixed area, and so the result follows by Gromov's compactness for holomorphic curves. □

Let (p, γ) and (p', γ') be two critical points of W. We describe how to define a completed version $\hat{R}_{(p,\gamma),(p',\gamma')}$ of the Hom complex (1.5) used in the case of finitely many critical points. Fix $N > 0$. For every Q as in (1.6), corresponding to a sequence $(p, \gamma), (p_1, \gamma_1), \ldots, (p_k, \gamma_k)$ of critical points, we say that $Q < N$ if

$$\sum_j \left| \int_{r(\gamma_{j+1}) - r(\gamma_j)} \Omega_I \right| < N.$$

It follows from Lemmas 2.6–2.8 that there are finitely many Q such that $Q < N$ and

$$R_Q := C((p, \gamma), (p_1, \gamma_1)) \otimes C((p_1, \gamma_1), (p_2, \gamma_2)) \otimes \cdots \otimes C((p_k, \gamma_k), (p', \gamma')) \neq 0.$$

Hence, the direct sum

$$R^{<N}_{(p,\gamma),(p',\gamma')} := \bigoplus_{Q<N} R_Q$$

contains only finitely many non-zero summands and all the categorical constructions described in §1.3 in the case of finitely many critical points make sense with $R^{<N}_{(p,\gamma),(p',\gamma')}$ and we obtain a truncated Fukaya-Seidel category $FS^{<N}(\widetilde{\mathcal{P}}_0, W)$ with Hom-complexes $R^{<N}_{(p,\gamma),(p',\gamma')}$. Finally, we define the Fukaya-Seidel category $FS(\widetilde{\mathcal{P}}_0, W)$ by taking the categorical limit $C \to +\infty$, with Hom complexes

$$\hat{R}_{(p,\gamma),(p',\gamma')} = \lim_{N \to +\infty} R^{<N}_{(p,\gamma),(p',\gamma')}.$$

2.6. 2-categories from holomorphic symplectic manifold. In the same way that Floer homology groups of pairs of Lagrangian submanifolds of a symplectic manifold can be realized as Hom spaces of the Fukaya category, it is natural to expect that the categories $FS(\widetilde{\mathcal{P}}_0, W)$ attached to pairs of holomorphic Lagrangian submanifolds of a holomorphic symplectic manifold can be realized as Hom categories of a 2-category.

CONJECTURE 2.9. *Let M be a holomorphic symplectic manifold, with complex structure I and holomorphic symplectic form Ω_I. The I-holomorphic Lagrangian submanifolds of M should form a linear 2-category, such that the category of morphisms* $\mathrm{Hom}(L_1, L_2)$ *between two objects L_1 and L_2 is the Fukaya-Seidel category of the holomorphic action functional on the infinite-dimensional space of paths between L_1 and L_2.*

We refer to the 2-category whose existence is asserted by Conjecture 2.9 as the *Fueter 2-category* of M, and we denote it by $\mathrm{Ft}(M)$, or $\mathrm{Ft}(M, \Omega_I)$ when we wish to make clear the complex structure and holomorphic symplectic form that are considered. We expect the Fueter 2-category to be independent on the choice of the auxiliary complex structures J and K (or to have only dependence on their asymptotic behaviour if M is non-compact), in the same way that the Fukaya category is a symplectic invariant and independent on auxiliary choices of almost-complex structures. If M is non-compact, one should also have a wrapped version of the Fueter 2-category, parallel to the wrapped Fukaya category [**1**].

For the simplest examples of holomorphic symplectic manifolds given as cotangent bundles T^*L of complex manifolds L, one expects the Hom category between the zero-section L and the graph L' of a holomorphic Morse function W on L to reproduce the Fukaya-Seidel category of (L, W), similarly to the way that the Floer homology between the zero-section and a perturbation of the zero-section in the cotangent bundle of a manifold reproduces Morse homology [**29, 32**].

CONJECTURE 2.10. *Let L be an exact Kähler manifold, W a holomorphic Morse function on L, and L' the graph of dW in T^*L. Then the Hom-category $\mathrm{Hom}_{\mathrm{Ft}(T^*L)}(L, L')$ in the Fueter 2-category of T^*L coincides with the Fukaya-Seidel category of (L, W).*

Conjectures 2.9 and 2.10 were independently formulated by Doan-Rezchikov [**23**]. In [**23**], Doan-Rezchikov introduce a flexible notion of hyperkähler geometry and start the analytic study of the compactness properties of solutions of the 3d Fueter equation. As an application, they obtain a first step towards Conjecture 2.10 by proving an isomorphism between moduli spaces of solutions of the 3d Fueter equation in T^*L and moduli spaces of pseudo-holomorphic curves in L.

REMARK 2.11. A "hyperkähler Floer theory" has been developed by Hohloch-Noetzel-Salamon [**47**], which is also based on a version of the Fueter equation. However, the proposal of [**47**] and Conjecture 2.9 are different: whereas the three complex structures I, J, K play symmetric roles in [**47**], it is not the case in Conjecture 2.9.

2.7. The Fueter 2-category and the Rozansky-Witten 2-category. As reviewed in §1, given a pair (X, W) where X is a Kähler manifold with $c_1(X) = 0$ and $W \colon X \to \mathbb{C}$ is a holomorphic Morse function, one can define the Fukaya-Seidel category $FS(X, W)$. It is a $\mathbb{Z}$-graded category which is not Calabi-Yau in general (the Serre functor is induced by the global monodromy around the critical points in the W-plane). Physically, $FS(X, W)$ is the category of boundary conditions in the A-model twist of the 2-dimensional $\mathcal{N} = (2, 2)$ Landau-Ginzburg theory defined by (X, W). Another category naturally attached to (X, W) is the category of matrix factorizations $MF(X, W)$, which is $\mathbb{Z}/2\mathbb{Z}$-graded, Calabi-Yau, and physically obtained as the category of boundary conditions in the B-model twist of the Landau-Ginzburg theory defined by (X, W). When (X, W) is the mirror of a Fano manifold Y, the Fukaya-Seidel category of (X, W) is the derived category of coherent sheaves on Y:

$$FS(X, W) = \mathrm{D^bCoh}(Y),$$

and the category of matrix factorizations of (X, W) is the Fukaya category of Y:

$$MF(X, W) = F(Y).$$

By construction, the Hom categories $\mathrm{Hom}_{\mathrm{Ft}(M,\Omega_I)}(L_1, L_2)$ in the Fueter 2-category $\mathrm{Ft}(M,\Omega_I)$ of a holomorphic symplectic manifold (M,Ω_I) are the Fukaya-Seidel categories of the holomorphic action functional on the spaces of paths between I-holomorphic Lagrangian submanifolds L_1 and L_2. If one replaces in this construction the Fukaya-Seidel category of the holomorphic action functional by the category of matrix factorizations of the holomorphic action functional, one obtains a 2-category $RW(M,\Omega_I)$. The proposal for such a 2-category attached to a holomorphic symplectic manifold was first formulated by Kapustin-Rozansky-Saulina [**56, 57**], who also argued that $RW(M,\Omega_I)$ should be the 2-category of boundary condition of the 3-dimensional Rozansky-Witten topological quantum field theory (TQFT) [**74**]. In general, a 2-category which is Calabi-Yau, as $RW(M,\Omega_I)$, is always the category of boundary conditions of a 3d TQFT [**69**][5]. By contrast, a 2-category which is not Calabi-Yau, as the Fueter 2-category, defines at best a *framed* 3d TQFT, which is only defined on framed manifolds of dimension ≤ 3 [**69**]. This is a 3d version of the maybe more familiar fact that the B-model (resp. A-model) of a holomorphic Morse function (X, W) (resp. of a Fano manifold Y) is a 2d TQFT whereas the A-model of (X, W) (resp. the B-model of Y) is only a framed 2d TQFT.

One expects that most of the structures existing for the categories $FS(X, W)$ and $MF(X, W)$ of a finite-dimensional Landau-Ginzburg model (X, W) have an analogue for the 2-categories $\mathrm{Ft}(M,\Omega_I)$ and $RW(M,\Omega_I)$ attached to a holomorphic symplectic manifold. For example, one can construct from $MF(X, W)$ a non-commutative Hodge structure [**59**], with Dolbeault data

$$\mathbb{H}^*(X, (\Omega_X^*, dW \wedge -)),$$

de Rham data given by the $\mathbb{C}((u))$-module

$$\mathbb{H}^*(X, (\Omega_X^*, ud + dW \wedge -))$$

and Betti data $H^*(X, X_{-\infty}, \mathbb{Z})$ where $X_{-\infty} = \{\mathrm{Re}W << 0\}$. The Riemann-Hilbert type isomorphism between de Rham and complexified Betti data is given by the exponential periods $\int_\gamma e^W \alpha$, where $\gamma \in H_*(X, X_{-\infty}, \mathbb{Z})$ are naturally classes of objects in $FS(X, W)$. Moreover, the Hodge filtration is encoded by a connection on the Rham data, with second order pole at $u = 0$, and space of flat sections given by the Betti data. Finally, the Stokes data of the irregular singularity of the connection at $u = 0$, controlling the jump of the asymptotic expansion of the periods $\int_\gamma e^{W/u}\alpha$ obtained by expressing γ as a linear combination of Lefschetz thimbles for $\mathrm{Re}(W/u)$, is determined by the counts of 2d BPS states of (X, W) (which are equal to Euler characteristics of the spaces of 2d BPS states reviewed in §1.2 for the construction of $FS(X, W)$).

It is conjectured in [**59**] that there is more generally an entirely categorical way to produce a non-commutative Hodge structure from a smooth proper dg-category C. For example, the Dolbeault data should be the Hochschild homology $HH(C)$ and the de Rham data should be the periodic cyclic homology $HC(C)$. We propose that this construction should have a 2-categorical analogue. For example, if C is a Calabi-Yau 2-category, as $RW(M,\Omega_I)$, one can define its Hochschild homology $HH(C)$ as the category obtained by evaluating the corresponding 3d TQFT on the

[5]We are neglecting here smoothness and compactness issues which could make the TQFT only partially defined.

circle S^1, and similarly its periodic cyclic homology $HC(C)$ as the category over $\mathbb{C}((u))$ obtained by S^1-equivariant compactification of the 3d TQFT over S^1.

CONJECTURE 2.12. *Let (M, Ω_I) be a holomorphic symplectic manifold. Then,*

(i) *the Hochschild homology of the 2-category $RW(M, \Omega_I)$ is the derived category of coherent sheaves on (M, I):*

$$HH(RW(M, \Omega_I)) = \mathrm{D^bCoh}(M, I).$$

(ii) *the periodic cyclic homology of the 2-category $RW(M, \Omega_I)$ is the category of DQ-modules* [**58, 62**] *obtained by deformation quantization of (M, Ω_I):*

$$HP(RW(M, \Omega_I)) = DQ(M, \Omega_I).$$

Conjecture 2.12(i) is the known expectation that the Rozansky-Witten theory compactified on S^1 is the B-model [**56, 57**], whereas the formulation of Conjecture 2.12(ii) might be new.

Finally, we describe how holomorphic Floer theory should be related to the Betti data of the hypothetical categorical non-commutative Hodge structure attached to $RW(\Omega_I, W)$. For every $u \in \mathbb{C}^*$, $\omega_u = \mathrm{Re}(u^{-1}\Omega_I)$ is a symplectic form on M, and so one can consider the Fukaya category of the symplectic manifold (M, ω_u). The Fukaya category is generally defined over a Novikov ring whose variable keeps track of the area of J_u-holomorphic curves. From now on, assume that for $u > 0$ small enough, the Fukaya category converges when we replace the Novikov parameter q^β by

$$e^{-\int_\beta (\omega_u + iB_u)} = e^{-\frac{1}{u}\int_\beta \Omega_I},$$

where $B_u := \mathrm{Im}(u^{-1}\Omega_I)$, and we denote by $F(M, \omega_u, B_u)$ the corresponding category. Viewed as a family over the formal punctured disk $\mathbb{C}((u))$, we obtain a category $F(M, \omega_u, B_u) \otimes \mathbb{C}((u))$ over $\mathbb{C}((u))$ where the contributions of J_u-holomorphic curves disappear because proportional to $e^{-\frac{1}{u}\int_\beta \Omega_I}$, which has zero Taylor expansion in u. Note that J_u-holomorphic disks only exist when the argument of u is in the complement of the countable set of arguments of complex numbers $\int_\beta \Omega_I$ indexed by relative homology classes $\beta \in H_2(M, L_1 \cup L_2, \mathbb{Z})$ between holomorphic Lagrangian submanifolds L_1 and L_2. The following conjecture is motivated by the known finite-dimensional story for $FS(X, W)$ and $MF(X, W)$.

CONJECTURE 2.13. *The Betti data for the non-commutative Hodge theory of $RW(M, \Omega_I)$ is induced by a Riemann-Hilbert type isomorphism*

$$DQ(M, \Omega_I) \simeq F(M, \omega_u, B_u) \otimes \mathbb{C}((u)). \tag{2.11}$$

Moreover, the Stokes data is induced by the holomorphic dependence of $F(M, \omega_u, B_u)$ in u and is determined by the counts of J_u-holomorphic curves entering the definition of the spaces of 2d BPS states of the holomorphic action functional described in §2.4 *for the construction of* $\mathrm{Ft}(M, \Omega_I)$.

A relation between deformation quantization and the Fukaya category is predicted from physics arguments by Kapustin in [**55**], and related works include [**8, 13, 70, 72, 82**]. Moreover, such a relation combined with the Stokes data interpretation of J_u-holomorphic curves is one of the main points of the work of Kontsevich-Soibelman [**66**] on non-perturbative quantization of (M, Ω_I) and the topic of resurgence. Conjecture 2.13 suggests a formulation of this relation in terms of non-commutative Hodge theory for $RW(M, \Omega_I)$.

2.8. Complex and holomorphic Atiyah-Floer conjectures. Holomorphic Floer theory is an example of infinite-dimensional complex Morse theory. In this section, we describe conjectural relations between holomorphic Floer theory and the two other examples of infinite-dimensional Morse theory briefly reviewed in §1.4 and given by complex-valued Chern-Simons theory and holomorphic Chern-Simons theory. These conjectures are respectively complex and holomorphic analogues of the Atiyah-Floer conjecture relating the two examples of infinite-dimensional Morse theory given by Lagrangian Floer theory and Chern-Simons theory (Floer homology for 3-manifolds) [**5**].

Let X be a compact 3-manifold with a choice of Heegaard splitting $X = X_1 \cup_\Sigma X_2$ along a compact surface Σ. Let $G_\mathbb{C}$ be the complexification of a compact Lie group. The character variety $M_{\Sigma,G_\mathbb{C}}$ parametrizing flat $G_\mathbb{C}$-connections on Σ is naturally a holomorphic symplectic manifold, and the locus L_{X_1} (resp. L_{X_2}) of $G_\mathbb{C}$ flat connections on Σ extending to X_1 (resp. X_2) is a holomorphic Lagrangian submanifold of $M_{\Sigma,G_\mathbb{C}}$. Holomorphic Floer theory applied to the pair of holomorphic Lagrangian submanifolds L_1 and L_2 conjecturally produces a Hom category $\mathrm{Hom}_{\mathrm{Ft}(M_{\Sigma,G_\mathbb{C}})}(L_{X_1}, L_{X_2})$ in the Fueter 2-category of $M_{\Sigma,G_\mathbb{C}}$. On the other hand, complex-valued Chern-Simons theory conjecturally attaches to the 3-manifold X a category $CS(X)$ constructed from counts of flat $G_\mathbb{C}$-connections on X, and of solutions to the Kapustin-Witten and Haydys-Witten equations on $X \times \mathbb{R}$ and $X \times \mathbb{R}^2$.

CONJECTURE 2.14. *Let X be a compact 3-manifold with a choice of Heegaard splitting $X = X_1 \cup_\Sigma X_2$ along a compact surface Σ, and let $G_\mathbb{C}$ be the complexification of a compact Lie group. Then, there exists an equivalence of categories*

$$\mathrm{Hom}_{\mathrm{Ft}(M_{\Sigma,G_\mathbb{C}})}(L_{X_1}, L_{X_2}) \simeq CS(X).$$

Let X be a compact Calabi-Yau 3-fold degenerating to the union $X_1 \cup_\Sigma X_2$ of two Fano 3-folds transversally glued along a common anticanonical $K3$ surface Σ. The moduli space $M_{\Sigma,\gamma}$ of stable holomorphic vector bundles on Σ with Chern classes γ is naturally a holomorphic symplectic manifold, and the locus L_{X_1} (resp. L_{X_2}) of holomorphic vector bundles on Σ extending to X_1 (resp. X_2) is a holomorphic Lagrangian submanifold of $M_{\Sigma,\gamma}$. Holomorphic Floer theory applied to the pair of holomorphic Lagrangian submanifolds L_1 and L_2 conjecturally produces a Hom category $\mathrm{Hom}_{\mathrm{Ft}(M_{\Sigma,\gamma})}(L_{X_1}, L_{X_2})$ in the Fueter 2-category of $M_{\Sigma,\gamma}$. On the other hand, holomorphic Chern-Simons theory conjecturally attaches to the Calabi-Yau 3-fold X and to the Chern classes γ a category $HCS(X,\gamma)$ constructed from counts of stable holomorphic vector bundles on X, and of G_2-instantons and $Spin(7)$-instantons on $X \times \mathbb{R}$ and $X \times \mathbb{R}^2$.

CONJECTURE 2.15. [6] *Let X be a compact Calabi-Yau 3-fold degenerating to the union $X_1 \cup_\Sigma X_2$ of two Fano 3-folds transversally glued along a common anticanonical $K3$ surface Σ. Then, there exists an equivalence of categories*

$$\mathrm{Hom}_{\mathrm{Ft}(M_{\Sigma,\gamma})}(L_{X_1}, L_{X_2}) \simeq HCS(X,\gamma).$$

[6]Conjecture 2.15 was suggested to the author by Richard Thomas. The uncategorified statement relating counts of holomorphic vector bundles and Lagrangian intersection numbers is discussed in [**25**, §4].

REMARK 2.16. As for the usual Atiyah-Floer conjecture, one expects Conjectures 2.14 and 2.15 to follow from an adiabatic dimensional reduction of the higher dimensional gauge theoretic equations.

3. 2d $\mathcal{N} = (2,2)$ theories and holomorphic Floer theory

In §2, we introduced holomorphic Floer theory as a Landau-Ginzburg model with an infinite dimensional target space. In this section, we explain that holomorphic Floer theory is actually relevant to the study of usual 2d $\mathcal{N} = (2,2)$ field theories such as Landau-Ginzburg models with finite dimensional target spaces. The conjectures that we formulate here will motivate our conjectures on the relation between holomorphic Floer theory and DT invariants in §4 (see Remark 4.7).

3.1. General 2d $\mathcal{N} = (2,2)$ theories. Let $\mathcal{T}$ be a massive 2-dimensional $\mathcal{N} = (2,2)$ field theory with n vacua. We denote by $\mathrm{Br}(\mathcal{T})$ its A_∞-category of boundary conditions (or branes) as considered in [**38**]. When $\mathcal{T}$ is the Landau-Ginzburg model defined by a holomorphic function $W: X \to \mathbb{C}$ on a Kähler manifold X, then $\mathrm{Br}(\mathcal{T})$ is the Fukaya-Seidel category of (X, W). The goal of this section is to argue that for general $\mathcal{T}$, the category of boundary conditions $\mathrm{Br}(\mathcal{T})$ can be extracted from a Fueter 2-category constructed from the moduli space of deformations of $\mathcal{T}$.

Let $\mathcal{N}_\mathcal{T}$ be the generically semi-simple Frobenius manifold parametrizing deformations of $\mathcal{T}$ [**20**, **26**]. In particular, the tangent bundle $T\mathcal{N}_\mathcal{T}$ has a structure of sheaf of commutative algebras over $\mathcal{N}_\mathcal{T}$, coming from the identification of the deformations of $\mathcal{T}$ with the the chiral ring. The natural quotient map

$$\mathrm{Sym}(T\mathcal{N}_\mathcal{T}) \to T\mathcal{N}_\mathcal{T}$$

induces an inclusion

$$\mathcal{L} = \mathrm{Spec}(T\mathcal{N}_\mathcal{T}) \hookrightarrow T^*\mathcal{N}_\mathcal{T} = \mathrm{Spec}(\mathrm{Sym}(T\mathcal{N}_\mathcal{T})).$$

The assumption that $\mathcal{T}$ is massive with n vacua implies that the spectral cover $\mathcal{L} \to \mathcal{N}_\mathcal{T}$ is generically of degree n. Furthermore, the inclusion $\mathcal{L} \subset T^*\mathcal{N}_\mathcal{T}$ naturally realizes $\mathcal{L}$ as a holomorphic Lagrangian submanifold of the holomorphic symplectic variety $T^*\mathcal{N}_\mathcal{T}$ by [**6**, Corollary 1.9]. On the other hand, let $F_\mathcal{T} \subset T^*\mathcal{N}_\mathcal{T}$ be the holomorphic Lagrangian submanifold given by the cotangent fiber over the point of $\mathcal{N}_\mathcal{T}$ corresponding to $\mathcal{T}$. The intersection $F_\mathcal{T} \cap \mathcal{L}$ consists of n points in natural correspondence with the n vacua of $\mathcal{T}$, see Figure 3.1.

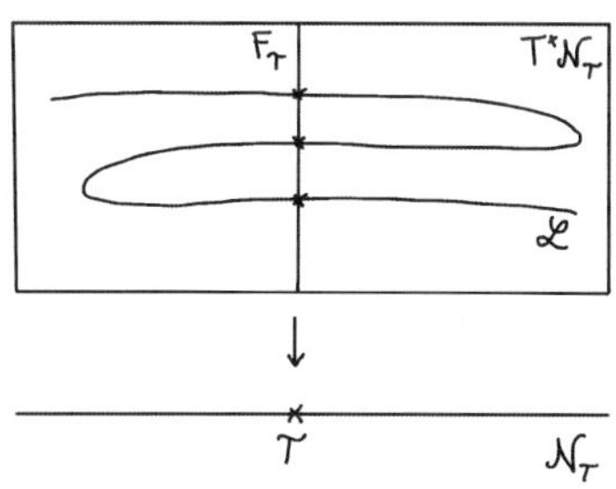

FIGURE 3.1. The spectral cover $\mathcal{L}$ and the cotangent fiber $F_\mathcal{T}$.

Conjecture 3.1. *The A_∞-category of boundary conditions* $\mathrm{Br}(\mathcal{T})$ *of a massive 2-dimensional* $\mathcal{N} = (2,2)$ *field theory* $\mathcal{T}$ *is equivalent to the Hom-category in the Fueter 2-category* $\mathrm{Ft}(T^*\mathcal{N}_\mathcal{T})$ *of between the spectral cover* $\mathcal{L}$ *and the cotangent fiber* $F_\mathcal{T}$*:*

$$\mathrm{Br}(\mathcal{T}) \simeq \mathrm{Hom}_{\mathrm{Ft}(T^*\mathcal{N}_\mathcal{T})}(\mathcal{L}, F_\mathcal{T}). \tag{3.1}$$

Remark 3.2. Moving the point $\mathcal{T}$ in $\mathcal{N}_\mathcal{T}$, that is moving the cotangent fiber $F_\mathcal{T}$, the A_∞-categories in (3.1) should form a "perverse sheaf of categories" over $\mathcal{N}_\mathcal{T}$, and so fit in the theory of perverse schobers [**53, 54**].

3.2. Surface defects and spectral networks. We also formulate below a version of Conjecture 3.1 in the context of theories of class $\mathcal{S}$. Let C be a Riemann surface (possibly with punctures), and B the base of the Hitchin fibration on the moduli space of $SL_n(\mathbb{C})$ Higgs bundles on C. For every point $b \in B$, we have the corresponding spectral curve $\Sigma_b \subset T^*C$. When Σ_b is smooth, the projection $\Sigma_b \to C$ is a ramified cover of degree n.

In physics language, Σ_b parametrizes 2-dimensional $\mathcal{N} = (2,2)$ field theories describing the "canonical surface defect" probing the vacuum $b \in B$ of the 4-dimensional $\mathcal{N} = 2$ of class $\mathcal{S}$ obtained by compactifying the 6-dimensional $\mathcal{N} = (2,0)$ superconformal field theory of type A_{n-1} on C [**35**, §7]. For every $x \in \Sigma_b$, we denote by $\mathcal{T}_{x,b}$ the corresponding 2-dimensional $\mathcal{N} = (2,2)$ field theory and by F_x the cotangent fiber of T^*C over x.

Conjecture 3.3. *For every* $b \in B$ *and* $x \in C$*, the* A_∞*-category of boundary conditions* $\mathrm{Br}(\mathcal{T}_{x,b})$ *of the canonical surface defect 2-dimensional* $\mathcal{N} = (2,2)$ *field theory* $\mathcal{T}_{x,b}$ *is equivalent to the Hom-category in the Fueter 2-category* $\mathrm{Ft}(T^*C)$ *of* T^*C *between the spectral curve* Σ_b *and the cotangent fiber* F_x*:*

$$\mathrm{Br}(\mathcal{T}_{x,b}) \simeq \mathrm{Hom}_{\mathrm{Ft}(T^*C)}(\Sigma_b, F_x). \tag{3.2}$$

Remark 3.4. Conjectures 3.1 and 3.3 are of a similar nature. The only difference is that in Conjecture 3.3, C is a space of particular deformations of $\mathcal{T}_{x,b}$, whereas in Conjecture 3.1, $\mathcal{N}_\mathcal{T}$ is the universal space of deformations of $\mathcal{T}$.

A non-trivial evidence for Conjecture 3.3 is provided by the M-theory realization of the BPS states of $\mathcal{T}_{x,b}$ and their description by spectral networks. Indeed, the M-theory realization of the BPS states of $\mathcal{T}_{x,b}$ is by BPS M2-branes with boundary on $F_x \cup \Sigma_b$, projecting onto spectral networks on C ending at x [**35**, §7]. BPS M2-branes are holomorphic disks with respect to a hyperkähler rotated complex structure prescribed by the phase of the central charge, and so are exactly the objects involved in the first step of the construction of $\mathrm{Hom}_{\mathrm{Ft}(T^*C)}(F_x, \Sigma_b)$. The second step involving solutions of the Fueter equations seems to be unexplored. It is an interesting question to find out what is the corresponding generalization in the description by spectral networks.

4. DT invariants from holomorphic Floer theory

After an overview of the geometry of Seiberg-Witten integrable systems in §4.1, we formulate in §4.2 our main conjecture relating holomorphic Floer theory and DT invariants occuring as BPS invariants of 4-dimensional $\mathcal{N} = 2$ field theories. We present a heuristic derivation of this conjecture in §4.3. Finally, we formulate further conjectures involving categories of line operators in §4.4.

4.1. Seiberg-Witten integrable systems. The low energy physics of a rank r 4-dimensional $\mathcal{N}=2$ field theory (without dynamical gravity) $\mathcal{T}$ is controlled by its Seiberg-Witten complex integrable system

$$\pi \colon M \longrightarrow B,$$

where B is the Coulomb branch of $\mathcal{T}$ on $\mathbb{R}^4$ and M is the Coulomb branch of $\mathcal{T}$ on $\mathbb{R}^3 \times S^1$. While B is a complex manifold of dimension r, the space M is a hyperkähler manifold of complex dimension $2r$ [**78**]. The low-energy effective theory of $\mathcal{T}$ on $\mathbb{R}^3 \times S^1$ is the 3d $\mathcal{N}=4$ sigma-model with target M.

We denote by I the complex structure on M in which π is holomorphic and by Ω_I the corresponding I-holomorphic symplectic form. As in §2.1, we denote by J and K the complex structures such that $\Omega_I = \omega_J + i\omega_K$, and for every $\zeta \in \mathbb{C}$, $|\zeta|=1$, we set $J_\zeta := (\mathrm{Re}\,\zeta)J + (\mathrm{Im}\,\zeta)K$. The fibers $F_b := \pi^{-1}(b)$ over points $b \in B$ in the complement of the discriminant locus Δ of π are r-dimensional abelian varieties and I-holomorphic Lagrangian submanifolds of (M, Ω_I), see Figure 4.1. The lattice of charges for states in the vacuum $b \in B \setminus \Delta$ is

$$\Gamma_b := \pi_2(M, F_b)$$

and for every $\gamma \in \Gamma_b$, we have a space BPS_γ^b of BPS states of charge γ in the vacuum b [**34, 37, 76–78**].

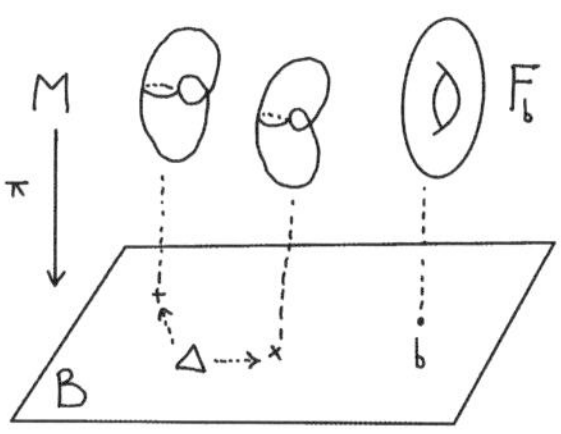

FIGURE 4.1. A rank 1 Seiberg-Witten integrable system $\pi : M \to B$, with discriminant locus Δ.

In the vacuum $b \in B \setminus \Delta$, the low-energy effective theory is a $U(1)^r$ gauge theory with electromagnetic charge lattice $H_1(F_b, \mathbb{Z})$. The natural map $\Gamma_b \to H_1(F_b, \mathbb{Z})$ is the projection of the lattice of charges on the lattice of electromagnetic charges and its kernel is the lattice of flavour charges. As already mentioned in §0.2.2, in many cases, there is an associated 3-dimensional Calabi-Yau (CY3) triangulated category $\mathcal{C}$, the base $B \setminus \Delta$ maps to the space of Bridgeland stability conditions on $\mathcal{C}$, the central charge at a point $b \in B \setminus \Delta$ is given by

$$Z_\gamma(b) = \int_\gamma \Omega_I,$$

and BPS_γ^b is mathematically realized as a cohomological DT invariant of $\mathcal{C}$ counting b-stable objects of class γ.

EXAMPLE 4.1. When $\mathcal{T}$ is a class $\mathcal{S}$ theory, obtained by compactifying the 6-dimensional $\mathcal{N}=(2,0)$ superconformal field theory of ADE type G on a Riemann surface C (possibly with punctures), the corresponding Seiberg-Witten complex integrable system is closely related to the Hitchin integrable system on the moduli space of G-Higgs bundles on C. Typically, the Seiberg-Witten integrable system is a self-dual fiberwise finite quotient of an Hitchin integrable system [**22**, **34**, **37**]. In particular, the base B is isomorphic to $\mathbb{C}^r$ as a complex manifold.

EXAMPLE 4.2. By considering a 4-dimensional $\mathcal{N}=2$ theory obtained by an appropriate compactification of a six-dimensional little string theory on T^2, one can obtain an elliptic K3 surface as Seiberg-Witten integrable system [**49**]. In this case, the base B is a complex projective line.

From now on, we assume that B is simply-connected. This assumption covers in particular the cases of Hitchin systems and K3 surfaces that we just described.

4.2. DT invariants and holomorphic Floer theory. Let D be a cigar geometry: D is the closed unit disk with polar coordinates $r \in [0,1]$, $\theta \in \mathbb{R}/2\pi\mathbb{Z}$ and a metric of the form $ds^2 = dr^2 + f(r)d\theta^2$ for a function $f(r)$ such that there exists $0 < \epsilon << 1$ such that $f(r) \sim r^2$ for $r << \epsilon$ and $f(r)$ is equal to a constant ρ for $\epsilon \leq r \leq 1$. The map $r : D \to [0,1]$ is a S^1-bundle for $r \neq 0$ and the S^1-fiber shrinks to a point at $r=0$, see Figure 4.2.

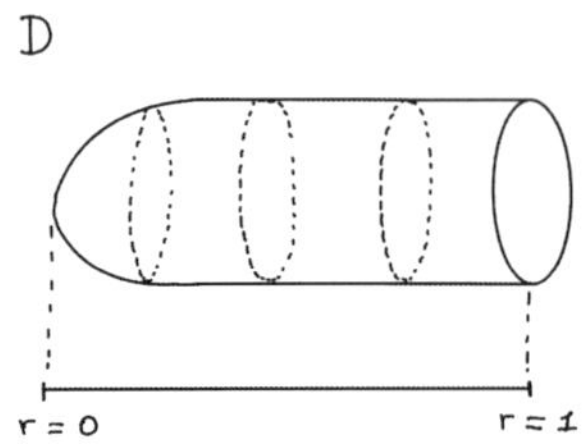

FIGURE 4.2. The cigar D.

We consider the 4d $\mathcal{N}=2$ theory $\mathcal{T}$ on $\mathbb{R}^2 \times D$ with a topological twist along D preserving half of the supersymmetries [**71**, §3.1]. In the limit where the radius of the circle S^1 in D shrinks to zero, the low energy description is the 3d sigma model on $\mathbb{R}^2 \times [0,1]$ with target M and a boundary condition at $r=0$ defined by the tip of the cigar. It is argued by Nekrasov-Witten [**71**, §3.1] that this boundary condition is defined by a I-holomorphic Lagrangian section S of $\pi \colon M \to B$. For example, for $b \in B \setminus \Delta$, the fiber $F_b = \pi^{-1}(b)$ is parametrized by a complexification of the holonomy around S^1 of the 4d low energy $U(1)^r$ gauge field, and the shrinking of S^1 at the tip of the cigar forces this holonomy to vanish. When $\mathcal{T}$ is a class S theory, M is a moduli space of Higgs bundles and S is the Hitchin section of the Hitchin fibration [**71**, §4.6].

In Conjecture 4.4 below, we propose that the BPS spectrum of $\mathcal{T}$ at a point $b \in B \setminus \Delta$ can be recovered from holomorphic Floer theory of (M, Ω_I) for the pair of I-holomorphic Lagrangian submanifolds S and F_b. To formulate this proposal, we first need some preliminary results on the space of paths $\mathcal{P}$ between S ad F_b. As S is a section of π and F_b is a fiber, the intersection $S \cap F_b$ consists of a single point p, see Figure 4.3:

$$S \cap F_b = \{p\}.$$

As in §2.2, we denote by $\mathcal{P}_0$ the connected component of $\mathcal{P}$ containing the constant path p.

LEMMA 4.3. *The fundamental group $\pi_1(\mathcal{P}_0)$ is naturally isomorphic to the charge lattice $\Gamma_b = \pi_2(M, F_b)$.*

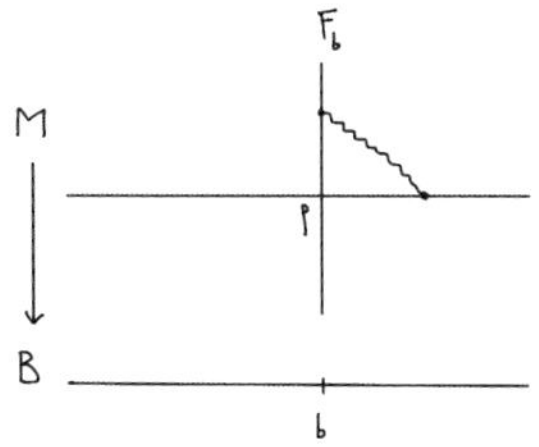

FIGURE 4.3. A path with endpoints on the section S and on the fiber F_b.

PROOF. Let w be a loop in $\mathcal{P}_0$ based at p:

$$\begin{aligned} w :[0,1] &\longrightarrow \mathcal{P}_0 \\ x &\mapsto (w_x : y \in [0,1] \mapsto w_x(y)), \end{aligned}$$

and $w_0(y) = w_0(y)$ for all $y \in [0,1]$, $w_x(0) \in S$ and $w_x(1) \in F_b$ for all $x \in [0,1]$. As $S \simeq B$ is assumed to be simply-connected, one can continuously deform w to contract the loop $x \mapsto w_x(0)$ in S, and so assume from now on without loss of generality that $w_x(0) = p$ for all $x \in [0,1]$.

The map $y \mapsto (x \mapsto w_x(y))$ is a one parameter family of loops in M, starting at $y = 0$ with the constant loop at p, and ending with the loop $x \mapsto w_x(1)$ in F_b. In particular, we obtain a map from a disk to M with boundary mapping to F_b, see Figure 4.4. The loop w is homotopically trivial in $\mathcal{P}_0$ if and only if this disk is homotopically trivial in M relatively to F_b. □

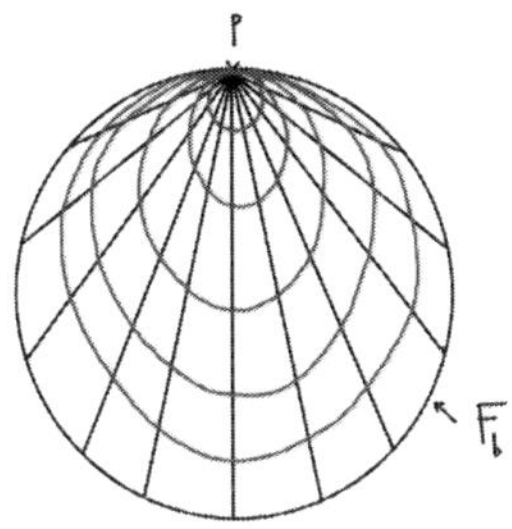

FIGURE 4.4. The one-parameter family of paths $x \mapsto (y \mapsto w_x(y))$, and the one-parameter family of loops $y \mapsto (x \mapsto w_x(y))$, forming a disk with boundary on F_b.

Let $\widetilde{\mathcal{P}_0} \to \mathcal{P}_0$ be the universal cover of $\mathcal{P}_0$. As the intersection $S \cap F_b$ consists of a single point p, the critical points of the holomorphic action functional on $\widetilde{\mathcal{P}_0}$ form a $\pi_1(\mathcal{P}_0)$-torsor by §2.3. By Lemma 4.3, $\pi_1(\mathcal{P}_0) = \Gamma_b$, and so we have a Γ_b-torsor of critical points. Fixing a reference critical point, we identify the set of critical points with Γ_b, and we denote by (p,γ) the critical point corresponding to $\gamma \in \Gamma_b$. By §2.4, holomorphic Floer theory attaches to any pair of critical points (p,γ) and (p,γ') a vector space $H((p,\gamma),(p,\gamma'))$ of 2d BPS states, see Figure 4.5. By Γ_b-equivariance of the holomorphic action functional on $\mathcal{P}_0$, the space $H((p,\gamma),(p,\gamma'))$ only depend on the difference $\gamma' - \gamma$. By definition, $H((p,0),(p,\gamma))$ is the cohomology of a complex generated by J_ζ-holomorphic curves in M with boundary on $S \cup F_b$, and with differential given by counts of solutions to the 3d ζ-Fueter equation asymptotic to these holomorphic curves, where

$$\zeta = \frac{\int_\gamma \Omega_I}{|\int_\gamma \Omega_I|}.$$

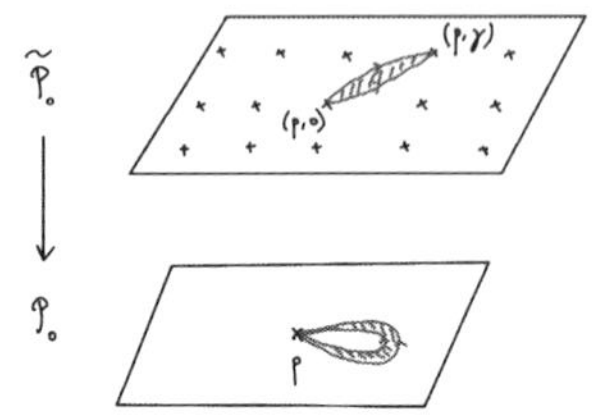

FIGURE 4.5. The lattice of critical points of W on $\widetilde{\mathcal{P}}_0$.

We can now state our main conjecture relating the BPS states of the 4d $\mathcal{N}=2$ theory $\mathcal{T}$ and the 2d BPS states of the holomorphic action functional.

CONJECTURE 4.4. *Let $\mathcal{T}$ be a 4d-dimensional $\mathcal{N}=2$ field theory. For every point $b \in B \setminus \Delta$ in the complement of the discriminant locus Δ of the Coulomb branch B of $\mathcal{T}$, and for every charge $\gamma \in \Gamma_b$, the space BPS_γ^b of BPS states of $\mathcal{T}$ of charge γ in the vacuum b is isomorphic to the space $H((p,0),(p,\gamma))$ of 2d BPS states of the holomorphic action functional on the space of paths between the holomorphic Lagrangian section S and the holomorphic Lagrangian fiber F_b of the Seiberg-Witten integrable system (M, Ω_I) of $\mathcal{T}$:*

$$BPS_\gamma^b \simeq H((p,0),(p,\gamma)).$$

When the 4d $\mathcal{N}=2$ theory $\mathcal{T}$ can be geometrically engineered by compactification of Type II string theory on a Calabi-Yau 3-fold, then the spaces of BPS states BPS_γ^b have a geometric realization as DT invariants, and so in this case Conjecture 4.4 predicts a relation between the DT theory of a Calabi-Yau 3-fold and the holomorphic Floer theory of the complex integrable system $\pi\colon M \to B$.

REMARK 4.5. In the decategorified limit, Conjecture 4.4 predicts a relation between BPS indices, or DT invariants, and counts of J_ζ-holomorphic curves in M with boundary on $S \cup F_b$. In fact, it follows from the proof of Lemma 4.3 that these J_ζ-holomorphic curves can be naturally viewed as J_ζ-holomorphic disks in M with boundary on the fiber F_b.

The counts of these disks are exactly the "quantum corrections" appearing in mirror symmetry: for every $\zeta \in \mathbb{C}$, $|\zeta| = 1$, the map $\pi\colon M \to B$ is a torus fibration in special Lagrangian on $(M, \omega_\zeta, J_\zeta)$, and so the Strominger–Yau–Zaslow picture of mirror symmetry [**83**] predicts that the mirror of $(M, \omega_\zeta, J_\zeta)$ is obtained by dualizing the torus fibration away from the singular fibers, and using counts of J_ζ-holomorphic disks with boundary on the torus fibers F_b to extend the complex structure of the mirror across the singular fibers [**31, 43, 63**]. For a general Calabi-Yau manifold M, the wall-crossing behaviour of these counts of disks as a function of b is based on the Lie algebra of vector fields preserving a holomorphic volume form on a complex torus [**63**]. Geometrically, the wall-crossing formula guarantees the consistency of the gluing procedure producing the mirror M.

The wall-crossing formula for DT invariants as a function of the stability is formally identical to the wall-crossing formula for counts of holomorphic disks when M is holomorphic symplectic: in this case, the wall-crossing transformations are symplectic and not just volume preserving as in the general Calabi-Yau situation [**65, 67, 68**]. This suggests a correspondence between DT invariants and counts of J_ζ-holomorphic disks in complex integrable system [**65**]. Another reason to expect this correspondence is the work of Gaiotto-Moore-Neitzke [**34, 37**] in which the hyperkähler metric on the Seiberg-Witten integrable system M is reconstructed from "quantum corrections" determined by the BPS spectrum. As M is self-mirror [**22**], compatibility between the SYZ mirror construction and the Gaiotto-Moore-Neitzke description of the hyperkähler metric requires a matching between counts of BPS states and counts of J_ζ-holomorphic curves. Conjecture 4.4 provides a categorification of this correspondence and is formulated in a way leading to a natural physics derivation presented in the next section. It would be very interesting to find more direct mathematical evidence for it.

Mathematically rigorous examples of correspondences between DT invariants and algebro-geometric versions of counts of holomorphic disks given by punctured Gromov–Witten invariants [**2, 4, 44, 45**] have been obtained for quiver DT invariants [**3, 10, 41, 42, 73**] and for DT invariants of local $\mathbb{P}^2$ [**9, 11, 12**]. In such correspondences, tropicalizations in B of holomorphic disks in M are interpreted as "attractor flow trees" in the space of stability conditions for DT invariants, see Figure 4.6 [**65, 85**]. We view these results as evidence for Conjecture 4.4.

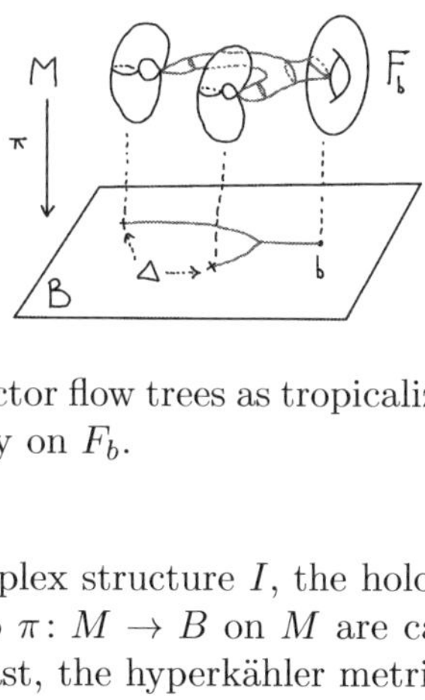

FIGURE 4.6. Attractor flow trees as tropicalization of holomorphic disks with boundary on F_b.

REMARK 4.6. The complex structure I, the holomorphic symplectic form Ω_I, and the I-holomorphic map $\pi\colon M \to B$ on M are canonically attached to the 4d $\mathcal{N} = 2$ theory $\mathcal{T}$. By contrast, the hyperkähler metric depends on the radius R of the circle S^1 on which one compactifies $\mathcal{T}$ to obtain a 3d $\mathcal{N} = 4$ sigma model. The fibers of π have size of order $1/R$. In the limit $R \to +\infty$, fibers collapse to the base: indeed, in this limit, the circle decompactifies, and the Coulomb branch M of $\mathcal{T}$ on $\mathbb{R}^3 \times S^1$ reduces to the Coulomb branch of $\mathcal{T}$ on $\mathbb{R}^4$. While one may expect that holomorphic Floer theory only depends on the I-holomorphic symplectic structure and not on the precise hyperkähler metric, it will be probably easier to define and study it in the collapsing limit $R \to +\infty$.

REMARK 4.7. There is a formal 2d-4d analogy between Conjecture 3.1 and Conjecture 4.4. In both cases, we have a space of parameters (the space $\mathcal{N}_\mathcal{T}$ of 2d $\mathcal{N} = (2,2)$ theories or the Coulomb branch of vacua of the 4d $\mathcal{N} = 2$ theory), a holomorphic symplectic manifold fibered over the space of parameters (the cotangent bundle $T^*\mathcal{N}_\mathcal{T} \to \mathcal{N}_\mathcal{T}$ or the Hitchin integrable system $\pi : M \to B$), and the space of (2d or 4d) BPS states at a given point of the parameter space is given by the holomorphic Floer theory of the holomorphic symplectic manifold for two holomorphic Lagrangian submanifolds given by the fiber of the fibration over this point and a section or multi-section (the spectral cover $\mathcal{L}$ or the Hitchin section S).

4.3. A physics derivation. In this section, we give a heuristic derivation of Conjecture 4.4 by compactification on a cigar geometry. We start with the 4d $\mathcal{N} = 2$ theory $\mathcal{T}$ on $\mathbb{R}^2 \times D$, where D is a cigar geometry reviewed at the beginning of §4.2. As boundary condition at the end of the cigar, one imposes to the theory to approach the vacuum $b \in B \setminus \Delta$. Finally, we restrict this system to the sector of charge $\gamma \in \Gamma_b$ and we look for the space V of half-supersymmetric states of lowest energy. As we assume that the system is in the vacuum u at the end of the cigar, non-trivial excitations should be localized near the tip of the cigar. But near the tip of the cigar, $\mathbb{R}^2 \times D$ looks like $\mathbb{R}^4$. By definition, BPS states of charge γ are the lowest energy half-supersymmetric states of the theory on $\mathbb{R}^4$ in the sector of charge γ, and so one concludes that $V \simeq BPS_\gamma^b$. In this picture, the BPS particles are just sitting at the tip of the cigar.

Now we take the limit where the circle S^1 in the cigar becomes very small. In this limit, the cigar D reduces to the interval $[0,1]$. The low-energy effective description of $\mathcal{T}$ compactified on a circle is the 3-dimensional $\mathcal{N} = 4$ sigma-model with target M. Therefore, our previous system reduces to a 3d $\mathcal{N} = 4$ sigma model of target M on $\mathbb{R}^2 \times [0,1]$ with appropriate boundary conditions at the ends of $[0,1]$. At $r = 0$, the boundary condition defined by the tip of the cigar is the section S, as reviewed in 4.2. On the other hand, at $r = 1$, the condition to be in the vacuum b translates into the boundary condition imposed by the fiber F_b.

In the limit where the interval $[0,1]$ also becomes very small, we obtain a 2d $\mathcal{N} = (2,2)$ sigma model on $\mathbb{R}^2$ with target the infinite-dimensional Kähler manifold $\mathcal{P}$ of paths $[0,1] \to M$ stretched between S and F_b. The central charge

$$Z_\gamma(b) = \int_\gamma \Omega_I$$

in the original 4d $\mathcal{N} = 2$ supersymmetry algebra induces a holomorphic superpotential on $\mathcal{P}$ which is exactly the holomorphic action functional W. Therefore, we obtain as low energy effective theory on $\mathbb{R}^2$ the 2d $\mathcal{N} = (2,2)$ Landau-Ginzburg model $(\mathcal{P}, W)$. Moreover, the sector of charge γ is obtained by considering only a pair of critical point of W differing by γ. By definition of holomorphic Floer theory, the space V of lowest energy supersymmetric states of this system is exactly the space $H((p,0),(p,\gamma))$.

In conclusion, Conjecture 4.4 follows from comparing the 4d $\mathcal{N} = 2$ theory $\mathcal{T}$ on $\mathbb{R}^2 \times D$ with the 2d $\mathcal{N} = (2,2)$ effective low energy on $\mathbb{R}^2$ describing the theory in the limit where the cigar D is small, see Figure 4.7

REMARK 4.8. It is clear from this argument that the assumption that $\mathcal{T}$ is a theory without dynamical theory is necessary. Indeed, a 4d $\mathcal{N} = 2$ supergravity theory, such as a Type II string compactification on a compact Calabi-Yau 3-fold, contains BPS black holes, and it is not possible to place a black hole at the tip of the cigar without destroying the local geometry.

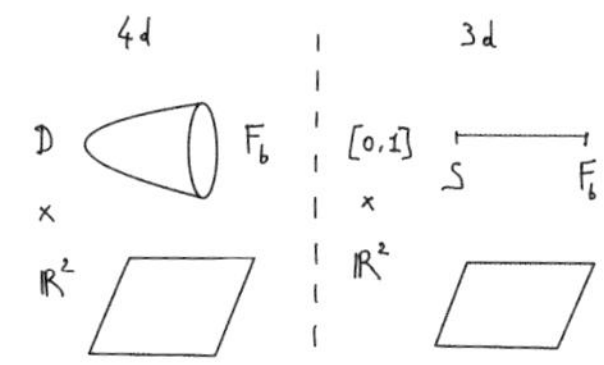

FIGURE 4.7. Dimensional reduction of the cigar to an interval.

REMARK 4.9. In the literature, compactification on a cigar geometry is often done in combination with a $\frac{1}{2}\Omega$-deformation along the S^1-rotations of the cigar [**71**]. It is essential for our purposes that the $\frac{1}{2}\Omega$-deformation parameter ϵ_1 is set to 0. For $\epsilon_1 = 0$, the possible supersymmetric boundary conditions at the boundary of the cigar are described by I-holomorphic Lagrangian submanifolds ((B,A,A)-branes) of M, and so we can take the fiber F_b as boundary condition. By contrast, for $\epsilon_1 \neq 0$, possible supersymmetric boundary conditions are J-holomorphic Lagrangian submanifolds ((A,B,A)-brane) and so the fiber F_b cannot be used (correspondingly, the boundary condition corresponding to the tip of the cigar deforms the (B,A,A)-brane given by the section S to a (A,B,A)-brane given by the locus of opers for class S theories [**71**]). On a related note, it would be interesting to understand precisely the relation with the limit $\epsilon_1 \to 0$ of the twistorial topological string of Cecotti-Neitzke-Vafa which is defined using $\epsilon_1 \neq 0$ [**19**]. Finally, Balasubramanian-Teschner consider in [**7**, §7.3] a family of (A,B,A)-branes which in the limit $\epsilon_1 \to 0$ reduce to the (B,A,A)-branes given by the fibers F_b and so it would also be interesting to see if this could be used to give a continuous deformation of our set-up (see also [**48**, Eq 5.6] and references there for further generalizations of these (A, B, A) branes in terms of Fenchel-Nielsen type spectral coordinates).

4.4. Holomorphic Floer theory and categories of line operators. In this section, we formulate conjectures about the categories attached by holomorphic Floer theory to the pairs of holomorphic Lagrangian submanifolds (S, F_b) and (S, S) in the Seiberg-Witten integrable system M. It follows from the definition in §2.5 of the Hom categories in the Fueter 2-category that the category

$$C(S, F_b) := \mathrm{Hom}_{\mathrm{Ft}(M,\Omega_I)}(S, F_b)$$

admits for every $\zeta \in \mathbb{C}^*$, $|\zeta| = 1$, an exceptional collections $(X_\gamma(\zeta))_{\gamma\in\Gamma_b}$ indexed by charges $\gamma \in \Gamma_b$. In other words, $C(S, F_b)$ is a deformation of the derived category of coherent sheaves $\mathrm{D^bCoh}(BT)$ on the classifying stack of the torus $T = \mathrm{Spec}\,\mathbb{C}[\Gamma_b] \simeq (\mathbb{C}^*)^n$. The Hom spaces in $C(F, S_b)$ are constructed as in §2.5 from the complexes of BPS states $C((p, \gamma), (p, \gamma'))$.

Formally, $X_\gamma(\zeta)$ is the Lefschetz thimble of gradient flow lines of $\mathrm{Re}(\zeta^{-1}W)$ emanating from the critical point (p, γ) of the holomorphic action functional W on the space of paths $\widetilde{\mathcal{P}}_0$. The object $X_\gamma(\zeta)$ should be viewed as the ζ-supersymmetric IR line defect of charge γ in the vacuum b [**21**, **36**]. Moreover, the expectation value $\langle X_\gamma(\zeta)\rangle$ of the IR line operator in the "conformal limit" in the sense of [**33**] should be formally given by the corresponding infinite-dimensional period integral:

$$\langle X_\gamma(\zeta)\rangle = \int_{X_\gamma(\zeta)} e^{\mathrm{Re}(\zeta^{-1}W)} vol,$$

where *vol* is "the volume form" on $\widetilde{\mathcal{P}}_0$. When ζ varies, the (infinite) exceptional collection $(X_\gamma(\zeta))$ changes by mutations determined by the complexes of BPS states. In particular, the infinite-dimensional period integrals $\langle X_\gamma(\zeta)\rangle$ are solutions of the Riemann-Hilbert problems formulated in terms of BPS/DT invariants in [**15**, **33**].

Working with an appropriate wrapped version of the Fueter 2-category, one should be able to make sense of the category

$$C(S,S) := \mathrm{Hom}_{\mathrm{Ft}(M,\Omega_I)}(S,S).$$

Similarly to what happens in the usual wrapped Fukaya category, one expects $C(S,S)$ to have for every $\zeta\in\mathbb{C}$, $|\zeta|=1$, an exceptional collection $(L_m(\zeta))_m$ indexed by the infinite set of intersection points between S and a wrapped perturbation of S. Moreover, the composition of Hom categories in the Fueter 2-category should define a structure of monoidal category on $C(S,S)$, and a structure of module category over $C(S,S)$ on $C(S,F_b)$. Acting with $C(S,S)$ on the object $X_0(\zeta)$ of $C(S,F_b)$ obtained for $\gamma=0$ induces a functor $RG_b(\zeta): C(S,S)\longrightarrow C(S,F_b)$.

CONJECTURE 4.10. *The monoidal category $C(S,S)$ is the category of supersymmetric UV line operators of the 4d $\mathcal{N}=2$ theory $\mathcal{T}$ as in* [**21**, **36**].

When $\mathcal{T}$ is a class S theory, M is a moduli space of Higgs bundles. Let M_B be the corresponding character variety: M_B is an affine algebraic variety, which as a complex manifold is isomorphic to M with complex structure J. In this case, Conjecture 4.10 predicts that $C(S,S)$ is a monoidal categorification of the algebra of regular functions $\mathbb{C}[M_B]$ of M_B. Moreover, the objects $L_m(\zeta)$ should be categorical lifts of the canonical basis of theta functions predicted by mirror symmetry [**44**, **45**].

CONJECTURE 4.11. *The functor $RG_b(\zeta): C(S,S)\longrightarrow C(S,F_b)$ is a categorification of the map in* [**21**, **36**] *expanding UV line operators in terms of IR line operators.*

In other words, Conjecture 4.11 predicts that $RG_b(\zeta)$ should be a categorification of the map expressing theta functions in terms of the cluster-like coordinates $\langle X_\gamma(\zeta)\rangle$.

Acknowledgment

The author thanks Tom Bridgeland and Richard Thomas for discussions at an early stage in 2017. The author also thanks the organizers of the String Math 2022 conference at the University of Warsaw, July 11-15, 2022, and of the MSRI Workshop New Four-Dimensional Gauge Theories, October 24-28 2022, for invitation to give talks where this work was presented.

References

[1] Mohammed Abouzaid and Paul Seidel, *An open string analogue of Viterbo functoriality*, Geom. Topol. **14** (2010), no. 2, 627–718, DOI 10.2140/gt.2010.14.627. MR2602848

[2] Dan Abramovich, Qile Chen, Mark Gross, and Bernd Siebert, *Punctured logarithmic maps*, preprint `arXiv:2009.07720`, 2020.

[3] Hülya Argüz and Pierrick Bousseau, *Quivers and curves in higher dimension*, preprint `arXiv:2308.07270`, 2023.

[4] Hülya Argüz and Mark Gross, *The higher-dimensional tropical vertex*, Geom. Topol. **26** (2022), no. 5, 2135–2235, DOI 10.2140/gt.2022.26.2135. MR4520304

[5] Michael Atiyah, *New invariants of* 3- *and* 4-*dimensional manifolds*, The mathematical heritage of Hermann Weyl (Durham, NC, 1987), Proc. Sympos. Pure Math., vol. 48, Amer. Math. Soc., Providence, RI, 1988, pp. 285–299, DOI 10.1090/pspum/048/974342. MR974342

[6] Michèle Audin, *Symplectic geometry in Frobenius manifolds and quantum cohomology*, J. Geom. Phys. **25** (1998), no. 1-2, 183–204, DOI 10.1016/S0393-0440(97)00026-0. MR1611969

[7] Aswin Balasubramanian and Jörg Teschner, *Supersymmetric field theories and geometric Langlands: the other side of the coin*, String-Math 2016, Proc. Sympos. Pure Math., vol. 98, Amer. Math. Soc., Providence, RI, 2018, pp. 79–105, DOI 10.1090/pspum/098/01723. MR3821750

[8] Kai Behrend and Barbara Fantechi, *Gerstenhaber and Batalin-Vilkovisky structures on Lagrangian intersections*, Algebra, arithmetic, and geometry: in honor of Yu. I. Manin. Vol. I, Progr. Math., vol. 269, Birkhäuser Boston, Boston, MA, 2009, pp. 1–47, DOI 10.1007/978-0-8176-4745-2_1. MR2641169

[9] Pierrick Bousseau, *A proof of N. Takahashi's conjecture for* $(\mathbb{P}^2, E)$ *and a refined sheaves/Gromov-Witten correspondence*, preprint `arXiv:1909.02992`, 2019.

[10] Pierrick Bousseau, *The quantum tropical vertex*, Geom. Topol. **24** (2020), no. 3, 1297–1379, DOI 10.2140/gt.2020.24.1297. MR4157555

[11] Pierrick Bousseau, *Scattering diagrams, stability conditions, and coherent sheaves on* $\mathbb{P}^2$, J. Algebraic Geom. **31** (2022), no. 4, 593–686. MR4484550

[12] Pierrick Bousseau, Pierre Descombes, Bruno Le Floch, and Boris Pioline, *BPS Dendroscopy on Local* $\mathbb{P}^2$, preprint `arXiv:2210.10712`, 2022.

[13] C. Brav, V. Bussi, D. Dupont, D. Joyce, and B. Szendrői, *Symmetries and stabilization for sheaves of vanishing cycles*, J. Singul. **11** (2015), 85–151, DOI 10.5427/jsing.2015.11e. With an appendix by Jörg Schürmann. MR3353002

[14] Tom Bridgeland, *Stability conditions on triangulated categories*, Ann. of Math. (2) **166** (2007), no. 2, 317–345, DOI 10.4007/annals.2007.166.317. MR2373143

[15] Tom Bridgeland, *Riemann-Hilbert problems from Donaldson-Thomas theory*, Invent. Math. **216** (2019), no. 1, 69–124, DOI 10.1007/s00222-018-0843-8. MR3935038

[16] Tom Bridgeland, *Geometry from Donaldson-Thomas invariants*, Integrability, quantization, and geometry II. Quantum theories and algebraic geometry, Proc. Sympos. Pure Math., vol. 103, Amer. Math. Soc., Providence, RI, [2021] ©2021, pp. 1–66, DOI 10.1090/pspum/103.2/01851. MR4285692

[17] Tom Bridgeland and Ivan Smith, *Quadratic differentials as stability conditions*, Publ. Math. Inst. Hautes Études Sci. **121** (2015), 155–278, DOI 10.1007/s10240-014-0066-5. MR3349833

[18] Tom Bridgeland and Valerio Toledano Laredo, *Stability conditions and Stokes factors*, Invent. Math. **187** (2012), no. 1, 61–98, DOI 10.1007/s00222-011-0329-4. MR2874935

[19] Sergio Cecotti, Andrew Neitzke, and Cumrun Vafa, *Twistorial topological strings and a* tt^* *geometry for* $\mathcal{N} = 2$ *theories in* 4*d*, Adv. Theor. Math. Phys. **20** (2016), no. 2, 193–312, DOI 10.4310/ATMP.2016.v20.n2.a1. MR3541845

[20] Sergio Cecotti and Cumrun Vafa, *On classification of* $N = 2$ *supersymmetric theories*, Comm. Math. Phys. **158** (1993), no. 3, 569–644. MR1255428

[21] Clay Córdova and Andrew Neitzke, *Line defects, tropicalization, and multi-centered quiver quantum mechanics*, J. High Energy Phys. **9** (2014), 099, front matter+65, DOI 10.1007/JHEP09(2014)099. MR3267980

[22] Richard Derryberry, *Stacky dualities for the moduli of Higgs bundles*, Adv. Math. **368** (2020), 107152, 55, DOI 10.1016/j.aim.2020.107152. MR4085143

[23] Aleksander Doan and Semon Rezchikov, *Holomorphic Floer theory and the Fueter equation*, `arXiv:2210.12047`, 2022.

[24] Simon Donaldson and Ed Segal, *Gauge theory in higher dimensions, II*, Surveys in differential geometry. Volume XVI. Geometry of special holonomy and related topics, Surv. Differ. Geom., vol. 16, Int. Press, Somerville, MA, 2011, pp. 1–41, DOI 10.4310/SDG.2011.v16.n1.a1. MR2893675

[25] S. K. Donaldson and R. P. Thomas, *Gauge theory in higher dimensions*, The geometric universe (Oxford, 1996), Oxford Univ. Press, Oxford, 1998, pp. 31–47. MR1634503

[26] Boris Dubrovin, *Geometry of* 2*D topological field theories*, Integrable systems and quantum groups (Montecatini Terme, 1993), Lecture Notes in Math., vol. 1620, Springer, Berlin, 1996, pp. 120–348, DOI 10.1007/BFb0094793. MR1397274

[27] Huijun Fan, Tyler Jarvis, and Yongbin Ruan, *The Witten equation, mirror symmetry, and quantum singularity theory*, Ann. of Math. (2) **178** (2013), no. 1, 1–106, DOI 10.4007/annals.2013.178.1.1. MR3043578

[28] Andreas Floer, *Morse theory for Lagrangian intersections*, J. Differential Geom. **28** (1988), no. 3, 513–547. MR965228
[29] Andreas Floer, *Witten's complex and infinite-dimensional Morse theory*, J. Differential Geom. **30** (1989), no. 1, 207–221. MR1001276
[30] Kenji Fukaya, *Morse homotopy, A^∞-category, and Floer homologies*, Proceedings of GARC Workshop on Geometry and Topology '93 (Seoul, 1993), Lecture Notes Ser., vol. 18, Seoul Nat. Univ., Seoul, 1993, pp. 1–102. MR1270931
[31] Kenji Fukaya, *Multivalued Morse theory, asymptotic analysis and mirror symmetry*, Graphs and patterns in mathematics and theoretical physics, Proc. Sympos. Pure Math., vol. 73, Amer. Math. Soc., Providence, RI, 2005, pp. 205–278, DOI 10.1090/pspum/073/2131017. MR2131017
[32] Kenji Fukaya and Yong-Geun Oh, *Zero-loop open strings in the cotangent bundle and Morse homotopy*, Asian J. Math. **1** (1997), no. 1, 96–180, DOI 10.4310/AJM.1997.v1.n1.a5. MR1480992
[33] Davide Gaiotto, *Opers and TBA*, preprint `arXiv:1403.6137`, 2014.
[34] Davide Gaiotto, Gregory W. Moore, and Andrew Neitzke, *Four-dimensional wall-crossing via three-dimensional field theory*, Comm. Math. Phys. **299** (2010), no. 1, 163–224, DOI 10.1007/s00220-010-1071-2. MR2672801
[35] Davide Gaiotto, Gregory W. Moore, and Andrew Neitzke, *Wall-crossing in coupled 2d-4d systems*, J. High Energy Phys. **12** (2012), 082, front matter + 166, DOI 10.1007/JHEP12(2012)082. MR3045271
[36] Davide Gaiotto, Gregory W. Moore, and Andrew Neitzke, *Framed BPS states*, Adv. Theor. Math. Phys. **17** (2013), no. 2, 241–397. MR3250763
[37] Davide Gaiotto, Gregory W. Moore, and Andrew Neitzke, *Wall-crossing, Hitchin systems, and the WKB approximation*, Adv. Math. **234** (2013), 239–403, DOI 10.1016/j.aim.2012.09.027. MR3003931
[38] Davide Gaiotto, Gregory Moore, and Edward Witten, *Algebra of the Infrared: String Field Theoretic Structures in Massive $N = (2,2)$ Field Theory In Two Dimensions*, preprint `arXiv:1506.04087`, 2015.
[39] Benjamin Gammage and Justin Hilburn, *Betti Tate's thesis and the trace of perverse schobers*, preprint `arXiv:2210.06548`, 2022.
[40] Benjamin Gammage, Justin Hilburn, and Aaron Mazel-Gee, *Perverse schobers and 3d mirror symmetry*, preprint `arXiv:2202.06833`, 2022.
[41] Mark Gross and Rahul Pandharipande, *Quivers, curves, and the tropical vertex*, Port. Math. **67** (2010), no. 2, 211–259, DOI 10.4171/PM/1865. MR2662867
[42] Mark Gross, Rahul Pandharipande, and Bernd Siebert, *The tropical vertex*, Duke Math. J. **153** (2010), no. 2, 297–362, DOI 10.1215/00127094-2010-025. MR2667135
[43] Mark Gross and Bernd Siebert, *From real affine geometry to complex geometry*, Ann. of Math. (2) **174** (2011), no. 3, 1301–1428, DOI 10.4007/annals.2011.174.3.1. MR2846484
[44] Mark Gross and Bernd Siebert, *Intrinsic mirror symmetry*, preprint `arXiv:1909.07649`, 2019.
[45] Mark Gross and Bernd Siebert, *The canonical wall structure and intrinsic mirror symmetry*, Invent. Math. **229** (2022), no. 3, 1101–1202, DOI 10.1007/s00222-022-01126-9. MR4462625
[46] Andriy Haydys, *Fukaya-Seidel category and gauge theory*, J. Symplectic Geom. **13** (2015), no. 1, 151–207, DOI 10.4310/JSG.2015.v13.n1.a5. MR3338833
[47] Sonja Hohloch, Gregor Noetzel, and Dietmar A. Salamon, *Floer homology groups in hyperkähler geometry*, New perspectives and challenges in symplectic field theory, CRM Proc. Lecture Notes, vol. 49, Amer. Math. Soc., Providence, RI, 2009, pp. 251–261, DOI 10.1090/crmp/049/09. MR2555940
[48] Lotte Hollands, Philipp Rüter, and Richard J. Szabo, *A geometric recipe for twisted superpotentials*, J. High Energy Phys. **12** (2021), Paper No. 164, 90, DOI 10.1007/jhep12(2021)164. MR4369437
[49] Kenneth Intriligator, *Compactified little string theories and compact moduli spaces of vacua*, Phys. Rev. D (3) **61** (2000), no. 10, 106005, 6, DOI 10.1103/PhysRevD.61.106005. MR1790780
[50] Dominic Joyce, *Holomorphic generating functions for invariants counting coherent sheaves on Calabi-Yau 3-folds*, Geom. Topol. **11** (2007), 667–725, DOI 10.2140/gt.2007.11.667. MR2302500

[51] Dominic Joyce and Yinan Song, *A theory of generalized Donaldson-Thomas invariants*, Mem. Amer. Math. Soc. **217** (2012), no. 1020, iv+199, DOI 10.1090/S0065-9266-2011-00630-1. MR2951762

[52] M. Kapranov, M. Kontsevich, and Y. Soibelman, *Algebra of the infrared and secondary polytopes*, Adv. Math. **300** (2016), 616–671, DOI 10.1016/j.aim.2016.03.028. MR3534842

[53] Mikhail Kapranov and Vadim Schechtman, *Perverse Schobers*, preprint `arXiv:1411.2772`, 2014.

[54] Mikhail Kapranov, Yan Soibelman, and Lev Soukhanov, *Perverse schobers and the Algebra of the Infrared*, preprint `arXiv:2011.00845`, 2020.

[55] Anton Kapustin, *A-branes and noncommutative geometry*, preprint `arXiv:hep-th/0502212`, 2005.

[56] Anton Kapustin and Lev Rozansky, *Three-dimensional topological field theory and symplectic algebraic geometry II*, Commun. Number Theory Phys. **4** (2010), no. 3, 463–549, DOI 10.4310/CNTP.2010.v4.n3.a1. MR2771578

[57] Anton Kapustin, Lev Rozansky, and Natalia Saulina, *Three-dimensional topological field theory and symplectic algebraic geometry. I*, Nuclear Phys. B **816** (2009), no. 3, 295–355, DOI 10.1016/j.nuclphysb.2009.01.027. MR2522724

[58] Masaki Kashiwara and Pierre Schapira, *Deformation quantization modules* (English, with English and French summaries), Astérisque **345** (2012), xii+147. MR3012169

[59] L. Katzarkov, M. Kontsevich, and T. Pantev, *Hodge theoretic aspects of mirror symmetry*, From Hodge theory to integrability and TQFT tt*-geometry, Proc. Sympos. Pure Math., vol. 78, Amer. Math. Soc., Providence, RI, 2008, pp. 87–174, DOI 10.1090/pspum/078/2483750. MR2483750

[60] Ahsan Khan. *Categorical Aspects of BPS States.* PhD thesis, Rutgers The State University of New Jersey, School of Graduate Studies, 2021.

[61] Ahsan Khan and Gregory Moore, *Categorical Wall-Crossing in Landau-Ginzburg Models*, preprint `arXiv:2010.11837`, 2020.

[62] Maxim Kontsevich, *Deformation quantization of algebraic varieties*, Lett. Math. Phys. **56** (2001), no. 3, 271–294, DOI 10.1023/A:1017957408559. EuroConférence Moshé Flato 2000, Part III (Dijon). MR1855264

[63] Maxim Kontsevich and Yan Soibelman, *Affine structures and non-Archimedean analytic spaces*, The unity of mathematics, Progr. Math., vol. 244, Birkhäuser Boston, Boston, MA, 2006, pp. 321–385, DOI 10.1007/0-8176-4467-9_9. MR2181810

[64] Maxim Kontsevich and Yan Soibelman, *Stability structures, motivic Donaldson-Thomas invariants and cluster transformations*, preprint `arXiv:0811.2435`, 2008.

[65] Maxim Kontsevich and Yan Soibelman, *Wall-crossing structures in Donaldson-Thomas invariants, integrable systems and mirror symmetry*, Homological mirror symmetry and tropical geometry, Lect. Notes Unione Mat. Ital., vol. 15, Springer, Cham, 2014, pp. 197–308, DOI 10.1007/978-3-319-06514-4_6. MR3330788

[66] Maxim Kontsevich and Yan Soibelman, *Holomorphic Floer Theory: brane quantization, exponential integrals and resurgence*, In preparation, 2022.

[67] Yu-Shen Lin, *Correspondence theorem between holomorphic discs and tropical discs on K3 surfaces*, J. Differential Geom. **117** (2021), no. 1, 41–92, DOI 10.4310/jdg/1609902017. MR4195752

[68] Wenxuan Lu, *Instanton Correction, Wall Crossing And Mirror Symmetry Of Hitchin's Moduli Spaces*, ProQuest LLC, Ann Arbor, MI, 2011. Thesis (Ph.D.)–Massachusetts Institute of Technology. MR2982375

[69] Jacob Lurie, *On the classification of topological field theories*, Current developments in mathematics, 2008, Int. Press, Somerville, MA, 2009, pp. 129–280. MR2555928

[70] Borislav Mladenov, *Formality of differential graded algebras and complex lagrangian submanifolds*, preprint `arXiv:2007.05498`, 2020.

[71] Nikita Nekrasov and Edward Witten, *The omega deformation, branes, integrability and Liouville theory*, J. High Energy Phys. **9** (2010), 092, i, 82, DOI 10.1007/JHEP09(2010)092. MR2776942

[72] Jonathan P. Pridham, *Quantisation of derived Lagrangians*, Geom. Topol. **26** (2022), no. 6, 2405–2489, DOI 10.2140/gt.2022.26.2405. MR4521246

[73] Markus Reineke and Thorsten Weist, *Refined GW/Kronecker correspondence*, Math. Ann. **355** (2013), no. 1, 17–56, DOI 10.1007/s00208-012-0778-0. MR3004575

[74] L. Rozansky and E. Witten, *Hyper-Kähler geometry and invariants of three-manifolds*, Selecta Math. (N.S.) **3** (1997), no. 3, 401–458, DOI 10.1007/s000290050016. MR1481135

[75] Günter Schwarz, *Hodge decomposition—a method for solving boundary value problems*, Lecture Notes in Mathematics, vol. 1607, Springer-Verlag, Berlin, 1995, DOI 10.1007/BFb0095978. MR1367287

[76] N. Seiberg and E. Witten, *Electric-magnetic duality, monopole condensation, and confinement in $N = 2$ supersymmetric Yang-Mills theory*, Nuclear Phys. B **426** (1994), no. 1, 19–52, DOI 10.1016/0550-3213(94)90124-4. MR1293681

[77] N. Seiberg and E. Witten, *Monopoles, duality and chiral symmetry breaking in $N = 2$ supersymmetric QCD*, Nuclear Phys. B **431** (1994), no. 3, 484–550, DOI 10.1016/0550-3213(94)90214-3. MR1306869

[78] N. Seiberg and E. Witten, *Gauge dynamics and compactification to three dimensions*, The mathematical beauty of physics (Saclay, 1996), Adv. Ser. Math. Phys., vol. 24, World Sci. Publ., River Edge, NJ, 1997, pp. 333–366. MR1490862

[79] Paul Seidel, *Vanishing cycles and mutation*, European Congress of Mathematics, Vol. II (Barcelona, 2000), Progr. Math., vol. 202, Birkhäuser, Basel, 2001, pp. 65–85. MR1905352

[80] Paul Seidel, *Fukaya categories and Picard-Lefschetz theory*, posted on 2008, viii+326, DOI 10.4171/063. MR2441780

[81] Ivan Smith, *Quiver algebras as Fukaya categories*, Geom. Topol. **19** (2015), no. 5, 2557–2617, DOI 10.2140/gt.2015.19.2557. MR3416110

[82] Jake P. Solomon and Misha Verbitsky, *Locality in the Fukaya category of a hyperkähler manifold*, Compos. Math. **155** (2019), no. 10, 1924–1958, DOI 10.1112/s0010437x1900753x. MR4010429

[83] Andrew Strominger, Shing-Tung Yau, and Eric Zaslow, *Mirror symmetry is T-duality*, Nuclear Phys. B **479** (1996), no. 1-2, 243–259, DOI 10.1016/0550-3213(96)00434-8. MR1429831

[84] Donghao Wang, *The Complex Gradient Flow equation and Seidel's Spectral Sequence*, preprint arXiv:2209.02810, 2022.

[85] Qiang Wang, *Wall-crossing structures and application to* SU(3) *Seiberg-Witten integrable systems*, J. Geom. Phys. **157** (2020), 103834, 15, DOI 10.1016/j.geomphys.2020.103834. MR4131444

[86] Edward Witten, *Supersymmetry and Morse theory*, J. Differential Geometry **17** (1982), no. 4, 661–692 (1983). MR683171

[87] Edward Witten, *Analytic continuation of Chern-Simons theory*, Chern-Simons gauge theory: 20 years after, AMS/IP Stud. Adv. Math., vol. 50, Amer. Math. Soc., Providence, RI, 2011, pp. 347–446, DOI 10.1090/amsip/050/19. MR2809462

Department of Mathematics, University of Georgia, Athens, Georgia 30605
Email address: Pierrick.Bousseau@uga.edu

Proceedings of Symposia in Pure Mathematics
Volume **107**, 2024
https://doi.org/10.1090/pspum/107/01945

Giant Gravitons and non-conformal vacua in twisted holography

Kasia Budzik

ABSTRACT. Twisted holography relates the two-dimensional chiral algebra subsector of $\mathcal{N} = 4$ SYM to the B-model topological string theory on the deformed conifold $SL(2, \mathbb{C})$. We review the relevant aspects of the duality and its two generalizations: the correspondence between determinant operators and "Giant Graviton" branes and the extension to non-conformal vacua of the chiral algebra.

1. Introduction and summary

1.1. Twisted holography. Twisting supersymmetric quantum field theories is the procedure of passing to the cohomology of a fermionic supercharge $\mathcal{Q}$. The operation produces a consistent subsector of the SQFT which is easier to study and may allow for exact computations. In the twisted theory several simplifications occur:

- Twisting restricts to protected (BPS) operators. Certain quantities, eg. correlation functions, might become independent of various parameters like the coupling constant.
- Dependence on certain spacetime coordinates drops out in cohomology. The fermionic supercharges satisfy anticommutation relations of the schematic form

$$\{Q, \tilde{Q}\} \sim P. \tag{1.1}$$

  The momenta that appear in the image of the twisting supercharge $\{\mathcal{Q}, \bullet\} \sim P$, are by definition exact in cohomology. Since momenta generate spacetime translations, dependence on the corresponding spacetime coordinates drops out in the twisted theory. There might be many twists available depending on the spacetime dimension and the number of supersymmetry (see eg. [**35**,**36**]). For example, if all of the spacetime momenta become exact we have *the topological twist*, and if half – *the holomorphic twist.*
- Twisted theories are endowed with extra mathematical structures and symmetries, which become visible at the level of cohomology [**19**,**20**,**39**,**40**,**56**,**58**]. However, this topic is beyond the scope of this note.

For the supersymmetric theories that have a holographic dual, a natural question is whether there exists an analogous twisting procedure on the gravity side. It was proposed in [**27**] that the dual operation corresponds to turning on a non-zero value for a bosonic ghost field[1] of the supergravity theory. The twisted type IIB supergravity was conjectured to be equivalent to the BCOV theory.[2] Remarkably, twisted supergravity can be quantized to all orders in perturbation theory, despite the theory being non-renormalizable [**23, 26**].

Applying the twisting procedures to both sides of a holographic duality should therefore produce an easier and more tractable duality, where many simplifications occur.[3] In addition, objects that appear after twisting are typically more well-defined mathematically (eg. vertex algebras, BCOV theory). One could hence hope for a more mathematical formulation of holography, at least at the twisted level. It has been proposed that twisted holography is captured by the ubiquitous mathematical duality known as Koszul duality [**24, 28, 29, 53, 63**].

The main example[4] of twisted holography in this note is the duality between a 2d chiral algebra $\mathcal{A}_N$ and the topological B-model on $SL(2,\mathbb{C})$ proposed in [**25**]. The chiral algebra is a subsector of $\mathcal{N}=4$ SYM obtained by the "$Q+S$" twist of [**6**].[5] One can arrive at the dual geometry $SL(2,\mathbb{C})$ by considering a stack of D1-branes in the topological B-model on $\mathbb{C}^3$, whose worldvolume theory is precisely the chiral algebra $\mathcal{A}_N$ (in analogy to the original derivation by Maldacena [**47**]). The stack of branes deforms the complex structure $\mathbb{C}^3 \setminus \mathbb{C}$ to $SL(2,\mathbb{C})$. According to [**25**], B-model on $SL(2,\mathbb{C})$ can also be seen as a twist of type IIB string theory on $\mathrm{AdS}_5 \times S^5$, where $SL(2,\mathbb{C}) \cong \mathrm{AdS}_3 \times S^3 \subset \mathrm{AdS}_5 \times S^5$. Therefore, the above duality can be regarded as a subsector of the standard $\mathrm{AdS}_5/\mathrm{CFT}_4$ correspondence.

At the twisted level, many aspects of the duality simplify:

- The dependence on the 't Hooft coupling $g^2_{\mathrm{YM}}N$ drops out. As a consequence, the computations on the chiral algebra side are essentially free theory computations (when considering BRST closed operators). The only coupling constant is the rank of the gauge group N, which gets mapped to the string coupling $g_s \sim 1/N$. Therefore, the large N computations on the gauge theory side can be readily matched with perturbative topological string theory expansion.
- The closed string field theory of the topological B-model is the Kodaira-Spencer (BCOV) theory [**11, 12**]. This theory has been shown to admit a unique quantization in perturbation theory [**26, 32**].
- D-branes in the topological B-model are holomorphic submanifolds of the target Calabi-Yau. In particular, D1-branes that will be considered in this note are holomorphic curves in $SL(2,\mathbb{C})$.

[1]The role of the bosonic ghost field in the BV formalism is to gauge a local supersymmetry. Coupling a supersymmetric field theory (worldvolume theory of a brane) to the background where a bosonic ghost Ψ is non-zero is equivalent to adding the corresponding supercharge Q_Ψ to the BRST charge of the supersymmetric field theory.

[2]The stronger conjecture is that twisted type IIB superstring theory is equivalent to topological B-model.

[3]There are also examples (holomorphic twist of $\mathcal{N}=1$ SYM and SQCD) where the dual of the original theory is not known, yet one can conjecture the holographic dual of the twisted theory [**19**].

[4]Other examples of twisted supergravity/holography include [**15, 29–31, 34, 37, 38, 41, 45, 49–51, 54, 55, 57, 64**].

[5]For an earlier work on the holographic dual of the chiral algebra subsector of $\mathcal{N}=4$ SYM see [**14**].

1.2. Summary of results. In [**25**], the holographic dictionary between single trace operators of the chiral algebra $\mathcal{A}_N$ and modifications of boundary conditions of the Kodaira-Spencer theory on $SL(2,\mathbb{C})$ was proposed. In the following sections, we review two generalization studied in [**17, 18**]:

(1) The correspondence between determinant operators and "Giant Graviton" D1-branes.
(2) The duality between the chiral algebra $\mathcal{A}_N$ in non-conformal vacua and topological B-model on asymptotically $SL(2,\mathbb{C})$ geometries.

As part of the standard AdS/CFT correspondence, an insertion of a determinant operator in $\mathcal{N}=4$ SYM is dual to a Giant Graviton D3-brane, which asymptotically wraps an $\mathbb{R}_+ \times S^3$ inside $\mathrm{AdS}_5 \times S^5$. In twisted holography, determinant operators in the chiral algebra $\mathcal{A}_N$ are dual to D1-branes[6] that asymptotically wrap $\mathbb{C}^* \cong \mathbb{R}_+ \times S^1$ inside $SL(2,\mathbb{C}) \cong \mathrm{AdS}_3 \times S^3$.

When considering insertions of multiple determinants, there might be different brane configurations (eg. connected or disconnected) with the same boundary behavior. Correlation functions of multiple determinants have large N saddles [**43**]. Using a spectral curve construction, to each large N saddle we can associate a holomorphic curve in $SL(2,\mathbb{C})$, which we conjecture is the support of the dual D1-brane [**17**]. The conjecture is tested by various holographic computations summarized in the next section.

As the next step in studying twisted holography, in [**18**] it was proposed that the duality can be extended to non-conformal vacua of the chiral algebra $\mathcal{A}_N$. The holographic dual geometries are deformations of $SL(2,\mathbb{C})$, originating from the backreaction of a stack of non-coincident D1-branes in the topological B-model on $\mathbb{C}^3$.

This is a twisted analog of the duality between the Coulomb branch of $\mathcal{N}=4$ SYM and the (asymptotically $\mathrm{AdS}_5 \times S^5$) multi-center supergravity solutions obtained from a near-horizon limit of a stack of non-coincident D3-branes [**13, 33, 44, 60, 61**].

Non-conformal (translation-invariant) vacua of a chiral algebra are a novel observable and in [**18**] it was conjectured that they correspond to the notion of *the associated variety* of the chiral algebra.[7] In case of chiral algebras of $\mathcal{N}=2$ SCFTs, we expect the 2d vacua to descent from the Higgs branch vacua of the 4d parent theory.[8] The conjecture is therefore in agreement with [**7**] which identified the Higgs branch of an $\mathcal{N}=2$ SCFT with the associated variety of its chiral algebra subsector.

1.3. Future directions. Twisted holography setup provides an excellent setting for computations that might be challenging in the full physical duality and for making more precise connections with mathematics.

One natural next step is to study even heavier operators of order N^2, such as $(\det Z)^N$, which are expected to produce finite deformations of the geometry [**46**].

---

[6]The twisting procedure is expected to reduce D3-branes to D1-branes [**52**].

[7]Chiral algebras are referred to as vertex algebras in the math literature.

[8]In a setting without conformal symmetry, the chiral algebra subsector can be defined using a version of the Ω-deformation [**21, 42, 48**]. We expect, but not prove, that this procedure is compatible with the Higgs branch vevs.

In the dual picture, one can consider a stack of N D1-branes wrapping $\mathbb{C}^*$ at $b = c = 0$ inside $SL(2,\mathbb{C})$ given by $ad - bc = 1$. The backreacted geometry can be described in patches [**16**].

The coordinates b, c remain holomorphic and coincide in all patches. In the patch $a \neq 0$, the coordinate a gets deformed to

$$\begin{aligned} \tilde{a}_b &= a\, e^{-\frac{\bar{c}}{b}\frac{1}{|b|^2+|c|^2}}, && \text{in } b \neq 0 \text{ patch} \\ \tilde{a}_c &= a\, e^{\frac{\bar{b}}{c}\frac{1}{|b|^2+|c|^2}}, && \text{in } c \neq 0 \text{ patch} \end{aligned} \tag{1.2}$$

with coordinate transition

$$\frac{\tilde{a}_c}{\tilde{a}_b} = e^{\frac{1}{bc}}. \tag{1.3}$$

Analogously, in the $d \neq 0$ patch:

$$\begin{aligned} \tilde{d}_b &= d\, e^{\frac{\bar{c}}{b}\frac{1}{|b|^2+|c|^2}}, && \text{in } b \neq 0 \text{ patch} \\ \tilde{d}_c &= d\, e^{-\frac{\bar{b}}{c}\frac{1}{|b|^2+|c|^2}}, && \text{in } c \neq 0 \text{ patch} \end{aligned} \tag{1.4}$$

with coordinate transition

$$\frac{\tilde{d}_c}{\tilde{d}_b} = e^{-\frac{1}{bc}}. \tag{1.5}$$

And finally, the coordinate transformation between $a \neq 0$ and $d \neq 0$ patches:

$$\tilde{a}_b \tilde{d}_b = 1 + bc. \tag{1.6}$$

It would be interesting to perform holographic computations in this setup. One of the challenges is the analysis of determinant correlation functions when the number of determinants is of order N.

Finally, we mention some other possible future research directions:

- It would be interesting to find the D3-branes in the physical 10d theory that descend to the proposed holomorphic curves in $SL(2,\mathbb{C})$.
- It could be conductive to consider determinant operators of adjoint as well as fundamental fields, which holographically translates to including space-filling branes.
- Mathematically, the conjecture of [**17**] can be stated as a correspondence between solutions of matrix equations (2.11) and holomorphic curves in $SL(2,\mathbb{C})$. For genus $g = 0$, this map was shown to be one-to-one (up to some non-physical redefinitions). It would be interesting to analyze the higher genus cases.
- Determinant modifications, reviewed in section 2.3.3, form a module for the infinite N global symmetry algebra identified in [**25**]. It could be useful to study this algebra in more detail. For example, a classification of its modules could provide a classification of objects in the topological B-model on $SL(2,\mathbb{C})$.

2. Giant Gravitons and determinants

In this section, we review the correspondence between determinant operators and Giant Graviton D1-branes in the twisted holography setup which was studied in [**17**].

2.1. Determinant correlation functions. The chiral algebra $\mathcal{A}_N$ is a gauged $\beta\gamma$ system of symplectic bosons X, Y valued in the adjoint representation of $U(N)$. It is convenient to define the linear combination

$$Z(u;z) \equiv X(z) + uY(z). \tag{2.1}$$

Then the symplectic boson OPE is[9]

$$Z(u;z)^a_b \, Z(v;w)^c_d \sim \delta^a_d \delta^c_b \frac{1}{N} \frac{u-v}{z-w}. \tag{2.2}$$

We are interested in correlation functions of determinant operators:

$$\mathcal{D}(m;u;z) \equiv \det\left(m + Z(u;z)\right), \quad m \in \mathbb{C}. \tag{2.3}$$

An insertion of a determinant of this form corresponds to a Giant Graviton brane wrapping a 1-dimensional complex curve, which approaches the boundary of AdS_3 at a point z along the line $b = au - m + \mathcal{O}(a^{-1})$, where $ad - bc = 1$ are the coordinates of $SL(2,\mathbb{C})$.[10] When considering insertions of multiple determinants $\mathcal{D}(m_i;u_i;z_i)$ there might be many brane configurations with the same asymptotics controlled by parameters m_i, u_i, z_i.

A method of computing correlations functions of determinant operators in the large N limit was presented in [**43**]. Following their prescription[11] (also implemented in [**10, 22, 62**]), we fermionize the determinants:

$$\left\langle \prod_i^k \mathcal{D}(m_i;u_i;z_i) \right\rangle = \int [\mathrm{d}\psi \mathrm{d}\bar{\psi}] \left\langle \prod_i^k e^{\bar{\psi}_i (m_i + Z(u_i;z_i))\psi^i} \right\rangle, \tag{2.4}$$

where $\psi^i, \bar{\psi}_i$ are auxiliary (anti)fundamental fermions and $\mathrm{d}\psi\mathrm{d}\bar{\psi} \equiv \prod_i^k \mathrm{d}\psi^i \mathrm{d}\bar{\psi}_i$. The expectation value on the right hand side stands for the chiral algebra path integral. Since the action for the symplectic bosons is free: $N \int \mathrm{Tr}\, X \bar{\partial} Y$, we can easily integrate them out:

$$\int [\mathrm{d}\psi \mathrm{d}\bar{\psi}] \, e^{-\frac{1}{2N}\sum_{i\neq j} \frac{u_i - u_j}{z_i - z_j} \bar{\psi}_i \psi^j \bar{\psi}_j \psi^i + \sum_i m_i \bar{\psi}_i \psi^i}. \tag{2.5}$$

To deal with the above integral, we perform the Hubbard-Stratonovich transformation by introducing auxiliary bosonic variables ρ^i_j for $i \neq j$ (and set $\rho^i_i \equiv m_i$):

$$\frac{1}{Z_\rho} \int [\mathrm{d}\psi \mathrm{d}\bar{\psi}][\mathrm{d}\rho] \, e^{\frac{N}{2}\sum_{i\neq j} \frac{z_i - z_j}{u_i - u_j} \rho^i_j \rho^j_i + \sum_{i,j} \rho^i_j \bar{\psi}_i \psi^i}, \tag{2.6}$$

where

$$Z_\rho = \int [\mathrm{d}\rho] \, e^{\frac{N}{2}\sum_{i\neq j} \frac{z_i - z_j}{u_i - u_j} \rho^i_j \rho^j_i} \tag{2.7}$$

and the integral $\int [\mathrm{d}\rho]$ involves only the off-diagonal components of ρ (and an appropriate contour).

---

[9]We include the N^{-1} factor in the OPE and also put N in front of single-trace operators. With this choice of conventions, the ribbon diagrams of genus g contribute at order N^{2-2g} in the 't Hooft expansion.

[10]The boundary behavior of a D1-brane dual to a determinant can be derived by adding a probe D1'-brane transverse to the stack of N D1-branes and finding its image in the backreacted geometry (see [**17**]).

[11]Notice that the tree level computations of [**43**] give an exact answer in the chiral algebra (when considering BRST closed operators) since the 't Hooft coupling dependence drops out in this subsector.

After integrating out the fermions, the ρ integral takes a form suitable for a large N saddle-point approximation:

$$\frac{1}{Z_\rho}\int [\mathrm{d}\rho]\, e^{NS[\rho]}, \tag{2.8}$$

with the action

$$S[\rho] = \frac{1}{2}\sum_{i\neq j}\frac{z_i - z_j}{u_i - u_j}\rho^i_j\rho^j_i + \log\det\rho. \tag{2.9}$$

The saddle point equations are

$$\frac{z_i - z_j}{u_i - u_j}\rho^i_j + (\rho^{-1})^i_j = 0, \quad i \neq j. \tag{2.10}$$

They can be rewritten in a simple form as a matrix equation:

$$[\zeta, \rho] + [\mu, \rho^{-1}] = 0, \tag{2.11}$$

where

- ρ is a $k \times k$ matrix, whose diagonal elements are fixed to be $\rho^i_i = m_i$ and the off-diagonal components are the variables we are solving for,
- ζ is a diagonal $k \times k$ matrix, whose diagonal elements are the positions z_i of determinants on the boundary of AdS_3,
- μ is a diagonal $k \times k$ matrix, whose diagonal elements u_i control the linear combinations of the symplectic bosons X, Y employed in determinants.

2.2. Spectral curve construction. In **[17]**, a spectral curve construction for Giant Graviton D1-branes was proposed. To each saddle ρ, we can associate a holomorphic curve in $SL(2, \mathbb{C})$ defined as a spectral curve of a system of commuting matrices:

$$\begin{aligned} B(a) &= a\mu - \rho \\ C(a) &= a\zeta + \rho^{-1} \\ D(a) &= a\zeta\mu + \rho^{-1}\mu - \zeta\rho. \end{aligned} \tag{2.12}$$

The spectral curve $\mathcal{S}_\rho$ is defined as a set of points (a, b, c, d), where $a \in \mathbb{C}$ and b, c, d are simultaneous eigenvalues of $B(a), C(a), D(a)$.

The matrices are defined in such a way that:

- they commute when ρ satisfies the saddle point equations (2.11). Therefore, for each saddle ρ, they can be simultaneously diagonalized,
- they satisfy
$$aD(a) - B(a)C(a) = 1, \tag{2.13}$$
which constraints the simultaneous eigenvalues to lay inside the locus $ad - bc = 1$. As a result, the spectral curve is a holomorphic curve inside $SL(2, \mathbb{C})$,
- the boundary behavior of the spectral curve matches the expected boundary behavior of a D1-brane dual to k insertions of determinants $\mathcal{D}(m_i; u_i; z_i)$ in the boundary theory. Explicitly, when $a \to \infty$, the eigenvalues of matrices $B(a)$ and $C(a)$ approach
$$\begin{aligned} b_i &= au_i - m_i + \dots \\ c_i &= az_i + p_i + \dots, \end{aligned} \tag{2.14}$$

where $p_i \equiv [\rho^{-1}]^i_i$. This means that the spectral curve reaches the holographic boundary at k points z_i, with an asymptotic behavior controlled by u_i and m_i as expected above.

2.3. Holographic checks. The spectral curve conjecture was tested by matching several chiral algebra and BCOV computations [**17**]. Here, we summarize three of them.

2.3.1. *Determinant correlation functions with a single trace.* First, we compare the large N expectation value of a single-trace operator in the presence of multiple determinants with a holographic Witten diagram computation.

A similar calculation to the one in section 2.1 yields

$$\left\langle N\,\mathrm{Tr}\,Z(u;z)^n \prod_i^k \mathcal{D}(u_i;z_i;m_i)\right\rangle = \frac{1}{Z_\rho}\int [\mathrm{d}\rho] e^{NS[\rho]}\left[-N\,\mathrm{Tr}_{k\times k}\left(-\rho^{-1}\frac{\mu-u}{\zeta-z}\right)^n\right]. \tag{2.15}$$

The ρ integral is controlled by the same large N saddles (2.11). At a given saddle $\rho=\rho^*$, it is equal to evaluating $e^{NS[\rho^*]}$ times

$$-N\,\mathrm{Tr}_{k\times k}\left(-\rho^{-1}\frac{\mu-u}{\zeta-z}\right)^n\Bigg|_{\rho=\rho^*}. \tag{2.16}$$

The holographic computation corresponds to a closed string propagating from a boundary insertion $N\,\mathrm{Tr}\,Z(u;z)^n$ and interacting with a Giant Graviton brane. The BCOV calculation accords to integrating a bulk-to-boundary propagator sourced by $N\,\mathrm{Tr}\,Z(u;z)^n$ along the brane:

$$\int_{\mathcal{S}} \partial^{-1}\alpha_n(u;z). \tag{2.17}$$

It was shown in [**17**] that (2.16) can be rewritten in the above form as an integral over the spectral curve $\mathcal{S}=S_{\rho^*}$.

2.3.2. *Matching actions.* An insertion of a determinant operator in the boundary can be engineered in the bulk by placing a probe D1'-brane transverse to the stack of N D1-branes. The auxiliary fermions $\psi^i, \bar\psi_i$, $i=1,\ldots,N$ introduced in (2.4) can be interpreted as open strings between the probe brane and the stack [**43**]. The auxiliary variables ρ^{-1} behave as fermion bilinears.

It would be very interesting to match the ρ action (2.9) with the action of open strings on the dual Giant Graviton D1-brane. In [**17**], the derivatives of the actions with respect to the parameters m_i, u_i, z_i were matched. In particular, the derivative of the saddle action (2.9) with respect to m_i equals

$$p_i \equiv \frac{\partial S}{\partial m_i} = [\rho^{-1}]^i_i. \tag{2.18}$$

The worldvolume theory of a D1-brane is a $u(1)$-gauged $\beta\gamma$ system,[12] where the two fields, β and γ, correspond to fluctuations in the two transverse directions, which we can take to be b and c. The action of the worldvolume theory of a brane

[12] Not to be confused with the chiral algebra $\mathcal{A}_N$ which is a $\mathrm{u}(N)$-gauged $\beta\gamma$ system, originally appearing as a worldvolume theory of a stack of N D1-branes.

$\mathcal{S}$ parametrized by a is[13]

$$\int_{\mathcal{S}} \beta\bar{\partial}\gamma \wedge \frac{\mathrm{d}a}{a}. \tag{2.19}$$

Close to the brane we can expand the two fields into modes:

$$\beta(a) = \sum_n \beta_n a^n, \qquad \gamma(a) = \sum_n \gamma_n a^n. \tag{2.20}$$

The asymptotic behaviour of a brane supported at the spectral curve $\mathcal{S} = \mathcal{S}_\rho$ was found to be (2.14). Near one of the asymptotic boundaries one can then identify the zero-modes as:

$$\beta_0 = -m_i, \qquad \gamma_0 = p_i. \tag{2.21}$$

These zero-modes are conjugate with respect to the action (2.19).

The above considerations give a match of variables p_i defined as conjugate to m_i. The derivatives of the actions with respect to u_i and z_i can be matched in a similar but more cumbersome way.

2.3.3. *Modifications of determinants.* Excitations of Giant Graviton D1-branes correspond to determinant modifications [**4, 5, 8, 9**] such as[14]

$$\det X = \frac{1}{N!}\varepsilon\varepsilon(X, X, \dots X, X) \mapsto \frac{1}{N!}\varepsilon\varepsilon(X, X, \dots, X, Y^2). \tag{2.22}$$

One can create BRST-closed modifications of determinants by acting with the modes of the global symmetry algebra of the chiral algebra $\mathcal{A}_N$ in the infinite N limit. This algebra was identified on both sides of the duality in [**25**]. On the chiral algebra side, the global symmetry algebra is defined by certain modes of the BRST-closed single-trace operators:[15]

$$a^{(n)}_{p,q} \sim \oint \mathrm{d}z\, z^{p+\frac{n}{2}} \oint \mathrm{d}u\, u^{q+\frac{n}{2}}\ \mathrm{Tr}\, Z(u;z)^n, \quad |p| \le \frac{n}{2} - 1,\ |q| \le \frac{n}{2}. \tag{2.23}$$

For example, the lowest modes generate an su(2) subalgebra:

$$a^{(2)}_{0,-1} \sim \oint \mathrm{d}z\ \mathrm{Tr}\, X^2(z), \quad a^{(2)}_{0,0} \sim \oint \mathrm{d}z\ \mathrm{Tr}\, XY(z), \quad a^{(2)}_{0,1} \sim \oint \mathrm{d}z\ \mathrm{Tr}\, Y^2(z). \tag{2.24}$$

The global symmetry algebra, when acting on determinants, produces BRST-closed modifications, eg.:

$$\left[a^{(2)}_{0,1}, \det X(0)\right] \sim \varepsilon\varepsilon(X, \dots, X, Y). \tag{2.25}$$

In principle, different modes can produce BRST-equivalent operators. The BRST-inequivalent modifications can be identified by studying the following matrix of two-point functions:

$$\left\langle \left[a^{(n)}_{-p,-q}, \det Y(\infty)\right] \left[a^{(m)}_{p,q}, \det X(0)\right] \right\rangle \Big|_{N\to\infty}. \tag{2.26}$$

---

[13] The denominator comes from contracting with the $SL(2,\mathbb{C})$ volume form, which can be written as $\Omega = \frac{\mathrm{d}a\mathrm{d}b\mathrm{d}c}{a}$ in the $a \neq 0$ patch.

[14] We use a schematic notation:

$$\varepsilon\varepsilon(Z_1, \dots, Z_N) \equiv \varepsilon_{i_1 \dots i_N} \varepsilon^{j_1 \dots j_N} (Z_1)^{i_1}_{j_1} \dots (Z_N)^{i_N}_{j_N}.$$

[15] There are also three other types of generators of the global symmetry algebra (see [**25**]) but we focus on this one.

The two types of inequivalent determinant modifications were found to be:

$$[a^{(n)}_{p,p-1}, \det X(0)] \sim n\,\varepsilon\varepsilon(X,\dots,Y^{1-2p}) \qquad , p \leq \frac{1}{2} \tag{2.27}$$

$$\begin{aligned}[a^{(n)}_{p,p+1}, \det X(0)] \sim\ & n\,\varepsilon\varepsilon(X,\dots,Y^{-1-2p}\partial X) \\ & + n\,\varepsilon\varepsilon(X,\dots,\partial^2 Y^{-3-2p}) \qquad , p < -\frac{1}{2}.\end{aligned} \tag{2.28}$$

On the holographic side, (the bosonic part of) the global symmetry algebra acts by holomorphic divergence-free vector fields on $SL(2,\mathbb{C})$. The modes of dual global symmetry algebra create excitations of the dual Giant Graviton brane.

The brane dual to the insertions of $\det X(0)$ and $\det Y(\infty)$ is

$$g = \begin{pmatrix} a & 0 \\ 0 & \frac{1}{a} \end{pmatrix} \subset SL(2,\mathbb{C}). \tag{2.29}$$

Likewise, we find two types of brane excitations

$$\delta g = \begin{pmatrix} a & \delta b \\ 0 & \frac{1}{a} \end{pmatrix}, \quad \delta b \sim n a^{1-2p} \tag{2.30}$$

$$\delta g = \begin{pmatrix} a & 0 \\ \delta c & \frac{1}{a} \end{pmatrix}, \quad \delta c \sim n a^{-1-2p}, \tag{2.31}$$

which can be matched with the two types of determinant modifications (2.27) and (2.28).

3. Non-conformal vacua

In this section, we review the twisted duality in the presence of non-zero vevs of the chiral algebra correlation functions.

3.1. Coulomb branch geometries. The translation-invariant vacua of the chiral algebra $\mathcal{A}_N$ can be parametrized by eigenvalues (x_i, y_i) of X and Y of multiplicity $N_i = \alpha_i N$, $i = 1, \dots, n$.

The dual geometries arise from separating the stack of N D1-branes in $\mathbb{C}^3$. The eigenvalues x_i and y_i are identified with the positions $x = x_i$, $y = y_i$ of N_i branes, where (x, y, z) are the coordinates in $\mathbb{C}^3$.

The backreaction of the stack of non-coincident branes was computed in [**18**]. The resulting geometries can be described using 2^n patches of $\mathbb{C}^3$, where for each $i = 1, \dots, n$ either $x - x_i$ or $y - y_i$ is non-zero. The coordinates x and y remain holomorphic and coincide in all patches. The coordinate z gets deformed to a new holomorphic coordinate z_I in a patch I. The holomorphic coordinate transformation on the intersection of two patches I and I' which differ by the i-th choice only ($y \neq y_i$ in I and $x \neq x_i$ in I') is

$$z_I = z_{I'} + \frac{\alpha_i}{(x - x_i)(y - y_i)}. \tag{3.1}$$

Let us consider two "extremal" patches: I_0 where $y - y_i$ are all non-zero and I_∞ where $x - x_i$ are all non-zero. The coordinate transformation on the intersection takes the form

$$z_0 - z_\infty = \sum_{i=1}^{n} \frac{\alpha_i}{(x - x_i)(y - y_i)} = \sum_{k\geq 0}\sum_{l\geq 0} \frac{1}{x^{k+1} y^{l+1}} \sum_{i=1}^{n} \alpha_i x_i^k y_i^l. \tag{3.2}$$

The first term of the sum ($k = l = 0$) describes the standard $SL(2,\mathbb{C})$ geometry obtained by a backreaction of coincident branes:[16]

$$z_0 - z_\infty = \frac{1}{xy}. \tag{3.3}$$

The subsequent terms $\frac{1}{x^{k+1}y^{l+1}}$ are deformations of the $SL(2,\mathbb{C})$ geometry. The coefficients can be identified with the vevs of single-trace operators in the vacuum parametrized by (x_i, y_i):

$$\langle \mathrm{Tr} X^k Y^l \rangle = \sum_{i=1}^{n} N_i x_i^k y_i^l. \tag{3.4}$$

3.2. Higgs branch conjecture. The Coulomb branch of $\mathcal{N} = 4$ SYM is simply $\mathbb{R}^{6N}/S_N$, parametrized by the commuting vevs of the six adjoint scalars $\vec{\Phi}$. In an $\mathcal{N} = 2$ language, it decomposes into the $\mathcal{N} = 2$ Coulomb branch and the $\mathcal{N} = 2$ Higgs branch, $\mathbb{C}^N/S_N$ and $\mathbb{C}^{2N}/S_N$, respectively. We expect the translation-invariant vacua of the chiral algebra $\mathcal{A}_N$ to descend from the Higgs branch vacua.

More generally, the Higgs branch of $\mathcal{N} = 2$ SCFTs has been conjectured to correspond to the notion of the associated variety of its chiral algebra subsector [**7**].[17]

Motivated by the above considerations, in [**18**] it was conjectured that the space of translation-invariant vacua of a 2d chiral algebra corresponds to the associated variety of the chiral algebra (see [**1**] for the definition). The conjecture can be verified in the case of gauged $\beta\gamma$ systems, which arise from Lagrangian $\mathcal{N} = 2$ SCFTs.

References

[1] Tomoyuki Arakawa, *A remark on the C_2-cofiniteness condition on vertex algebras*, Math. Z. **270** (2012), no. 1-2, 559–575, DOI 10.1007/s00209-010-0812-4. MR2875849

[2] Tomoyuki Arakawa, *Associated varieties and Higgs branches (a survey)*, Vertex algebras and geometry, Contemp. Math., vol. 711, Amer. Math. Soc., [Providence], RI, [2018] ©2018, pp. 37–44, DOI 10.1090/conm/711/14300. MR3831567

[3] Tomoyuki Arakawa, *Representation theory of W-algebras and Higgs branch conjecture*, Proceedings of the International Congress of Mathematicians—Rio de Janeiro 2018. Vol. II. Invited lectures, World Sci. Publ., Hackensack, NJ, 2018, pp. 1263–1281. MR3966808

[4] Vijay Balasubramanian, David Berenstein, Bo Feng, and Min-xin Huang, *D-branes in Yang-Mills theory and emergent gauge symmetry*, J. High Energy Phys. **3** (2005), 006, 51, DOI 10.1088/1126-6708/2005/03/006. MR2151986

[5] Vijay Balasubramanian, Min-xin Huang, Thomas S. Levi, and Asad Naqvi, *Open strings from $\mathcal{N} = 4$ super Yang-Mills*, J. High Energy Phys. **8** (2002), no. 37, 19, DOI 10.1088/1126-6708/2002/08/037. MR1942154

[6] Christopher Beem, Madalena Lemos, Pedro Liendo, Wolfger Peelaers, Leonardo Rastelli, and Balt C. van Rees, *Infinite chiral symmetry in four dimensions*, Comm. Math. Phys. **336** (2015), no. 3, 1359–1433, DOI 10.1007/s00220-014-2272-x. MR3324147

[7] Christopher Beem and Leonardo Rastelli, *Vertex operator algebras, Higgs branches, and modular differential equations*, J. High Energy Phys. **8** (2018), 114, front matter+70, DOI 10.1007/jhep08(2018)114. MR3861208

[8] David Berenstein, *Shape and holography: studies of dual operators to giant gravitons*, Nuclear Phys. B **675** (2003), no. 1-2, 179–204, DOI 10.1016/j.nuclphysb.2003.10.004. MR2022667

[9] David Berenstein and Samuel E. Vázquez, *Integrable open spin chains from giant gravitons*, J. High Energy Phys. **6** (2005), 059, 17, DOI 10.1088/1126-6708/2005/06/059. MR2158308

[16] In coordinates $w_0 = z_0 x$, $w_\infty = z_\infty y$, we get the familiar $SL(2,\mathbb{C})$ relation: $w_0 y - w_\infty x = 1$.

[17] See also [**59**] and the reviews [**2**, **3**].

[10] David Berenstein and Shannon Wang, *BPS coherent states and localization*, J. High Energy Phys. **8** (2022), Paper No. 164, 35, DOI 10.1007/jhep08(2022)164. MR4470159

[11] M. Bershadsky, S. Cecotti, H. Ooguri, and C. Vafa, *Holomorphic anomalies in topological field theories*, Nuclear Phys. B **405** (1993), no. 2-3, 279–304, DOI 10.1016/0550-3213(93)90548-4. MR1240687

[12] M. Bershadsky, S. Cecotti, H. Ooguri, and C. Vafa, *Kodaira-Spencer theory of gravity and exact results for quantum string amplitudes*, Comm. Math. Phys. **165** (1994), no. 2, 311–427. MR1301851

[13] Adel Bilal and Chong-Sun Chu, *D3-brane(s) in* $\mathrm{AdS}_5 \times S^5$ *and* $\mathcal{N} = 4, 2, 1$ *SYM*, Nuclear Phys. B **547** (1999), no. 1-2, 179–200, DOI 10.1016/S0550-3213(99)00116-9. MR1694564

[14] Federico Bonetti and Leonardo Rastelli, *Supersymmetric localization in* AdS_5 *and the protected chiral algebra*, J. High Energy Phys. **8** (2018), 098, front matter+39, DOI 10.1007/jhep08(2018)098. MR3866291

[15] Ilka Brunner, Ioannis Lavdas, and Ingmar Saberi, *Holomorphic boundary conditions for topological field theories via branes in twisted supergravity* (2021).`arXiv:2110.15257` [hep-th]

[16] Kasia Budzik and Davide Gaiotto, *Unpublished*, (2022).

[17] Kasia Budzik and Davide Gaiotto, *Giant gravitons in twisted holography*, J. High Energy Phys. **10** (2023), 131, DOI 10.1007/JHEP10(2023)131.`arXiv:2106.14859` [hep-th]

[18] Kasia Budzik and Davide Gaiotto, *Twisted holography without conformal symmetry*, J. High Energy Phys. **12** (2023), 102, DOI 10.1007/JHEP12(2023)104.`arXiv:2211.01419` [hep-th]

[19] Kasia Budzik, Davide Gaiotto, Justin Kulp, Brian R. Williams, Jingxiang Wu, and Matthew Yu, *Semi-Chiral Operators in 4d* $N = 1$ *Gauge Theories*, (2023) `arXiv:2306.01039` [hep-th].

[20] Kasia Budzik, Davide Gaiotto, Justin Kulp, Jingxiang Wu, and Matthew Yu, *Feynman diagrams in four-dimensional holomorphic theories and the Operatope*, J. High Energy Phys. **7** (2023), Paper No. 127, 38, DOI 10.1007/jhep07(2023)127. MR4618055

[21] Dylan Butson, *Equivariant localization in factorization homology and applications in mathematical physics II: Gauge theory applications*, (2020) `arXiv:2011.14978` [math.RT].

[22] Gaoli Chen, Robert de Mello Koch, Minkyoo Kim, and Hendrik J. R. Van Zyl, *Structure constants of heavy operators in ABJM and ABJ theory*, Phys. Rev. D **100** (2019), no. 8, 086019, 16, DOI 10.1103/physrevd.100.086019. MR4032035

[23] Kevin Costello, *M-theory in the Omega-background and 5-dimensional non-commutative gauge theory*, (2016) `arXiv:1610.04144` [hep-th].

[24] Kevin Costello, *Holography and Koszul duality: the example of the* $M2$ *brane*, (2017) `arXiv:1705.02500` [hep-th].

[25] Kevin Costello and Davide Gaiotto, *Twisted Holography*, (2018) `arXiv:1812.09257` [hep-th].

[26] Kevin Costello and Si Li, *Quantization of open-closed BCOV theory, I*, (2015) `arXiv:1505.06703` [hep-th].

[27] Kevin Costello and Si Li, *Twisted supergravity and its quantization*, (2016) `arXiv:1606.00365` [hep-th].

[28] Kevin Costello and Natalie M. Paquette, *Twisted supergravity and Koszul duality: a case study in* AdS_3, Comm. Math. Phys. **384** (2021), no. 1, 279–339, DOI 10.1007/s00220-021-04065-3. MR4252878

[29] Kevin Costello and Natalie M. Paquette, *Celestial holography meets twisted holography: 4d amplitudes from chiral correlators*, J. High Energy Phys. **10** (2022), 193.

[30] Kevin Costello, Natalie M. Paquette, and Atul Sharma, *Top-down holography in an asymptotically flat spacetime*, Phys. Rev. Lett. **130** (2023), no. 6, Paper No. 061602, 7, DOI 10.1103/physrevlett.130.061602. MR4550721

[31] Kevin Costello and Brian R. Williams, *Twisted heterotic/type I duality*, (2021) `arXiv:2110.14616` [hep-th].

[32] Kevin J. Costello and Si Li, *Quantum BCOV theory on Calabi-Yau manifolds and the higher genus B-model*, (2012) `arXiv:1201.4501` [math.QA].

[33] Michael R. Douglas and Washington Taylor, *Branes in the bulk of Anti-de Sitter space*, PUPT-1806, RU-98-32, (1998) `arXiv:hep-th/9807225`.

[34] Richard Eager and Fabian Hahner, *Maximally twisted eleven-dimensional supergravity*, Comm. Math. Phys. **398** (2023), no. 1, 59–88, DOI 10.1007/s00220-022-04516-5. MR4544182

[35] Chris Elliott and Pavel Safronov, *Topological twists of supersymmetric algebras of observables*, Comm. Math. Phys. **371** (2019), no. 2, 727–786, DOI 10.1007/s00220-019-03393-9. MR4019918

[36] Chris Elliott, Pavel Safronov, and Brian R. Williams, *A taxonomy of twists of supersymmetric Yang-Mills theory*, Selecta Math. (N.S.) **28** (2022), no. 4, Paper No. 73, 124, DOI 10.1007/s00029-022-00786-y. MR4468561

[37] Davide Gaiotto and Jacob Abajian, *Twisted M2 brane holography and sphere correlation functions*, (2020) arXiv:2004.13810 [hep-th].

[38] Davide Gaiotto and Miroslav Rapčák, *Miura operators, degenerate fields and the M2-M5 intersection*, J. High Energy Phys. **1** (2022), Paper No. 086, 78, DOI 10.1007/jhep01(2022)086. MR4403458

[39] Niklas Garner and Natalie M. Paquette, *Mathematics of String Dualities*, PoS **TASI2021** (2023), 007.

[40] Owen Gwilliam and Brian R. Williams, *Higher Kac-Moody algebras and symmetries of holomorphic field theories*, Adv. Theor. Math. Phys. **25** (2021), no. 1, 129–239, DOI 10.4310/ATMP.2021.v25.n1.a4. MR4320072

[41] Nafiz Ishtiaque, Seyed Faroogh Moosavian, and Yehao Zhou, *Topological holography: the example of the D2-D4 brane system*, SciPost Phys. **9** (2020), no. 2, Paper No. 017, 63, DOI 10.21468/scipostphys.9.2.017. MR4210690

[42] Saebyeok Jeong, *SCFT/VOA correspondence via Ω-deformation*, J. High Energy Phys. **10** (2019), 171, 22, DOI 10.1007/jhep10(2019)171. MR4055858

[43] Yunfeng Jiang, Shota Komatsu, and Edoardo Vescovi, *Structure constants in $\mathcal{N} = 4$ SYM at finite coupling as worldsheet g-function*, J. High Energy Phys. **7** (2020), 037, 206, DOI 10.1007/jhep07(2020)037. MR4138059

[44] Per Kraus, Finn Larsen, and Sandip P. Trivedi, *The Coulomb branch of gauge theory from rotating branes*, J. High Energy Phys. **3** (1999), Paper 3, 27, DOI 10.1088/1126-6708/1999/03/003. MR1691064

[45] Songyuan Li and Jan Troost, *Pure and twisted holography*, J. High Energy Phys. **3** (2020), 144, 21, DOI 10.1007/jhep03(2020)144. MR4089978

[46] Hai Lin, Oleg Lunin, and Juan Maldacena, *Bubbling AdS space and 1/2 BPS geometries*, J. High Energy Phys. **10** (2004), 025, 68, DOI 10.1088/1126-6708/2004/10/025. MR2116025

[47] Juan Maldacena, *The large N limit of superconformal field theories and supergravity*, Adv. Theor. Math. Phys. **2** (1998), no. 2, 231–252, DOI 10.4310/ATMP.1998.v2.n2.a1. MR1633016

[48] Jihwan Oh and Junya Yagi, *Chiral algebras from Ω-deformation*, J. High Energy Phys. **8** (2019), 143, 26, DOI 10.1007/jhep08(2019)143. MR4013415

[49] Jihwan Oh and Yehao Zhou, *A domain wall in twisted* M*-theory*, SciPost Phys. **11** (2021), no. 4, Paper No. 077, 29, DOI 10.21468/scipostphys.11.4.077. MR4336165

[50] Jihwan Oh and Yehao Zhou, *Feynman diagrams and Ω-deformed M-theory*, SciPost Phys. **10** (2021), no. 2, Paper No. 029, 51, DOI 10.21468/scipostphys.10.2.029. MR4237705

[51] Jihwan Oh and Yehao Zhou, *Twisted holography of defect fusions*, SciPost Phys. **10** (2021), no. 5, Paper No. 105, 38, DOI 10.21468/scipostphys.10.5.105. MR4262332

[52] Hirosi Ooguri and Cumrun Vafa, *Knot invariants and topological strings*, Nuclear Phys. B **577** (2000), no. 3, 419–438, DOI 10.1016/S0550-3213(00)00118-8. MR1765411

[53] Natalie M. Paquette and Brian R. Williams, *Koszul duality in quantum field theory*, Confluentes Math. **14** (2022), no. 2, 87–138. MR4561878

[54] Surya Raghavendran, Ingmar Saberi, and Brian R. Williams, *Twisted eleven-dimensional supergravity*, Comm. Math. Phys. **402** (2023), no. 2, 1103–1166, DOI 10.1007/s00220-023-04745-2. MR4627318

[55] Surya Raghavendran and Philsang Yoo, *Twisted S-Duality*, (2019) arXiv:1910.13653 [math-ph].

[56] Ingmar Saberi and Brian R. Williams, *Twisted characters and holomorphic symmetries*, Lett. Math. Phys. **110** (2020), no. 10, 2779–2853, DOI 10.1007/s11005-020-01319-4. MR4146954

[57] Ingmar Saberi and Brian R. Williams, *Twisting pure spinor superfields, with applications to supergravity*, (2021) arXiv:2106.15639 [math-ph].

[58] Ingmar Saberi and Brian R. Williams, *Superconformal algebras and holomorphic field theories*, Ann. Henri Poincaré **24** (2023), no. 2, 541–604, DOI 10.1007/s00023-022-01224-7. MR4548528

[59] Jaewon Song, Dan Xie, and Wenbin Yan, *Vertex operator algebras of Argyres-Douglas theories from M5-branes*, J. High Energy Phys. **12** (2017), 123, front matter+35, DOI 10.1007/jhep12(2017)123. MR3756667

[60] A. A. Tseytlin and S. Yankielowicz, *Free energy of $N = 4$ super Yang-Mills in Higgs phase and non-extremal* D3*-brane interactions*, Nuclear Phys. B **541** (1999), no. 1-2, 145–162, DOI 10.1016/S0550-3213(98)00762-7. MR1682406

[61] Yi-Yen Wu, *A Note on AdS / SYM correspondence on the Coulomb branch*, JHU-TIPAC-98009 (1998) arXiv:hep-th/9809055.

[62] Peihe Yang, Yunfeng Jiang, Shota Komatsu, and Jun-Bao Wu, *D-branes and orbit average*, SciPost Phys. **12** (2022), no. 2, Paper No. 055, 45, DOI 10.21468/scipostphys.12.2.055. MR4378189

[63] Keyou Zeng, *Twisted Holography and Celestial Holography from Boundary Chiral Algebra*, Commun. Math. Phys. **405** (2024), no. 1, 19, DOI 10.1007/s00220-023-04917-0.arXiv:2302.06693 [hep-th]

[64] Yehao Zhou, *Twisted Holography: The Examples of 4d and 5d Chern-Simons Theories*, Ph.D. thesis, U. Waterloo (main), 2022.

Perimeter Institute for Theoretical Physics, Waterloo, Ontario N2L 2Y5, Canada
Email address: kbudzik@perimeterinstitute.ca

Proceedings of Symposia in Pure Mathematics
Volume **107**, 2024
https://doi.org/10.1090/pspum/107/01946

Introduction to topological symmetry in QFT

Daniel S. Freed

Abstract. We publicize the quantum framework for symmetry that is developed in our joint paper [*Topological symmetry in quantum field theory*, arXiv.org/2209.07471] with Greg Moore and Constantin Teleman. We include additional motivation and an application to a selection rule for line defects in 4-dimensional gauge theories.

This conference proceedings is a preview (appearing in arrears) of joint work with Greg Moore and Constantin Teleman [**FMT**], to which we refer for further exposition, further development, more examples, and for extensive references to the literature, since we omit citations here.

As a first introduction to our framework for topological symmetries, consider the simpler setting of isometries in Riemannian geometry. Let M be a Riemannian manifold. In differential geometry we define the notions of (i) a Lie group G, and (ii) the action of G on M by isometries. These are both fairly straightforward definitions. What is more difficult (Myers-Steenrod theorem) is to prove that the group of *all* isometries of M has a canonical Lie group structure. Our main definitions for topological symmetries in quantum field theory are analogous to (i) and (ii); we do not attempt to give structure to the collection of all topological symmetries of a quantum field theory.[1]

Here are the main points:

(1) We define separately (i) abstract symmetry data, and (ii) a concrete realization of abstract symmetry data on a quantum field theory. The data (i)

This material is based upon work supported by the National Science Foundation under Grant Number DMS-2005286 and by the Simons Foundation Award 888988 as part of the Simons Collaboration on Global Categorical Symmetries. It is also based upon work supported by the National Science Foundation under Grant No. 1440140, while the author was in residence at the Mathematical Sciences Research Institute in Berkeley, California, during the fall semester of 2022.

[1]There are well-developed mathematical foundations for *topological* field theory, but the geometric development of general quantum field theories is at a much more nascent phase. So while we treat topological field theories rigorously, we take a more heuristic approach to nontopological quantum field theories. Therefore, we do not even attempt a definition of topological symmetries in a nontopological quantum field theory.

is a *quiche*[2] (σ,ρ) in which σ is a *topological* field theory and ρ is a *topological* boundary theory. The data of the action (ii) of (σ,ρ) on a quantum field theory F belongs to nontopological field theory if F is not topological.

(2) Topological symmetry is treated as a *quantum* object: the pair (σ,ρ). This goes beyond the usual coupling of a theory to a background gauge field, which treats symmetry in a *classical* context. We illustrate the power of this perspective by means of an example at the end of this note.

(3) Our definitions are inspired by an analogy that we explicate below:

$$\begin{aligned}\text{an algebra and right module } (A,R) &\leftrightsquigarrow \text{a quiche } (\sigma,\rho)\\ \text{element of } A &\leftrightsquigarrow \text{defect in } (\sigma,\rho)\end{aligned}$$

Our framework applies to internal topological symmetries of quantum field theories. It includes homotopical symmetries: higher groups, 2-groups, etc.

To reinforce the power of separating abstract symmetry from its concrete action, we begin with an illustration from representation theory. Then we recount the key definitions from [**FMT**]. We end with an application of our framework to 4-dimensional gauge theory, or more to properly 4-dimensional quantum field theories with a certain symmetry: the derivation of a selection rule for line defects.

Motivation: Computations in representations of $\mathfrak{sl}_2\mathbb{R}$.

2-dimensional representation. Set

$$h=\begin{pmatrix}1&0\\0&-1\end{pmatrix}\qquad e=\begin{pmatrix}0&1\\0&0\end{pmatrix}\qquad f=\begin{pmatrix}0&0\\1&0\end{pmatrix}\tag{1}$$

These matrices form a basis of the Lie algebra $\mathfrak{sl}_2\mathbb{R}$. Straightforward addition and multiplication of 2×2 matrices verifies the identity

$$\frac{1}{2}h^2+ef+fe=\frac{1}{2}h^2+h+2fe\tag{2}$$

In fact, both sides equal the scalar matrix $\begin{pmatrix}3/2&0\\0&3/2\end{pmatrix}$.

3-dimensional representation. Here the matrices that represent the elements (1) of $\mathfrak{sl}_2\mathbb{R}$ are

$$h'=\begin{pmatrix}2&0&0\\0&0&0\\0&0&-2\end{pmatrix}\qquad e'=\begin{pmatrix}0&1&0\\0&0&2\\0&0&0\end{pmatrix}\qquad f'=\begin{pmatrix}0&0&0\\2&0&0\\0&1&0\end{pmatrix}\tag{3}$$

One can compute—perhaps not quite as easily—that the identity (2) holds:

$$\frac{1}{2}(h')^2+e'f'+f'e'=\frac{1}{2}(h')^2+h'+2f'e'\tag{4}$$

Again both sides are scalar matrices, namely 4 times the identity matrix.

[2]This strange terminology for an object that encodes abstract *qu*antum symmetries evokes a pair (σ,ρ) in which ρ is the crust and σ the filling; imagine the sandwich on the left of Figure 3 made open-faced by omitting the right "piece of bread" $\widetilde{F}$. One can (and one did!) point out that a *quiche* is *not* an open-faced sandwich. (Google 'tartine'.)

An infinite dimensional representation. The Lie group $\mathrm{SL}_2\mathbb{R}$ acts on the projective line $\mathbb{RP}^1$ as fractional linear transformations, effectively through its quotient $\mathrm{PSL}_2\mathbb{R}$. For each complex number $\lambda \in \mathbb{C}$, there is an induced action on λ-differentials $\phi(x)(dx)^\lambda$. The infinitesimal action of the Lie algebra $\mathfrak{sl}_2\mathbb{R}$ is by the first-order differential operators

$$\begin{aligned} \tilde{h}&: \phi \longmapsto -2x\phi' - 2\lambda\phi \\ \tilde{e}&: \phi \longmapsto -\phi' \\ \tilde{f}&: \phi \longmapsto \ x^2\phi' + 2\lambda x\phi \end{aligned} \tag{5}$$

Simple calculus manipulations verify the identity (2), which here says that the ostensibly second-order differential operator on each side of

$$\frac{1}{2}\tilde{h}^2 + \tilde{e}\tilde{f} + \tilde{f}\tilde{e} = \frac{1}{2}\tilde{h}^2 + \tilde{h} + 2\tilde{f}\tilde{e} \tag{6}$$

is the scalar operator that multiplies ϕ by $4\lambda^2 - 2\lambda$.

The universal enveloping algebra. Let $\mathcal{U}(\mathfrak{sl}_2\mathbb{R})$ be the unital associative algebra generated by $\mathfrak{sl}_2\mathbb{R}$ subject to the relation generated by equating commutators in $\mathcal{U}(\mathfrak{sl}_2\mathbb{R})$ with the Lie bracket in $\mathfrak{sl}_2\mathbb{R}$:

$$xy - yx = [x, y], \qquad x, y \in \mathfrak{sl}_2\mathbb{R}. \tag{7}$$

Any linear representation of the Lie algebra $\mathfrak{sl}_2\mathbb{R}$ extends to a left module for this *universal enveloping algebra* $\mathcal{U}(\mathfrak{sl}_2\mathbb{R})$. A special case of (7) is the relation

$$ef = fe + h \tag{8}$$

in $\mathcal{U}(\mathfrak{sl}_2\mathbb{R})$, where $e, f, h \in \mathfrak{sl}_2\mathbb{R}$ are the matrices (1). Substitute (8) into the left hand side of (2) to immediately obtain the right hand side of (2). That's it! This simple manipulation in the abstract algebra $\mathcal{U}(\mathfrak{sl}_2\mathbb{R})$ proves that the relation (2) holds in *every* linear representation of $\mathfrak{sl}_2\mathbb{R}$, and so in particular immediately proves (4) and (6). (Observe that the expression $h^2/2 + ef + fe$ lies in the center of $\mathcal{U}(\mathfrak{sl}_2\mathbb{R})$, and therefore necessarily acts as a scalar in every irreducible representation, as we also saw in the three representations considered above. However, this centrality is not the main point we are making here.)

The framework we introduce for symmetry in quantum field theory enables similar universal computations in an abstract quiche (σ, ρ), and these computations imply relations in every quantum field theory on which (σ, ρ) acts. Computations are made with defects in the topological field theory (σ, ρ).

Motivation: Algebras.

We first remark that the word 'symmetry' is usually used for a transformation of a mathematical object that preserves structure, so in particular is invertible. But if, say, a finite group G acts linearly on a complex vector space V, then one can take linear combinations of the linear operators corresponding to elements of G to endow V with the structure of a module over the group algebra $\mathbb{C}[G]$. In this way one can talk about an "algebra of symmetries", though the algebra contains many non-units (noninvertible elements). As an extreme case, we might now call the action of $0 \in \mathbb{C}[G]$ a "symmetry" of V. Whether or not such "noninvertible symmetries" merit this nomenclature, it is these algebras of symmetries that motivate our definitions in field theory.

Abstract algebras of symmetries. Let A be an algebra, and for simplicity suppose that the ground field is $\mathbb{C}$. For our motivational exposition it suffices to assume that A and the modules that follow are finite dimensional. We do not use a topology on A. Let R be the right *regular* A-module, i.e., the vector space underlying the algebra A together with the right module structure $R \times A \to R$ given by multiplication. Then an abstract algebra of symmetries is the pair (A, R).

Concrete algebras of symmetries. Let V be a complex vector space. The action of (A, R) on V is specified by the data of a pair (L, θ): L is a left A-module and θ is an isomorphism of vector spaces

$$\theta\colon R \otimes_A L \xrightarrow{\;\cong\;} V \tag{9}$$

Since R is the right *regular* module, the tensor product in the domain of (9) is $A \otimes_A L$, and there is a natural isomorphism of this mixing of A and L with the vector space that underlies the left module L. Passing from a module to its underlying vector space is such a natural process that we do not usually bother to articulate the isomorphism (9). But in the field theory context there is no operation analogous to 'underlying vector space'; Example 16 below suggests such an operation is not possible. Elements of $A = \mathrm{End}_A(R)$ act as linear operators on V via the isomorphism θ.

A generalization. We can relax the condition that R be the *regular* module and consider (A, R) with *any* right module. Then this very general notion of abstract symmetry allows many possibilities. For example, if $A = \mathbb{C}$ is the ground field as an algebra and R is any vector space, then an (A, R)-action on a vector space V decomposes V as a tensor product $R \otimes L$ for some vector space L. For general (A, R), the algebra $\mathrm{End}_A(R)$ acts on $R \otimes_A L$ and so too on V via the isomorphism θ.

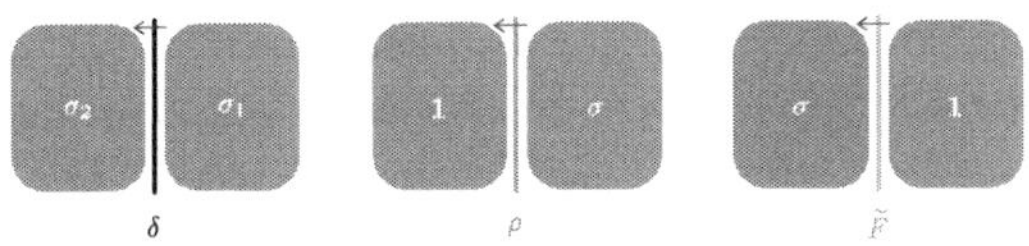

FIGURE 1. A domain wall, a right boundary theory, and a left boundary theory

Quiches and their action on quantum field theories.

Nomenclature: from algebra to field theory. A Wick-rotated field theory is a linear representation of a bordism category: this is the elevator speech version of Graeme Segal's axiom system. In this sense a Wick-rotated field theory is analogous to a linear representation of a Lie group, or to a left module over an algebra.[3] We adopt terminology from these analogs. A field theory has a (spacetime) dimension and a collection of background fields that determine the domain bordism category; specifying these is analogous to specifying which Lie group or algebra we are representing. But in this note we often leave these important specifications implicit. Suppose $\sigma, \sigma_1, \sigma_2$ are field theories (of the same dimension on the same background

[3]Suppose A is an algebra and M is a right A-module. Then one can contemplate an algebra A' that acts on M on the left and commutes with the A-action. For example, one often considers the commutant $A' = \mathrm{End}_A(M)$. In the analogy with field theory, A is the bordism category, the module M is the Wick-rotated field theory, and A' is the quiche, defined in the next few sections.

fields). We use the following nomenclature for the domain walls and boundaries in Figure 1:

$$\begin{array}{rcl} \text{domain wall } \delta\colon \sigma_1 \to \sigma_2 & \rightsquigarrow & (\sigma_2, \sigma_1)\text{-bimodule} \\ \text{right boundary theory } \rho\colon \sigma \to \mathbb{1} & \rightsquigarrow & \text{right } \sigma\text{-module} \\ \text{left boundary theory } \widetilde{F}\colon \mathbb{1} \to \sigma & \rightsquigarrow & \text{left } \sigma\text{-module} \end{array} \tag{10}$$

Here $\mathbb{1}$ is the tensor unit theory: $\mathbb{1} \otimes F \cong F$ for all theories F. Furthermore, we use the notation

$$\rho \otimes_\sigma \widetilde{F} \tag{11}$$

for the dimensional reduction F of σ along the interval pictured in Figure 2, a presentation of F that we sometimes call a *sandwich*: the bulk theory σ is sandwiched between the right boundary theory ρ and the left boundary theory $\widetilde{F}$.

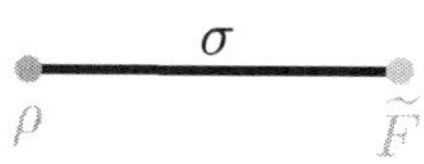

FIGURE 2. The interval used in the dimensional reduction (11)

Abstract field theoretic symmetry data: the quiche. The nonnegative integer n in the following is the dimension of field theories on which the quiche (σ, ρ) acts.

DEFINITION 12. Fix $n \in \mathbb{Z}^{\geq 0}$. Then an *$n$-dimensional quiche* is a pair (σ, ρ) in which σ is an $(n+1)$-dimensional topological field theory and ρ is a right topological σ-module.

It is the topological nature of the field theory σ and its right boundary theory ρ that make this definition appropriate for topological symmetries. One can relax the definition and only require that σ be a once-categorified n-dimensional topological field theory; in its role as symmetries of an n-dimensional field theory, it is not necessary to evaluate σ on arbitrary $(n+1)$-manifolds.

REMARK 13. In the motivational case of algebras, there is a distinguished right module: the regular module. Similarly, there are classes of topological field theories σ for which we can define the notion of a regular σ-module. The examples in this note are of this type.

The most basic quiche is associated to a finite group G and any nonnegative integer n. Let σ be free $(n+1)$-dimensional gauge theory with gauge group G, and let ρ be the right boundary theory which sums over sections of the restriction of a principal G-bundle to the boundary. (This is sometimes called a Dirichlet boundary theory: the fluctuating principal G-bundle is trivialized on the boundary.) The detailed example at the end of this note is a homotopical variant in which G is the homotopical abelian group BA, the classifying space of a finite abelian group A.

Action of a quiche on a quantum field theory. The following definition is illustrated in Figure 3.

DEFINITION 14. Let (σ, ρ) be an n-dimensional quiche, and let F be an n-dimensional field theory. A *(σ, ρ)-module structure* on F is a pair $(\widetilde{F}, \theta)$ in which

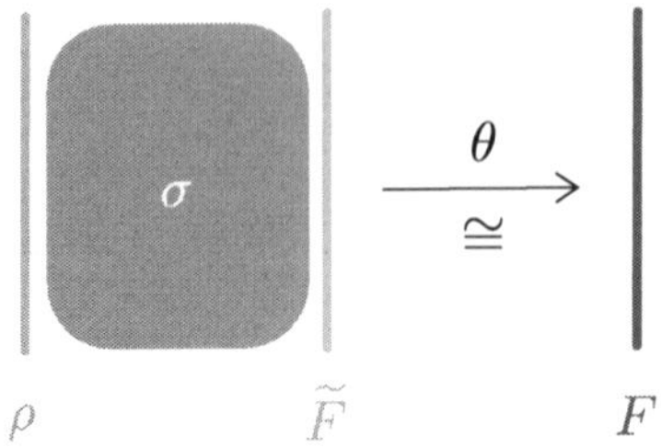

FIGURE 3. (σ, ρ)-module data on a field theory F

$\widetilde{F}$ is a left σ-module and θ is an isomorphism

$$\theta\colon \rho \otimes_\sigma \widetilde{F} \xrightarrow{\;\cong\;} F \tag{15}$$

of absolute n-dimensional theories.

Neither F nor $\widetilde{F}$ is assumed to be topological. (See footnote [1] for the mathematical status that we adopt in this note and in [**FMT**] for nontopological field theories.)

EXAMPLE 16 (center symmetry in gauge theory). Let H be a Lie group and suppose $A \subset H$ is a finite subgroup of the center of H. Then A is abelian. Set $\overline{H} = H/A$. (As an example, take $H = \mathrm{SU}_2$ and $A = \mu\!\!\!/_2 = \{\pm 1\} \subset \mathrm{SU}_2$ the center; then $\overline{H} \cong \mathrm{SO}_3$.) Suppose F is an H-gauge theory in some dimension n, and assume that F carries the homotopical symmetry of the group BA, the classifying space of A. For example, the theory may have matter fields that are neutral under the central subgroup $A \subset H$, in which case it has BA-symmetry. A map into BA classifies a principal A-bundle, and there is a bilinear pairing $\{\text{principal } A\text{-bundles}\} \times \{H\text{-connections}\} \to \{H\text{-connections}\}$: multiplication $A \times H \to H$ is a homomorphism of Lie groups since A is central. We assume that the quantum field theory F has this action as a symmetry. (For example, this action does not alter the curvature, so pure Yang-Mills theory has this symmetry.) In the physics literature this is termed a "1-form symmetry". It is usually expressed by introducing a new background field—an A-gerbe[4]—and extending the theory F to a theory with this *background/classical* field. In [**FMT**] we advocate for the *quantum* quiche picture: σ is the quantum theory that sums over A-gerbes, ρ is the boundary theory that trivializes an A-gerbe, and $\widetilde{F}$ is an $\overline{H}$-gauge theory. Observe that no details about the quantum field theory F are used to describe the symmetry; we only have to know that it carries a BA-symmetry. In particular, our characterization of F as "an H-gauge theory" is not necessary, and furthermore this characterization only pertains to the description of the theory, not to the abstract theory: the possibility of isomorphisms between H-gauge theories and quantum field theories that lack a fluctuating H-connection obstructs the statement "The theory F is an H-gauge theory".

Quotients.

[4] An A-gerbe is a geometric representative of a cohomology class in $H^2(-;A)$; it arises here as the obstruction to lifting a principal $\overline{H}$-bundle to a principal H-bundle. For $H = \mathrm{SU}_2$ and $A = \mu\!\!\!/_2$ this obstruction is the second Stiefel-Whitney class.

Augmentations of algebras. We begin as before with motivation from algebra.

DEFINITION 17. Let A be an algebra over $\mathbb{C}$. An *augmentation* of A is an algebra homomorphism $\epsilon_A\colon A \to \mathbb{C}$ from A to the ground field.

An augmentation gives the ground field $\mathbb{C}$ the structure of an A-module $\mathbb{C}_{\epsilon_A}$, which we take to be a right module: set $\lambda \cdot a = \lambda\, \epsilon_A(a)$ for $\lambda \in \mathbb{C}_{\epsilon_A}$, $a \in A$. Let R be the right regular module, and let V be a vector space equipped with an (A, R)-action, including the structure isomorphism (9) that recovers the underlying vector space V. *Define* the vector space

$$Q := \mathbb{C}_{\epsilon_A} \otimes_A L \tag{18}$$

Then Q is by definition the quotient of V by the action of (A, R) using the augmentation ϵ_A.

EXAMPLE 19. Let G be a finite group and $A = \mathbb{C}[G]$ its complex group algebra. A character χ of A determines the augmentation that sends $g \mapsto \chi(g)$, $g \in G$. In particular, there is a canonical augmentation from the unit character: $\chi(g) = 1$ for all $g \in G$. Let S be a set with a (left) G-action, and let $V = \mathbb{C}\langle S\rangle$ be the free vector space on S. Then for the unit character, Q in (18) can be identified with $\mathbb{C}\langle S/G\rangle$, the free vector space on the quotient of S by the G-action. Any character χ determines a complex line bundle $\mathcal{L} \to S/\!/G$ over the groupoid (stack) quotient $S/\!/G$. In this case Q can be identified with the vector space of sections of $\mathcal{L} \to S/\!/G$. Note that at each $s \in S$ the stabilizer subgroup $G_s \subset G$ acts on the fiber $\mathcal{L}_s$ by a character. If that character is nontrivial, then all sections vanish at s.

Augmentations do not always exist and, as in Equation 19, they may not be unique.

Quotients in field theory. This often goes under the name 'gauging a symmetry'. Let (σ, ρ) be an n-dimensional quiche for some nonnegative integer n. Just as the notion of a *regular* module only makes sense in special contexts (Equation 13), so too does the notion of an *augmentation* of (σ, ρ). In fully local topological field theory, one special context occurs when the codomain $(n+1)$-category of the theory σ is the Morita category of algebras in some n-category. In that case we have the notions of both the regular module and augmentations: an augmentation is an algebra map to the tensor unit. We realize the augmentation ϵ as a right σ-module. Then if ϵ^L is its left adjoint—a particular dual left σ-module—the sandwich $\rho \otimes_\sigma \epsilon^L$ is the trivial theory. Augmentations may not exist, whereas the regular module of an algebra always exists. For the gauge theory of a finite group G introduced after Equation 13, an augmentation is the right boundary theory ϵ which does not have any additional fluctuating fields: there is no constraint on the bulk fluctuating G-bundle, hence the quantization sums over all G-bundles. As mentioned, this is the process of quotienting, or gauging, (by) the G-symmetry. The quotient theory along the augmentation ϵ is by definition

$$F/\sigma := \epsilon \otimes_\sigma \widetilde{F}, \tag{20}$$

as illustrated in Figure 4.

Line defects in 4-dimensional theories with BA-symmetry.

What follows is an extended example to illustrate the power of the quantum quiche framework. Our exposition is inspired by the papers [**GMN, AST**]. We first

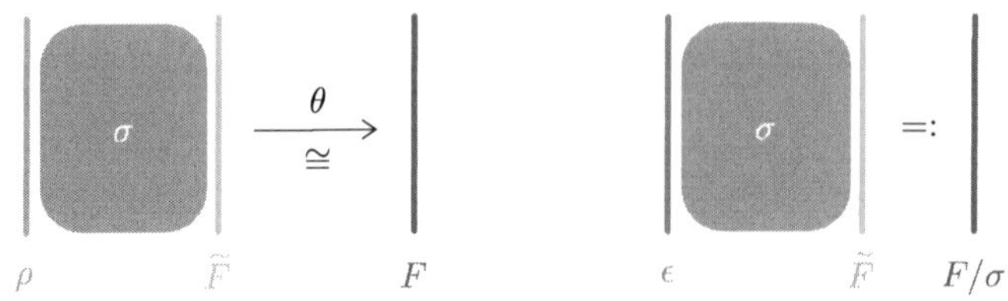

FIGURE 4. On the left, the structure isomorphism of the (σ, ρ)-module structure on F; on the right, the definition of the quotient theory F/σ

state the result, which is a selection rule for line[5] defects. Then we briefly introduce *finite homotopy theories*, an important general class of topological field theories. We conclude with a derivation of the selection rule.

Setup and the selection rule. As in Equation 16, let H be a Lie group and suppose $A \subset H$ is a finite subgroup of its center. Suppose F is a 4-dimensional field theory with BA-symmetry; we refer to it as an H-gauge theory. (But see the comment at the end of Equation 16.) For every pair (A', q) consisting of a subgroup $A' \subset A$ and a quadratic function $q\colon A' \to \mathbb{C}^\times$ there is an associated[6] H/A'-gauge theory $F_{A',q}$. We compute a selection rule for line defects in this theory. As we explain below, the category of line defects is a sum of categories indexed by

(21) $$(m, e) \in A \times A^\vee,$$

where $A^\vee = \mathrm{Hom}(A, \mathbb{C}^\times)$ is the group of characters—the Pontrjagin dual group—to A. Now the quadratic function $q\colon A' \to \mathbb{C}^\times$ has an associated bicharacter

(22) $$b_q\colon A' \times A' \to \mathbb{C}^\times,$$

and so by transposition a homomorphism

(23) $$\epsilon_q\colon A' \to (A')^\vee.$$

The selection rule for line defects in the theory $F_{A',q}$ is:

(24) $$\boxed{\begin{aligned} &m \in A' \\ &e\big|_{A'} = \epsilon_q(m)^{-1} \end{aligned}}$$

For later use: The quadratic function q gives rise to a *Pontrjagin square* cohomology operation

(25) $$\mathcal{P}_q\colon H^2(X; A') \longrightarrow H^4(X; \mathbb{C}^\times)$$

[5]Affine geometry: point, line, plane. Differential geometry: point, curve, surface. 'Curve defects' anyone?

[6]Below we define the theory $F_{A',q}$ from F using only the BA-symmetry of F; see (33) and Figure 5. The descriptions as gauge theories are not only superfluous but also do not have any intrinsic meaning. Furthermore, an important feature of our approach, as opposed to that in [**GMN, AST**], is that we give a *local* construction of $F_{A',q}$ from F. Characterizing a quantum field theory by cataloging its line defects is not a local procedure. Here we *compute* the selection rule on line defects from the local description of the theory, and moreover our computation takes place entirely in the topological field theory that encodes the BA-symmetry.

on any topological space X, and it too is a quadratic function:

$$\mathcal{P}_q(x_1 + x_2) = \mathcal{P}_q(x_1) \cdot \mathcal{P}_q(x_2) \cdot \beta_q(x_1, x_2), \qquad x_1, x_2 \in H^2(X; A'), \tag{26}$$

where $\beta_q \colon H^2(X;A') \times H^2(X;A') \to H^4(X;\mathbb{C}^\times)$ is the symmetric bihomomorphism defined by the cup product and the bicharacter (22) on coefficients.

Finite homotopy theories. These topological field theories are constructed by a finite, homotopical version of the Feynman path integral, hence are amenable to computations using techniques from algebraic topology. Some boundaries and defects in these theories have semiclassical descriptions, so are similarly susceptible to straightforward computation. The data that defines a finite homotopy theory is a pair $(\mathcal{X}, \lambda)$ consisting of a π-finite space[7] $\mathcal{X}$ and a cocycle λ on $\mathcal{X}$; the theory exists in any dimension. In our example we use cocycles for singular cohomology with coefficients in $\mathbb{C}^\times$.

Set $(\mathcal{X}, \lambda) = (B^2A, 1)$, where B^2A is an Eilenberg-MacLane space $K(A,2)$ and 1 is the identity cocycle that represents the identity element in $H^4(B^2A;\mathbb{C}^\times)$. We construct a 5-dimensional topological field theory σ. Let M be a closed manifold of dimension ≤ 4; we briefly describe the quantization $\sigma(M)$ for $\dim M = 4,3,2$. Consider the mapping space

$$\mathcal{X}^M = \mathrm{Map}(M, B^2A). \tag{27}$$

This mapping space is the space of "fluctuating fields" of the theory on M; its quantizations (29)–(31) proceed via a homotopical form of integration. Eilenberg-MacLane spaces classify singular cohomology, from which it follows that

$$\begin{aligned} \pi_0(\mathcal{X}^M) &= H^2(M;A) \\ \pi_1(\mathcal{X}^M) &= H^1(M;A) \\ \pi_2(\mathcal{X}^M) &= H^0(M;A) \end{aligned} \tag{28}$$

Also, the mapping space $\mathcal{X}^M = \mathrm{Map}(M, B^2A)$ is a product of Eilenberg-MacLane spaces: all k-invariants vanish. For $\dim M = 4$ the vector space $\sigma(M)$ is the space of locally constant complex functions on $\mathcal{X}^M$:

$$\sigma(M) = \mathrm{Fun}_{\mathrm{flat}}(\mathcal{X}^M) = \mathrm{Fun}(\pi_0\mathcal{X}^M), \qquad \dim M = 4. \tag{29}$$

Notice that this quantization only uses the set $\pi_0\mathcal{X}^M$. For $\dim M = 3$ the linear category $\sigma(M)$ is the category of flat complex vector bundles on $\mathcal{X}^M$:

$$\sigma(M) = \mathrm{Vect}_{\mathrm{flat}}(\mathcal{X}^M) = \mathrm{Vect}(\pi_{\le 1}\mathcal{X}^M), \qquad \dim M = 3. \tag{30}$$

Let $\mathbb{V}$ denote the linear category of finite dimensional complex vector spaces and linear maps. Notice that the quantization (30) only uses the fundamental groupoid $\pi_{\le 1}\mathcal{X}^M$; the category $\sigma(M)$ is equivalent to the category of complex vector bundles over that groupoid, i.e., the category of functors $\pi_{\le 1}\mathcal{X}^M \to \mathbb{V}$. Now a $\mathbb{V}$*-module* is a complex linear category, and a *flat* $\mathbb{V}$*-bundle* over a topological space is a local system of $\mathbb{V}$-modules. For $\dim M = 2$ the linear 2-category $\sigma(M)$ is the 2-category of flat $\mathbb{V}$-bundles on $\mathcal{X}^M$:

$$\sigma(M) = \mathrm{Cat}_{\mathrm{flat}}(\mathcal{X}^M) = \mathrm{Cat}(\pi_{\le 2}\mathcal{X}^M), \qquad \dim M = 2. \tag{31}$$

---

[7]The π-finiteness condition—$\mathcal{X}$ has finitely many path components; there exists $Q \in \mathbb{Z}^{\ge 0}$ such that $\pi_q(\mathcal{X}, x) = 0$ for all $q > Q$, $x \in \mathcal{X}$; and $\pi_q(\mathcal{X}, x)$ is a finite group for all $q \in \mathbb{Z}^{\ge 0}$, $x \in \mathcal{X}$—is only needed to define the partition function in the top dimension. Without that assumption we obtain a once-categorified field theory. Our example $\mathcal{X} = B^2A$ *is* π-finite, so defines a full topological field theory.

Let $\mathcal{C}$ denote the linear 2-category of suitable complex linear categories. Again the quantization (31) depends only on a truncation of the mapping space, here the fundamental 2-groupoid $\pi_{\le 2}\mathcal{X}^M$, and (31) is the 2-category of functors $\pi_{\le 2}\mathcal{X}^M \to \mathcal{C}$. Thus to each object of $\pi_{\le 2}\mathcal{X}^M$ is assigned a linear category, to each 1-morphism is assigned a linear functor, and to each 2-morphism is assigned a natural transformation of functors. In particular, an elements of $\pi_2(\mathcal{X}^M, \phi)$ acts as an automorphism of the identity functor of the category attached to $\phi \in \mathcal{X}^M$.

There are semiclassical data that specify some special topological boundary theories and topological defects in a finite homotopy theory. The semiclassical data of a topological boundary theory is a (π-finite) map $f\colon \mathcal{Y} \to \mathcal{X}$ together with a suitable cochain on $\mathcal{Y}$; for a topological defect the codomain is an iterated free loop space of $\mathcal{X}$. A regular boundary theory has $\mathcal{Y} = *$ a point; an augmentation has $f = \mathrm{id}_{\mathcal{X}}$. We defer to [**FMT**, Appendix A] for details.

Proof of the selection rule. Recall the setup in Example 16. Namely, A is a finite abelian group, σ is the 5-dimensional finite homotopy theory constructed from $(B^2A, 1)$, and ρ is the right regular boundary theory that quantizes a basepoint $* \to B^2A$. Furthermore, F is a 4-dimensional quantum field theory with BA-symmetry, and $(\widetilde{F}, \theta)$ is the data that expresses the BA-symmetry. A special case already mentioned is gauge theory: H is a Lie group that contains A as a subgroup of its center, then set $\overline{H} = H/A$, and suppose F as an H-gauge theory with BA-symmetry, in which case $\widetilde{F}$ is an $\overline{H}$-gauge theory (obtained by coupling H-connections to A-gerbes).

As stated in the text following (31), semiclassical data for a right boundary theory is a π-finite map $\mathcal{Y} \to B^2A$ together with a cocycle μ on $\mathcal{Y}$ for a class in $H^4(\mathcal{Y}; \mathbb{C}^\times)$. If $A' \subset A$ is a subgroup, then $\mathcal{Y} = B^2A' \to B^2A$ is a π-finite map. There is an isomorphism

$$H^4(B^2A'; \mathbb{C}^\times) \cong \{\text{quadratic functions } q\colon A' \longrightarrow \mathbb{C}^\times\} \tag{32}$$

by a classical computation of Eilenberg-MacLane: the quadratic function q maps to the Pontrjagin square $\mathcal{P}_q(\iota) \in H^4(B^2A'; \mathbb{C}^\times)$ of the tautological class $\iota \in H^2(B^2A'; A')$. Hence a pair (A', q) determines a right topological boundary theory $R_{A',q}$ for σ. Furthermore, the sandwich

$$F_{A',q} := R_{A',q} \otimes_\sigma \widetilde{F} \tag{33}$$

is the 4-dimensional quantum field theory which is a twisted quotient by BA', as illustrated in Figure 5. In the language of gauge theories, $R_{A',q} \otimes_\sigma \widetilde{F}$ is a theory with gauge group H/A'; the quadratic form determines topological terms in the action of that gauge theory.

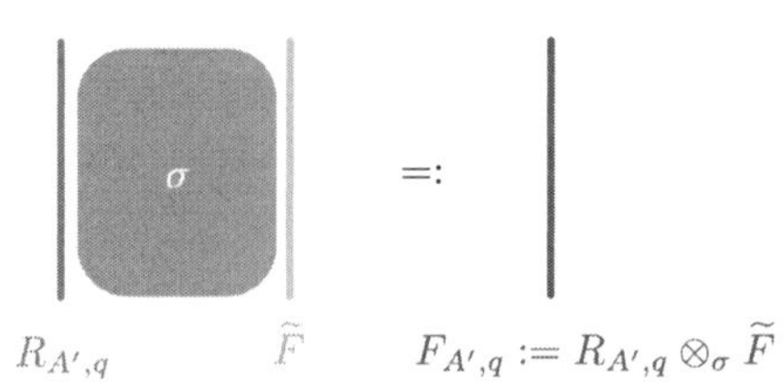

FIGURE 5. The q-twisted quotient of F by BA'

Fix A', q. We investigate line defects in the theory $R_{A',q} \otimes_{\sigma} \widetilde{F}$. The computation is local, but for ease of language assume given a 4-manifold M and a 1-dimensional submanifold $C \subset M$ on which we "wrap" defects. In the sandwich picture, these are defects supported on the 2-dimensional submanifold $[0,1] \times C \subset [0,1] \times M$; see Figure 6. The 1-category of line defects[8] supported on $C \subset M$ is the inverse limit $\varprojlim_{\epsilon \to 0} \mathrm{Hom}\big(1, \sigma(L_{\epsilon})\big)$, where in the sandwich picture L_{ϵ} is the Cartesian product of $[0,1]$ and the linking 2-sphere of size ϵ of $C \subset M$ at some $p \in C$. To compute σ on $L_{\epsilon} \approx [0,1] \times S^2$, cut along $\{1/2\} \times S^2$ and so express $\sigma(L_{\epsilon})$ as the 1-category

$$\mathrm{Hom}\big(\mathcal{V}_{A',q}, \mathcal{V}_{\widetilde{F}}(\epsilon)\big) \tag{34}$$

between objects in the 2-category $\sigma(S^2)$.[9] Observe that $\mathcal{V}_{A',q}$ is topological—the top half-cylinder Γ in Figure 6 lies in the domain of the topological field theory $(\sigma, R_{A',q})$—so no $\lim\limits_{\epsilon \to 0}$ is required. The limit is required for $\mathcal{V}_{\widetilde{F}}(\epsilon)$, but the derivation of the selection rule (24) uses only the top half-cylinder Γ, so we do not encounter the limit in the sequel.

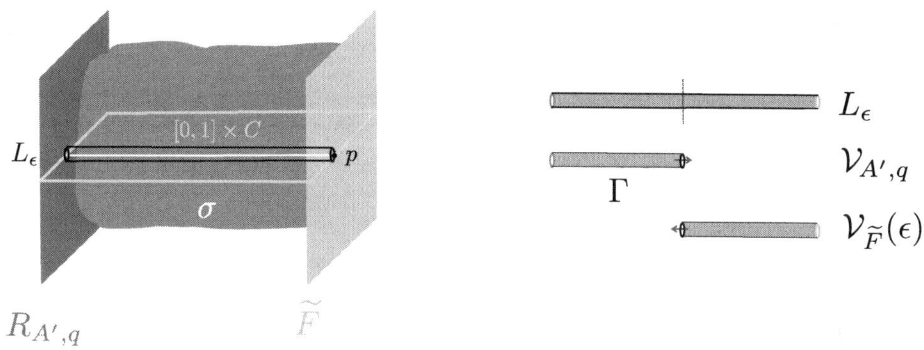

FIGURE 6. The 1-category of line defects

REMARK 35. This slicing of L_{ϵ} is the maneuver that sequesters the topological computation from the nontopological part of the theory. Such sequestration maneuvers on defects are what the sandwich presentation of the theory make possible.

Next, we compute the 2-category $\sigma(S^2)$ along which we factor in (34). From (28) we have

$$\begin{aligned} \pi_0\big(\mathrm{Map}(S^2, B^2A)\big) &= H^2(S^2; A) \cong A \\ \pi_1\big(\mathrm{Map}(S^2, B^2A)\big) &= H^1(S^2; A) = 0 \\ \pi_2\big(\mathrm{Map}(S^2, B^2A)\big) &= H^0(S^2; A) \cong A \end{aligned} \tag{36}$$

Therefore, according to (31) the quantization $\sigma(S^2)$ of $\mathcal{M}$ is the 2-category of flat $\mathbb{V}$-bundles $\mathcal{V} \to \pi_{\le 2}\mathcal{M}$. The computation (36) implies that for each $m \in \pi_0\mathcal{M}$ we

[8]The global statement on C is correct if the normal bundle to $C \subset M$ is framed. In any case, it is the local statement at a point of C that is our focus here.

[9]To glue we must reverse the depicted arrow of time at the boundary of the top half-cylinder Γ. The finite semisimplicity of $\sigma(S^2)$ makes this a straightforward operation.

have a linear category $\mathcal{V}^{(m)}$ equipped with an action of $\pi_2 \cong A$ by automorphisms of the identity functor, hence $\mathcal{V}^{(m)}$ decomposes as

$$\mathcal{V}^{(m)} = \bigoplus_e \mathcal{V}^{(m,e)} \cdot e, \qquad e \in H^0(S^2; A)^\vee \cong A^\vee. \tag{37}$$

In the gauge theory context, m and e are called discrete magnetic and electric fluxes.

As preparation for the final step of the computation, fix a cocycle μ_q that represents the class in $H^4(B^2A'; \mathbb{C}^\times)$ that corresponds to q under the isomorphism (32). Suppose $f_1, f_2\colon S^2 \to B^2A'$ are maps whose homotopy classes are $m_1, m_2 \in A'$, respectively. Then

$$\big\langle (f_1 \times f_2)^*(\mu_q) \,,\, [S^2 \times S^2] \big\rangle = \beta_q(m_1, m_2), \tag{38}$$

follows from an application of (26) with $x_i = f_i^*(\iota)$, where $\iota \in H^2(B^2A'; A')$ is the tautological class.

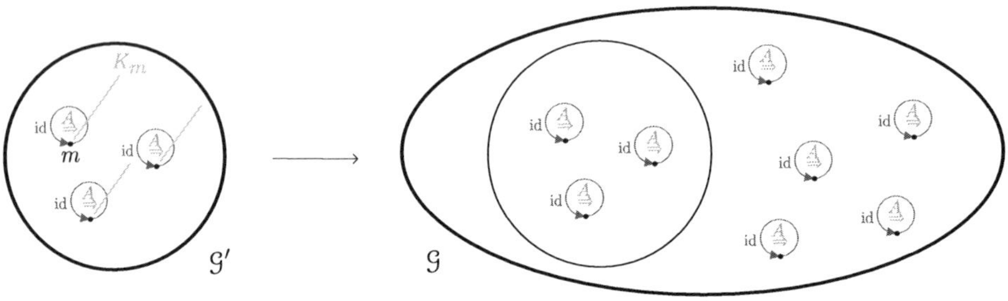

FIGURE 7. The map $\mathcal{G}' \longrightarrow \mathcal{G}$ of 2-groupoids; see (41)

Finally, we compute the value of $(\sigma, R_{A',q})$ on the half-cylinder Γ in Figure 6; the result is an object $\mathcal{V}_{A',q} = \bigoplus_{m,e} \mathcal{V}_{A',q}^{(m,e)} \cdot e$ in the 2-category $\sigma(S^2)$. Note that Γ is a morphism $\emptyset^2 \to S^2$ in the bordism category; the $R_{A',q}$-coloring of the left boundary has the effect of "coning off" that boundary component. Intuitively, $\Gamma\colon \emptyset^2 \to S^2$ is the bordism that computes the value of the boundary theory $R_{A',q}$ on S^2. The semiclassical value of this bordism is the correspondence

$$\begin{array}{ccc} & \big(\mathrm{Map}(S^2, B^2A'), \tau^2(\mu_q)\big) & \\ {}^{p_0}\swarrow & & \searrow^{p_1} \\ * & & \mathrm{Map}(S^2, B^2A) \end{array} \tag{39}$$

in which the cocycle $\tau^2(\mu_q)$ is the transgression of μ_q; its cohomology class lies in $H^2\big(\mathrm{Map}(S^2, B^2A'); \mathbb{C}^\times\big)$. Represent $\tau^2(\mu_q)$ as a flat $\mathbb{V}$-line bundle

$$\mathcal{K} \to \mathrm{Map}(S^2, B^2A').$$

Formally, the quantization of Γ is

$$\mathcal{V}_{A',q} = (p_1)_*\big(\mathcal{K} \otimes p_0^*(\underline{\mathbb{V}})\big) = (p_1)_*(\mathcal{K}), \tag{40}$$

where $\underline{\mathbb{V}} \to *$ is the trivial $\mathbb{V}$-line bundle. We compute $(p_1)_*(\mathcal{K})$ using the map

$$\mathcal{G}' := \pi_{\le 2}\, \mathrm{Map}(S^2, B^2A') \xrightarrow{\;j\;} \pi_{\le 2}\, \mathrm{Map}(S^2, B^2A) =: \mathcal{G} \tag{41}$$

of fundamental 2-groupoids, as depicted in Figure 7. (Recall the homotopy groups (36) and the fact that all k-invariants vanish. Hence we have strict models in which (41) is an inclusion.) Our task is to compute the pushforward $j_*(\mathcal{K}) \to \mathcal{G}$ of the $\mathbb{V}$-line bundle $\mathcal{K} \to \mathcal{G}'$. First, it follows from (38) that for $m \in A'$, the action of $\pi_2(\mathcal{G}', m) \cong A'$ on $\mathcal{K}_m$ is via the character $\epsilon_q(m)$, where recall the definition (23) of ϵ_q. Now since j induces an inclusion $A' \hookrightarrow A$ on π_0, it follows that $\mathcal{V}_{A',q}^{(m,e)} = 0$ if $m \notin A'$. This proves the first of the selection rules (24). For the second, observe that for $m \in A'$, the pushforward of $\mathcal{K}_m$ yields a $\mathbb{V}$-module over $\mathcal{G}$ supported at $m \in A' \subset A$ on which the action of $\pi_2(\mathcal{G}, m) \cong A$ is by the induced representation $\mathrm{Ind}_{A'}^{A}\bigl(\epsilon_q(m)\bigr)$. The latter is the sum of 1-dimensional representations: characters of A that restrict to $\epsilon_q(m)$ on A' appear with multiplicity one. Therefore, the $\mathbb{V}$-bundle $\mathcal{V}_{A',q} \to \mathcal{G}$ has support on $A' \subset A$ and the fiber at $m \in A'$ is a sum of $\mathbb{V}$-lines indexed by characters of A whose restriction to A' is $\epsilon_q(m)$. It only remains to observe that the inverse in the second selection rule in (24) is due to the fact that $\mathcal{V}_{A',q}$ is in the domain of (34), so appears dualized.

Acknowledgments

The author warmly thanks Greg Moore and Constantin Teleman for their collaboration and their feedback on this note.

References

[AST] Ofer Aharony, Nathan Seiberg, and Yuji Tachikawa, *Reading between the lines of four-dimensional gauge theories*, J. High Energy Phys. **8** (2013), 115, front matter+41, DOI 10.1007/JHEP08(2013)115. MR3106323 ↑99, 100

[FMT] Daniel S. Freed, Gregory W. Moore, and Constantin Teleman, *Topological symmetry in quantum field theory*, arXiv:2209.07471. 93, 94, 98, 102

[GMN] Davide Gaiotto, Gregory W. Moore, and Andrew Neitzke, *Framed BPS states*, Adv. Theor. Math. Phys. **17** (2013), no. 2, 241–397. MR3250763 ↑99, 100

Department of Mathematics, Harvard University, Cambridge, Massachusetts 02138
Email address: dafr@math.harvard.edu

Proceedings of Symposia in Pure Mathematics
Volume **107**, 2024
https://doi.org/10.1090/pspum/107/01947

Superconformal algebras for the Schoen Calabi–Yau manifold

Mateo Galdeano

Abstract. We revisit a recent proposal for the worldsheet description of string theory compactifications on special holonomy manifolds obtained via connected sums: the geometric construction corresponds to a diamond of inclusions of worldsheet algebras. We present new evidence for the proposal by considering compactifications on the Schoen Calabi–Yau manifold.

Contents

1. Introduction

The purpose of this note is to provide further evidence in favour of a proposal of [**29**] regarding the worldsheet formulation of string theory compactifications on connected sum manifolds.

Connected sum manifolds are obtained by gluing two Asymptotically Cylindrical (ACyl) open manifolds along isomorphic asymptotic ends. This technique has proved extremely fruitful to construct compact manifolds of special holonomy [**10, 14–16, 31, 43, 47**] which have then been used to study string compactifications, see for example [**6, 7, 9, 34, 35**] and references therein.

From the point of view of the worldsheet, a string theory compactification on a special holonomy manifold corresponds to an extended algebra of chiral symmetries [**25, 48, 51**]. This can be understood from the fact that covariantly constant forms give rise to new worldsheet symmetries and their associated currents [**37–39**].

2020 *Mathematics Subject Classification.* Primary 81R10, 81Q70; Secondary 81T30, 81T33.

This work was supported by a scholarship from the Mathematical Institute, University of Oxford, as well as the Deutsche Forschungsgemeinschaft (DFG, German Research Foundation) under Germany's Excellence Strategy – EXC 2121 "Quantum Universe" – 390833306.

This correspondence is further enhanced when the compactifying manifold can be described in terms of a connected sum. The features of this geometric structure are reflected in the worldsheet in terms of a diamond of inclusions of chiral algebras [**29**]. This has been verified for Twisted Connected Sum (TCS) [**28**], Extra Twisted Connected Sum (ETCS) and Generalized Connected Sum (GCS) manifolds [**29**].

In this note we check that this proposal also holds for the Schoen Calabi–Yau manifold [**50**]. The connected sum structure of this manifold was obtained in [**9**] by applying M-theory/heterotic duality to TCS G_2-manifold compactifications. Consequently, we will find that the associated diamond of algebras presents some similarities with that of the TCS.

This note is organised as follows: we first introduce in section 2 some necessary background on connected sums and worldsheet algebras together with a summary of the proposal of [**29**]. In section 3 we discuss as a warm-up case—and with the purpose of setting up notation—the example of TCS G_2-manifolds. This case was already analysed in [**28**], but here we have modified the presentation to follow [**29**].

Section 4 contains the most important results of this note. We begin with a review of the connected sum structure of the Schoen manifold following [**9**] before introducing the corresponding diamond of chiral algebras. We proceed to study the diamond in detail, looking at the the relationship between the algebra at the bottom and the intersection of the algebras at the lateral tips. We then study automorphisms of the diamond and relate them to mirror symmetry maps, before summarizing our results and concluding in section 5.

2. Background and general discussion

2.1. Special holonomy manifolds via connected sums. The Twisted Connected Sum (TCS) construction was introduced by Kovalev [**43**] and later extended by Corti, Haskins, Nordström and Pacini [**14, 15**] to provide new examples of compact manifolds with G_2-holonomy. This technique has been further generalized within the G_2-holonomy setting to the Extra Twisted Connected Sum (ETCS) construction [**16, 31, 47**], and several alternative versions of these "connected sum" constructions have been proposed with the aim of producing new manifolds with holonomy SU(3) [**19**], SU(4) [**20**] or Spin(7) [**10, 21**].[1]

We now describe the general ideas underlying all these constructions, and refer the reader to the original papers as well as [**44**] for the technical details. First of all, we introduce some basic definitions. We say a manifold M *asymptotes* the cylinder $\mathbb{R}^+ \times N$ if there exists a compact submanifold of M whose complement is diffeomorphic to such cylinder. In that case, we call M an *Asymptotically Cylindrical* (ACyl) manifold with *cross-section* N and we represent the asymptotic behaviour with an arrow

$$M \longrightarrow \mathbb{R}^+ \times N. \tag{2.1}$$

Furthermore, we require the metric and certain covariantly constant forms on M—which will be determined by the holonomy of the ACyl manifold—to asymptote those of $\mathbb{R}^+ \times N$ at a certain rate. The technical details of the analysis involved in this asymptotic behaviour will not be relevant for us, hence we omit them. We

[1]The Generalized Connected Sum (GCS) of Spin(7)-holonomy manifolds proposed in [**10**] is supported by strong evidence from string theory compactifications [**10, 18, 29**]. Nevertheless, a formal mathematical proof of its general validity is still missing in the literature.

parametrize the $\mathbb{R}^+$ factor by a coordinate t and represent the asymptotic behaviour of the tensors by an arrow. In particular, for the metric we require

$$g_M \longrightarrow g_{M,\infty} = \mathrm{d}t^2 + g_N, \tag{2.2}$$

where the subscript ∞ refers to the fact that the equality holds in the limit $t \longrightarrow \infty$.

We want to obtain a compact manifold M with holonomy equal to G. To this end, we consider two ACyl manifolds $M_\pm$ with cross-sections $N_\pm$. Both $M_\pm$ must have holonomy contained in G, so they are both equipped with torsion-free G-structures. We also require the cross-sections $N_\pm$ to be isomorphic. We call these ACyl manifolds the *building blocks* of the construction.

The next step is to glue the two building blocks along their isomorphic asymptotic ends to produce a compact manifold M. Parametrizing the directions of the cylinders $\mathbb{R}^+_\pm$ by $t_\pm$, we truncate both cylinders and introduce boundaries at $t_\pm = t_0 + 1$. We can then define an isomorphism F_{t_0} between $[t_0, t_0+1]_+ \times N_+$ and $[t_0, t_0+1]_- \times N_-$ as follows

$$F_{t_0} : [t_0, t_0+1]_+ \times N_+ \ni (t, n) \longmapsto (2t_0 + 1 - t, \phi(n)) \in [t_0, t_0+1]_- \times N_-, \tag{2.3}$$

where ϕ is an isomorphism between N_+ and N_-. We call F_{t_0} the *gluing map*, it identifies the asymptotic ends of $M_\pm$ and produces a compact manifold M. The area where the gluing is performed is usually denominated the *neck* region of M, and we will denote it by $M_+ \cap M_-$. The gluing map is chosen such that the G-structures of $M_\pm$ can be patched together into a globally well-defined G-structure on M.

As we have truncated $M_\pm$ at $t_\pm = t_0 + 1$, the G-structure on M has non-trivial torsion localized in the neck region. The key point of the construction is that through some careful analysis it can be shown that for a sufficiently long neck—equivalently, for large enough t_0—it is possible to deform the G-structure to a torsion-free one. This completes the construction. Once again, the technical aspects of the analysis are unimportant for our purposes and we will not discuss them.

2.2. Worldsheet algebras for special holonomy manifolds. We now focus on worldsheet aspects of string theory compactifications on special holonomy manifolds. For the purpose of this note, we will consider supersymmetric type II backgrounds given by a direct product of a special holonomy manifold M and Minkowski spacetime. In practice, we will ignore the spacetime factor and study these compactifications in terms of an $\mathcal{N} = (1, 1)$ non-linear sigma model with target M. From the worldsheet point of view, this is described by a two-dimensional chiral superconformal field theory. Throughout this note we restrict ourselves to one of the chiral components, say the holomorphic one, and study the underlying superconformal algebra.

The special holonomy of the target manifold M has important consequences for the sigma model. Recall that reduced holonomy can be equivalently formulated as the existence of certain covariantly constant forms on the manifold [**42**]. It was observed by Howe and Papadopoulos [**37–39**] that each of these covariantly constant forms leads to an additional (classical) non-linear chiral symmetry on the sigma model action. The conserved (super-)currents associated to these symmetries are constructed directly from the covariantly constant forms.

At the level of the quantum worldsheet theory, this implies the existence of additional chiral operators corresponding to the covariantly constant forms on M. These operators come in supersymmetric pairs: to each p-form we can associate an operator of conformal weight $\frac{p}{2}$ and its supersymmetric partner of weight $\frac{p+1}{2}$. Furthermore, these operators retain some features reminiscent of their geometric origin, as we will motivate in more detail in an example below.

We thus find that the original underlying $\mathcal{N} = 1$ (super-)Virasoro algebra, given by the (holomorphic) stress-tensor T and the chiral supersymmetry current G, can be extended by additional operators to obtain a new $\mathcal{N} = 1$ W-algebra [**4, 25**–**27, 48, 51**]. This means that we can associate to each special holonomy manifold M a particular chiral algebra $\mathcal{A}(M)$ that captures the underlying chiral symmetries upon compactification.

We illustrate this with an example: consider a Kähler n-fold, that is, a $2n$-dimensional manifold with $\mathrm{U}(n)$-holonomy. This manifold comes equipped with a covariantly constant Hermitian 2-form ω_n, which gives rise to a weight 1 superprimary multiplet (J_n^3, G_n^3). This extends the original $\mathcal{N} = 1$ Virasoro algebra to an $\mathcal{N} = 2$ Virasoro algebra and reflects the fact that a compactification on a Kähler manifold preserves $\mathcal{N} = 2$ supersymmetry.

The most relevant case for us will be when the holonomy is further reduced to $\mathrm{SU}(n)$, which corresponds to a Calabi–Yau n-fold. In this situation we have an additional covariantly constant complex n-form known as the holomorphic volume form Ω_n. This results in two additional real superprimary multiplets of conformal weight $\frac{n}{2}$ corresponding to the real and imaginary part of Ω_n. We denote them by (A_n, C_n) and (B_n, D_n), respectively. These further extend the $\mathcal{N} = 2$ Virasoro algebra of the Kähler case to the *Odake algebra* [**48**], which we denote by Od_n.

For $n \geq 3$ the Odake algebra is associative modulo the following *singular* fields and their descendants (which we call *null* fields)

$$N_n^1 = \partial A_n - {:}J_n^3 B_n{:}, \qquad N_n^2 = \partial B_n + {:}J_n^3 A_n{:}, \tag{2.4}$$

where colons represent normal ordering. Singular fields are descendants which are annihilated by all positive modes of the generators of the algebra. They have zero norm, so they do not contribute to correlation functions and they must be quotiented out of our extended algebras. This is a subtle point in the analysis of this note, as many of the statements we check are true "up to null fields".

The case $n = 2$ of the Odake algebra is very special as it corresponds to K3 surfaces, which are also hyper-Kähler manifolds. Some of the geometric features of these manifolds are indeed present in Od_2: recall there exists a triple of complex structures I, J, K on the K3 surface satisfying the identities

$$I^2 = J^2 = K^2 = -\mathrm{Id}, \qquad IJ = -JI = K. \tag{2.5}$$

Each of them has a corresponding Hermitian form $\{\omega^I, \omega^J, \omega^K\}$, and these give rise to the operators $\{J_2^3, A_2, B_2\}$.[2] In addition to (2.4) the following fields are also singular

$$N_2^3 = {:}J_2^3 J_2^3{:} - {:}A_2 A_2{:}, \qquad N_2^4 = {:}A_2 A_2{:} - {:}B_2 B_2{:}, \tag{2.6}$$

[2]Note the SU(2) forms are given in terms of the Hermitian forms by

$$\omega_2 = \omega^I, \qquad \Omega_2 = \omega^J + i\,\omega^K.$$

this is a manifestation of the first identity of (2.5) at the algebra level. Another example of the correlation between geometry and algebra can be seen as follows: the Od_2 algebra is actually isomorphic to the (little) $\mathcal{N} = 4$ Virasoro algebra, obtained by extending $\mathcal{N} = 1$ Virasoro by an SU(2) Kac–Moody current superalgebra at level $k = 1$. This is in direct correspondence with the SU(2)-structure of the K3 surface, and indeed the OPEs of the Kac–Moody current reproduce—in an appropriate sense—the second identity of (2.5).

We conclude this section mentioning some additional W-algebras that will be relevant for the discussion. The extended algebras associated to manifolds of holonomy G_2 and Spin(7) were introduced by Shatashvili and Vafa [**51**]. We call these the Shatashvili–Vafa G_2 and Spin(7) algebras, and denote them by $\mathrm{SV}^{\mathrm{G}_2}$ and $\mathrm{SV}^{\mathrm{Spin}(7)}$ respectively. The $\mathrm{SV}^{\mathrm{G}_2}$ algebra has a superprimary multiplet (P, K) of weight $\frac{3}{2}$ corresponding to the *associative* 3-form φ and a multiplet (X, M) of weight 2 corresponding to the *coassociative* 4-form ψ. We will not make use of the $\mathrm{SV}^{\mathrm{Spin}(7)}$ algebra, but we mention that it has a multiplet of weight 2 corresponding to the *Cayley* 4-form Ψ.

Finally, when the target manifold is either $\mathbb{R}$ or $\mathbb{S}^1$—which corresponds to the case of trivial holonomy—we can parametrize it by a coordinate t. The form dt is covariantly constant and the associated currents are a free Majorana–Weyl fermion ψ_t and a $\widehat{\mathfrak{u}}(1)$ current $j_t = i\partial_t$. These generate what we call the *Free algebra*, denoted by Fr^1. The $\mathcal{N} = 1$ Virasoro algebra can be found as a subalgebra of Fr^1, with its generators given by

$$(2.7) \qquad G^t = :j_t\psi_t:, \qquad T^t = \frac{1}{2} :(\partial\psi_t\psi_t + j_t j_t): .$$

We will keep reusing this notation to describe the Virasoro algebra inside Fr^1. When n copies of $\mathbb{R}$ or $\mathbb{S}^1$ are considered, we obtain n copies of the Free algebra $\mathrm{Fr}^n = \mathrm{Fr}^1 \oplus \overset{n}{\cdots} \oplus \mathrm{Fr}^1$.

2.3. Worldsheet algebras for connected sum manifolds. When the manifold M has a connected sum structure, the correspondence between the holonomy of M and its associated W-algebra—which we denote by $\mathcal{A}(M)$—can be further extended. It has been proposed [**28, 29**] that the piecewise decomposition of M has its counterpart at the level of worldsheet superconformal algebras, so that chiral algebras can be associated to the building blocks $M_\pm$ as well as the neck region $M_+ \cap M_-$ and arranged into a diamond of inclusions that matches the embeddings in geometry.

The argument for the proposal goes as follows: we consider an open submanifold U of the manifold M and study the algebra $\mathcal{A}(U)$ of chiral operators associated to U. The covariantly constant forms on M restrict to covariantly constant forms on U, so the corresponding operators also belong to $\mathcal{A}(U)$. Nevertheless, there can be covariantly constant forms on U which are not well-defined over the whole of M. Equivalently, the holonomy of U might be further reduced than the holonomy of M. This means the chiral algebras and the corresponding submanifolds follow the opposite inclusion pattern

$$(2.8) \qquad U \subset M \quad \Leftrightarrow \quad \mathcal{A}(U) \supset \mathcal{A}(M).$$

We point out that the algebra inclusion in (2.8) might not be unique, reflecting the additional geometric freedom available in the submanifold U. Consider now a different open submanifold $V \subset M$, which again satisfies $\mathcal{A}(V) \supset \mathcal{A}(M)$, and the overlap of the two open patches $U \cap V$. Following the reasoning of (2.8), we can find $\mathcal{A}(M)$ as a subalgebra of $\mathcal{A}(U \cap V)$ via the inclusion $\mathcal{A}(U) \subset \mathcal{A}(U \cap V)$ or the inclusion $\mathcal{A}(V) \subset \mathcal{A}(U \cap V)$. Since $\mathcal{A}(M)$ represents the chiral symmetries of the whole manifold, both realizations of the inclusion should agree: this restricts some of the freedom present in the embedding (2.8). The same argument can be repeated with further open patches of M and in the end the embedding of $\mathcal{A}(M)$ should be uniquely fixed, as there should be no ambiguity in identifying the chiral symmetries of the theory.

Connected sum manifolds constitute a very particular case since they are completely described by just two open submanifolds—the building blocks $M_{\pm}$—glued along their intersection—the neck region $M_+ \cap M_-$. Following the previous arguments, we should be able to arrange the chiral algebras associated to M, $M_{\pm}$ and $M_+ \cap M_-$ in a diamond of inclusions uniquely determined by the geometry of the connected sum construction. We display this in fig. 1.

This line of reasoning is a motivation rather than a proof, and the validity of this proposal has to be checked on a case-by-case basis. This has been done so far for TCS G_2-manifolds [**28**], ETCS G_2-manifolds and GCS Spin(7)-manifolds [**29**]. We present in section 4.3 a new check of the proposal for the Schoen manifold.

The gluing map of a connected sum is an isomorphism of $M_+ \cap M_-$ that provides a well-defined G-structure on the compact manifold M. An analogous structure should also be present for the diamond and this is indeed the case: the gluing map can be translated into an automorphism of $\mathcal{A}(M_+ \cap M_-)$ that preserves $\mathcal{A}(M)$.

We now comment on the chiral algebra corresponding to the intersection of the lateral tips of the diamond $\mathcal{A}(M_+) \cap \mathcal{A}(M_-)$. By construction, $\mathcal{A}(M)$ is contained inside this algebra and we might wonder if this inclusion is strict. If that was the case, this would mean that the set of chiral symmetries of M could be further enhanced beyond those associated to a generic manifold with G-holonomy. On the other hand, an agreement between these algebras would indicate that the connected sum construction considered produces—at least from a worldsheet point of view—generic manifolds of G-holonomy. An analytical proof of this equality is currently beyond reach as the vacuum characters of extended W-algebra intersections are poorly understood. Nevertheless, we provided in [**29**] numerical checks for the connected sums mentioned above up to a certain level in the vacuum modules. Similar checks are presented in this note for the Schoen manifold.

Finally, we would like to point out that the diamond of algebras can be used to study mirror symmetry for connected sum manifolds M. A mirror map in geometry corresponds to a certain algebra automorphism of the associated chiral algebra $\mathcal{A}(M)$. However, if M admits a connected sum decomposition, a mirror map respecting the connected sum structure should correspond to an automorphism of the whole diamond of inclusions. This was studied in [**29**] for GCS Spin(7)-manifolds. Even though the only automorphism of $\mathrm{SV}^{\mathrm{Spin}(7)}$ is the identity, it was found that non-trivial automorphisms of the whole diamond exist and these were used to propose new GCS mirror maps beyond the ones described in [**8**]. The validity of these mirror maps was explicitly checked for GCS Spin(7)-manifolds given by Joyce orbifolds [**40**] admitting an SYZ fibration [**52**].

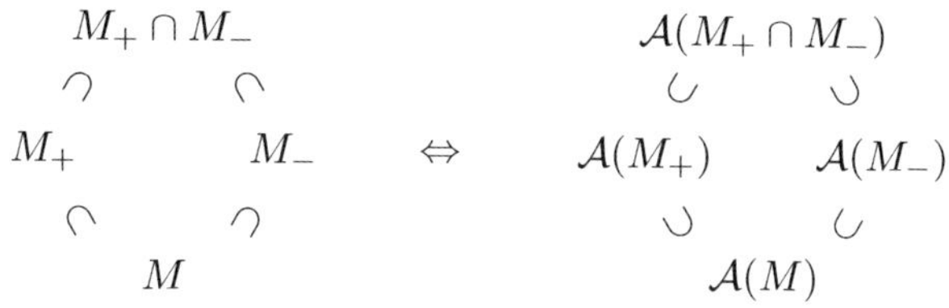

FIGURE 1. Diamond of submanifold embeddings and diamond of subalgebra inclusions for a generic connected sum manifold.

3. Review of superconformal algebras for TCS G_2-manifolds

In this section we briefly recall the Twisted Connected Sum construction (TCS) of G_2-holonomy manifolds [**14, 15, 43**] and the corresponding diamond of superconformal algebras [**28, 29**]. This serves the twofold purpose of establishing some notation that will be used throughout the rest of the note as well as presenting the construction of the TCS diamond as a warm-up case for the Schoen manifold.

3.1. Geometry of the TCS construction. As discussed in section 2.1, the idea of the TCS construction is to glue two Asymptotically Cylindrical (ACyl) manifolds—called *building blocks*—along their isomorphic asymptotic ends. This is done via a *gluing map* that identifies the *cross-sections* of the cylinders and produces a compact manifold M with a long *neck* region. This manifold has a well-defined G_2-structure that becomes torsion-free after a certain deformation.

Firstly, we need to introduce the concept of Asymptotically Cylindrical (ACyl) Calabi–Yau n-fold (CY_n). This is an open Calabi–Yau n-fold which asymptotes a cylinder with cross-section given by a Calabi–Yau $(n-1)$-fold times a circle $\mathbb{S}^1$, that is

$$\text{ACyl CY}_n \longrightarrow \mathbb{R}^+ \times \mathbb{S}^1 \times \mathrm{CY}_{n-1}. \tag{3.1}$$

Denoting by (t, θ) the coordinates of $\mathbb{R}^+ \times \mathbb{S}^1$, the metric g_n, Hermitian form ω_n and holomorphic volume form Ω_n of the ACyl CY_n satisfy the following asymptotic relations

$$g_n \longrightarrow g_{n,\infty} = \mathrm{d}t^2 + \mathrm{d}\theta^2 + g_{n-1}, \tag{3.2}$$

$$\omega_n \longrightarrow \omega_{n,\infty} = \mathrm{d}t \wedge \mathrm{d}\theta + \omega_{n-1}, \tag{3.3}$$

$$\Omega_n \longrightarrow \Omega_{n,\infty} = (\mathrm{d}\theta - i\mathrm{d}t) \wedge \Omega_{n-1}, \tag{3.4}$$

with the notation introduced in section 2.1.

For a TCS manifold, each of the building blocks consist on an ACyl Calabi–Yau 3-fold times a circle, $M_\pm = \mathbb{S}^1_{\xi\pm} \times \text{ACyl CY}_{3\pm}$. We call these circles the *external* circles and parametrize them by coordinates $\xi_\pm$. The cross-sections of these building blocks are given by $N_\pm = \mathrm{CY}_{2\pm} \times \mathbb{S}^1_{\theta\pm} \times \mathbb{S}^1_{\xi\pm}$, where the circles $\mathbb{S}^1_{\theta\pm}$ coming from the $\mathrm{CY}_{3\pm}$ are called *internal* circles and we denote their coordinates by $\theta_\pm$. Thus, the building blocks satisfy

$$M_\pm = \mathbb{S}^1_{\xi\pm} \times \text{ACyl CY}_{3\pm} \longrightarrow \mathbb{R}^+_\pm \times \mathbb{S}^1_{\theta\pm} \times \mathbb{S}^1_{\xi\pm} \times \mathrm{CY}_{2\pm} = \mathbb{R}^+_\pm \times N_\pm. \tag{3.5}$$

Each of the building blocks admits a G_2-structure that can be described in terms of an *associative* 3-form. We denote these forms by $\varphi_\pm$ and define them in terms of the forms on the Calabi–Yau 3-folds and the external circles

$$\varphi_\pm = \mathrm{d}\xi_\pm \wedge \omega_{3\pm} + \mathrm{Re}(\Omega_{3\pm}). \tag{3.6}$$

Parametrizing the $\mathbb{R}^+_\pm$ factors by coordinates $t_\pm$, these forms follow the asymptotic behaviour

$$\varphi_\pm \longrightarrow \mathrm{d}\xi_\pm \wedge \omega^I_\pm + \mathrm{d}\theta_\pm \wedge \omega^J_\pm + \mathrm{d}t_\pm \wedge \omega^K_\pm + \mathrm{d}\xi_\pm \wedge \mathrm{d}t_\pm \wedge \mathrm{d}\theta_\pm, \tag{3.7}$$

where we have made explicit the hyper-Kähler structure of the K3 surfaces $\mathrm{CY}_{2\pm}$ identifying

$$\omega_{2\pm} = \omega^I_\pm, \qquad \Omega_{2\pm} = \omega^J_\pm + i\,\omega^K_\pm. \tag{3.8}$$

The gluing map F between the two asymptotic ends identifies the external circle on M_+ with the internal circle on M_- and vice-versa. In order to compensate this "twist", the K3 surfaces are identified via a *hyper-Kähler matching*: this is an isometry of the K3 surfaces that effectively rotates the hyper-Kähler structures. At the level of differential forms, the effect of the gluing map F is as follows

$$F^*(\mathrm{d}t_-) = -\mathrm{d}t_+, \qquad F^*(\mathrm{d}\xi_-) = \mathrm{d}\theta_+, \qquad F^*(\mathrm{d}\theta_-) = \mathrm{d}\xi_+, \tag{3.9}$$

$$F^*(\omega^K_-) = -\omega^K_+, \qquad F^*(\omega^I_-) = \omega^J_+, \qquad F^*(\omega^J_-) = \omega^I_+. \tag{3.10}$$

Note the gluing map identifies the asymptotic ends of the associative 3-forms $\varphi_\pm$ (3.7) and provides a well-defined G_2-structure to the compact manifold M. We denote the corresponding associative 3-form by φ.

3.2. Chiral algebra perspective. We now summarize how the diamond of algebras corresponding to the TCS construction can be obtained, showing explicitly the inclusions of the most important generators. The guiding principle for the algebra inclusions are the asymptotic geometric relations we have just described. The algebra at the top of the diamond corresponds to the neck region, which has the form $\mathbb{R}^+ \times \mathbb{S}^1 \times \mathbb{S}^1 \times \mathrm{CY}_2$. This immediately implies that the associated algebra is $\mathrm{Fr}^3 \oplus \mathrm{Od}_2$. The neck region can be described from the point of view of $R^+_+ \times N_+$ or $R^+_- \times N_-$, to be completely explicit we follow the $R^+_+ \times N_+$ description here.

The lateral tips of the diamond correspond to the building blocks $M_\pm = \mathbb{S}^1_{\xi\pm} \times$ ACyl $\mathrm{CY}_{3\pm}$ so we have to associate algebras $\left(\mathrm{Fr}^1 \oplus \mathrm{Od}_3\right)_\pm$ to them. The right ansatz for the inclusion $\left(\mathrm{Fr}^1 \oplus \mathrm{Od}_3\right)_+ \subset \mathrm{Fr}^3 \oplus \mathrm{Od}_2$ is provided by the asymptotic behaviour of the forms (3.3), (3.4), which translates into the language of chiral operators as

$$J^3_{3+} = J^3_2 + {:}\psi_t\,\psi_\theta{:}, \qquad A_{3+} + iB_{3+} = {:}(\psi_\theta - i\psi_t)(A_2 + iB_2){:}\,. \tag{3.11}$$

Since M has G_2-holonomy, the algebra at the bottom of the diamond must be $\mathrm{SV}^{\mathrm{G}_2}$. The inclusion $\mathrm{SV}^{\mathrm{G}_2} \subset \left(\mathrm{Fr}^1 \oplus \mathrm{Od}_3\right)_+$ was already studied in [**25**], and the relation (3.6) provides the correct ansatz for our presentation

$$P = {:}\psi_\xi\, J^3_3{:} + A_3. \tag{3.12}$$

Combining this with (3.11) we obtain the ansatz for the inclusion $\mathrm{SV}^{\mathrm{G}_2} \subset \mathrm{Fr}^3 \oplus \mathrm{Od}_2$, which matches the expectation from (3.7)

$$P = {:}\psi_\xi J^3_2{:} + {:}\psi_\theta\, A_2{:} + {:}\psi_t\, B_2{:} + {:}\psi_\xi\, \psi_t\, \psi_\theta{:}\,. \tag{3.13}$$

$$
\begin{array}{ccccccc}
 & \mathbb{R}^+ \times \mathbb{S}^1_\theta \times \mathbb{S}^1_\xi \times \mathrm{CY}_2 & & & & \mathrm{Fr}^3 \oplus \mathrm{Od}_2 & \\
 \cap & & \cap & & \cup & & \cup \\
\mathbb{S}^1_{\xi+} \times \mathrm{CY}_{3+} & & \mathbb{S}^1_{\xi-} \times \mathrm{CY}_{3-} & \Leftrightarrow & \left(\mathrm{Fr}^1 \oplus \mathrm{Od}_3\right)_+ & & \left(\mathrm{Fr}^1 \oplus \mathrm{Od}_3\right)_- \\
 \cap & & \cap & & \cup & & \cup \\
 & \mathrm{TCS\ G}_2 & & & & \mathrm{SV}^{\mathrm{G}_2} &
\end{array}
$$

FIGURE 2. Diamond of submanifold embeddings and diamond of subalgebra inclusions for a TCS G_2 manifold.

The TCS gluing map (3.9), (3.10) naturally provides an ansatz for an automorphism of $\mathrm{Fr}^3 \oplus \mathrm{Od}_2$, given as follows:

$$
(3.14) \qquad \left(\psi_t, \psi_\theta, \psi_\xi, B_2, J_2^3, A_2\right) \longmapsto \left(-\psi_t, \psi_\xi, \psi_\theta, -B_2, A_2, J_2^3\right),
$$

with the same action on the superpartners. This is indeed an automorphism and it can be checked that it leaves the $\mathrm{SV}^{\mathrm{G}_2}$ algebra at the bottom of the diamond invariant. Applying this automorphism to $\left(\mathrm{Fr}^1 \oplus \mathrm{Od}_3\right)_+$ we obtain an explicit description of the remaining algebra $\left(\mathrm{Fr}^1 \oplus \mathrm{Od}_3\right)_-$ and its inclusion in the diamond

$$
(3.15) \qquad J^3_{3-} = J^3_2 - {:}\psi_t\, \psi_\xi{:}, \qquad A_{3-} + iB_{3-} = {:}(\psi_\xi + i\psi_t)(J^3_2 - iB_2){:}\,.
$$

We thus conclude that the diamond associated to a TCS manifold is well-defined. We present it in fig. 2.

4. Superconformal algebras for the Schoen Calabi–Yau manifold

In this section we provide new results regarding the diamond of worldsheet algebras associated to the Schoen manifold [**50**]. The Schoen manifold is a Calabi–Yau 3-fold with Hodge numbers $\left(h^{1,1}, h^{2,1}\right) = (19, 19)$ and it has been extensively studied together with its quotients in the context of F-theory and heterotic compactifications, see for example [**3, 11–13, 17, 32, 46**]. The Schoen manifold is also known in the literature as the split bicubic because it can be realized via a complete intersection with configuration matrix

$$
(4.1) \qquad \left[\begin{array}{c|cc} \mathbb{CP}^1 & 1 & 1 \\ \mathbb{CP}^2 & 3 & 0 \\ \mathbb{CP}^2 & 0 & 3 \end{array}\right].
$$

We will not need a detailed understanding of the Schoen manifold as a CICY, but it will be useful to keep in mind that the Schoen manifold can be described as a fibration over $\mathbb{CP}^1$ whose generic fiber is given by two elliptic surfaces (i.e. two 2-tori).

We are interested in the Schoen manifold as a connected sum. This was explicitly described in [**9**] using the duality between M-theory on a K3 surface and heterotic string theory on a 3-torus and we present a summary of the main ideas in section 4.1. We also present a complementary description of the Schoen manifold as an orbifold resolution in section 4.2. We then move to a description of the worldsheet chiral algebras, introducing the corresponding diamond in section 4.3 and discussing its genericity in section 4.4. We conclude in section 4.5 with a study of automorphisms of the diamond which include a realization of SYZ mirror symmetry in the worldsheet.

4.1. The Schoen manifold as a connected sum. We summarize the discussion of [**9**] leading to the connected sum decomposition of the Schoen manifold and refer the reader to the original paper for further details. The key idea is a fiberwise application of the duality between M-theory on a K3 surface and heterotic string theory on a 3-torus [**1, 2, 33**].

Consider an M-theory compactification of a TCS G_2-manifold[3] and focus on its building blocks, $M_\pm = \mathbb{S}^1_{\xi\pm} \times \text{ACyl CY}_{3\pm}$. The standard procedure to construct ACyl Calabi–Yau 3-folds [**14, 15**] is to start from a closed Kähler 3-fold given by a K3 fibration over $\mathbb{CP}^1$ (satisfying certain properties) and then remove a point from the $\mathbb{CP}^1$ as well as the K3 fiber over that point. We can thus think of each ACyl $\text{CY}_{3\pm}$ as a fibration: the fiber is a K3 surface $\text{CY}_{2\pm}$ and the base $B_\pm$ is given by $\mathbb{CP}^1$ minus a point. These fibrations realize the asymptotic cylindrical behaviour (3.5), where $(t_\pm, \theta_\pm)$ are the asymptotic coordinates of $B_\pm \longrightarrow \mathbb{R}^+_\pm \times \mathbb{S}^1_{\theta\pm}$.

For technical reasons, we assume that the $\text{CY}_{2\pm}$ fibers are themselves given by an elliptic fibration over $\mathbb{CP}^1$ with a well-defined section, and we further assume that the elliptic curves are holomorphically fibered over the $\mathbb{CP}^1$. These assumptions imply that when the $\text{CY}_{2\pm}$ are fibered over $B_\pm$ to give the ACyl $\text{CY}_{3\pm}$, the Hermitian forms $\omega_{2\pm} = \omega^I_\pm$ remain constant whereas the holomorphic volume forms $\Omega_{2\pm} = \omega^J_\pm + i\,\omega^K_\pm$ vary over the fibration.

We can now consider the dual heterotic picture by applying M-theory/heterotic duality to the K3 fibers $\text{CY}_{2\pm}$. These are then replaced by 3-torus fibers $\mathbb{T}^3_\pm$ while keeping the base $B_\pm$ and the external circle $\mathbb{S}^1_{\xi\pm}$ intact. This duality establishes a relation between the circles $\mathbb{S}^1_{a\pm}$ of the $\mathbb{T}^3_\pm$ fibers and the Hermitian forms $\omega^a_\pm$ of the $\text{CY}_{2\pm}$ via a period decomposition, where $a \in \{I, J, K\}$. In particular, the $\mathbb{T}^3_\pm$ fiber factorizes as $\mathbb{T}_{f\pm} \times \mathbb{S}^1_{I\pm}$ in the following way: the $\mathbb{S}^1_{I\pm}$ circle—which is related to the Hermitian forms $\omega^I_\pm$ that remain constant in the $\text{CY}_{2\pm}$ fibers—is trivially fibered over $B_\pm$, whereas the torus $\mathbb{T}_{f\pm} = \mathbb{S}^1_{J\pm} \times \mathbb{S}^1_{K\pm}$ is non-trivially fibered over $B_\pm$ and gives rise to a new open Calabi–Yau 2-fold.

Therefore, after M-theory/heterotic duality we obtain the dual building blocks $\widetilde{M}_\pm = \mathbb{S}^1_{\xi\pm} \times \mathbb{S}^1_{I\pm} \times \widetilde{\text{CY}}_{2\pm}$, where the $\widetilde{\text{CY}}_{2\pm}$ factors are given by the $\mathbb{T}_{f\pm}$ fibration over $B_\pm$. These building blocks are again asymptotically cylindrical with a cross-section given by a product of five circles:

$$\widetilde{M}_\pm = \mathbb{S}^1_{\xi\pm} \times \mathbb{S}^1_{I\pm} \times \widetilde{\text{CY}}_{2\pm} \longrightarrow \mathbb{R}^+_\pm \times \mathbb{S}^1_{\theta\pm} \times \mathbb{S}^1_{\xi\pm} \times \mathbb{S}^1_{I\pm} \times \mathbb{S}^1_{J\pm} \times \mathbb{S}^1_{K\pm} = \mathbb{R}^+_\pm \times \widetilde{N}_\pm. \tag{4.2}$$

In what follows we denote by $(x_{I\pm}, x_{J\pm}, x_{K\pm})$ the coordinates on the circles $\mathbb{S}^1_{I\pm} \times \mathbb{S}^1_{J\pm} \times \mathbb{S}^1_{K\pm}$. Since the glued TCS manifold has G_2-holonomy, the duality implies that the compact manifold $\widetilde{M}$ obtained after gluing $\widetilde{M}_\pm$ will be a Calabi–Yau 3-fold. We therefore want to define an SU(3)-structure on each building block $\widetilde{M}_\pm$, this can be done mimicking the ACyl CY asymptotic relations (3.3) and (3.4)

$$\widetilde{\omega}_{3\pm} = d\xi_\pm \wedge dx_{I\pm} + \widetilde{\omega}_{2\pm}, \qquad \widetilde{\Omega}_{3\pm} = (dx_{I\pm} - i d\xi_\pm) \wedge \widetilde{\Omega}_{2\pm}, \tag{4.3}$$

where $(\widetilde{\omega}_{2\pm}, \widetilde{\Omega}_{2\pm})$ are the Hermitian and holomorphic volume forms on the $\widetilde{\text{CY}}_{2\pm}$. Their correct asymptotic behaviour is again provided by the duality, for example $\widetilde{\omega}_{2\pm,\infty}$ can be read off directly from (3.7) following the identification between

[3]We follow the notation introduced in section 3.1.

Hermitian forms and circles,

$$\widetilde{\omega}_{2\pm} \longrightarrow \widetilde{\omega}_{2\pm,\infty} = \mathrm{d}\theta_\pm \wedge \mathrm{d}x_{J\pm} + \mathrm{d}t_\pm \wedge \mathrm{d}x_{K\pm}, \tag{4.4}$$

$$\widetilde{\Omega}_{2\pm} \longrightarrow \widetilde{\Omega}_{2\pm,\infty} = (\mathrm{d}x_{J\pm} - i\mathrm{d}\theta_\pm) \wedge (\mathrm{d}x_{K\pm} - i\mathrm{d}t_\pm). \tag{4.5}$$

Combining (4.4) and (4.5) with (4.3) we find the asymptotic behaviour of the SU(3)-structure with respect to the neck region

$$\widetilde{\omega}_{3\pm} \longrightarrow \widetilde{\omega}_{3\pm,\infty} = \mathrm{d}\xi_\pm \wedge \mathrm{d}x_{I\pm} + \mathrm{d}\theta_\pm \wedge \mathrm{d}x_{J\pm} + \mathrm{d}t_\pm \wedge \mathrm{d}x_{K\pm}, \tag{4.6}$$

$$\widetilde{\Omega}_{3\pm} \longrightarrow \widetilde{\Omega}_{3\pm,\infty} = (\mathrm{d}x_{I\pm} - i\mathrm{d}\xi_\pm) \wedge (\mathrm{d}x_{J\pm} - i\mathrm{d}\theta_\pm) \wedge (\mathrm{d}x_{K\pm} - i\mathrm{d}t_\pm). \tag{4.7}$$

We have fully described the building blocks and the neck region of the dual manifold for the heterotic compactification. It remains to identify the gluing map $\widetilde{F}$ that can be used to fuse $\widetilde{M}_\pm$ together to obtain the compact dual manifold $\widetilde{M}$.

The gluing map can be identified from the TCS gluing map F. The gluing of the base and the external circle are exactly as in (3.9), where the internal and external circles are interchanged, $\mathbb{S}^1_{\theta\pm} \leftrightarrow \mathbb{S}^1_{\xi\pm}$. The gluing of the $\mathbb{T}^3_\pm$ fibers is obtained from (3.10) identifying the Hermitian forms with the corresponding circles. This shows the trivially fibered circle is exchanged with one of the $\mathbb{T}_{f\pm}$ circles, $\mathbb{S}^1_{I\pm} \leftrightarrow \mathbb{S}^1_{J\pm}$, whereas the leftover circles reverse orientation upon gluing. Explicitly, the action of the gluing map on the forms is given by

$$\widetilde{F}^*(\mathrm{d}t_-) = -\mathrm{d}t_+, \qquad \widetilde{F}^*(\mathrm{d}\xi_-) = \mathrm{d}\theta_+, \qquad \widetilde{F}^*(\mathrm{d}\theta_-) = \mathrm{d}\xi_+, \tag{4.8}$$

$$\widetilde{F}^*(\mathrm{d}x_{K-}) = -\mathrm{d}x_{K+}, \qquad \widetilde{F}^*(\mathrm{d}x_{I-}) = \mathrm{d}x_{J+}, \qquad \widetilde{F}^*(\mathrm{d}x_{J-}) = \mathrm{d}x_{I+}. \tag{4.9}$$

It is immediate to check from (4.6) and (4.7) that the gluing map $\widetilde{F}$ identifies the asymptotic ends of the Hermitian forms $\widetilde{\omega}_{3\pm}$ and the holomorphic volume forms $\widetilde{\Omega}_{3\pm}$. Thus, we obtain a well-defined SU(3)-structure on $\widetilde{M}$ and denote its associated forms by $(\widetilde{\omega}_3, \widetilde{\Omega}_3)$.

In the limit of large neck and after a small deformation the original TCS manifold has G_2-holonomy, by duality we expect that the same happens for the connected sum we have just described. Therefore, in the appropriate limit the manifold $\widetilde{M}$ obtained upon gluing is a Calabi–Yau 3-fold.

In [**9**] the authors make use of a Mayer–Vietoris sequence to compute the Hodge numbers of $\widetilde{M}$. They find $(h^{1,1}, h^{2,1}) = (19, 19)$, which uniquely identifies the connected sum Calabi–Yau 3-fold to be the Schoen manifold.

An important observation is that all M-theory compactifications on TCS G_2-manifolds with the technical assumptions we have considered—such as having elliptically fibered K3 fibers—are dual to heterotic compactifications on the same Calabi–Yau manifold. This might seem surprising at first, however the duality also specifies a choice of vector bundles over the Schoen manifold that depends on the TCS manifold and distinguishes the dual heterotic theories [**9**]. For the purpose of this note, we just focus on the connected sum decomposition of the Schoen manifold.

4.2. The Schoen manifold as an orbifold resolution. The Schoen manifold can also be described as a crepant resolution of a toroidal orbifold of the form $\mathbb{T}^6/(\mathbb{Z}_2 \times \mathbb{Z}_2)$ [**41**]. In order to describe the associated diamond of algebras it will be useful to establish a connection between the toroidal description and the connected sum construction, and we do so following [**9**] again.[4]

[4]Note however that our conventions differ slightly from those of [**9**].

	x^1	x^2	x^3	x^4	x^5	x^6
α	$+$	$+$	$-$	$-$	$-$	$-$
β	b_1-	$-$	b_3+	$+$	b_5-	$-$

TABLE 1. Action of $\mathbb{Z}_2 \times \mathbb{Z}_2$ on $\mathbb{T}^6$. The choice of parameters $(b_1, b_3, b_5) = (0, 0, \frac{1}{2})$ produces the Schoen manifold after resolution. The $\pm$ entries correspond to a global $\pm$ sign whereas the $\frac{1}{2}-$ entries correspond to $x^i \mapsto -x^i + \frac{1}{2}$.

We parametrize $\mathbb{T}^6$ with coordinates x^i ranging from 0 to 1, with $i = 1, \ldots, 6$. We choose complex coordinates

$$z^1 = x^1 + ix^2, \qquad z^2 = x^3 + ix^4, \qquad z^3 = x^5 + ix^6, \tag{4.10}$$

and define the following SU(3)-structure on the torus[5]

$$\widetilde{\omega_3} = \frac{i}{2}\left(\mathrm{d}z^1 \wedge \mathrm{d}\bar{z}^1 + \mathrm{d}z^2 \wedge \mathrm{d}\bar{z}^2 + \mathrm{d}z^3 \wedge \mathrm{d}\bar{z}^3\right), \qquad \widetilde{\Omega_3} = i\,\mathrm{d}z^1 \wedge \mathrm{d}z^2 \wedge \mathrm{d}z^3. \tag{4.11}$$

The quotient group $\mathbb{Z}_2 \times \mathbb{Z}_2$ has two generators α, β whose action on the coordinates can be found in table 1 for the choice $(b_1, b_3, b_5) = (0, 0, \frac{1}{2})$. The singularities of the orbifold $\mathbb{T}^6/(\mathbb{Z}_2 \times \mathbb{Z}_2)$ can be resolved to obtain a Calabi–Yau manifold with Hodge numbers $(h^{1,1}, h^{2,1}) = (19, 19)$, we thus find the Schoen manifold.

We can identify the connected sum decomposition even before the resolution: the orbifold coordinate x_5 ranges within the interval $[0, 1/4]$ and we can pull the orbifold apart along this direction. The generic fiber is a product of five circles $(\mathbb{S}^1)^5$, as corresponds to the neck region, and the building blocks correspond to the ends $\widetilde{M}_+ = \left((\mathbb{S}^1)^5 \times \mathbb{R}\right)/\mathbb{Z}_2^\alpha$ at $x_5 = 0$ and $\widetilde{M}_- = \left((\mathbb{S}^1)^5 \times \mathbb{R}\right)/\mathbb{Z}_2^\beta$ at $x_5 = 1/4$. The quotient groups at these ends act non-trivially on the $\widetilde{\mathrm{CY}}_{2\pm}$ factors while leaving the trivial circles intact. As a result, we can identify the orbifold coordinates with their corresponding asymptotic coordinates as follows

$$\widetilde{M}_+ : \quad x^1 = \xi_+, \quad x^2 = x_{I+}, \quad x^3 = \theta_+, \quad x^4 = x_{J+}, \quad x^5 = t_+, \qquad x^6 = x_{K+}, \tag{4.12}$$

$$\widetilde{M}_- : \quad x^1 = \theta_-, \quad x^2 = x_{J-}, \quad x^3 = \xi_-, \quad x^4 = x_{I-}, \quad x^5 = -t_-, \quad x^6 = -x_{K-}. \tag{4.13}$$

This provides a coordinate description of the Schoen orbifold. Note in particular that the Hermitian and holomorphic volume forms from (4.11) agree with those of (4.3) in the limits (4.6), (4.7) upon the identifications (4.12) and (4.13).

We conclude with a brief comment about alternative choices of parameters (b_1, b_3, b_5). Donagi and Wendland [**22**] obtained a complete classification of Calabi–Yau 3-folds arising from orbifolds of $\mathbb{T}^6$ whose quotient group is given by an abelian extension of $\mathbb{Z}_2 \times \mathbb{Z}_2$. For the particular case where the group is precisely $\mathbb{Z}_2 \times \mathbb{Z}_2$, in addition to the Schoen manifold we also have the choices $(0, 0, 0)$, $(0, \frac{1}{2}, \frac{1}{2})$ and $(\frac{1}{2}, \frac{1}{2}, \frac{1}{2})$.[6] We point out how the last two cases can be pulled apart along the

[5]Here we are rotating the standard holomorphic volume form by a factor of i, this is done simply to match the conventions we have inherited from the definition of ACyl CY manifolds.

[6]The first one is the Vafa–Witten orbifold [**54**] and the second is a Borcea–Voisin manifold [**5, 55**] with Hodge numbers $(h^{1,1}, h^{2,1}) = (11, 11)$ [**24**].

coordinates x^5, x^3 and x^5, x^3, x^1. This suggests that these manifolds might also admit a connected sum description.

4.3. Chiral algebra perspective. In this section we describe the diamond of chiral algebras corresponding to the connected sum description of the Schoen manifold. The procedure is similar to that of section 3.2: the geometry of the manifolds determines the corresponding chiral algebras, and the asymptotic relations between the forms on these manifolds provide a natural ansatz for the algebra inclusions in the diamond. We have verified that the ansätze we present indeed define the desired algebra inclusions.[7] This involves a careful treatment of the Odake null fields, see (2.4).

As motivated in sections 2.2 and 2.3, we will associate an Odake algebra Od_n to each Calabi–Yau n-fold and a Free algebra Fr^1 to each $\mathbb{S}^1$ or $\mathbb{R}$. We will describe inclusions of the Odake algebra through the explicit expressions of the operators (G_n, J^3_n, A_n, B_n), this is enough since the expressions of the remaining operators are implied by the OPEs.[8]

The neck region of the Schoen manifold is of the form $\mathbb{R}^+ \times (\mathbb{S}^1)^5$, so the associated algebra is simply Fr^6. We denote the Free algebras Fr^1_i and their operators by (ψ_i, j_i) with $i = 1, \ldots, 6$, matching the coordinates of the orbifold description (4.10).

The building blocks are of the form $\widetilde{M}_\pm = \mathbb{S}^1_{\xi\pm} \times \mathbb{S}^1_{I\pm} \times \widetilde{\mathrm{CY}}_{2\pm}$ so the corresponding algebras on the lateral tips of the diamond are $\left(\mathrm{Fr}^2 \oplus \mathrm{Od}_2\right)_\pm$. Each Free algebra is identified with a Free algebra in Fr^6 via the coordinate identifications (4.12) and (4.13). The Od_2 factors are given by a free field realization motivated by the asymptotic behaviour of the Hermitian and holomorphic volume forms (4.4), (4.5), together with (4.12) and (4.13) again. For $\left(\mathrm{Fr}^2 \oplus \mathrm{Od}_2\right)_+ \subset \mathrm{Fr}^6$ we have

$$\begin{aligned} \mathrm{Fr}^1_{\xi+} &= \mathrm{Fr}^1_1, \qquad \mathrm{Fr}^1_{I+} = \mathrm{Fr}^1_2, \\ G_{2+} &= G^3 + G^4 + G^5 + G^6, \\ J^3_{2+} &= :\psi_3\psi_4: + :\psi_5\psi_6:, \\ A_{2+} + iB_{2+} &= :(\psi_4 - i\psi_3)(\psi_6 - i\psi_5):, \end{aligned} \tag{4.14}$$

whereas for $\left(\mathrm{Fr}^2 \oplus \mathrm{Od}_2\right)_- \subset \mathrm{Fr}^6$ we obtain

$$\begin{aligned} \mathrm{Fr}^1_{\xi-} &= \mathrm{Fr}^1_3, \qquad \mathrm{Fr}^1_{I-} = \mathrm{Fr}^1_4, \\ G_{2-} &= G^1 + G^2 + G^5 + G^6, \\ J^3_{2-} &= :\psi_1\psi_2: + :\psi_5\psi_6:, \\ A_{2-} + iB_{2-} &= :(\psi_2 - i\psi_1)(-\psi_6 + i\psi_5): . \end{aligned} \tag{4.15}$$

The algebra at the bottom of the diamond must be Od_3 since the Schoen manifold is a Calabi–Yau 3-fold. It is well-known that Od_3 can be embedded into $\mathrm{Fr}^2 \oplus \mathrm{Od}_2$, recall (3.11) and see [**28**]. The precise embedding in our case is motivated by the

[7] All the algebra computations in this note have been performed with the help of the package *OPEdefs* by Thielemans [**53**].

[8] The explicit OPEs of all the operators involved can be found in the appendices of [**30**].

$$
\begin{array}{ccccc}
 & \mathbb{R}^+ \times \left(\mathbb{S}^1\right)^5 & & & \mathrm{Fr}^6 & \\
 \rotatebox{-45}{$\supset$} & & \rotatebox{45}{$\subset$} & & \cup \quad \cup & \\
\mathbb{S}^1_{\xi+} \times \mathbb{S}^1_{I+} \times \widetilde{\mathrm{CY}}_{2+} & & \mathbb{S}^1_{\xi-} \times \mathbb{S}^1_{I-} \times \widetilde{\mathrm{CY}}_{2-} & \Leftrightarrow & \left(\mathrm{Fr}^2 \oplus \mathrm{Od}_2\right)_+ \quad \left(\mathrm{Fr}^2 \oplus \mathrm{Od}_2\right)_- \\
 \cap & & \cap & & \cup \quad \cup \\
 & \widetilde{\mathrm{CY}}_3 & & & \mathrm{Od}_3
\end{array}
$$

FIGURE 3. Diamond of submanifold embeddings and diamond of subalgebra inclusions for the Schoen manifold.

SU(3)-structure on the building blocks (4.3). For $\mathrm{Od}_3 \subset \left(\mathrm{Fr}^2 \oplus \mathrm{Od}_2\right)_+$ we find

$$
\begin{aligned}
G_3 &= G^1 + G^2 + G_{2+}, \\
J_3^3 &= :\psi_1\psi_2: + J_{2+}^3, \\
A_3 + iB_3 &= :(\psi_2 - i\psi_1)\,(A_{2+} + iB_{2+}):,
\end{aligned}
\tag{4.16}
$$

whereas for the embedding $\mathrm{Od}_3 \subset \left(\mathrm{Fr}^2 \oplus \mathrm{Od}_2\right)_-$ we have

$$
\begin{aligned}
G_3 &= G^3 + G^4 + G_{2-}, \\
J_3^3 &= :\psi_3\psi_4: + J_{2-}^3, \\
A_3 + iB_3 &= :(\psi_4 - i\psi_3)\,(A_{2-} + iB_{2-}): .
\end{aligned}
\tag{4.17}
$$

Combining (4.16) with (4.14) and (4.17) with (4.15) we obtain the same free field realization of Od_3 inside Fr^6, showing that the diamond closes in the way of fig. 3. Furthermore, this realization agrees with the ansatz suggested by the SU(3)-structure on the toroidal orbifold (4.11)

$$
\begin{aligned}
G_3 &= G^1 + G^2 + G^3 + G^4 + G^5 + G^6, \\
J_3^3 &= :\psi_1\psi_2: + :\psi_3\psi_4: + :\psi_5\psi_6:, \\
A_3 + iB_3 &= :(\psi_2 - i\psi_1)(\psi_4 - i\psi_3)(\psi_6 - i\psi_5): .
\end{aligned}
\tag{4.18}
$$

Finally, we can study the automorphism of the diamond corresponding to the gluing map (4.8), (4.9). Identifying the coordinates of the orbifold with those of $\widetilde{M}_+$ via (4.12), we obtain

$$
(\psi_5, j_5) \longmapsto (-\psi_5, -j_5)\,, \quad (\psi_3, j_3) \longmapsto (\psi_1, j_1)\,, \quad (\psi_1, j_1) \longmapsto (\psi_3, j_3)\,, \tag{4.19}
$$

$$
(\psi_6, j_6) \longmapsto (-\psi_6, -j_6)\,, \quad (\psi_4, j_4) \longmapsto (\psi_2, j_2)\,, \quad (\psi_2, j_2) \longmapsto (\psi_4, j_4)\,. \tag{4.20}
$$

This is an automorphism of Fr^6 that acts as the identity on the bottom algebra Od_3 and maps $\left(\mathrm{Fr}^2 \oplus \mathrm{Od}_2\right)_+$ to $\left(\mathrm{Fr}^2 \oplus \mathrm{Od}_2\right)_-$ and vice-versa, as can be seen from (4.18), (4.14) and (4.15).

The structure of the diamond and the gluing automorphism are analogous to those of the TCS discussed in section 3.2. This should not come as a surprise since the connected sum structure of the Schoen manifold is obtained from that of the TCS manifold via M-theory/heterotic duality.

4.4. Is the construction generic? We now address the relationship between the algebra Od_3 at the bottom of the diamond and the intersection $\left(\mathrm{Fr}^2 \oplus \mathrm{Od}_2\right)_+ \cap \left(\mathrm{Fr}^2 \oplus \mathrm{Od}_2\right)_-$ of the algebras at the lateral tips. We expect these two algebras to be equal. The reason is that this worldsheet algebra statement would correspond in geometry to the statement that the gluing procedure of section 4.1 produces a manifold with holonomy precisely equal to SU(3)—and we know this is indeed the case for the Schoen manifold.

We know from the diamond that $\mathrm{Od}_3 \subset \left(\mathrm{Fr}^2 \oplus \mathrm{Od}_2\right)_+ \cap \left(\mathrm{Fr}^2 \oplus \mathrm{Od}_2\right)_-$, so the equality could be verified by checking that both algebras have the same number of independent operators of each conformal weight. Equivalently, we could show the agreement between the vacuum module characters of both chiral algebras, defined as

$$\chi(\mathcal{A}) = \mathrm{tr}\left(q^{L_0 - \frac{c}{24}}\right), \tag{4.21}$$

where the trace is taken over the vacuum module and $q = e^{2\pi i \tau}$. The character of the Odake algebra with $n = 3$ can be found in [**49**],

$$\chi(\mathrm{Od}_3) = \sum_{m\in\mathbb{Z}} \frac{(1-q)\, q^{m^2+m-\frac{1}{2}-\frac{9}{24}}}{\left(1+q^{m-\frac{1}{2}}\right)\left(1+q^{m+\frac{1}{2}}\right)} \prod_{n\in\mathbb{N}} \frac{\left(1+q^{n-\frac{1}{2}}\right)^2}{(1-q^n)^2}. \tag{4.22}$$

The character of the algebras $\left(\mathrm{Fr}^2 \oplus \mathrm{Od}_2\right)_\pm$ is also well-known, since Od_2 is isomorphic to the $\mathcal{N} = 4$ Virasoro algebra whose character can be found in [**23**]. Unfortunately, we do not have an analytic expression for the character of the intersection algebra $\left(\mathrm{Fr}^2 \oplus \mathrm{Od}_2\right)_+ \cap \left(\mathrm{Fr}^2 \oplus \mathrm{Od}_2\right)_-$. Therefore, we can only verify our conjecture numerically.

Expanding the character χ in powers of q, we can read off the number of independent fields at level h as the coefficient of the $q^{h-\frac{c}{24}}$ term. From (4.22) we find

$$\chi(\mathrm{Od}_3) = q^{-\frac{3}{8}}\left(1+q+4q^{3/2}+5q^2+6q^{5/2}+10q^3+18q^{7/2}+27q^4+36q^{9/2}+\dots\right). \tag{4.23}$$

We have verified that the number of independent fields of the intersection algebra agrees with those of Od_3 up to level 9/2, providing support to our conjecture.

4.5. Automorphisms and mirror symmetry. In section 4.1 we have described the Schoen manifold as a $\mathbb{T}^3$ fibration. It is well-known this actually constitutes an SYZ fibration [**52**] and we can obtain a mirror symmetry map by performing three T-dualities along the circles in the fiber. As the Schoen manifold is self-dual [**36**], the mirror map in this case maps the Schoen manifold to itself.

Mirror symmetry can also be described from the worldsheet algebra perspective. Two mirror manifolds provide equivalent superconformal field theories upon compactification, so the underlying chiral algebras must remain the same. Therefore, a mirror map in geometry corresponds to an automorphism of the associated W-algebra.

For the Odake algebra, the *Mirror symmetry* automorphism $\mathbf{M}$ is given by

$$\mathbf{M}: \left(J^3_n, G^3_n, B_n, D_n\right) \longmapsto \left(-J^3_n, -G^3_n, -B_n, -D_n\right). \tag{4.24}$$

We will also need the *Phase* automorphism $\mathbf{Ph}^\pi$, that has the geometrical interpretation of a rotation of the holomorphic volume form by an angle π,[9]

$$\mathbf{Ph}^\pi : (A_n, B_n, C_n, D_n) \longmapsto (-A_n, -B_n, -C_n, -D_n). \tag{4.25}$$

[9]We will encounter both $\mathbf{M}$ and $\mathbf{M} \circ \mathbf{Ph}^\pi$ as the automorphisms associated to mirror maps. The appearance of the $\mathbf{Ph}^\pi$ factor depends on the dimension of the Calabi–Yau manifold and whether the supersymmetric cycle is calibrated by $\mathrm{Re}(\Omega_n)$ or $\mathrm{Im}(\Omega_n)$.

Similarly, performing a T-duality along a circle has a corresponding automorphism in the associated Free algebra, that we also call the *T-duality* automorphism $\mathbf{T}$,

$$\mathbf{T} : (\psi, j) \longmapsto (-\psi, -j)\,. \tag{4.26}$$

We now want to consider the mirror map of the Schoen manifold $\widetilde{M}$ obtained by T-dualising the $\mathbb{T}^3$ fiber. Since $\widetilde{M}$ is constructed by gluing two building blocks $\widetilde{M}_\pm$ with a $\mathbb{T}^3$ fibration, the $\widetilde{M}$ mirror map should give rise to a mirror map for each of the building blocks $\widetilde{M}_\pm$. Furthermore, these maps must agree over the neck region since the T-duality can be performed consistently over the whole $\widetilde{M}$.

At the level of the chiral algebras, this implies that the mirror map must be an automorphism of the whole diamond of algebras associated to the Schoen manifold. Therefore, we must be able to describe the mirror map by an automorphism of the top algebra of the diamond that reduces to automorphisms of the algebras at the lateral tips and the bottom of the diamond.

This automorphism is easily found from the orbifold perspective, where the top algebra Fr^6 can be directly associated to the orbifold coordinates. In our notation, the $\mathbb{T}^3$ fiber corresponds to the orbifold coordinates (x^2, x^4, x^6). Therefore, we define the mirror automorphism by a composition of three $\mathbf{T}$ maps along the corresponding Free algebras: $\mathbf{T}_2 \circ \mathbf{T}_4 \circ \mathbf{T}_6$.

We now study the restriction of this automorphism to the subalgebras of the diamond. For the algebras in the lateral tips $\left(\mathrm{Fr}^2 \oplus \mathrm{Od}_2\right)_\pm$, we obtain a $\mathbf{T}$ map along the Free algebras associated to the circles $\mathbb{S}^1_{I\pm}$ and a $\mathbf{M}$ map on the Od_2. For the bottom Od_3 algebra, we obtain the automorphism $\mathbf{M} \circ \mathbf{Ph}^\pi$.

Therefore, $\mathbf{T}_2 \circ \mathbf{T}_4 \circ \mathbf{T}_6$ provides an automorphism of the diamond that reduces to mirror symmetry automorphisms for the different Odake algebras, reproducing the SYZ mirror map at the level of chiral algebras.

These automorphisms can be understood geometrically via the orbifold picture, as they arise from T-dualities along toroidal supersymmetric fibers. Indeed, it is not hard to see from (4.5) together with (4.12) and (4.13) that the tori $\mathbb{S}^1_4 \times \mathbb{S}^1_6$ and $\mathbb{S}^1_2 \times \mathbb{S}^1_6$ are calibrated by $\mathrm{Re}(\widetilde{\Omega}_{2+})$ and $\mathrm{Re}(\widetilde{\Omega}_{2-})$, respectively. Therefore, these tori constitute special Lagrangian submanifolds of $\widetilde{\mathrm{CY}}_{2\pm}$ [**42**] and T-duality should give rise to a mirror automorphism in Od_2. Analogously, the $\mathbb{T}^3$ fiber is calibrated by $\mathrm{Re}(\widetilde{\Omega}_3)$, see (4.11), so it is a special Lagrangian submanifold of $\widetilde{M}$ as expected.

Finally, we discussed in section 4.2 the existence of other Calabi–Yau orbifolds that can be pulled along different coordinates. If these orbifolds do admit connected sum structures, their associated diamonds of chiral algebras will be the same as the one of the Schoen manifold. This motivates a more general study of the automorphisms of our diamond, as these orbifolds could admit supersymmetric fibrations along directions different from the ones we have just presented. A similar analysis was performed for $\mathbb{T}^8$ orbifolds describing GCS Spin(7)-manifolds in [**29**].

We investigate which combinations of $\mathbf{T}_i$ maps preserve the diamond of fig. 3 and descend to automorphisms of the subalgebras. We find that the allowed combinations are determined by the terms in the holomorphic volume form $\widetilde{\Omega}_3$, which can be read off (4.11). Geometrically, these correspond to 3-tori of $\mathbb{T}^6$ calibrated by either $\mathrm{Re}(\widetilde{\Omega}_3)$ or $\mathrm{Im}(\widetilde{\Omega}_3)$. We have collected the automorphisms and their action on the diamond in tables 2 and 3.

Coordinates	$(2,4,6)$	$(2,3,5)$	$(1,4,5)$	$(1,3,6)$
Fr^6	$\mathbf{T}_2 \circ \mathbf{T}_4 \circ \mathbf{T}_6$	$\mathbf{T}_2 \circ \mathbf{T}_3 \circ \mathbf{T}_5$	$\mathbf{T}_1 \circ \mathbf{T}_4 \circ \mathbf{T}_5$	$\mathbf{T}_1 \circ \mathbf{T}_3 \circ \mathbf{T}_6$
$(\mathrm{Fr}^2 \oplus \mathrm{Od}_2)_+$	$\mathbf{T}_2 \circ \mathbf{M}$	$\mathbf{T}_2 \circ \mathbf{M}$	$\mathbf{T}_1 \circ \mathbf{M} \circ \mathbf{Ph}^\pi$	$\mathbf{T}_1 \circ \mathbf{M} \circ \mathbf{Ph}^\pi$
$(\mathrm{Fr}^2 \oplus \mathrm{Od}_2)_-$	$\mathbf{T}_4 \circ \mathbf{M}$	$\mathbf{T}_3 \circ \mathbf{M} \circ \mathbf{Ph}^\pi$	$\mathbf{T}_4 \circ \mathbf{M}$	$\mathbf{T}_3 \circ \mathbf{M} \circ \mathbf{Ph}^\pi$
Od_3	$\mathbf{M} \circ \mathbf{Ph}^\pi$	$\mathbf{M} \circ \mathbf{Ph}^\pi$	$\mathbf{M} \circ \mathbf{Ph}^\pi$	$\mathbf{M} \circ \mathbf{Ph}^\pi$

TABLE 2. Automorphisms of the diamond restricting to $\mathbf{M} \circ \mathbf{Ph}^\pi$ in Od_3.

Coordinates	$(2,4,5)$	$(2,3,6)$	$(1,4,6)$	$(1,3,5)$
Fr^6	$\mathbf{T}_2 \circ \mathbf{T}_4 \circ \mathbf{T}_5$	$\mathbf{T}_2 \circ \mathbf{T}_3 \circ \mathbf{T}_6$	$\mathbf{T}_1 \circ \mathbf{T}_4 \circ \mathbf{T}_6$	$\mathbf{T}_1 \circ \mathbf{T}_3 \circ \mathbf{T}_5$
$(\mathrm{Fr}^2 \oplus \mathrm{Od}_2)_+$	$\mathbf{T}_2 \circ \mathbf{M} \circ \mathbf{Ph}^\pi$	$\mathbf{T}_2 \circ \mathbf{M} \circ \mathbf{Ph}^\pi$	$\mathbf{T}_1 \circ \mathbf{M}$	$\mathbf{T}_1 \circ \mathbf{M}$
$(\mathrm{Fr}^2 \oplus \mathrm{Od}_2)_-$	$\mathbf{T}_4 \circ \mathbf{M} \circ \mathbf{Ph}^\pi$	$\mathbf{T}_3 \circ \mathbf{M}$	$\mathbf{T}_4 \circ \mathbf{M} \circ \mathbf{Ph}^\pi$	$\mathbf{T}_3 \circ \mathbf{M}$
Od_3	$\mathbf{M}$	$\mathbf{M}$	$\mathbf{M}$	$\mathbf{M}$

TABLE 3. Automorphisms of the diamond restricting to $\mathbf{M}$ in Od_3.

We find that, in addition to the automorphism associated to the SYZ fibration of the Schoen manifold, there are seven additional potential mirror maps. It would be interesting to explicitly check which ones are realized by supersymmetric fibrations of the Calabi–Yau orbifolds from [**22**].

5. Conclusion

In this note we have discussed connected sum manifolds from the point of view of their associated superconformal algebras. After introducing some necessary background, we have revisited the proposal of [**29**] that assigns a unique diamond of algebra inclusions to a connected sum manifold. This diamond reflects the piecewise structure of the manifold and provides a counterpart to the geometry at the level of chiral algebras.

Then, we have summarized the case of TCS G_2-manifolds—discussed in [**28, 29**] in a different format—before presenting new evidence for the proposal. In particular, we have described in detail the realization of the diamond associated to the Schoen Calabi–Yau manifold, studying different algebra inclusions and analysing its automorphisms. To this end, we have used the connected sum decomposition first described in [**9**].

There are several future directions that would prove interesting to pursue. First of all, it would be illuminating to test the diamond proposal on additional manifolds admitting a connected sum decomposition. The "doubling constructions" of Doi and Yotsutani [**19–21**] would be natural candidates. Note that for the case of Calabi–Yau 3-folds [**19**] this construction recovers some of the Borcea–Voisin manifolds [**5, 55**]—including the Schoen manifold—so a first step would be verifying if the Schoen diamond extends to this more general setting.

We could also follow the ideas of section 4.1 and apply different string dualities to known compactifications on connected sum manifolds. In this way we could obtain new descriptions of connected sum manifolds and their associated diamonds. For example, one could try to study the Calabi–Yau 4-fold dual to a TCS manifold via M-theory/F-theory duality [**9**].

Testing the proposal on manifolds beyond connected sums is another enticing possibility. By the arguments of section 2.3, for a manifold described by several open patches it should still be possible to identify uniquely its underlying chiral algebra in terms of the algebras associated to the open patches.

On a different note, our analysis in section 4.5 shows several candidates for mirror maps of Calabi–Yau 3-folds obtained from toroidal orbifolds [**22**]. At the moment, it is an open question whether these are realised in geometry or not.

Finally, it would be extremely interesting to relate the discussion in this note to the chiral de Rham complex [**45**]. This would allow us to make a connection between the vast literature in the subject and our diamond proposal, perhaps shedding some new light on our results.

Acknowledgments

I would like to thank my collaborator Marc-Antoine Fiset for introducing me into this line of research and for his contributions to some of the results presented here. I would also like to thank Xenia de la Ossa and Enrico Marchetto for helpful comments and discussions. Finally, I am very grateful to the organisers of String Math 2022 for putting together a fantastic conference and for the opportunity to write this contribution.

References

[1] B. S. Acharya, $N = 1$ *heterotic/M-theory duality and Joyce manifolds*, Nuclear Phys. B **475** (1996), no. 3, 579–596, DOI 10.1016/0550-3213(96)00326-4. MR1408389

[2] Bobby Samir Acharya and Edward Witten, *Chiral fermions from manifolds of G(2) holonomy*, arXiv:hep-th/0109152v1, 2001.

[3] Fabio Apruzzi, Fridrik Freyr Gautason, Susha Parameswaran, and Marco Zagermann, *Wilson lines and Chern-Simons flux in explicit heterotic Calabi-Yau compactifications*, J. High Energy Phys. **2** (2015), 183, front matter+45, DOI 10.1007/JHEP02(2015)183. MR3321270

[4] R. Blumenhagen, *Covariant construction of* $N = 1$ *super W-algebras*, Nuclear Phys. B **381** (1992), no. 3, 641–669, DOI 10.1016/0550-3213(92)90494-V. MR1181628

[5] Ciprian Borcea, *K3 surfaces with involution and mirror pairs of Calabi-Yau manifolds*, Mirror symmetry, II, AMS/IP Stud. Adv. Math., vol. 1, Amer. Math. Soc., Providence, RI, 1997, pp. 717–743, DOI 10.1090/amsip/001/28. MR1416355

[6] Andreas P. Braun, Sebastjan Cizel, Max Hübner, and Sakura Schäfer-Nameki, *Higgs bundles for M-theory on* G_2*-manifolds*, J. High Energy Phys. **3** (2019), 199, 69, DOI 10.1007/jhep03(2019)199. MR3940806

[7] Andreas P. Braun and Michele Del Zotto, *Mirror symmetry for* G_2*-manifolds: twisted connected sums and dual tops*, J. High Energy Phys. **5** (2017), 080, front matter+28, DOI 10.1007/jhep05(2017)080. MR3662851

[8] Andreas P. Braun, Suvajit Majumder, and Alexander Otto, *On mirror maps for manifolds of exceptional holonomy*, J. High Energy Phys. **10** (2019), 204, 37, DOI 10.1007/jhep10(2019)204. MR4051050

[9] Andreas P. Braun and Sakura Schäfer-Nameki, *Compact, singular* G_2*-holonomy manifolds and M/heterotic/F-theory duality*, J. High Energy Phys. **4** (2018), 126, front matter+40, DOI 10.1007/jhep04(2018)126. MR3801148

[10] Andreas P. Braun and Sakura Schäfer-Nameki, Spin(7)*-manifolds as generalized connected sums and 3d* $\mathcal{N} = 1$ *theories*, J. High Energy Phys. **6** (2018), 103, front matter+40, DOI 10.1007/jhep06(2018)103. MR3831764

[11] Volker Braun, Yang-Hui He, Burt A. Ovrut, and Tony Pantev, *A heterotic standard model*, Phys. Lett. B **618** (2005), no. 1-4, 252–258, DOI 10.1016/j.physletb.2005.05.007. MR2148973

[12] Volker Braun, Yang-Hui He, Burt A. Ovrut, and Tony Pantev, *The exact MSSM spectrum from string theory*, J. High Energy Phys. **5** (2006), 043, 14, DOI 10.1088/1126-6708/2006/05/043. MR2231497

[13] Philip Candelas, Xenia de la Ossa, Yang-Hui He, and Balázs Szendrői, *Triadophilia: a special corner in the landscape*, Adv. Theor. Math. Phys. **12** (2008), no. 2, 429–473. MR2399316

[14] Alessio Corti, Mark Haskins, Johannes Nordström, and Tommaso Pacini, *Asymptotically cylindrical Calabi-Yau 3-folds from weak Fano 3-folds*, Geom. Topol. **17** (2013), no. 4, 1955–2059, DOI 10.2140/gt.2013.17.1955. MR3109862

[15] Alessio Corti, Mark Haskins, Johannes Nordström, and Tommaso Pacini, G_2*-manifolds and associative submanifolds via semi-Fano 3-folds*, Duke Math. J. **164** (2015), no. 10, 1971–2092, DOI 10.1215/00127094-3120743. MR3369307

[16] Diarmuid Crowley, Sebastian Goette, and Johannes Nordström, *An analytic invariant of* G_2 *manifolds*, `arXiv:1505.02734`, 2015.

[17] Gottfried Curio and Dieter Lüst, *A class of* $N = 1$ *dual string pairs and its modular superpotential*, Internat. J. Modern Phys. A **12** (1997), no. 32, 5847–5865, DOI 10.1142/S0217751X97003066. MR1492334

[18] Mirjam Cvetič, Jonathan J. Heckman, Thomas B. Rochais, Ethan Torres, and Gianluca Zoccarato, *Geometric unification of Higgs bundle vacua*, Phys. Rev. D **102** (2020), no. 10, 106012, 34, DOI 10.1103/physrevd.102.106012. MR4188436

[19] Mamoru Doi and Naoto Yotsutani, *Doubling construction of Calabi-Yau threefolds*, New York J. Math. **20** (2014), 1203–1235. MR3291616

[20] Mamoru Doi and Naoto Yotsutani, *Doubling construction of Calabi-Yau fourfolds from toric Fano fourfolds*, Commun. Math. Stat. **3** (2015), no. 3, 423–447, DOI 10.1007/s40304-015-0066-x. MR3397147

[21] Mamoru Doi and Naoto Yotsutani, *Gluing construction of compact Spin(7)-manifolds*, J. Math. Soc. Japan **71** (2019), no. 2, 349–382, DOI 10.2969/jmsj/77007700. MR3943852

[22] Ron Donagi and Katrin Wendland, *On orbifolds and free fermion constructions*, J. Geom. Phys. **59** (2009), no. 7, 942–968, DOI 10.1016/j.geomphys.2009.04.004. MR2536855

[23] Tohru Eguchi and Anne Taormina, *Character formulas for the* $N = 4$ *superconformal algebra*, Phys. Lett. B **200** (1988), no. 3, 315–322, DOI 10.1016/0370-2693(88)90778-2. MR926866

[24] Sergio Ferrara, Jeffrey A. Harvey, Andrew Strominger, and Cumrun Vafa, *Second-quantized mirror symmetry*, Phys. Lett. B **361** (1995), no. 1-4, 59–65, DOI 10.1016/0370-2693(95)01074-Z. MR1362446

[25] José M. Figueroa-O'Farrill, *Extended superconformal algebras associated with manifolds of exceptional holonomy*, Phys. Lett. B **392** (1997), no. 1-2, 77–84, DOI 10.1016/S0370-2693(96)01506-7. MR1435225

[26] José M. Figueroa-O'Farrill and Stany Schrans, *Extended superconformal algebras*, Phys. Lett. B **257** (1991), no. 1-2, 69–73, DOI 10.1016/0370-2693(91)90860-S. MR1100860

[27] José M. Figueroa-O'Farrill and Stany Schrans, *The conformal bootstrap and super* W *algebras*, Internat. J. Modern Phys. A **7** (1992), no. 3, 591–617, DOI 10.1142/S0217751X92000260. MR1142948

[28] Marc-Antoine Fiset, *Superconformal algebras for twisted connected sums and* G_2 *mirror symmetry*, J. High Energy Phys. **12** (2018), 011, front matter+27, DOI 10.1007/jhep12(2018)011. MR3908231

[29] Marc-Antoine Fiset and Mateo Galdeano, *Superconformal algebras for generalized* Spin(7) *and* G_2 *connected sums*, J. High Energy Phys. **10** (2021), Paper No. 137, 29, DOI 10.1007/jhep10(2021)137. MR4339789

[30] Mateo Galdeano, *The geometry and superconformal algebras of string compactifications with a G-structure*, Ph.D. thesis, Oxford University, 2022.

[31] Sebastian Goette, Johannes Nordström, and Don Zagier, *Nu-invariants of extra-twisted connected sums*, `arXiv:2010.16367`, 2020.

[32] Rajesh Gopakumar and Sunil Mukhi, *Orbifold and orientifold compactifications of F-theory and M-theory to six and four dimensions*, Nuclear Phys. B **479** (1996), no. 1-2, 260–284, DOI 10.1016/0550-3213(96)00460-9. MR1429832

[33] Sergei Gukov, Shing-Tung Yau, and Eric Zaslow, *Duality and fibrations on* G_2 *manifolds*, Turkish J. Math. **27** (2003), no. 1, 61–97. MR1975332

[34] James Halverson and David R. Morrison, *The landscape of M-theory compactifications on seven-manifolds with* G_2 *holonomy*, J. High Energy Phys. **4** (2015), 047, front matter+33, DOI 10.1007/JHEP04(2015)047. MR3351322

[35] Jonathan J. Heckman, Craig Lawrie, Ling Lin, and Gianluca Zoccarato, *F-theory and dark energy*, Fortschr. Phys. **67** (2019), no. 10, 1900057, 27, DOI 10.1002/prop.201900057. MR4024826

[36] Shinobu Hosono, Masa-Hiko Saito, and Jan Stienstra, *On the mirror symmetry conjecture for Schoen's Calabi-Yau 3-folds*, Integrable systems and algebraic geometry (Kobe/Kyoto, 1997), World Sci. Publ., River Edge, NJ, 1998, pp. 194–235. MR1672045

[37] P. S. Howe and G. Papadopoulos, *A note on holonomy groups and sigma models*, Phys. Lett. B **263** (1991), no. 2, 230–232, DOI 10.1016/0370-2693(91)90591-D. MR1117678

[38] P. S. Howe and G. Papadopoulos, *W-symmetries of a class of $d = 2$, $N = 1$ supersymmetric sigma models*, Phys. Lett. B **267** (1991), no. 3, 362–365, DOI 10.1016/0370-2693(91)90946-N. MR1128063

[39] P. S. Howe and G. Papadopoulos, *Holonomy groups and W-symmetries*, Comm. Math. Phys. **151** (1993), no. 3, 467–479. MR1207260

[40] D. D. Joyce, *Compact 8-manifolds with holonomy* Spin(7), Invent. Math. **123** (1996), no. 3, 507–552, DOI 10.1007/s002220050039. MR1383960

[41] Dominic D. Joyce, *Compact Riemannian 7-manifolds with holonomy G_2. I, II*, J. Differential Geom. **43** (1996), no. 2, 291–328, 329–375. MR1424428

[42] Dominic D. Joyce, *Riemannian holonomy groups and calibrated geometry*, Oxford Graduate Texts in Mathematics, vol. 12, Oxford University Press, Oxford, 2007. MR2292510

[43] Alexei Kovalev, *Twisted connected sums and special Riemannian holonomy*, J. Reine Angew. Math. **565** (2003), 125–160, DOI 10.1515/crll.2003.097. MR2024648

[44] Thibault Langlais, *Analysis and spectral theory of neck-stretching problems*, `arXiv:2301.03513`, 2023.

[45] Fyodor Malikov, Vadim Schechtman, and Arkady Vaintrob, *Chiral de Rham complex*, Comm. Math. Phys. **204** (1999), no. 2, 439–473, DOI 10.1007/s002200050653. MR1704283

[46] David R. Morrison and Cumrun Vafa, *Compactifications of F-theory on Calabi-Yau threefolds. II*, Nuclear Phys. B **476** (1996), no. 3, 437–469, DOI 10.1016/0550-3213(96)00369-0. MR1412112

[47] Johannes Nordström, *Extra-twisted connected sum G_2-manifolds*, Ann. Global Anal. Geom. **64** (2023), no. 1, Paper No. 2, 80, DOI 10.1007/s10455-023-09893-1. MR4601966

[48] Satoru Odake, *Extension of $N = 2$ superconformal algebra and Calabi-Yau compactification*, Modern Phys. Lett. A **4** (1989), no. 6, 557–568, DOI 10.1142/S021773238900068X. MR1016545

[49] Satoru Odake, *Character formulas of an extended superconformal algebra relevant to string compactification*, Internat. J. Modern Phys. A **5** (1990), no. 5, 897–914, DOI 10.1142/S0217751X90000428. MR1035398

[50] Chad Schoen, *On fiber products of rational elliptic surfaces with section*, Math. Z. **197** (1988), no. 2, 177–199, DOI 10.1007/BF01215188. MR923487

[51] S. L. Shatashvili and C. Vafa, *Superstrings and manifolds of exceptional holonomy*, Selecta Math. (N.S.) **1** (1995), no. 2, 347–381, DOI 10.1007/BF01671569. MR1354601

[52] Andrew Strominger, Shing-Tung Yau, and Eric Zaslow, *Mirror symmetry is T-duality*, Nuclear Phys. B **479** (1996), no. 1-2, 243–259, DOI 10.1016/0550-3213(96)00434-8. MR1429831

[53] Kris Thielemans, *An Algorithmic approach to operator product expansions, W algebras and W strings*, Ph.D. thesis, Leuven U., 1994.

[54] Cumrun Vafa and Edward Witten, *On orbifolds with discrete torsion*, J. Geom. Phys. **15** (1995), no. 3, 189–214, DOI 10.1016/0393-0440(94)00048-9. MR1316330

[55] Claire Voisin, *Miroirs et involutions sur les surfaces $K3$* (French), Astérisque **218** (1993), 273–323. Journées de Géométrie Algébrique d'Orsay (Orsay, 1992). MR1265318

Fachbereich Mathematik, Universität Hamburg, Bundesstrasse 55, 20146, Hamburg, Germany

Email address: `mateo.galdeano@uni-hamburg.de`

Proceedings of Symposia in Pure Mathematics
Volume **107**, 2024
https://doi.org/10.1090/pspum/107/01948

Langlands duality for skein modules of 3-manifolds

David Jordan

Abstract. I introduce new Langlands duality conjectures concerning skein modules of 3-manifolds, which we have made recently with David Ben-Zvi, Sam Gunningham, and Pavel Safronov. I recount some historical motivation and some recent special cases where the conjecture is confirmed. The proofs in these cases combine the representation theory of double affine Hecke algebras and a new 1-form symmetry structure on skein modules related to electric-magnetic duality. This note is an expansion of my talk given at String Math 2022 in Warsaw, and is submitted to the String Math 2022 Proceedings publication.

1. Introduction

In his 1967 letter to André Weil, Robert Langlands proposed a mysterious conjectural correspondence – now known as Langlands reciprocity – between what are now called **automorphic** representations of a simple algebraic group G and **Galois** representations to its *Langlands dual* group ${}^LG$. Let us fix a number field $\mathbb{F}$ with ring of adeles $\mathbb{A}_\mathbb{F}$. Automorphic representations of G are, loosely speaking, certain $G(\mathbb{A}_\mathbb{F})$-representations which may be realised inside $L^2(G(\mathbb{A}_\mathbb{F})/G(\mathbb{F}))$, and hence are closely related to automorphic forms. The Langlands dual group ${}^LG$ is another simple algebraic group obtained from G by interchanging root data. Galois representations to ${}^LG$ are group homomorphisms from the absolute Galois group $\Gamma(\overline{\mathbb{F}}/\mathbb{F})$ to ${}^LG(\mathbb{C})$. Among the many remarkable features of Langlands reciprocity already evident is that the objects it relates are of *a priori* very different nature. Langlands reciprocity and its many consequences and relatives – including variants for function fields and local fields – became collectively known as **arithmetic Langlands duality**.

Decades after Langlands circulated his conjectures, Beilinson and Drinfeld discovered a new kind of Langlands duality – also conjectural – taking place between certain moduli spaces of bundles over a smooth projective curve X. Their proposal, now known as the **de Rham geometric Langlands duality**, asserts an equivalence between different algebro-geometric categories living over two such moduli spaces. On the automorphic side lies a certain category of $\mathcal{D}$-modules – i.e. systems of polynomial differential equations – on a moduli space of holomorphic G-bundles on X. On the Galois side lies a category of coherent sheaves – i.e., finite quotients of maps between vector bundles – on a moduli space of flat ${}^LG$-bundles (i.e. ${}^LG$-bundles equipped with a flat connection 1-form) on X. As in the arithmetic setting, the geometric Langlands duality relates objects of *a priori* very different nature, and bearing no elementary relationship.

The thread connecting the arithmetic and geometric conjectures passes through a deep series of analogies – known as Weil's Rosetta stone – which relate the arithmetic of function fields to the geometry of complex curves. One imagines the ring of adeles $\mathbb{A}_{\mathbb{F}}$ of a function field $\mathbb{F}$ to be a complex curve, with the primes as points, and hence regards $G(\mathbb{A}_{\mathbb{F}})$ as giving "local coordinates" of a G-bundle in the formal neighborhood of each prime; on the geometric side this corresponds to the specification of a G-bundle by its Taylor expansion at each point of a curve. One thereby situates the category of $\mathcal{D}$-modules on Bun_G on the geometric side opposite to a category of ℓ-adic sheaves on the arithmetic side; these in turn are a kind of categorification of automorphic forms, via the "sheaf-to-function" correspondence for function fields. On the other side of Langlands duality, the absolute Galois group is well-understood to be an arithmetic analog of the fundamental group of the curve.

Beilinson and Drinfeld's conjectures were later deformed in the language of twisted $\mathcal{D}$-modules by Feigin, Frenkel and Gaitsgory, following a proposal of Stoyanovsky. In their **quantum de Rham geometric Langlands duality** a new symmetry emerges which is not present in the arithmetic or geometric settings: the distinction between automorphic and Galois sides evaporates, and both moduli spaces which appear are those of holomorphic G-bundles (respectively, ${}^LG$-bundles), while the categories appearing are both categories of (Ψ- or ${}^L\Psi$-twisted, respectively) $\mathcal{D}$-modules. Here the twisting depends on a possibly infinite complex parameter $\Psi \in \mathbf{CP}^1$, and ${}^L\Psi = -1/(n_G\Psi)$, where $n_G \in \{1,2,3\}$ is the lacing number of G. The classical geometric Langlands conjecture is the special case $\Psi = \infty, {}^L\Psi = 0$.

A physical manifestion of quantum geometric Langlands duality was discovered by Kapustin and Witten, as an instance of **S-duality**, also known as Montonen–Olive duality. This duality relates certain low-energy approximations called "twists" of $\mathcal{N} = 4$ super Yang-Mills quantum field theory in four dimensions. S-duality is a sort of non-abelian Fourier transform, whose existence is predicted by $\mathcal{M}$-theory, and which can be understood as a generalisation of electric-magnetic duality in Maxwell theory. In this telling G and ${}^LG$ each appear as gauge groups of distinct theories, and the parameters $\Psi, {}^L\Psi$ each give $\mathbb{CP}^1$ charts on the space of twists. S-duality asserts an equivalence of physical theories,

$$\mathcal{Z}_{G,\Psi} \simeq \mathcal{Z}_{{}^LG,{}^L\Psi},$$

and Kapustin and Witten explained how to derive the categories appearing in both the classical and quantum geometric Langlands conjectures as categories of surface operators $\mathcal{Z}_{G,\Psi}(X), \mathcal{Z}_{{}^LG,{}^L\Psi}(X)$. This reformulation has been very influential in both mathematics and physics communities, and motivates much of the discussion which follows.

Ben-Zvi and Nadler proposed a further combinatorial incaranation of the geometric Langlands duality, now known as the **Betti geometric Langlands duality**. The key idea is to apply the Riemann–Hilbert correspondence on both sides of de Rham conjectures. On the Galois side, to a de Rham local system on X is associated its monodromy homomorphism $\pi_1(X) \to {}^LG$. On the automorphic side, to a $\mathcal{D}$-module on $\mathrm{Bun}_G(X)$ is associated its sheaf of flat sections over $\mathrm{Bun}_G(X)$. The Betti geometric Langlands duality conjectures assert an equivalence between a suitable category of coherent sheaves on the character variety – the moduli space of homomorphisms $\pi_1(X) \to {}^LG$ – and a category of sheaves of vector spaces on

$\mathrm{Bun}_G(X)$. In particular the latter is conjectured, like the former, to depend only on the topological surface Σ underlying the algebraic curve X.

The Galois side of the Betti correspondence was subsequently q-deformed in the framework of quantum groups in our works with Brochier, Ben-Zvi, Snyder and Safronov, where it was called the quantum character theory TFT, and given the structure of a fully extended four-dimensional TFT (with divergent partition function, what is sometimes called a once-categorified 3-dimensional TQFT, or a (3+1)-TQFT). The essential idea in these constructions was that the Betti Galois moduli spaces may be regarded as mapping stacks $\mathrm{Maps}(\Sigma, BG)$ into the classifying stack BG, with $\mathrm{Coh}(BG) = Rep(G)$. The stack BG has a canonical 2-shifted symplectic structure, whose deformation quantization BG_q is given algebraically by the braided tensor category $\mathrm{Rep}_q(G)$. Using the language of higher Morita theory and factorization homology, it was possible to make formal sense of a functor "$\mathrm{Maps}(\Sigma, BG_q)$", and to establish that this extends to a 4-dimensional TQFT with divergent partition function.

The purpose of this note is to share a recent conjecture we have made jointly with David Ben–Zvi, Sam Gunningham and Pavel Safronov, which envisions Langlands duality between the skein modules of closed oriented 3-manifolds. The G-skein module $\mathrm{Sk}_{G,q}(M)$ of a closed, oriented 3-manifold is the formal linear span of G-labelled ribbon graphs, modulo local relations modelled on the ribbon braided tensor category $\mathrm{Rep}_q(G)$ of representations of the quantum group $U_q\mathfrak{g}$; skein categories are defined in a similar spirit (see Section 2 for precise definitions and some examples). In its simplest form, our conjecture states:

CONJECTURE 1.1. *Let G be a semisimple algebraic group, and let ${}^LG$ denote its Langlands dual group. Let M be a closed, oriented 3-manifold, suppose that $\Psi \in \mathbb{C}^\times$ is transcendental, and let $q = e^{i\Psi}$ and ${}^Lq = e^{i^L\Psi}$. Then we have a linear isomorphism,*

$$\mathrm{Sk}_{G,q}(M) \cong \mathrm{Sk}_{{}^LG,{}^Lq}(M).$$

See Section 4 for the statement of various refinements and strengthenings of the conjecture, and Section 5 for proofs in a few special families of examples.

REMARK 1.2. It was conjectured by Edward Witten and proved in our work with Gunningham and Safronov that the skein modules are in fact finite-dimensional under the above assumptions, so that our conjecture can be rephrased as an equality of natural numbers:

$$\dim \mathrm{Sk}_{G,q}(M) = \dim \mathrm{Sk}_{{}^LG,{}^Lq}(M).$$

Moreover our method of proof established that the dimension is independent of q so long as it is taken to be transcendental (it is expected to suffice to require only that q is not root of unity). Hence the precise specification of parameters q and ${}^Lq$ in the conjecture is somewhat superfluous, and is included only to give the reader a sense for the parallel with de Rham quantum geometric Langlands conjectures.

In particular, to confirm the conjecture as stated it is enough to compute the generic dimension of the skein module for G and for ${}^LG$. In fact, all the cases where we can confirm the conjecture are by independently computing both dimensions. Because of the transcendental relationship between q and ${}^Lq$, a canonical isomorphism of vector spaces (e.g. one intertwining the action of the diffeomorphism group on each side) is likely to depend analytically, rather than algebraically, on the parameter q.

The following table summarises the many instances of Langlands duality recounted so far. Note the symmetry shared between the quantum de Rham and quantum skein module formulations.

Shorthand	Object of study	Automorphic/A side		Galois/B side
Arithmetic	Number field F	$\{V \subseteq L^2(G(\mathbb{A}_{\mathrm{F}})/G(\mathbb{F}))\}$	$\leftrightarrow$	$\{\rho : \Gamma(\overline{\mathrm{F}}/\mathrm{F}) \to {}^LG\}$
Classical				
de Rham	Complex curve X	$\mathcal{D}(\mathrm{Bun}_G(X))$	$\leftrightarrow$	$\mathrm{Coh}(\mathrm{Loc}_{{}^LG}(X))$
Betti	Real surface Σ	$\mathrm{Shv}(\mathrm{Bun}_G(\Sigma))$	$\leftrightarrow$	$\mathrm{Coh}(\mathrm{Ch}_{{}^LG}(\Sigma))$
Quantum				
de Rham	Complex curve X	$\mathcal{D}_\Psi(\mathrm{Bun}_G(X))$	$\leftrightarrow$	$\mathcal{D}_{{}^L\Psi}(\mathrm{Bun}_{{}^LG}(X))$
Betti	Real surface Σ	$\mathrm{Shv}_q(\mathrm{Bun}_G(\Sigma))$	$\leftrightarrow$	$\mathrm{Coh}_q(\mathrm{Ch}_G(\Sigma))$
Skein	3-manifold	$\mathrm{Sk}_{G,q}(M)$	$\cong$	$\mathrm{Sk}_{{}^LG,{}^Lq}(M)$

The most basic motivation for the conjecture is that the ${}^LG$-skein module of a 3-manifold M specialises at $q = 1$ to the algebra of functions on the ${}^LG$-character variety of M, and the skein category of a surface Σ, specialises at $q = 1$ to the ${}^LG$-character stack of a surface Σ. Indeed, a ${}^LG$-labelled ribbon graph defines in a straightforward way a holomorphic function on the ${}^LG$-character variety, essentially by sending a ${}^LG$-local system E to the trace of an associated bundle living along the support of the graph: these functions, and their deformations, are called Wilson loop observables. It is then natural to hope – though far from automatic – that the one-parameter family of quantum deformations given by the skein module indeed coincides with the one-parameter family of twists in the Kapustin–Witten theories. The rich structure of extended topological field theory enjoyed by skein modules – they are expected to coincide with the value of the quantum character theory TFT on 3-manifolds, and hence to describe the Kapustin–Witten state space – as well as their natural interpretation as q-deformed Wilson loop operators lends support for this hope.

Another source of motivation for our conjecture comes from the deep analogies of Mazur, Kapranov, and Reznikov, relating 3-manifold topology to algebraic number theory, and from the subsequent development of *arithmetic quantum field theory*. In this "MKR dictionary" – which builds on Weil's Rosetta stone – one imagines a number field (more precisely its ring $\mathbb{O}$ of integers) to be a closed 3-manifold; one treats ideals in $\mathbb{O}$ as links, and prime ideals as knots. Units in $\mathbb{O}$ are treated as embedded surfaces, field extensions are treated as branched covers, homology groups as ideal class groups, etc. Building on this dictionary, the impetus of arithmetic field theory is to compute arithmetic invariants of number fields as if they were partition functions for a quantum field theory. This idea was explored first by Kim, who formulated an arithmetic analog of 3D Chern-Simons TQFT as a mechanism for constructing L-functions, and taken up more recently by Ben-Zvi, Sakellaridis, and Venkatesh who have proposed to study various aspects of arithmetic Langlands duality using ideas from Kapustin and Witten's 4D TQFT.

It is perhaps remarkable in hindsight given the MKR dictionary that our conjecture so significantly post-dates the geometric Langlands conjectures. Just as skein theory attaches a vector space to each 3-manifold, arithmetic Langlands duality attaches vector spaces – the space of automorphic forms, and the space of algebraic functions on the arithmetic character variety, on each side of the duality

– to each number field. The geometric Langlands dualities differ in two key respects: they involve complex curves (↔ function fields) as opposed to 3-manifolds (↔ number fields), and they replace vector space-valued invariants with categorical ones. This of course makes Langlands duality for skein modules more elementary to formulate, and to falsify, than geometric Langlands duality, since it reduces to a statement about equality of integer dimensions rather than about equivalences of ∞-categories. From a more physical perspective, it is interesting to note that Kapustin and Witten's expressed quantum geometric Langlands duality for *2-dimensional manifolds* using a *four-dimensional* QFT, but largely left aside the intervening question of mathematically describing the Hilbert spaces attached to *3-manifolds*.

This sense of anachronism can be partly explained by posing the following two natural and (to our knowledge) unanswered questions concerning Langlands duality in the arithmetic and physical settings. The answers to each question would fill in some more of the dots between our conjectures and the canon of Langlands duality which I have surveyed above. On the side of arithmetic:

QUESTION 1.3. What sense, if any, can be made of a **quantum arithmetic Langlands duality**?

In particular, what role does the symplectic structure on character varieties, and its deformation quantization, play in number theory, and what is the arithmetic meaning of the deformation parameter q? On the physical side:

QUESTION 1.4. What is the Hilbert space attached to a closed oriented 3-manifold by the Kapustin–Witten A-side twist at $\Psi = 0$?

The answer to this question would sit across Langlands duality from the ${}^LG$-character variety of M, and should have an A-side flavour involving symplectic/contact geometry, Fukaya categories, Floer theory, etc. One may contemplate possible answers either by degenerating our understanding for generic Ψ (as a skein module), or by extrapolating up in dimension from the case of surfaces, where the A-side twist at $\Psi = 0$ is a category of $\mathcal{D}$-modules on $\mathrm{Bun}_G(X)$.

One among many complications which arise when contemplating this A-side twist at $\Psi = 0$ for 3-manifolds is that while S-duality predicts an equivalence of theories $\mathcal{Z}_{G,\Psi} \simeq \mathcal{Z}_{{}^LG,{}^L\Psi}$, it nevertheless *interchanges* the Dirchlet and Nahm pole boundary conditions. The skein module is, in some sense by definition, the orbit of the Dirichlet boundary condition by the Wilson loop operators of the theory; and for irrational non-zero Ψ (equivalently, q not a root of unity) we anticipate that these orbits will coincide. We however do not expect such a coincidence at $\Psi = 0, \infty$, and therefore one must formulate the Nahm pole boundary condition on 3-manifolds in mathematical terms.

References. Many of the conjectures recalled in this introduction were circulated first as private correspondence, were never published, or were published only much later as folklore. For this reason, I have avoided in-line citations throughout the introduction, even in those cases which appeared promptly in print, such as Kapustin and Witten's paper [**22**] on electric-magnetic duality and Ben-Zvi and Nadler's paper [**7**] introducing Betti geometric Langlands conjectures. For a more thorough recounting of Langlands duality in its arithmetic, geometric, and physical manifestions, and for an introduction to arithmetic topology, I recommend:

- *The genesis of the Langlands program* [**27**], an LMS Lecture series which covers Robert Langlands's biography, his mathematics, and his letter to Andre Weil,
- *Between electric-magnetic duality and the Langlands program* [**8**], an unpublished series of lecture notes, on the relations between geometric Langlands, arithmetic Langlands, and electric-magnetic duality.
- *Knots and primes* [**26**], a Springer textbook introducing the foundations of arithemetic topology.

Acknowledgements. I thank my collaborators David Ben-Zvi, Sam Gunningham, Pavel Safronov, Monica Vazirani, and Haiping Yang for our many discussions surveyed here, and for their encouragement to share our conjectures and work in progress in this forum. I also thank Dan Freed, Du Pei and Edward Witten for several enlightening discussions and correspondences regarding electric-magnetic duality, and Minhyong Kim for inspiring discussions about arithmetic quantum field theory. Finally, I thank the organisers of String Math 2022 for a great conference, and for the opportunity to prepare these notes.

2. Quantum groups and their skein theory

The construction of skein modules and skein categories depends on the specification of a cp-rigid ribbon braided tensor category[1] $\mathcal{A}$. This means that $\mathcal{A}$ is a locally presentable abelian category with enough compact projective objects, equipped with a bilinear monoidal functor $\otimes : \mathcal{A} \times \mathcal{A} \to \mathcal{A}$, a braiding natural isomorphism σ, and a ribbon element ν, in which all compact projective objects are dualizable.

The idea will be that we can encode all the algebraic structure of a cp-rigid ribbon braided tensor category in a diagrammatic form, using the *graphical calculus* of Reshetikhin and Turaev. These diagrams and all relations between them take place in a decorated cylindrical ball $D^2 \times I$, but we can then allow our diagrams to permeate an entire oriented 3-manifold, imposing only locally the relations coming from the graphical calculus in some ball. The resulting vector space is called the skein module. We may additionally allow these diagrams to end at boundary surface components of our 3-manifold; this leads to the notion of a skein category, and of the induced functor between skein categories.

2.1. Typical examples of ribbon braided tensor categories. Our main examples of interest will be the categories $\mathrm{Rep}_q(G)$ attached[2] to an algebraic group G at transcendental values of the parameter q. This means we consider the category of locally finite-dimensional representations of the quantum group $U_q(\mathfrak{g})$ attached to the Lie algebra $\mathfrak{g}$ of G, and within it we consider a tensor subcategory consisting of those $U_q(\mathfrak{g})$-representations, all of whose weights lie in the weight lattice of G. We recall that the weight lattice of G lies between those of the adjoint and the simply connected group of $\mathfrak{g}$, and hence is determined by specifying $Z(G)$, the quotient of

[1] For experts: equivalently, we may consider instead the subcategory of compact projective objects, which will be a (not-necessarily-unital!) ribbon braided monoidal tensor category in which all objects are dualizable. We note that unless $\mathcal{A}$ is semisimple, the compact-projective objects do not form an abelian category, but instead an additive idempotent complete category.

[2] More precisely the specification of ribbon structure depends on choosing a dth root of q, where d is the determinant of the Cartan matrix, but we will suppress this point from the notation.

the weight lattice of G by the root lattice; or equivalently by $\pi_1(G)$, the quotient of the coweight lattice by the coroot lattice of G.

EXAMPLE 2.1. We will treat the examples $\mathrm{Rep}_q(\mathrm{SL}_2)$ and $\mathrm{Rep}_q(\mathrm{PGL}_2)$ in detail throughout these notes. We recall that $\mathrm{Rep}_q(\mathrm{SL}_2)$ is a semisimple category with simples $\mathrm{V}(0), \mathrm{V}(1), \mathrm{V}(2), \ldots$ with each $\mathrm{V}(k)$ having dimension $k+1$ and highest weight k. The category $\mathrm{Rep}_q(\mathrm{PGL}_2)$ is the full subcategory of $\mathrm{Rep}_q(\mathrm{SL}_2)$ consisting of direct sums of the irreducible representations $\mathrm{V}(0), \mathrm{V}(2), \mathrm{V}(4), \ldots$.

Let us recall some further standard algebraic facts about these two categories. First, in the case of $\mathrm{Rep_q}(\mathrm{SL}_2)$, we recall that the defining two-dimensional representation $\mathrm{V}(1)$ is a *tensor generator*: every representation $\mathrm{V}(k)$ appears as a direct summand in the kth tensor power $\mathrm{V}(1)^{\otimes k}$. Secondly, we have that the Hom spaces $\mathrm{Hom}(\mathrm{V}(1)^{\otimes k}, \mathrm{V}(1)^{\otimes \ell})$ are spanned by repeated compositions of the evaluation and coevalution maps depicted in the first term of the RHS below. In particular, we can express the braiding in terms of the evaluation, coevaluation and identity maps:

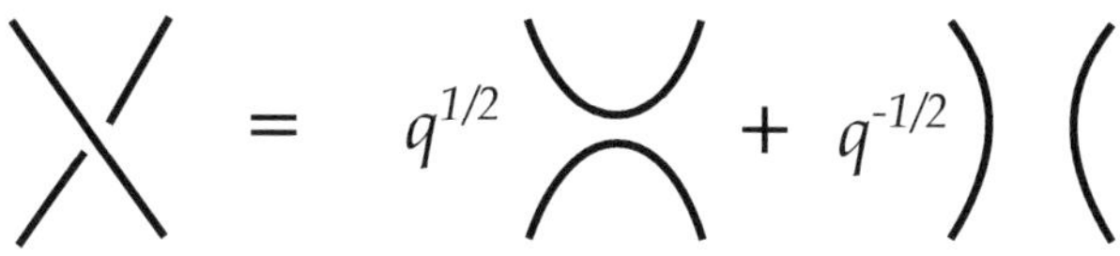

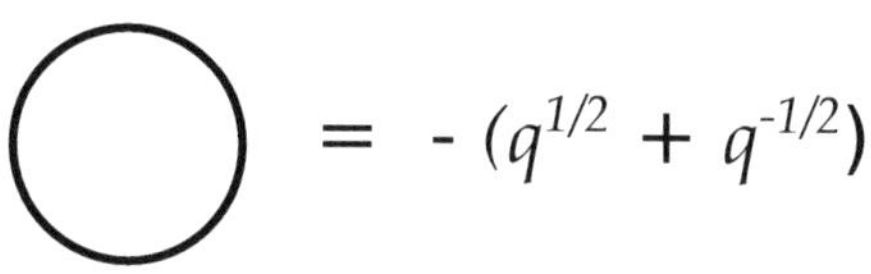

We have omitted the source and target $\mathrm{V}(1)$ from the generators. Let us note that the evaluation and coevaluation ought usually to involve $\mathrm{V}(1)^*$, but that we have identified $\mathrm{V}(1) \cong \mathrm{V}(1)^*$. There are two natural candidates for such isomorphisms, and it is customary to favor the skew-symmetric pairing, hence the negative sign in the formula for the value of the circle, i.e. the quantum dimension of $\mathrm{V}(1)$.

EXAMPLE 2.2. For $\mathrm{Rep_q}(\mathrm{PGL}_2)$, the situation is similar, but somewhat more complicated. It is again easy to show that $\mathrm{Rep_q}(\mathrm{PGL}_2)$ is tensor generated by the generator $\mathrm{V}(2)$. Recalling the isomorphism $\mathrm{V}(2) \otimes \mathrm{V}(2) \cong \mathrm{V}(0) \oplus \mathrm{V}(2) \oplus \mathrm{V}(4)$, we now have a three-dimensional space $\mathrm{End}(\mathrm{V}(2) \otimes \mathrm{V}(2))$, spanned by the three diagrams on the RHS below (corresponding to the projection onto the $V(2)$ component, onto the $V(0)$ component, and the identity morphism, respectively). In particular, we may express the braiding:

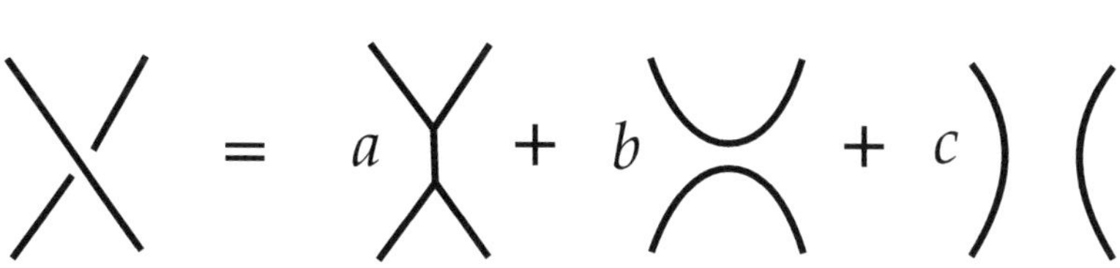

for suitable constants $a, b, c \in \mathbf{C}(q)$ which can be determined by direct computation in the given basis. Here again we have omitted the label $\mathrm{V}(2)$ from the generators.

EXAMPLE 2.3. For the family of groups $G = \mathrm{SL}_N$, we also have that $\mathrm{Rep_q}(\mathrm{SL}_N)$ is tensor generated by the defining representation $\mathrm{V}(\omega_1)$ associated to the first fundamental weight of SL_N. It is possible to give a similar explicit diagrammatic description for Hom spaces of tensor powers of $\mathrm{V}(\omega_1)$, using either MOY calculus [**28**] or the spiders/webs descriptions of [**12, 25**], to which we refer for details.

2.2. Graphical calculus, skein modules and skein categories. The key fact we will use about ribbon braided tensor categories is that they admit a *graphical calculus*. We have already seen some examples of graphical calculus above; let us recall what it means more precisely.

DEFINITION 2.4 (Sketch. See Figure 1 and [**13**, Section 4.2]). Let Σ be an oriented surface.

- An **$\mathcal{A}$-labeling of** Σ is the data, X, of an oriented embedding of finitely many disjoint disks $x_1, \ldots, x_n \colon \mathbb{D}^2 \to \Sigma$ labeled by objects $V_1, \ldots, V_n$ of $\mathcal{A}$. We denote by $\vec{x_i}$ the x axis sitting inside each disk x_i, and denote $\vec{X} = \cup_i \vec{x_i}$.
- A ribbon graph has "ribbons" connecting "coupons". A coupon is simply a zero-simplex in $\Sigma \times I$, equipped with the choice of a normal vector. A ribbon is an embedded 1-simplex in $\Sigma \times I$, whose endpoints each lie either in the set X, or at some coupon, and which is equipped with a normal vector along it. We furthermore require that the tangent vector of each embedded simplex points directly up when ending at $X \times \{0, 1\}$ or else in the direction of the coupon's normal vector when ending at a coupon. Finally, we require a cyclic ordering of the incoming and outgoing ribbons at each coupon.
- An **$\mathcal{A}$-coloring** of a ribbon graph is a labelling of each ribbon by an object of $\mathcal{A}$, and of each coupon by a morphism from the (ordered) tensor product of incoming edges to the (ordered) tensor product of outgoing edges.
- We say that an $\mathcal{A}$-colored ribbon graph Γ is **compatible with an $\mathcal{A}$-labeling** if $\partial\Gamma = \vec{X}$, and denote by $\mathrm{Rib}_{\mathcal{A}}(M, X)$ the vector space with basis the $\mathcal{A}$-colored ribbon graphs on M compatible with X.

REMARK 2.5. The graphical calculus for ribbon tensor categories has appeared in the literature with many differing but essentially equivalent formulations. The formulation we present here is adapted from our presentation in [**17**], however the customary "ribbons" and "coupons" formulation has been replaced with a topologically equivalent formulation which will be more compatible with simplicial homology arising later in the paper.

Let us now describe the defining relations of skein modules. Consider the 3-ball written as the cylinder $\mathbb{D}^2 \times I$ over the 2-disk $\mathbb{D}^2$, and fix a labeling $X \cup Y$ with disks $X = (x_1, V_1), \ldots, (x_n, V_n)$ embedded in $\mathbb{D}^2 \times \{0\}$ and $Y = (y_1, W_1), \ldots (y_m, W_m)$ embedded in $\mathbb{D}^2 \times \{1\}$. Then we have a well-defined surjection,

$$\mathrm{Rib}_{\mathcal{A}}(\mathbb{D}^2 \times I, X \cup Y) \to \mathrm{Hom}_{\mathcal{A}}(V_1 \otimes \cdots \otimes V_n, W_1 \otimes \cdots \otimes W_m),$$

called the Reshetikhin–Turaev evaluation map, see [**32**]. In essence this map takes a ribbon graph in a cylinder, and a generic projection, and interprets each crossing as a braiding, and each coupon as a morphism between each tensor product, and computes the corresponding composition: that this is well-defined independent of

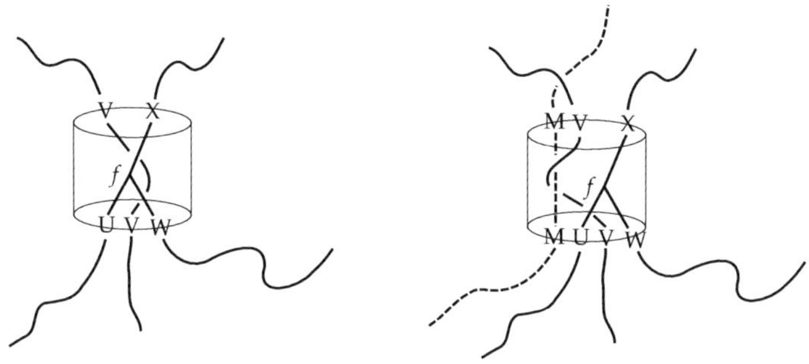

FIGURE 1. At left: the skein module is spanned by $\mathcal{A}$-labelled ribbon diagrams; relations; the skein relations state that the diagram within the cylinder can be replaced by any other diagram whose RT-evaluation gives the same element of $\mathrm{Hom}(U \otimes V \otimes W, V \otimes X)$. At right: a defect is introduced along the dotted line, and a braided module category is inserted along it (see Section 3). The skein module now includes labels and relations from the braided tensor category away from the defect, and from its braided module category along the defect.

crossings is perhaps the most fundamental algebraic property of ribbon categories. We will call the kernel of this map the **skein relations** between X and Y.

DEFINITION 2.6. Let M be an oriented 3-manifold equipped with a decomposition of its boundary $\partial M \cong \mathrm{rev}\,\Sigma_{in} \coprod \Sigma_{out}$, and $\mathcal{A}$-labelings X_{in} of Σ_{in} and X_{out} of Σ_{out}.

- The **relative $\mathcal{A}$-skein module** $\mathrm{Sk}_{\mathcal{A}}(M, X_{in}, X_{out})$ is the vector space spanned by isotopy classes of $\mathcal{A}$-colored ribbon graphs in M compatible with $X_{in} \cup X_{out}$, taken modulo isotopy and the skein relations between X_{in} and X_{out} determined by any oriented ball $\mathbb{D}^2 \times I \subset M^3$.
- When $\partial M = \emptyset$ (hence $\partial\Gamma = \emptyset$), we call this the **$\mathcal{A}$-skein module**, and denote it by $\mathrm{Sk}_{\mathcal{A}}(M)$.

DEFINITION 2.7. Let Σ be an oriented surface. The **skein category** $\mathrm{SkCat}_{\mathcal{A}}(\Sigma)$ of Σ has:

- As its objects, $\mathcal{A}$-labelings of Σ.
- As the 1-morphisms from X to Y the relative $\mathcal{A}$-skein module of $(\Sigma \times [0,1], X, Y)$.

The composition of 1-morphisms is by concatenation of relative skeins in the evident sense.

EXAMPLE 2.8. In the cases $\mathcal{A} = \mathrm{Rep_q}(\mathrm{SL}_2)$ and $\mathcal{A} = \mathrm{Rep_q}(\mathrm{PGL}_2)$ the rather intricate definition of the skein module involving labelled ribbon graphs simplifies considerably, owing to the simplifications highlighted in Example 2.1. In both cases, one can dispense with labels and orientations on strands: for labels, one takes $\mathrm{V}(1)$ or $\mathrm{V}(2)$, respectively as the label always (and thus omits it), and one further applies the self-duality of $\mathrm{V}(1)$, respectively $\mathrm{V}(2)$ to eliminate orientation data.

[3]Here we assume without loss of generality that $\mathbb{D}^2 \times \{0\} \subset \Sigma \times \{0\}$ and $\mathbb{D}^2 \times \{1\} \subset \Sigma \times \{1\}$.

For coupons, in the $\mathrm{Rep_q}(\mathrm{SL}_2)$ case, Schur Weyl duality implies that all morphisms can be expressed in terms of the braidings, so no coupons are necessary at all. For $\mathrm{Rep_q}(\mathrm{PGL}_2)$, we require only the trivalent coupon from Example (2.2).

2.3. Relations to deformation quantization. The starting point for studying skein modules using geometric representation theory comes from the following description of skein modules and skein categories at $q = 1$.

DEFINITION 2.9. The G-character stack $\underline{\mathrm{Ch}}_G(X)$ of a topological space X is the moduli stack of flat G-bundles on X, equivalently, it is the quotient stack obtained by quotienting the G**-representation variety** $\mathrm{Hom}(\pi_1(X), G)$, by the conjugation action of G. The G**-character variety** $\mathrm{Ch}_G(X)$ is the GIT quotient, i.e. the set of closed G-orbits on the representation variety.

The following theorem follows[4] from the main results of [**6**, **13**].

THEOREM 2.10. *The idempotent completion (respectively, free cocompletion) of the skein category* $\mathrm{SkCat}_{G,q=1}(\Sigma)$ *is naturally equivalent to the category of vector bundles (respectively quasi-coherent sheaves) on the G-character stack* $\underline{\mathrm{Ch}}_G(\Sigma)$*)*.

Moreover, in the case $G = \mathrm{SL}_2$, the endomorphism algebra of the empty skein object (known as the skein algebra) is a flat deformation of the algebra of functions on the SL_2-character variety of the surface. A similar statement is expected for an arbitrary simple group G. Here, flatness in the parameter q means that Hom spaces defined in $\mathrm{SkCat}_G(\Sigma)$ are actually free $\mathbf{C}[q, q^{-1}]$-modules; informally, one says then that the vector space supporting the skein module is "constant" in q and that only the algebra structure changes as a function of q: in such cases, we obtain a Poisson bracket on the $q = 1$ limit, and we say that the skein algebra/skein category is a deformation quantization of the character variety/character stack. The resulting Poisson bracket quantizes the Atiyah–Bott/Goldman/Fock-Rosly Poisson bracket (a single Poisson bracket on the character variety independently constructed by each grouping of authors in three different ways). One has a similar interpretation for the $q = 1$ limit of $\mathrm{Sk}_G(M)$ for any 3-manifold. The following is proved for $G = \mathrm{SL}_2$ in [**10**, **31**], but is expected to hold for any group G.

THEOREM 2.11. *The skein module* $\mathrm{Sk}_{\mathrm{SL}_2,q=1}(M)$ *of an oriented 3-manifold* M *is naturally isomorphic to the algebra* $\mathcal{O}(\mathrm{Ch}_{\mathrm{SL}_2}(M))$ *of algebraic functions on the* SL_2*-character variety of* M.

However, when we move the quantum parameter q away from one, skein modules behave very differently than skein categories in elementary terms. While skein categories of surfaces (more precisely the relative skein algebras which appear as their Hom spaces) depend *flatly* on the quantum parameter q, quite the opposite is true for skein modules of closed 3-manifolds. We have noted that at $q = 1$ the skein module of a closed 3-manifold may be identified with the algebra of algebraic functions on the G-character variety of M – this is in particular typically an infinite-dimensional vector space. By contrast, we have

[4]More precisely, Cooke proves that the assignment $\mathrm{SkCat}_{G,q}(\Sigma)$ satisfies *excision*, a universal cutting and gluing formula which characterises factorization homology, while Ben-Zvi, Francis and Nadler proved that the assignment $\Sigma \mapsto \mathcal{QC}(\underline{\mathrm{Ch}}_G(\Sigma))$ satisfies the same universal property of excision. Hence the constructions coincide, noting that by construction both assign the category $\mathrm{Rep}(G)$ to the disk, $\mathbb{D}^2$.

THEOREM 2.12 ([**17**]). *Let $q \in \mathbb{C}^\times$ be generic, and let M be a closed, oriented 3-manifold. Then the G-skein module of M is finite-dimensional.*

This statement first appeared as a conjecture of Edward Witten in 2014, and came largely as a surprise to the skein theory community: because of the many known results asserting that skein algebras and skein categories of surfaces deform flatly, it was generally expected that skein modules of 3-manifolds should as well. With the benefit of hindsight and modern approaches to deformation quantization – specifically the framework of shifted symplectic structures [**24**], [**29**] – the finiteness asserted in the theorem is not at all surprising.

The character stacks of surfaces, which are deformed by skein categories, are symplectic (what is now called "0-shifted symplectic"), where as the character stacks of 3-manifolds have a very different structure: they are what is now called (-1)-shifted symplectic. One way to understand (-1)-shifted symplectic stacks is by a shifted analog of Darboux coordinates: whereas Darboux coordinates equip a 0-shifted symplectic stack with a local symplectomorphism to the unique symplectic vector space in that dimension, Darboux coordinates for (-1)-shifted stack identify the neighborhood of any point with the intersection of two Lagrangian subspaces of the symplectic vector space.

The finite-dimensionality of skein modules can be understood in this context from the fact that the notion of deformation quantization for 0-shifted and (-1)-shifted symplectic structures is very different in nature. Whereas the familiar notion for 0-shifted structures involves a non-commutative algebra, with a flatness criterion on the underlying vector space of the quantization, the quantization of a (-1)-shifted symplectic structure defines a perverse sheaf on the classical variety, whose global sections describes the deformation. There are (at least) two general avenues to construct this perverse sheaf: an algebraic construction of deformation quantizations and their tensor products, following Kashiwara and Schapira [**23**], or a geometric construction following [**3**], [**30**].

In the algebraic construction, we suppose that the (-1)-shifted structure really does arise as the intersection of two Lagrangian subvarieties L_1, L_2, and that the symplectic variety X has a deformation quantization into some non-commutative algebra A (or sheaf of algebras), and that the Lagrangians each quantize to holonomic (left, and right, respectively) modules M_1, M_2 over A. In this case, Kashiwara and Schapira show that the non-commutative intersection – i.e. the relative tensor product $M_1 \otimes_A M_2$ – defines a perverse sheaf over the classical intersection $L_1 \cap L_2$. A careful analysis of the sections of this perverse sheaf in the setting of skein modules yielded the finiteness result of [**17**].

A related geometric construction of Abouzaid and Manolescu [**1**] of complexified instanton Floer homology also hinges on the construction of the character stack of a 3-manifold, as an intersection of two standard Lagrangians in the character stack of a surface of some genus g. The Brav–Bussi–Ben-Bassat–Joyce–Pridham quantization of such a (-1)-shifted symplectic stack is controlled by a perverse sheaf of *vanishing cycles*, and in good situations this will have, indeed, a finite dimensional space of global sections.

REMARK 2.13. Work in progress by Gunningham and Safronov gives an identification of the Kashiwara–Schapira sheaf and the Brav–Bussi–Ben-Bassat–Joyce sheaf in great generality, including in our setting of skein modules, thus unifying the two approaches.

We might be tempted to understand the finiteness obtained in these two constructions as follows: generically, two Lagrangian submanifolds should intersect transversely, but of course in practice they may not. However, when we quantize, even if the classical Lagrangians fail to intersect transversely – giving a positive dimensional character variety, and hence an infinite-dimensional algebra of functions – at the quantum level, and at generic parameters, they exhibit the generic behavior.

Both the Kashiwara–Schapira and the Abouzaid–Manolescu quantizations are by construction not only vector spaces but chain complexes; we have so far only constructed on their cohomology in degree zero. This suggests that the skein module also has a natural derived enhancement, with finite-dimensional cohomologies in each degree, and satisfying a Langlands duality akin to 1.1. We will not discuss this derived enhancement in the present note.

2.4. Skein topological field theory. Given a 3-manifold M with oriented boundary $\partial M = \Sigma_+ \sqcup \Sigma_-$, we obtain a functor,

$$\mathrm{Sk}_{\mathcal{A}}(M) : \mathrm{SkCat}(\Sigma_-)^{op} \otimes \mathrm{SkCat}(\Sigma_+) \to \mathrm{Vect},$$

called the skein bimodule of M, which assigns to a pair of skein objects the relative skeins in M compatible with the given marking at the boundary.

THEOREM 2.14 (Walker, unpublished). *The assignments of surfaces and 3-manifolds to their skein category and skein (bi)modules, respectively, defines a functor from the category* Cob^3 *of surfaces and 3-cobordisms, to the category of categories and categorical bimodules.*

It was shown in [**13**] that skein categories satisfy a universal property called excision, which implies that the skein category coincides with the categorical factorization homology of Ayala–Francis [**2**], as was studied in this context in [**4, 5**]. The excision property for skein modules should be regarded as expressing the existence of an extension of skein theory all the way down to the point: the $(3,2)$-TFT axioms allow us to cut 3-manifolds along surfaces, and excision allows us to cut surfaces up along embedded 1-manifolds, and then those 1-manifolds up along 0-manifolds.

However, while it is known that skein modules define a $(3,2)$-TFT after Walker, and that excision for skein categories implies a $(2,1,0)$-TFT via factorization homology, I am not aware of a rigorous construction of the desired $(3,2,1,0)$-TFT in the framework of fully extended topological field theories. In [**9**], we have constructed a $(3,2,1,0)$-TFT from the same input, of a cp-rigid braided tensor category, however it remains to identify the value of this theory on 3-manifolds with the skein module, as one expects. We will nevertheless largely pass notions between the skein-theoretic and fully extended formulations of the theory in Sections 3.

3. Gradings and twists of skein modules

It is an elementary observation that the ribbon braided tensor category $\mathrm{Rep}_q(G)$ is naturally graded (as an abelian monoidal category) by the Pontryagin dual $Z(G)^\vee$ of the centre $Z(G)$ of the group G: the central elements of G define grouplike central elements of the quantum group $U_q(\mathfrak{g})$, and by Schur's lemma, any simple representation uniquely determines a character of the centre. The group-like property implies that the representations tensor compatibly with the group structure.

Because the skein relations all take place locally in a ball, the central grading on $\mathrm{Rep}_q(G)$ extends naturally to a grading by $H_1(M, Z(G)^\vee)$ on the G-skein module: given a G-labelled ribbon graph, we obtain a simplicial 1-cycle by replacing the label along each 1-simplex with its grading by $Z(G)^\vee$. We note that we always obtain a cycle in this way, since at coupons, there can only be a nonzero morphism if the sum of degrees is zero. If the manifold has boundary, and the skein is defined relative to the boundary, then the 1-cycle will give a relative $Z(G)^\vee$ 1-cycle, by similar reasoning. This assignment gives the grading of the G-skein module by $H_1(M, Z(G)^\vee)$, and this grading will play an important role in what follows.

We will see that skein modules carry another structure: they can be twisted by braided module categories inserted along codimension-two defects. The grading and twisting constructions are dual in a precise sense, which we will soon see. For now let us outline the construction of twisting of skein modules: the complete details are presented in the forthcoming [**18**], so we will be somewhat brief.

DEFINITION 3.1. A braided module category $\mathcal{M}$ for a braided tensor category $\mathcal{A}$ is a module category $\mathcal{M}$ for $\mathcal{A}$, together with a 'monodromy" natural automorphism,

$$\alpha : (X, m) \mapsto (\alpha_{X,m} : X \otimes m \to X \otimes m),$$

of the action functor of $\mathcal{A}$ on $\mathcal{M}$, which satisfies:

$$\alpha_{X\otimes Y,m} = (\sigma_{Y,X}\sigma_{X,Y} \otimes \mathrm{id}_M) \circ \alpha_{X,Y\otimes M} \circ (\mathrm{id}_X \otimes \alpha_{Y,M}).$$

REMARK 3.2. To motivate this definition, we should imagine that the braided tensor structure on $\mathcal{A}$ is encoded by the E_2-operad of little disks (see Figure 2), and that $\mathcal{M}$ sits at a single point (a co-dimension two defect) in the plane. The module action comes from bringing $\mathcal{A}$-disks to $\mathcal{M}$, and the natural isomorphism α is indeed the monodromy obtained by sliding those disks once counterclockwise around $\mathcal{M}$ before colliding.

DEFINITION 3.3. The E_2-monoidal bicategory $\mathrm{BrMod}(\mathcal{A})$ has as objects braided module categories, as its 1-morphisms braided module functors (i.e. module functors (F, J) satisfying an evident compatibility with α), and as 2-morphisms the braided module natural transformations. The E_2-monoidal structure is given by sending the pair $\mathcal{M}_1$, $\mathcal{M}_2$ to the relative tensor product $\mathcal{M}_1 \otimes_{\mathcal{A}} \mathcal{M}_2$, with the resulting holonomy natural automorphism given as the composition of holonomies.

We will say that the braided module category $\mathcal{M}$ is dualizable if it is fully dualizable as an object of $\mathrm{BrMod}(\mathcal{A})$.

Given a fully dualizable braided module category, and a knot K in a 3-manifold M, we may define the skein module, $\mathrm{Sk}_{\mathcal{M}}(M, K)$ as the formal span of all $(\mathcal{A}, \mathcal{M})$-labelled ribbon graphs, modulo $(\mathcal{A}, \mathcal{M})$-skein relations. Here, an $(\mathcal{A}, \mathcal{M})$-labelled ribbon graph is a ribbon graph in M, equipped with a labelling by $\mathcal{A}$ in the open set $M \backslash K$, and with a labelling by $\mathcal{M}$ along K. The $(\mathcal{A}, \mathcal{M})$-skein relations are given by imposing within any open cylinder contained in $M \backslash K$ the Reshetikhin–Turaev evaluation relations of $\mathcal{A}$, and within any open cylinder intersecting K along $\{0\} \times I \subset D^2 \times I$ the Reshetikhin–Turaev evaluation relations of $\mathcal{M}$.

3.1. Higher form symmetry actions of abelian groups. The most important source of dualizable module categories for the present discussion are the *invertible* module categories, those for which the evalution and coevaluation are equivalences of braided module categories, identifying the dual braided module category as an inverse. These form an abelian group under tensor product, and we can use this additional structure to define twists not only along knots but along simplicial cycles in M. To motivate the construction, we recall now the framework of higher form symmetries in quantum field theory.

Recall that an ordinary action of an abelian group A on a $(d+1)$-dimensional QFT induces an action of $H_d(M,A)$ on the Hilbert space $Z(M)$ attached to any d-manifold M – in particular if M is closed and connected we have an action of the group $A \cong H_d(M,A)$ itself, on the vector space $Z(M)$.

In purely local/field-theoretic terms, the A-action arises because we redefine the field equations in the neighborhood of some d-cycle, requiring that fields transform according by the A-action as they pass through the cycle: in particular, this introduces a singularity to the field and so a field in the twisted sense is not a field in the ordinary sense. The action of $H_d(M,A)$ on $Z(M)$ which we highlighted arises by situating a d-cycle at the halfway point of trivial mapping cylinders $M \times I$, thereby modifying the notion of correlation functions.

In the literature on extended defects, such an action is called a 0-form symmetry, the "0" expressing that it is not yet an extended symmetry. In this context, we may alternatively contemplate the structure of k-form symmetry action of a group A on a $(d+1)$-dimensional QFT: in mathematical/functorial terms this means we consider an action of the $(k+1)$-group B^kA on the theory (here B^k denotes the k-fold delooping, a $k+1$-groupoid with unique object, and unique n=morphism for $n<k$ and with A as its group of $k+1$-morphisms). From such an extended action we obtain (among other things) an action of $H_{d-k}(M,A)$ on the Hilbert space $Z(M)$ attached to each d-manifold M. Again in physical terms, the data of a k-form symmetry action of the group A allows for modifications of the field equations, now along higher co-dimension submanifolds of M.

3.2. Twisting and grading together: the Picard groupoid. The homology groups $H_{d-k}(M,A)$ appearing above are *global* invariants of M – that is, they depend on the global topology of M, and they allow us to twist Hilbert spaces globally along some global cycle. It will be very useful in practice to give a more local enhancement of a k-form symmetry action, by introducing *twists* $Z(M)[\gamma]$ of the Hilbert space $Z(M)$ indexed by elements $\gamma \in H_{d-k-1}(M,A)$, and isomorphisms between homologous twists (thereby refining the action of H_{d-k}). Very roughly, introduction of twists allows us to consider cycles in M^{d+1} relative to the boundary, and therefore to "cut up" the $d-k$-cycles in $H_{d-k}(M,A)$ into more basic pieces.

In order to simplify notation, we will fix $d=3$ and $k=1$ in what follows; however it is straightforward to modify the general notions for arbitrary d and k, and also to introduce higher groupoids facilitating twists in higher codimension.

Recall that a simplicial 1-cycle γ on M with values in A is by definition an oriented graph embedded into M, with edges labelled by elements of A, and with the property that at each vertex the sum of all labels of incoming edges equals to the sum of all labels of outgoing edges.

DEFINITION 3.4. Fix an oriented 3-manifold M. The Picard groupoid $H_{1,2}(M,A)$ has as its objects the abelian group of simplicial 1-cycles $C_1(M,A)$, with symmetric monoidal product given by addition. Morphisms from γ to γ' are homology classes of 2-chains η satisfying $d\eta = \gamma' - \gamma$.

It follows immediately from the definition that the group of isomorphism classes of objects of $H_{1,2}(M)$ is canonically isomorphic to $H_1(M, A)$, and that the automorphism group of any object is canonically isomorphic to $H_2(M, A) \cong H_1(M, A^\vee)$. Hence, the data of a functor $H_{1,2}(M, A) \to \mathrm{Vect}$ consists of a vector space $F([\gamma])$ for each $[\gamma] \in H_1(M, A)$, each equipped with an action of the abelian group $H_1(M, A^\vee)$.

3.3. One-form actions on braided tensor categories. Recall from [**9**] that braided tensor categories form into a 4-category BrTens. This is the 4-category whose:

- objects are cp-rigid braided tensor categories,
- 1-morphisms from $\mathcal{A}$ to $\mathcal{B}$ are $\mathcal{A}$-$\mathcal{B}$-central algebras, i.e. cp-rigid tensor categories $\mathcal{C}$ equipped with braided tensor functors $\mathcal{A} \otimes \mathcal{B}^{op} \to \mathcal{Z}(\mathcal{C})$,
- 2-morphisms from $\mathcal{C}$ to $\mathcal{D}$ are $\mathcal{C}$-$\mathcal{D}$ bimodule categories, compatibly with the central action,
- 3-morphisms are bimodule functors, and
- 4-morphisms are natural transformations.

In particular, given a braided tensor category $\mathcal{A}$, its endomorphism 3-category consists of $\mathcal{A}$-$\mathcal{A}$-central algebras. The unit endomorphism is $\mathcal{A}$ itself with its regular action. We may then consider the endomorphisms 2-category of $\mathcal{A}$ thusly regarded, it is precisely the 2-category $\mathrm{BrMod}(\mathcal{A})$ of braided module categories we have met previously. See Figure 2 for a comparison with the notion in Definition 3.1

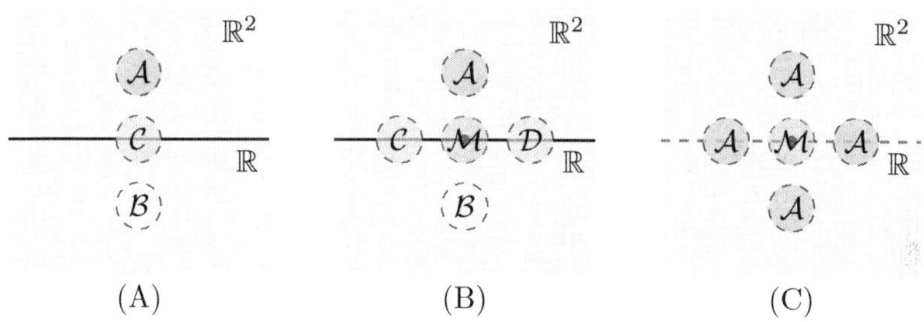

FIGURE 2. (A) depicts a pair of braided tensor categories $\mathcal{A}$ and $\mathcal{B}$, encoded by embeddings of disks $\mathbb{R}^2$, and a 1-morphism between them: an $\mathcal{A}$-$\mathcal{B}$-central algebra $\mathcal{C}$ encoded by embeddings of intervals $\mathbb{R}$ sandwiched between the $\mathcal{A}$- and $\mathcal{B}$-regions. (B) depicts a 2-morphism: a $\mathcal{C}$-$\mathcal{D}$-bimodule structure on $\mathcal{M}$ is encoded by bringing disks into the central point. (C) depicts the special case $\mathcal{A} = \mathcal{B} = \mathcal{C} = \mathcal{D}$, in which case we can erase the $\mathbb{R}$, and $\mathcal{M}$ sits on an isolated point in $\mathbb{R}^2$, and the braided module category structure comes from approaching $\mathcal{M}$ in any direction with $\mathcal{A}$-disks.

DEFINITION 3.5. A 1-form symmetry action of the abelian group A on the braided tensor category $\mathcal{A}$ is the data of an E_2-monoidal functor $A \to \mathrm{BrMod}(\mathcal{A})$, where A is regarded as a discrete symmetric monoidal 2-category.

More concretely, a 1-form symmetry action of A on $\mathcal{A}$ is the data of a braided module category $\mathcal{M}_a \in \mathrm{BrMod}(\mathcal{A})$ for each $a \in A$, and fixed equivalences $\phi_{a,b}$: $M_a \otimes_{\mathcal{A}} M_b \to M_{a+b}$ of braided module categories, satisfying an evident pentagon equation. Here are the primary two sources of examples we will use:

EXAMPLE 3.6. Let A be any subgroup of the group $Z_{gp}(\mathcal{A})$ of group-like Bernstein-central elements of $\mathcal{A}$, $Z_{gp}(\mathcal{A}) = \{a \in \mathrm{Aut}(\mathrm{id}_{\mathcal{A}}) \mid a_{V\otimes W} = a_V \otimes a_W\}$, and define $\mathcal{M}_a = \mathcal{A}$ to be the regular $\mathcal{A}$-module, but with monodromy isomorphism $\alpha_{a,m} = (\mathrm{id}_Y \otimes a) \circ \sigma_{m,Y}\sigma_{Y,m}$ twisted by α.

A particular case of this example arises when we take $\mathcal{A}$ to be $\mathrm{Rep_q}(\mathrm{SL}_2)$, and $A = Z(\mathrm{SL}_2) = \mathbb{Z}/2$. Then as $\mathcal{A}$-modules we have $\mathcal{M}_0 = \mathcal{M}_1 = \mathcal{A}$, so it remains only to describe the monodromy isomorphism; since each module is freely generated by $\mathbf{1}$, it is enough to specify $\alpha_{X,\mathbf{1}}$. Since the double braiding of $\mathbf{1}$ with X is trivial, this means that the monodromy isomorphism is trivial for $\mathcal{M}_0$ and is multiplication by ± 1 on a simple X of even/odd degree. See Figure 3 for a graphical/skein interpretation. In physical literature (motivated by the example of Maxwell theory), the resulting action is called the **electric 1-form symmetry**.

EXAMPLE 3.7. Let $\widetilde{\mathcal{A}}$ be any A-graded braided tensor extension of $\mathcal{A}$, and define $\mathcal{M}_a$ to be the degree a component $\mathrm{Rep_q}(\widetilde{G})_a$, with α and ϕ the restrictions of the double braiding and the multiplication functor on $\widetilde{\mathcal{A}}$. A particular case of this example arises when we take $\mathcal{A}$ to be $\mathrm{Rep_q}(\mathrm{PGL}_2)$, $A = \pi_1(\mathrm{PGL}_2) = \mathbb{Z}/2$, and $\widetilde{\mathcal{A}} = \mathrm{Rep_q}(\mathrm{SL}_2)$. Then we can let $\mathcal{M}_0 = \mathrm{Rep_q}(\mathrm{SL}_2)_{even} = \mathrm{Rep_q}(\mathrm{PGL}_2)$, and $\mathcal{M}_1 = \mathrm{Rep_q}(\mathrm{SL}_2)_{odd}$. In the physical literature, the resulting action is called the **magnetic 1-form symmetry**.

3.4. The twisted skein module. Let us now translate the abstract notions above into the formalism of skein modules, and ultimately linear algebra.

DEFINITION 3.8. Fix a 1-form action $\pi : \mathcal{A} \to \mathrm{BrMod}(\mathcal{A})$. For each simplicial cycle $\gamma \in C_1(M, A)$, the **twisted skein module** $\mathrm{TwSk}(M)[\gamma]$ is the span of all $(\mathcal{A}, \pi, \gamma)$-labelled ribbon graphs, modulo the $(\mathcal{A}, \pi, \gamma)$-skein relations.

Here, a $(\mathcal{A}, \pi, \gamma)$-labelled ribbon graph is a ribbon graph in M, with strands and coupons labelled by objects and morphisms of $\mathcal{A}$ on the complement of the support of γ, and with strands and coupons labelled by objects and morphisms of $\mathcal{M}_a$ along any 1-simplex of γ weighted by a. The skein relations are defined in the evident way, by treating separately cylinders which do not intersect the support of γ, cylinders which intersect a 1-simplex of γ, and cylinders which intersect a 0-simplex of γ. We omit complete definitions.

PROPOSITION 3.9 ([**18**]). *We have a well-defined functor,*

$$\mathrm{TwSk}(M) : H_{1,2}(M, Z(G)^\vee) \to \mathrm{Vect},$$

mapping each $\gamma \in Z_1(M, Z(G)^\vee)$ *to the* γ*-twisted skein module, and mapping each* $\eta : \gamma_1 \to \gamma_2$ *to an isomorphism induced by Reshetikhin-Turaev evaluation on a bounding disk.*

This means simply that for each choice of $\gamma \in C_1(M, A)$ we will define a twisted skein module $\mathrm{TwSk}(M)[\gamma]$, and for each $\eta \in C_2(M, A)/dC_3(M, A)$ with $d\eta = \gamma' - \gamma$ we define an isomorphism,

$$\mathrm{TwSk}(M)[\eta] : \mathrm{TwSk}(M)[\gamma] \to \mathrm{TwSk}(M)[\gamma'],$$

and that these isomorphisms will compose compatibly.

Finally, we can consider the two flavours of twisted skein modules coming from Example 2.1.

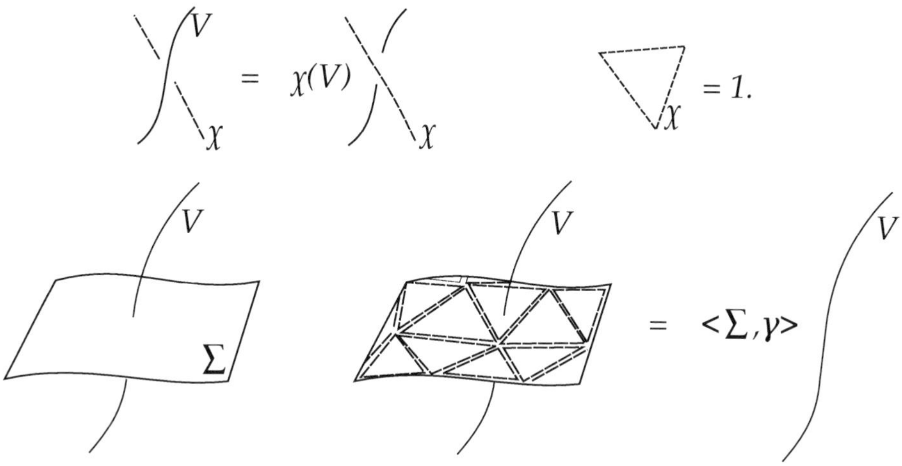

FIGURE 3. Top left: each character χ determines a braided module category $\mathcal{M}_\chi$, whose underlying category is $\mathcal{A}$, and where the double braiding is given by the central action of χ on each object $V \in \mathcal{A}$ marking a skein. Top right: this braided module category is invertible; closed loops contract trivially. At bottom: the pairing between some 2-cycle Σ and some skein is obtained by first condensing within Σ some network of triangles, and contracting each triangle through the skein. The result is multiplication by the Poincare pairing $\langle \Sigma, \gamma \rangle$ between the cycle Σ and the support 1-cycle γ of the skein. We note that invertibility of χ implies already that adjacent edges of the triangulation cancel pairwise, hence the resulting skein module is not twisted despite the introduction of defects, so that we indeed obtain an endomorphism.

EXAMPLE 3.10. Let $A = Z(G)$. We have seen in Example 3.6 that we have an action of $Z(G)$ on $\mathrm{Rep}_q(G)$ by 1-form symmetries; the resulting action by $H_2(M, Z(G))$ on $\mathrm{Sk}(M) = \mathrm{TwSk}(M)[0]$ allows us to decompose $\mathrm{Sk}(M)$ according to the characters $H_2(M, Z(G))^\vee \cong H_1(M, Z(G)^\vee)$. The skein-theoretic interpretation is depicted in Figure 3. The resulting graded decomposition is the elementary one observed at the start of Section 3. This grading, in its elementary form, has featured prominently in the computations of [**11**], [**15**], [**16**] and [**14**].

To our knowledge, the twisted skein modules $\mathrm{TwSk}(M)[\gamma]$ for non-trivial $\gamma \in H_1(M, Z(G))$ are new, and so do not have an *a priori* elementary description. However, it is easy to describe the twists in this case: the condition on defects says that when a 1-simplex of a ribbon graph labelled with a simple representation V passes through a 1-simplex of the defect with weight $z \in Z(G)$, it picks up a scalar factor equal to the evaluation of z on V.

EXAMPLE 3.11. Let $A = \pi_1(G)$. We have seen in Example 3.7 that we obtain an action of $\pi_1(G)$ on $\mathrm{Rep}_q(G)$ by 1-form symmetries. To our knowledge both the resulting action of $H_2(M, Z(G))$ and hence the resulting decomposition of $H_1(M, \pi_1(G)^\vee)$ on $\mathrm{Sk}(M)$, and the twisted skein modules are new, and so do not have an *a priori* elementary description. However, it is again easy to define twists in this case. Given a G^{sc}-labelled ribbon graph with simple labels, we can produce a 1-simplex in $C_1(M, \pi_1(G))$, by replacing each simple label V by its grading with respect to $\pi_1(G)$, regarding $\mathrm{Rep_q}(G^{sc})$ as a $\pi_1(G)$-graded extension of $\mathrm{Rep_q}(G)$. The space $\mathrm{TwSk}(M)[\gamma]$ is the span of all G^{sc}-labelled ribbon graphs, whose underlying 1-cocycle equals γ, and modulo only those skein relations also with underlying 1-cocycle γ.

For any group G, and for $A = Z$ or π_1, and for $a \in H_1(M, A^\vee)$ and $b \in H_1(M, A)$ we denote by $\mathrm{Sk}^{a,b}_{G,A}(M)$ the a-graded component of the b-twisted skein module. It follows from the definition that we have a natural linear map of vector spaces,

$$\Phi : \mathrm{Sk}_{G^{ad}}(M) \to \mathrm{Sk}_{G^{sc}}(M),$$

which regards a $\mathrm{Rep_q}(G^{ad})$-labelled ribbon graph as a $\mathrm{Rep_q}(G^{sc})$-labelled ribbon graph. An important observation is that while $\mathrm{Rep_q}(G^{ad}) \to \mathrm{Rep_q}(G^{sc})$ is fully faithful (so that in a ball this map is simply an isomorphism), it is *not* an isomorphism globally, and has both kernel and cokernel. The cokernel is clear: by construction the image of Φ is contained in the degree zero part of $\mathrm{Sk}_{G^{sc}}(M)$. The kernel is more subtle, but it stems from the fact that on the source, by definition, only skein relations which have trivial underlying 1-cycle are imposed, whereas in the target they are not. In the skein module $\mathrm{Sk}_{G^{sc}}(M)$, two ribbon graphs of degree zero can be related in a way that nevertheless involves elements of nontrivial degree. This is captured cleanly in the following

THEOREM 3.12 ([**18**]). *Let M be an oriented 3-manifold (not necessarily closed), and let $q \in \mathbb{C}^\times$ be an arbitrary complex number. Then we have a canonical isomorphism,*

$$\mathrm{Sk}^{a,b}_{G/Z(G),\pi_1}(M) \cong \mathrm{Sk}^{a,b}_{G,Z}(M).$$

Unpacking, it says that the map Φ above actually extends firstly to a map

$$\mathrm{Sk}_{G^{ad},\pi_1}(M) \to \bigoplus_{b \in H_1(M,Z(G))} \mathrm{Sk}^{0,b}_{G^{sc},Z}(M),$$

which is injective, and then again to a map,

$$\bigoplus_{a,b} \mathrm{Sk}^{a,b}_{G^{ad},\pi_1}(M) \to \bigoplus_{a,b} \mathrm{Sk}^{a,b}_{G^{sc},Z}(M),$$

which is an isomorphism. We note that the indices a and b appear on botht he LHS and RHS above. This is justified by the natural isomorphisms, $\pi_1(G^{ad}) \cong Z(G^{sc})^\vee$.

The reader familiar with the language of gauging of higher form symmetries may not find this surprising: Indeed, we have $G^{ad} = G/Z(G)$, hence $\mathrm{Rep_q}(G^{ad})$ is a gauging of the electric 1-form symmetry by $Z(G)$; gauging an electric 1-form symmetry induces on the new theory a dual magnetic 1-form symmetry, with an isomorphism like that above.

4. The bigraded refinement of the skein conjecture

In total, we have defined bi-graded skein module,

$$\mathrm{Sk}^{a,b}_{G^{sc},Z}(M), \qquad \text{for } a \in H_1(M, Z(G^{sc})^\vee), b \in H_1(M, Z(G^{sc})),$$

using the electric 1-form symmetry, and

$$\mathrm{Sk}^{a,b}_{G^{ad},\pi_1}(M), \qquad \text{for } a \in H_1(M, \pi_1), b \in H_1(M, \pi_1(G^{ad})^\vee),$$

using the magnetic 1-form symmetry, where in each case a denotes the grading degree, and b denotes the twist. Now recall that under Langlands/S-duality, the electric and magnetic 1-form symmetries are interchanged. Thus the same electric-magnetic duality principle which informs Conjecture 1.1 also predicts:

CONJECTURE 4.1 (Bigraded refinement of skeins conjecture). *Suppose that $q \in \mathbb{C}^\times$ is transcendental and that M is a closed oriented 3-manifold. Then we have an equality of dimensions,*

$$\dim \mathrm{Sk}^{a,b}_{G,Z}(M) = \dim \mathrm{Sk}^{b,a}_{{}^LG,\pi_1}(M).$$

Now suppose that G is simply connected of ADE type, so that ${}^L\mathfrak{g} = \mathfrak{g}$. In this case the Langlands dual group ${}^LG = G^{ad}$ has a second manifestation as a quotient $G/Z(G)$ of G by the electric 1-form symmetry we exploited to define the bi-grading. Hence we may apply Theorem 3.12, to obtain the following equivalent

CONJECTURE 4.2 (Reformulation of refined skein conjecture). *Suppose that G is simply connected of ADE type, that $q \in \mathbb{C}^\times$ is transcendental and that M is a closed oriented 3-manifold. Then we have an equality of dimensions,*

$$\dim \mathrm{Sk}^{a,b}_{G,Z}(M) = \dim \mathrm{Sk}^{b,a}_{G,Z}(M),$$

for all $a \in H_1(M, Z(G)^\vee), b \in H_1(M, Z(G))$.

REMARK 4.3. The reformulation of Conjecture 4.1 is very convenient: applying it to the case $G = \mathrm{SL}_N$, it means we can check the refined conjecture completely internally to either the Z-twists on SL_N or alternatively the π_1-twists on PGL_N. Because more is known about SL_N skein theory, we work with the Z-twists in the next section.

REMARK 4.4. The isomorphism in Theorem 3.12 is completely canonical, holds for arbitrary q and also for 3-manifolds with boundary: it is a very general principle, and in particular, there was an a priori natural alignment of the indices a and b appearing on either side of the asserted isomorphism. By contrast, the indices a and b do not themselves live in the same group naturally, but rather live in Pontryagin dual groups. Hence Conjecture 4.2 seems not to parse semantically.

Indeed, to make sense of Conjecture 4.2, one must choose an isomorphism $Z(G) \cong Z(G)^\vee$, though clearly the statement itself does not depend on the choice. This should be regarded as a further symptom of the fact that Langlands duality should be swapping $Z(G)$ with $\pi_1({}^LG)^\vee$, but that we are identifying $\pi_1({}^LG)^\vee$ with $Z(G)$ in the ADE case.

Recall that the untwisted skein modules in Conjecture 1.1 are obtained as sums over the untwisted graded components. Hence, Conjecture 1.1 follows from the following weakening of Conjecture 4.2.

CONJECTURE 4.5. *Suppose that G is simply connected of ADE type, that $q \in \mathbb{C}^\times$ is transcendental and that M is a closed oriented 3-manifold. Then we have an equality of dimensions,*

$$\dim \mathrm{Sk}^{a,0}_{G,Z}(M) = \dim \mathrm{Sk}^{0,a}_{G,Z}(M),$$

for all $a \in H_1(M, Z(G)^\vee)$.

5. Some evidence for the conjecture

Dimensions of skein modules are very difficult to compute in elementary terms. Indeed, this is underscored by the twenty five years which passed between the two fundamental examples $S^2 \times S^1$ [**21**] and T^3 [**11, 15**]. Since then a dam has broken and these dimensions are being computed using a combination of elementary and categorical techniques. This allows us now to prove our conjecture in a number of special cases. Recall that the Lie algebra $\mathfrak{sl}_N$ is fixed under Langlands duality; however on the group level, Langlands duality interchanges the simply connected form with the adjoint form on the dual group, so it is meaningful to contemplate Langlands duality for the pair $(\mathrm{SL}_N, \mathrm{PGL}_N)$.

We focus on two families of skein modules, where complete computations can be done. In the first family, we fix the Langlands dual pair to be $(\mathrm{SL}_2, \mathrm{PGL}_2)$, and consider mapping cylinders $\Sigma_g \times S^1$ of arbitrary genus g, while in the second family we instead fix the 3-manifold to be T^3, and vary the Langlands dual pair to be $(\mathrm{SL}_N, \mathrm{PGL}_N)$. For the first family, we recall the following:

THEOREM 5.1 ([**16**], [**14**]). *Suppose that the quantum parameter is generic. Then we have:*

$$\dim \mathrm{Sk}_{SL_2}(\Sigma_g \times S^1) = 2^{2g+1} + 2g - 1.$$

With respect to the electric 1-form symmetry action, we have:

$$\dim \mathrm{Sk}^{a,0}_{\mathrm{SL}_2}(\Sigma_g \times S^1) = \begin{cases} g+1, & a = (0,0) \\ g, & a = (1,0) \\ 1, & \textit{otherwise} \end{cases},$$

where the cases refer to the Künneth decomposition,

$$\begin{aligned} H_1(\Sigma_g \times S^1, \mathbb{Z}/2) &= H_0(\Sigma, \mathbb{Z}/2) \otimes H_1(S^1) \oplus H_1(\Sigma, \mathbb{Z}/2) \otimes H_0(S^1) \\ &= H_0(\Sigma, \mathbb{Z}/2) \oplus H_1(\Sigma, \mathbb{Z}/2). \end{aligned}$$

The first non-trivial confirmation of Langlands duality for skein module is then given by the following

THEOREM 5.2 ([**18**], to appear). *Suppose that the quantum parameter is generic. Then we have:*

$$\dim \mathrm{Sk}_{\mathrm{PGL}_2}(\Sigma_g \times S^1) = 2^{2g+1} + 2g - 1$$

In fact the proof is not by computing with $\mathrm{Sk}_{\mathrm{PGL}_2}$, but instead by confirming Conjecture 4.5 for $M = \Sigma_g \times S^1$ with $G = \mathrm{SL}_2$. More precisely we prove:

$$\dim \mathrm{Sk}^{0,a}_{\mathrm{SL}_2}(\Sigma_g \times S^1) = \begin{cases} g+1, & a = (0,0) \\ g, & a = (1,0) \\ 1, & \text{otherwise} \end{cases}.$$

The first case is, of course, already proved by Carrega and Gilmer, we repeat it only for emphasis. For the second case, we use the fact that the twisted skein category at $q = 1$ becomes the category of quasi-coherent sheaves on the *twisted character variety*. This is in fact a *smooth affine variety*, hence the Hochschild homology of its quantization to the skein category is identified with the middle homology group of the twisted character variety, which was shown to be g-dimensional by Hausel and Rodriguez-Villegas [**20**], following Hitchin.

Turning now to the second family of skein modules, we recall the following notation: $\mathcal{P}(d)$ denotes the partition number of d, i.e. the number of ways to write d as a sum of non-negative integers, $J_3 = C * \mu$ is the third Jordan totient function, obtained as the Möbius inversion of the cube function $C(n) = n^3$.

THEOREM 5.3 ([**19**], to appear). *Suppose that the quantum parameter is generic. Then we have:*

$$\dim \mathrm{Sk}_{SL_N}(T^3) = (\mathcal{P} * J_3)(N) = \sum_{N=d\cdot e\cdot f} \mathcal{P}(d)\cdot e^3 \cdot \mu(f).$$

With respect to the electric 1-form symmetry action we have

$$\mathrm{Sk}_{SL_N}(T^3)^{(a,b,c),0} = \mathrm{gcd(a,b,c,N)},$$

where we write (a,b,c) for an element of $(\mathbb{Z}/N)^3 = H_1(T^3,\mathbb{Z}/N)$.

This theorem is proved by first obtaining an equivalence between the skein category of the torus and the category of modules for a double affine Hecke algebra, subsequently giving an isomorphism between this double affine Hecke algebra and a smash product of a quantum torus and the symmetric group, and finally computing the Hochschild homology directly by combinatorial means.

Our second non-trivial confirmation of Langlands duality is given by the following

THEOREM 5.4 ([**18**]). *Suppose that the quantum parameter is generic, and that N is prime. Then we have:*

$$\dim \mathrm{Sk}_{PGL_N}(T^3) = (\mathcal{P} * J_3)(N) = \sum_{N=d\cdot e\cdot f} \mathcal{P}(d)\cdot e^3 \cdot \mu(f).$$

With respect to the magnetic 1-form symmetry action we have

$$\mathrm{Sk}_{\mathrm{PGL}_N}(T^3)^{(a,b,c),0} = \mathrm{gcd(a,b,c,N)} = 1.$$

Our proof is similar to that of Theorem 5.2. We instead verify Conjecture 4.5 for $M = T^3$ and $G = \mathrm{SL}_N$, by proving:

$$\dim \mathrm{Sk}^{0,(a,b,c)}_{\mathrm{SL}_N}(T^3) = 1, \quad \text{if } (a,b,c) \neq 0.$$

For this, we rely again on the fact that the twisted charac ter variety is smooth, and that its middle homology is known to be one-dimensional. Of course, we expect the theorem to hold when N is not prime, but in this case the twisted character variety is not smooth, so we can't easily apply the same techniques for the proof.

References

[1] Mohammed Abouzaid and Ciprian Manolescu, *A sheaf-theoretic model for* $\mathrm{SL}(2,\mathbb{C})$ *Floer homology*, J. Eur. Math. Soc. (JEMS) **22** (2020), no. 11, 3641–3695, DOI 10.4171/jems/994. MR4167016

[2] David Ayala and John Francis, *Factorization homology of topological manifolds*, J. Topol. **8** (2015), no. 4, 1045–1084, DOI 10.1112/jtopol/jtv028. MR3431668

[3] Oren Ben-Bassat, Christopher Brav, Vittoria Bussi, and Dominic Joyce, *A 'Darboux theorem' for shifted symplectic structures on derived Artin stacks, with applications*, Geom. Topol. **19** (2015), no. 3, 1287–1359, DOI 10.2140/gt.2015.19.1287. MR3352237

[4] David Ben-Zvi, Adrien Brochier, and David Jordan, *Integrating quantum groups over surfaces*, J. Topol. **11** (2018), no. 4, 874–917, DOI 10.1112/topo.12072. MR3847209

[5] David Ben-Zvi, Adrien Brochier, and David Jordan, *Quantum character varieties and braided module categories*, Selecta Mathematica (2018).

[6] David Ben-Zvi, John Francis, and David Nadler, *Integral transforms and Drinfeld centers in derived algebraic geometry*, J. Amer. Math. Soc. **23** (2010), no. 4, 909–966, DOI 10.1090/S0894-0347-10-00669-7. MR2669705

[7] David Ben-Zvi and David Nadler, *Betti geometric Langlands*, Algebraic geometry: Salt Lake City 2015, Proc. Sympos. Pure Math., vol. 97, Amer. Math. Soc., Providence, RI, 2018, pp. 3–41, DOI 10.1090/pspum/097.2/01. MR3821166

[8] David Ben-Zvi and Jackson Van Dyke, *Between electric-magnetic duality and the langlands program*, `https://web.ma.utexas.edu/users/vandyke/notes/langlands_sp21/langlands.pdf`, 2022.

[9] Adrien Brochier, David Jordan, and Noah Snyder, *On dualizability of braided tensor categories*, Compos. Math. **157** (2021), no. 3, 435–483, DOI 10.1112/s0010437x20007630. MR4228258

[10] Doug Bullock, *Rings of* $\mathrm{SL}_2(\mathbf{C})$*-characters and the Kauffman bracket skein module*, Comment. Math. Helv. **72** (1997), no. 4, 521–542, DOI 10.1007/s000140050032. MR1600138

[11] Alessio Carrega, *Nine generators of the skein space of the 3-torus*, Algebr. Geom. Topol. **17** (2017), no. 6, 3449–3460, DOI 10.2140/agt.2017.17.3449. MR3709652

[12] Sabin Cautis, Joel Kamnitzer, and Scott Morrison, *Webs and quantum skew Howe duality*, Math. Ann. **360** (2014), no. 1-2, 351–390, DOI 10.1007/s00208-013-0984-4. MR3263166

[13] Juliet Cooke, *Factorisation homology and skein categories of surfaces*, `arXiv:1910.02630`, 2019.

[14] Renaud Detcherry and Maxime Wolff, *A basis for the Kauffman skein module of the product of a surface and a circle*, Algebr. Geom. Topol. **21** (2021), no. 6, 2959–2993, DOI 10.2140/agt.2021.21.2959. MR4344875

[15] Patrick M. Gilmer, *On the Kauffman bracket skein module of the 3-torus*, Indiana Univ. Math. J. **67** (2018), no. 3, 993–998, DOI 10.1512/iumj.2018.67.7327. MR3820238

[16] Patrick M. Gilmer and Gregor Masbaum, *On the skein module of the product of a surface and a circle*, Proc. Amer. Math. Soc. **147** (2019), no. 9, 4091–4106, DOI 10.1090/proc/14553. MR3993800

[17] Sam Gunningham, David Jordan, and Pavel Safronov, *The finiteness conjecture for skein modules*, Inventiones mathematicae, **232.1** (2023): 301-363.

[18] Sam Gunningham, David Jordan, and Pavel Safronov, *One-form symmetries on skein modules*, To appear, 2023.

[19] Sam Gunningham, David Jordan, Monica Vazirani, and Haiping Yang, *Skeins on tori*, To appear, 2023.

[20] Tamás Hausel and Fernando Rodriguez-Villegas, *Mixed Hodge polynomials of character varieties*, Invent. Math. **174** (2008), no. 3, 555–624, DOI 10.1007/s00222-008-0142-x. With an appendix by Nicholas M. Katz. MR2453601

[21] Jim Hoste and Józef H. Przytycki, *The Kauffman bracket skein module of* $S^1 \times S^2$, Math. Z. **220** (1995), no. 1, 65–73, DOI 10.1007/BF02572603. MR1347158

[22] Anton Kapustin and Edward Witten, *Electric-magnetic duality and the geometric Langlands program*, Commun. Number Theory Phys. **1** (2007), no. 1, 1–236, DOI 10.4310/CNTP.2007.v1.n1.a1. MR2306566

[23] Masaki Kashiwara and Pierre Schapira, *Deformation quantization modules* (English, with English and French summaries), Astérisque **345** (2012), xii+147. MR3012169

[24] Hovhannes Khudaverdian and Theodore Voronov, *Higher Poisson Brackets and Differential Forms*, AIP Cong. Proc. **1079** (2008), DOI 10.1063/1.3043861.

[25] Greg Kuperberg, *Spiders for rank* 2 *Lie algebras*, Comm. Math. Phys. **180** (1996), no. 1, 109–151. MR1403861

[26] Masanori Morishita, *Knots and primes: An introduction to arithmetic topology*, Universitext, Springer, London, 2012, DOI 10.1007/978-1-4471-2158-9. MR2905431

[27] Julia Mueller and Freydoon Shahidi (eds.), *The genesis of the Langlands Program*, London Mathematical Society Lecture Note Series, vol. 467, Cambridge University Press, Cambridge, 2021. Edited by Julia Mueller and Freydoon Shahidi, DOI 10.1017/9781108591218. MR4274520

[28] Hitoshi Murakami, Tomotada Ohtsuki, and Shuji Yamada, *Homfly polynomial via an invariant of colored plane graphs*, Enseign. Math. (2) **44** (1998), no. 3-4, 325–360. MR1659228

[29] Tony Pantev, Bertrand Toën, Michel Vaquié, and Gabriele Vezzosi, *Shifted symplectic structures*, Publ. Math. Inst. Hautes Études Sci. **117** (2013), 271–328, DOI 10.1007/s10240-013-0054-1. MR3090262

[30] J. P. Pridham, *Deformation quantisation for (-1)-shifted symplectic structures and vanishing cycles* (2019).preprint, `arXiv:1508.07936`

[31] Józef H. Przytycki and Adam S. Sikora, *On skein algebras and* $\mathrm{Sl}_2(\mathbf{C})$*-character varieties*, Topology **39** (2000), no. 1, 115–148, DOI 10.1016/S0040-9383(98)00062-7. MR1710996

[32] Vladimir G. Turaev, *Quantum invariants of knots and 3-manifolds*, De Gruyter Studies in Mathematics, vol. 18, De Gruyter, Berlin, 2016. Third edition [of MR1292673], DOI 10.1515/9783110435221. MR3617439

School of Mathematics, University of Edinburgh, Edinburgh EH9 3FD, United Kingdom

Proceedings of Symposia in Pure Mathematics
Volume **107**, 2024
https://doi.org/10.1090/pspum/107/01949

Link homology from homological mirror symmetry

Elise LePage

ABSTRACT. We explain how to calculate link homology for a Lie algebra $\mathfrak{g}$ using the Fukaya category associated to a 2d A-model. Links are represented as configurations of particular A-branes and link homology is given by Homs between these A-branes. In the case of $\mathfrak{g} = \mathfrak{su}_2$, we explain how to explicitly construct projective resolutions of the relevant A-branes in terms of thimbles, whose algebra is known. This gives an explicit algorithm for computing Khovanov homology. This algorithm can be extended to all Lie algebras.

1. Quantum group invariants and categorification

Given a Lie algebra and a representation, one may define a certain polynomial link invariant, known as a quantum group invariant. The first example is the Jones polynomial, associated to the fundamental representation of $\mathfrak{su}_2$ and defined by Jones in 1984 [**9**]. Four years later, Witten gave a physical origin of the Jones polynomial using Chern-Simons theory for $\mathfrak{su}_2$, which allowed one to define a link invariant for any Lie algebra $\mathfrak{g}$ using Chern-Simons theory for $\mathfrak{g}$ [**14**].

A natural question to ask is whether one can categorify quantum group invariants to get new invariants of links. In 1999, Khovanov proposed a categorification for the case of $\mathfrak{su}_2$, now known as Khovanov homology [**10**]. Khovanov homology categorifies the Jones polynomial in the sense that it associates to a link a series of homology groups, which are themselves link invariants, whose Euler characteristic coincides with the Jones polynomial of the link. Khovanov and Rozansky extended this categorification to $\mathfrak{su}_n$, defining Khovanov-Rozansky homology [**11**]. In 2013, Webster proposed a categorification that works for all Lie algebras [**13**]. Even more recently, Aganagic proposed a solution based in physics using homological mirror symmetry [**1**], [**2**]. Her work gives the same categorification as Webster's work but has the advantage that the physical origin is clearer and calculations become very tractable. Furthermore, Aganagic's approach extends to Lie superalgebras. Although Aganagic defines link homology in a pair of mirror theories, here we focus only on the A-side of mirror symmetry.

2. The A-model

Consider a simple Lie algebra $\mathfrak{g}$. We want to define a Landau-Ginzburg theory associated to $\mathfrak{g}$. Towards this end, let $\mathcal{A} = \mathbb{C}^\times$ and $Y = \bigotimes_{a=1}^{rk\mathfrak{g}} Sym^{d_a}\mathcal{A}$ where

2020 *Mathematics Subject Classification.* Primary 57K18, 53D37.
Key words and phrases. Khovanov homology, mirror symmetry.

the d_a may take any integer value. We will see shortly that we will require special choices of d_a in order to get link invariants. It is helpful to think of $\mathcal{A}$ as an infinite cylinder and Y as the configuration space of colored points (with d_a points of each color) on the cylinder.

Let a_i be a series of marked points on $\mathcal{A}$, each colored by a representation of $\mathfrak{g}$ with highest weight μ_i. Denote the roots of $\mathfrak{g}$ by e_a. Let $y_{\alpha,a}$ be the coordinate on each copy of $\mathcal{A}$ in Y, labelled such that $(y_{1,a}, y_{2,a}, \dots, y_{d_a,a})$ are coordinates on $Sym^{d_a}\mathcal{A}$. Define a holomorphic function $W : Y \to \mathbb{C}$ by

$$W(y) = \lambda_0 W^0(y) + \sum_{a=1}^{rk\mathfrak{g}} \lambda_a W^a(y) \tag{2.1}$$

where $\lambda_0, \lambda_a \in \mathbb{C}$ are parameters and

$$W^0(y) = \sum_{a=1}^{rk\mathfrak{g}} \ln f_a(y), \qquad W^a(y) = \sum_{\alpha=1}^{d_a} \ln y_{\alpha,a} \tag{2.2}$$

with

$$f_a(y) = \prod_{\alpha=1}^{d_a} \frac{\prod_i (1 - a_i/y_{\alpha,a})^{\langle e_a,\mu_i\rangle}}{\prod_{(b,\beta)\neq(a,\alpha)} (1 - y_{\beta,b}/y_{\alpha,a})^{\langle e_a,e_b\rangle/2}}. \tag{2.3}$$

We will also need to equip Y with a choice of a top holomorphic form given by

$$\Omega = \bigwedge_{a=1}^{rk\mathfrak{g}} \bigwedge_{\alpha=1}^{d_a} \frac{dy_{\alpha,a}}{y_{\alpha,a}}. \tag{2.4}$$

We will consider the derived Fukaya-Seidel category $\mathscr{D}_Y$ associated to (Y, W). The objects of this category are A-branes supported on graded Lagrangian submanifolds of Y.

2.1. The gradings. Lagrangians on Y admit multiple gradings: a Maslov grading M depending on Ω and equivariant gradings J^a depending on W^a. An A-brane requires both a choice of a Lagrangian and a choice of gradings.

The Maslov grading comes from a choice of phase of Ω along a Lagrangian L. Let $\Omega^2 = |\Omega^2|e^{2i\varphi}$. Let $\tilde{Y}$ be a cover of Y on which 2φ is single-valued. Let $\tilde{L}$ be a lift of L to $\tilde{Y}$. Let $\tilde{L}[M]$ denote a Lagrangian on which the value of φ differs from its value on $\tilde{L}$ by $\pi i M$:

$$\varphi|_{\tilde{L}[M]} = \varphi|_{\tilde{L}} + \pi i M. \tag{2.5}$$

The equivariant gradings arise from the non-single-valuedness of W^a. More precisely, a shift in the equivariant gradings of $\tilde{L}$ changes the value of W on $\tilde{L}$ as follows:

$$W^a|_{\tilde{L}\{\vec{J}\}} = W^a|_{\tilde{L}} + 2\pi i J^a. \tag{2.6}$$

More details regarding the gradings can be found in [**2**].

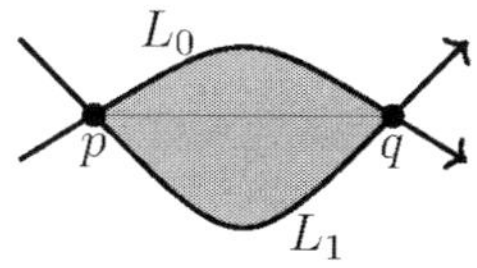

FIGURE 1. An example of a holomorphic disk contributing to Qp

2.2. Morphisms between A-branes. Morphisms between A-branes are defined via Floer theory. Let L_0 and L_1 be two A-branes. Then morphisms from L_0 to L_1 are given by

$$Hom_{\mathscr{D}_Y}(L_0, L_1) = HF^{0,0}(L_0, L_1) \tag{2.7}$$

where $HF^{0,0}$ denotes the Floer cohomology groups in Maslov and equivariant degrees zero. See [**5**] or [**12**] for a more detailed review of Floer theory. The groups $HF^{M,\vec{J}}(L_0, L_1)$ are generated by the intersection points of $\tilde{L}_0$ with $\tilde{L}_1[M]\{\vec{J}\}$ in $\tilde{Y}$ modulo the action of a differential Q. Equivalently, the intersection points are the degree $(M, \vec{J})$ intersection points of L_0 with L_1 in Y.

The Floer differential Q counts certain degree $(1, \vec{0})$ holomorphic disks between intersection points. Let $p, q \in L_0 \cap L_1$. Let $y : [0,1] \times \mathbb{R} \to Y$ be a holomorphic map such that $y(0,t) = L_0$, $y(1,t) = L_1$, $\lim_{t\to\infty}(s,t) = p$, and $\lim_{t\to-\infty}(s,t) = q$. We require that y^*W is a regular function. An example of the image in Y of one such disk is shown in Figure 1. Note that by the Riemann mapping theorem, we can just as well consider maps from the unit disk D to Y, which we will also call y. The Maslov and equivariant degrees of such maps are given by

$$M(y) = \frac{1}{\pi} \oint_{\partial D} y^* d\varphi \tag{2.8}$$

and

$$J^i(y) = -\frac{1}{2\pi i} \oint_{\partial D} y^* dW^i. \tag{2.9}$$

The degrees of y are related to the degrees of p and q by

$$M(q) - M(p) = M(y), \qquad J^a(q) - J^a(p) = J^a(y) \tag{2.10}$$

Disks only contribute to Q modulo reparametrization along t, so let $\#M(p,q,y)$ be the signed count of disks connecting p to q modulo this reparametrization.

DEFINITION 2.1. The Floer differential is defined by

$$Qp = \sum_{\substack{q \in L_0 \cap L_1 \\ M(y)=1, \vec{J}(y)=\vec{0}}} \#M(p,q,y)\, q. \tag{2.11}$$

From a physics perspective, the intersection points are ground states of the theory and the Floer differential counts instantons tunneling between ground states.

It follows immediately that

$$Hom_{\mathscr{D}_Y}(L_0, L_1[M]\{\vec{J}\}) = HF^{M,\vec{J}}(L_0, L_1). \tag{2.12}$$

Furthermore, one can use (2.8) and (2.9) to determine the relative degrees of any pair of intersection points regardless of whether y contributes to Q. In practice however, there exist simple combinatorial versions of these formulas, derived from the definitions given here, which are far easier to use [**2**].

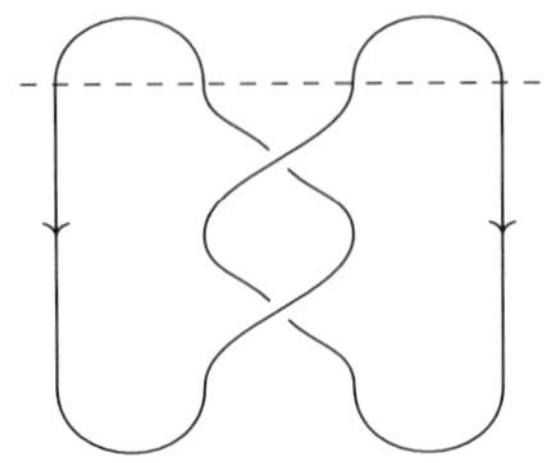

FIGURE 2. A presentation of the Hopf link as the plat closure a braid with a dashed line dividing it into "caps" and "cups".

3. Links as A-branes

We start with a presentation of an oriented knot or link as the plat closure of a braid β on $2d$ strands, such as the Hopf link shown in Figure 2. We then choose a simple Lie algebra $\mathfrak{g}$ and a representation of $\mathfrak{g}$ to color each strand of the knot or link.

We break the plat closure into two pieces: "cups" which are segments braided by β and "caps" which are unbraided segments. In Figure 2, the cups and caps are separated by a dashed line. These caps and cups will correspond to special A-branes in a particular A-model. The A-model we want is the one described above, associated to the same Lie algebra $\mathfrak{g}$ and with d marked points on $\mathcal{A}$ colored by representations chosen and d marked points colored by their dual representations. We require d_a to satisfy

$$\sum_{i=1}^{d} \mu_i = \sum_{a=1}^{rk\mathfrak{g}} d_a e_a \tag{3.1}$$

This data defines a pair (Y, W) and the associated derived Fukaya-Seidel category.

From now on, we will set $\mathfrak{g} = \mathfrak{su}_2$ with each strand colored by the fundamental representation of $\mathfrak{su}_2$. Here, the target space of our theory is $Y = Sym^d \mathcal{A}$ and the potential is

$$W(y) = \lambda_0 \ln f(y) + \lambda_1 \sum_{\alpha=1}^{d} \ln y_\alpha \tag{3.2}$$

with

$$f(y) = \prod_{\alpha=1}^{d} \frac{\prod_{i=1}^{2d}(1 - a_i/y_\alpha)}{\prod_{\beta\neq\alpha}(1 - y_\beta/y_\alpha)}. \tag{3.3}$$

The caps are then associated to straight line Lagrangians connecting pairs of punctures, such as those shown in Figure 3a. To obtain the cups, start with d figure-eight-shaped Lagrangians E_i enclosing pairs of adjacent punctures. We can use β to define a map $\mathscr{B}$ from Y to itself by braiding the punctures according to β. Denote the image of E_i under this map by $\mathscr{B}E_i$. The braided cups are associated to $\mathscr{B}E_{\mathcal{U}} = \mathscr{B}E_1 \times \ldots \times \mathscr{B}E_n$, the product of the braided figure-eight Lagrangians. An example of $\mathscr{B}E_{\mathcal{U}}$ for the Hopf link is shown in Figure 3b.

3.1. Link invariants. Denote the caps by $I_{\mathcal{U}} = I_1 \times \ldots \times I_n$.

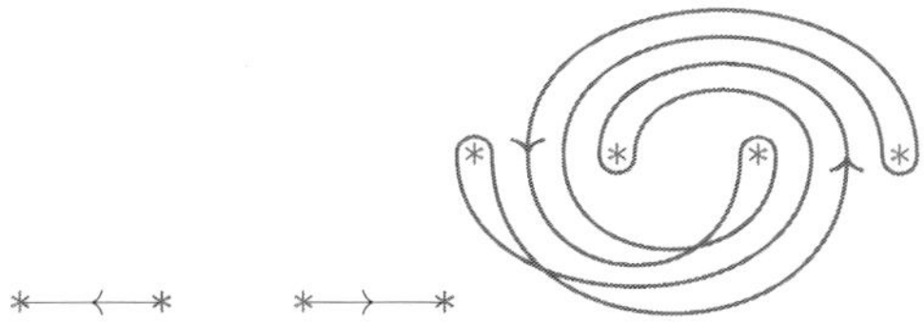

(A) Caps (B) Cups

FIGURE 3. Lagrangians corresponding to the caps and cups for the presentation of the Hopf link shown in Figure 2 colored by the fundamental representation of $\mathfrak{su}_2$ with $n = 2$

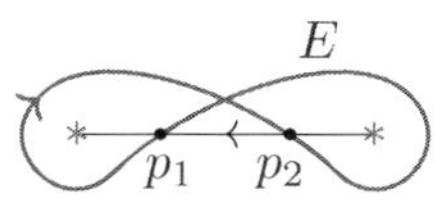

FIGURE 4. A-branes representing an unknot

THEOREM 3.1. *The homology groups*

$$Hom^{*,*}_{\mathscr{D}_Y}(\mathscr{B}E_{\mathcal{U}}, I_{\mathcal{U}}) = \bigoplus_{M\in\mathbb{Z}, \vec{J}\in\mathbb{Z}^2} Hom_{\mathscr{D}_Y}(\mathscr{B}E_{\mathcal{U}}, I_{\mathcal{U}}[M]\{\vec{J}\}) \tag{3.4}$$

are link invariants.

We will prove this theorem in Section 4 using techniques developed in that section.

THEOREM 3.2. *The homology groups*

$$Hom_{\mathscr{D}_Y}(\mathscr{B}E_{\mathcal{U}}, I_{\mathcal{U}}[M]\{J^0\})$$

of a link L coincide with the Khovanov homology groups $Kh^{i,j}(L)$ of the same link with the gradings related by

$$i = M + 2J^0 + i_0, \qquad j = 2J^0 + j_0 \tag{3.5}$$

with

$$i_0 = \frac{w}{2} - \frac{e}{2}, \qquad j_0 = d + \frac{3w}{2} - \frac{e}{2} \tag{3.6}$$

where d is half the number of strands of the braid, w is the writhe the braid, and e is the sum of the exponents of the braid group generators for the presentation chosen.

The two theorems are due to Aganagic [**2**] with the general statement of i_0 and j_0 given in [**4**].

EXAMPLE 3.3. Consider the simplest presentation of the unknot, shown in Figure 4, which has $d = 1$ and corresponds to the identity braid. There are two intersection points $p_1, p_2 \in E \cap I$. Let y be either of the two disks connecting p_1 to p_2. Using (2.10), we find

$$M(p_2) - M(p_1) = 2 \tag{3.7}$$

and

$$J(p_2) - J(p_1) = -1. \tag{3.8}$$

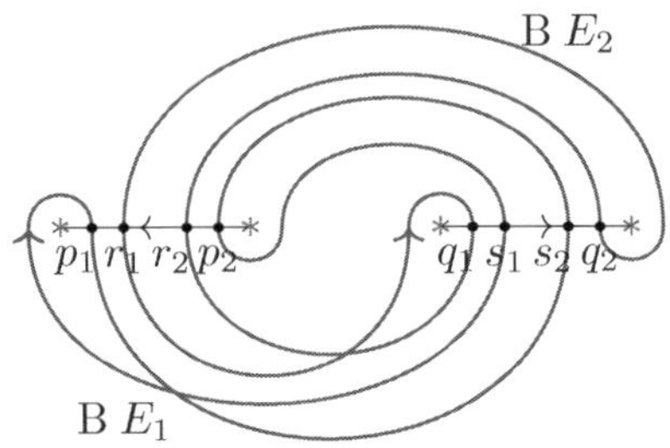

FIGURE 5. A-branes representing a Hopf link and corresponding to the braid presentation shown in Figure 2

The absolute degrees are chosen such that p_1 has $(M, J^0) = (0,0)$. Once this choice is made for the unknot, the absolute degrees for all other knots are fixed as explained in [**4**].

Because p_1 and p_2 do not have the correct degrees to be related by Q, we know that $Qp_1 = 0 = Qp_2$. Hence, we find

$$Hom^{M,J^0}_{\mathscr{D}_Y}(E,I) = \begin{cases} \mathbb{Z} & \textit{if } (M,J^0) = (0,0) \textit{ or } (M,J^0) = (2,-1), \\ 0 & \textit{otherwise.} \end{cases} \tag{3.9}$$

To relate this result to Khovanov homology, note that $w = 0$ and $e = 0$ so that $i_0 = 0$ and $j_0 = 1$. Using Theorem 3.2, we find

$$Kh^{i,j}(L) = \begin{cases} \mathbb{Z} & \textit{if } (i,j) = (0,1) \textit{ or } (i,j) = (0,-1), \\ 0 & \textit{otherwise,} \end{cases} \tag{3.10}$$

which agrees with the known Khovanov homology of the unknot.

EXAMPLE 3.4. As a second example, consider the Hopf link shown in Figure 5, which corresponds to the braid presentation shown in Figure 2. There are eight points in $\mathscr{B}E_{\mathcal{U}} \cap I_{\mathcal{U}}$. These points and their degrees are shown in Table 1.

TABLE 1. The Maslov degrees M and the equivariant degrees J^0 of the intersection points of the branes shown in Figure 5

	p_1q_1	p_1q_2	p_2q_1	p_2q_2	r_1s_1	r_1s_2	r_2s_1	r_2s_2
M	2	4	0	2	0	1	1	2
J^0	0	-2	0	0	1	0	0	-1

One finds the following action of the differential:

$$\begin{aligned} Qp_2q_1 &= r_1s_2 + r_2s_1, \\ Qr_1s_2 &= -p_1q_1 - p_2q_2, \\ Qr_2s_1 &= p_1q_1 + p_2q_2, \end{aligned}$$

giving

$$(3.11)\qquad Hom^{M,J^0}_{\mathscr{D}_Y}(E,I)=\begin{cases}\mathbb{Z} & if\ (M,J^0)\in\{(4,-2),(2,-1),(2,0),(0,1)\},\\ 0 & otherwise.\end{cases}$$

Looking at Figure 2, we see that this presentation has $w=2$ and $e=2$, so the necessary degree shifts to relate our result to Khovanov homology are $i_0=0$ and $j_0=4$. With these shifts, we recover

$$(3.12)\qquad Kh^{i,j}(L)=\begin{cases}\mathbb{Z} & if\ (i,j)\in\{(0,0),(0,2),(2,4),(2,6)\},\\ 0 & otherwise,\end{cases}$$

which is the known Khovanov homology of the positively oriented Hopf link.

4. An algebraic approach

In theory, one could always compute the homology groups $Hom^{*,*}_{\mathscr{D}_Y}(\mathscr{B}E_{\mathcal{U}},I_{\mathcal{U}})$ directly by finding the intersection points of $\mathscr{B}E_{\mathcal{U}}$ with $I_{\mathcal{U}}[M]\{\vec{J}\}$ and counting holomorphic disks between intersection points. However, there is no general method for counting these holomorphic disks. We explain an alternative method of computing $Hom^{*,*}_{\mathscr{D}_Y}(\mathscr{B}E_{\mathcal{U}},I_{\mathcal{U}})$, based in algebra rather than geometry, that is more suitable for calculation.

We will use the fact that $\mathscr{D}_Y$ is generated by Lefshetz thimbles to resolve $\mathscr{B}E_{\mathcal{U}}$ in terms of left thimbles. It will turn out that $I_{\mathcal{U}}$ are right thimbles, and so calculating Homs between a resolution of $\mathscr{B}E_{\mathcal{U}}$ and $I_{\mathcal{U}}$ is easy. A remarkable result is that there is a method for finding a resolution of any $\mathscr{B}E_{\mathcal{U}}$, giving us an algorithm $Hom^{*,*}_{\mathscr{D}_Y}(\mathscr{B}E_{\mathcal{U}},I_{\mathcal{U}})$ for any link [**4**].

4.1. Lefshetz thimbles. Let $\mathcal{C}$ be a critical point of $\operatorname{Re} W$. Each critical point has an associated left Lefshetz thimble $T_{\mathcal{C}}$ and an associated right Lefshetz thimble $I_{\mathcal{C}}$.

DEFINITION 4.1. The left Lefshetz thimble $T_{\mathcal{C}}$ is the set of all initial conditions for downward gradients flows of $\operatorname{Re} W$ that end on $\mathcal{C}$. The right Lefshetz thimble $I_{\mathcal{C}}$ is the set of all initial conditions for upward gradients flows of $\operatorname{Re} W$ that end on $\mathcal{C}$.

It is sufficient for us to know the isotopy classes of the thimbles in $\mathscr{D}_Y$. The left thimbles are isotopic to products of real line Lagrangians running from one end of the cylinder to the other. The right thimbles are isotopic to products of lines running between pairs of punctures. Examples of left and right thimbles are shown in Figure 6. The left thimbles are numbered such that T_i is a line passing between punctures a_i and a_{i-1}, and $T_{i_1\ldots i_d}$ is the product $T_{i_1}\times\ldots\times T_{i_d}$. The right thimbles are numbered similarly.

Left and right Lefshetz thimbles are dual in the sense that they only intersect if they are associated to the same critical point. Hence, we have

$$(4.1)\qquad Hom_{\mathscr{D}_Y}(T_{\mathcal{C}},I_{\mathcal{C}'})=\delta_{\mathcal{C},\mathcal{C}'}.$$

To compute link invariants, it will be necessary to compute Homs between the right thimbles as well. Because these thimbles are noncompact, we define Homs between them using wrapping. Let $T^{\varsigma}_{\mathcal{C}}$ be the right thimble obtained by replacing

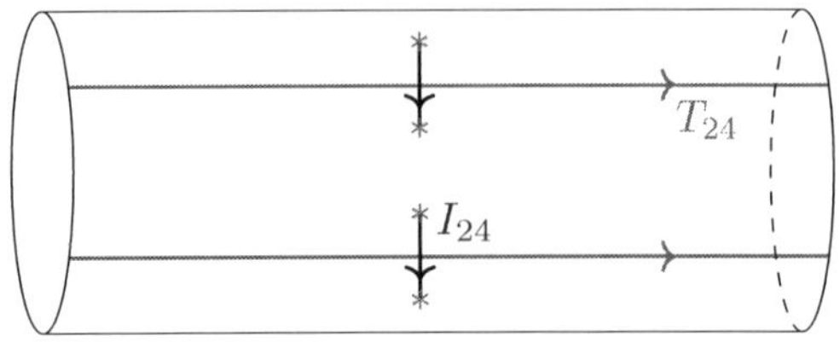

FIGURE 6. Examples of left and right thimbles in $\mathscr{D}_Y$

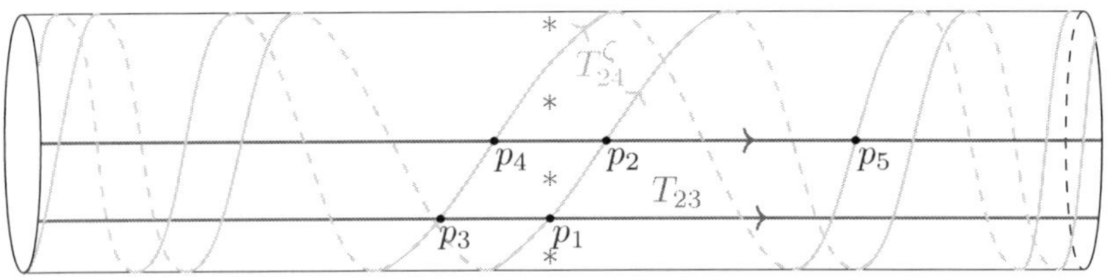

FIGURE 7. An example of a thimble T_{23} and a wrapped thimble T_{24}^{ζ} whose intersection points give $Hom_{\mathscr{D}_Y}(T_{24}, T_{23})$

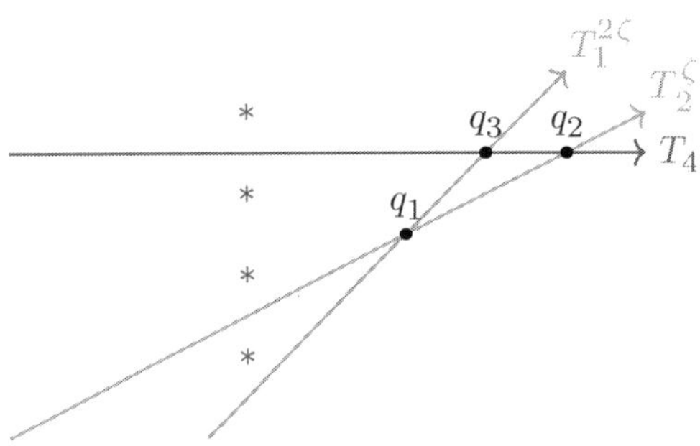

FIGURE 8. A holomorphic triangle giving the relation $q_2 \cdot q_1 = q_3$

W with $We^{-i\zeta}$ [**8**]. The effect of this replacement is shown in Figure 7. We then define morphisms between right thimbles as

$$Hom_{\mathscr{D}_Y}(T_C, T_{C'}) = HF^{0,0}(T_C^{\zeta}, T_{C'}). \tag{4.2}$$

Morphisms between left thimbles are defined analogously.

Morphisms compose via

$$Hom_{\mathscr{D}_Y}(T_j, T_k) \otimes Hom_{\mathscr{D}_Y}(T_i, T_j) \to Hom_{\mathscr{D}_Y}(T_i, T_k), \tag{4.3}$$

which, in terms of Floer theory, is

$$HF^{0,0}(T_j^{\zeta}, T_k) \otimes HF^{0,0}(T_i^{2\zeta}, T_j^{\zeta}) \to HF^{0,0}(T_i^{2\zeta}, T_k). \tag{4.4}$$

Composition of morphisms amounts to counting holomorphic triangles such as the one shown in Figure 8.

Morphisms between all left thimbles form a graded, associative algebra A. It is best described as a strand algebra on a cylinder with $2d$ red strands corresponding to the fixed positions of the punctures and d blue strands corresponding to the positions of the thimbles. The thimble T_i is represented by a blue strand between the ith and $(i-1)$th red strands. The thimble $T_{i_1 \dots i_d}$ is represented by the d blue strands representing $T_{i_1}, \dots, T_{i_d}$.

Consider a morphism from $T_{\mathcal{C}}$ to $T_{\mathcal{C}'}$ and its associated intersection point. The ordering of the strands at the base of the strand algebra cylinder corresponds to $T_{\mathcal{C}}$ and the ordering of the strands at the top of the strand algebra cylinder corresponds to $T_{\mathcal{C}'}$. The intersection point uniquely specifies how the strands connect the base of the cylinder to the top of the cylinder.

For example, the three intersection points $p_1p_4, p_2p_3, p_1p_5 \in Hom_{\mathscr{D}_Y}(T_{24}, T_{23})$ shown in Figure 7 correspond to the diagrams

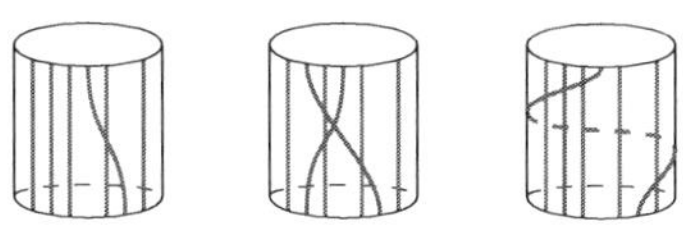

respectively.

Composition of morphisms is given by stacking the corresponding cylinders and rescaling. For example, the relation $q_2 \cdot q_1 = q_3$ shown in Figure 8 is given in terms of strand diagrams as

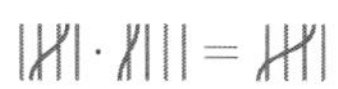

(4.5)

where we have omitted drawing the cylinders for convenience. For knots in $\mathbb{R}^3$, one will never encounter algebra elements that wrap behind the cylinder.

Altogether, the strand algebra is generated by certain "bits":

that have $J^0 = 0$, $J^0 = 1$, and $J^0 = -1$, respectively. The equivariant degree of each strand algebra element is the sum of the equivariant degrees of the "bits" that compose it.

It turns out that the branes in $\mathscr{D}_Y$ have slightly more structure than we have explained so far. Each brane comes equipped with a local system of modules of a graded algebra $\mathcal{B}$. A morphism between branes will also involve homomorphisms of their local systems at the corresponding intersection point. The local systems can be viewed as coming from a larger A-model $(\mathcal{Y}, \mathcal{W})$ where $\mathcal{Y}$ fibers over Y with $(\mathbb{C}^\times)^d$ fibers. Each thimble $T_{\mathcal{C}}$ corresponds to a thimble $\mathcal{T}_{\mathcal{C}}$ in $\mathcal{Y}$ where $T_{\mathcal{C}}$ and $\mathcal{T}_{\mathcal{C}}$ agree on Y. From this perspective, one can view homomorphisms of $\mathcal{B}$-modules as coming from the intersections of thimbles in the fibers over their intersection points in Y. For details, see [**3**].

For our purposes, it is sufficient to only consider the effect of the local systems on the thimble algebra. The effect is to add another generator, a dot with $J^0 = 1$ that may be placed on any blue strand.

The strand algebra contains several nontrivial relations that are proven in [**3**] by counting holomorphic disks in the derived Fukaya-Seidel category associated to

$(\mathcal{Y}, \mathcal{W})$. The relations are

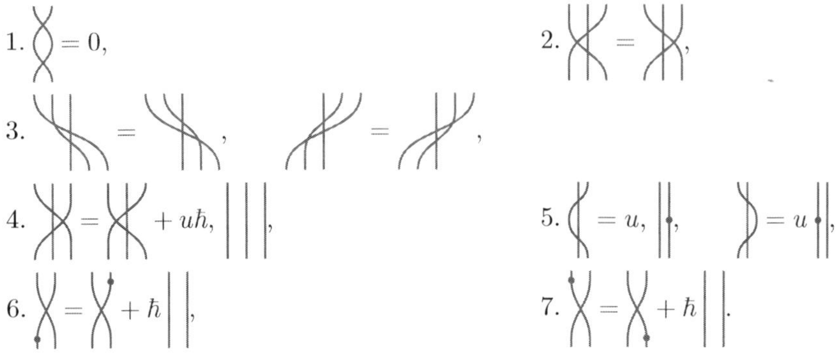

One may always rescale the algebra generators to set $u = 1$ and $\hbar = 1$, but we leave them in for convenience. With $u = 1$, these relations are the same as those for the KLRW algebra [**13**]. The fact that the thimble algebra coincides with the KLRW algebra is expected from homological mirror symmetry [**2**].

REMARK 4.2. Let Δ be the divisor of $\prod_{\alpha<\beta}(y_\alpha - y_\beta)$. The algebra at $\hbar = 0$ is the strand algebra on the space $Y_0 = Y \backslash \Delta$, the complement of the diagonal Δ in Y. The theory on Y_0 is similar to the theory on Y except that disks cannot pass through any point with $y_\alpha = y_\beta$, so every disk is a product of one-dimensional disks.

REMARK 4.3. Let D_a be the divisor of $\prod_{\alpha,i}(y_\alpha - a_i)$. The algebra at $u = 0$ and $\hbar = 0$ is the strand algebra on the space $Y_0 \backslash D_a$. In this theory, disks are not allowed to pass through the punctures. It corresponds to all branes being equipped with a trivial local system.

4.2. Resolutions of A-branes. Every A-brane in $\mathscr{D}_Y$ admits a resolution in terms of thimbles. Resolutions of one-dimensional branes can be easily written down from geometry. We will use these one-dimensional resolutions as building blocks to find resolutions of higher dimensional branes.

4.2.1. *Resolutions of one-dimensional Lagrangians.* We use mapping cones to construct of one-dimensional Lagrangians.

DEFINITION 4.4. Let (L_1, δ^{L_1}) and (L_2, δ^{L_2}) be chain complexes and let $f : L_1 \to L_2$ be a chain map. The cone over f is the complex $L_1[1] \oplus L_2$ equipped with the differential

$$\delta = \begin{pmatrix} \delta^{L_1} & 0 \\ f & -\delta^{L_2} \end{pmatrix} \tag{4.6}$$

In one dimension, the cone over $f : L_1 \to L_2$ amount to gluing $L_1[1]$ and L_2 at the point $p \in Hom_{\mathscr{D}_Y}(L_1, L_2)$. For example, consider the map  $: T_3\{-1\} \to T_2$. The cone over  is

$$T_3\{-1\} \longrightarrow T_2. \tag{4.7}$$

Geometrically, this cone is the resolution of the brane shown in Figure 9.

In $\mathscr{D}_{Y_0 \backslash D_a}$, resolutions of more complicated one-dimensional branes can be built by gluing thimbles along maps. For resolutions of branes in $\mathscr{D}_{Y_0}$ however, we need to add additional maps. To find a resolution of a brane in $\mathscr{D}_{Y_0}$, we start with a

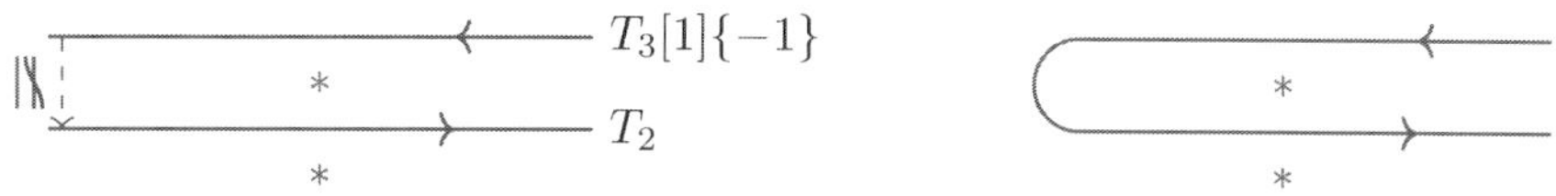

FIGURE 9. On the left, two thimbles with a map between them and on the right, the Lagrangian corresponding to the cone over that map

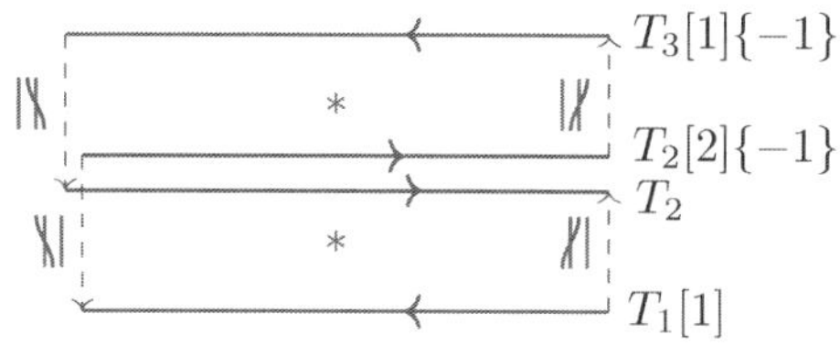

FIGURE 10. The thimbles and maps between them making up the resolution of the brane E

resolution of a brane in $\mathscr{D}_{Y_0\setminus D_a}$. Call the differential δ, and note that $\delta^2|_{u=0} = 0$. We then make an ansatz for the differential in $\mathscr{D}_{Y_0}$ by adding to δ all possible maps with dots as allowed by degree with arbitrary coefficients. We then use the condition $\delta^2 = 0$ to fix the coefficients. This uniquely determines a brane in $\mathscr{D}_{Y_0}$. Since the brane is one-dimensional, it is also a brane in $\mathscr{D}_Y$.

EXAMPLE 4.5. Consider the figure-eight brane E shown in Figure 4, which we refer to as E_2 from now on. We can break E_2 into thimbles and maps between them as shown in Figure 10. Assembling the thimbles and maps into a complex, we find

(4.8) 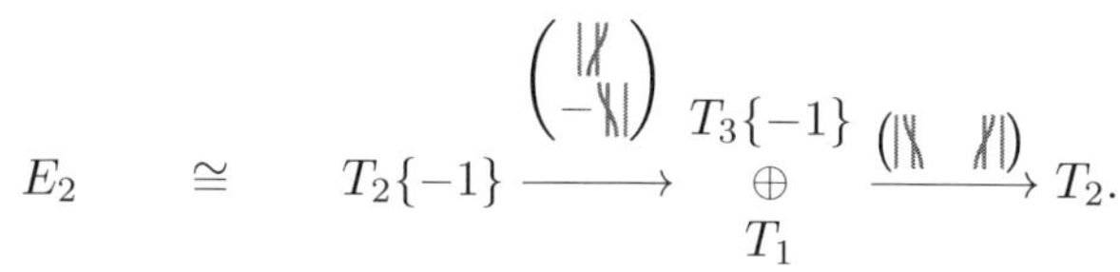

This differential squares to zero in $\mathscr{D}_Y$, so we are done.

As a check, note that applying $Hom^{*,*}_{\mathscr{D}_Y}(-, I_2)$ to E_2 recovers the correct Khovanov homology of the unknot. The resolution of E_2 contains two copies of T_2, so we find

$$Hom^{M,J^0}_{\mathscr{D}_Y}(E_2, I_2) = \begin{cases} \mathbb{Z} & \textit{if } (M, J^0) = (0,0) \textit{ or } (M, J^0) = (2,-1), \\ 0 & \textit{otherwise}, \end{cases} \tag{4.9}$$

agreeing with what we found previously.

EXAMPLE 4.6. Consider the brane $\mathscr{B}E_1$ shown in Figure 5. We can decompose $\mathscr{B}E_1$ as thimbles and maps between them as shown in Figure 11. Assembling these thimbles and maps into a chain complex gives

(4.10) 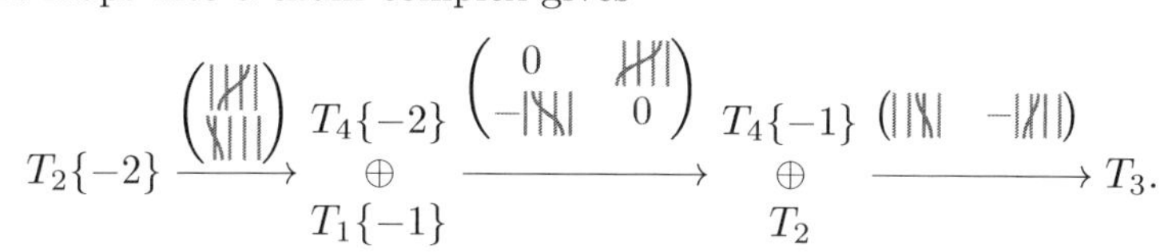

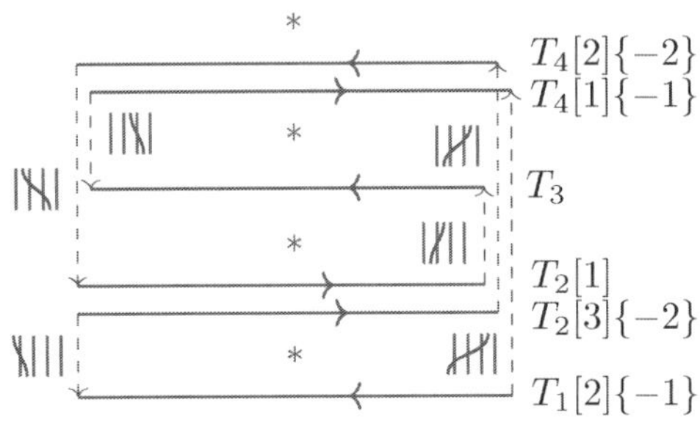

FIGURE 11. The thimbles and maps between them making up the resolution of the brane $\mathscr{B}E_1$ shown in Figure 5

This differential squares to zero when $u = 0$ but not when $u \neq 0$, so this complex is a resolution of the brane in $\mathscr{D}_{Y_0 \setminus D_a}$ but not in $\mathscr{D}_Y$. In order to find a resolution in $\mathscr{D}_Y$, we must add dotted correction terms. Adding in all possible dotted corrections with undetermined coefficients x_i, the complex becomes

(4.11)

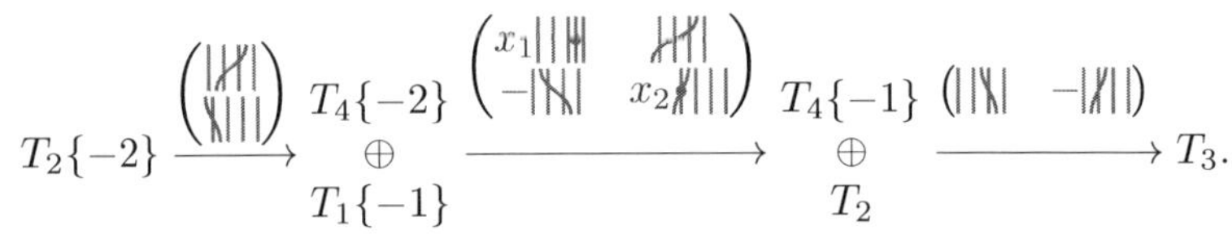

Requiring $\delta^2 = 0$ fixes $x_1 = -u$ and $x_2 = u$, giving us a resolution of $\mathscr{B}E_1$ in $\mathscr{D}_Y$. One can follow a similar procedure to resolve $\mathscr{B}E_2$.

4.2.2. *Resolutions of higher dimensional Lagrangians.* Given a d-dimensional Lagrangian L, we first break it into a product of one-dimensional Lagrangians L_i for $i = 1, \dots, d$. We then find a resolution of each one-dimensional Lagrangian using the methods just described. To construct a resolution L, we iteratively take the product of $L_1 \times \dots \times L_i$ with L_{i+1}, adding in extra maps to ensure $\delta^2 = 0$ at each step. We will briefly outline how this construction goes. More details can be found in [**4**].

Let δ^{L_i} be the differential on L_i. Start by considering the product $L_1 \times L_2$ where differential acts by

$$\delta_{\text{geo}}^{L_1 \times L_2}(\ell_1, \ell_2) = (\delta^{L_1}\ell_1, \ell_2) + (-1)^{M(\ell_1)}(\ell_1, \delta^{L_2}\ell_2). \tag{4.12}$$

This differential does not necessarily satisfy $\left(\delta_{\text{geo}}^{L_1 \times L_2}\right)^2 = 0$, so we add a correction term $\tilde{\delta}^{L_1 \times L_2}$ to get a differential

$$\delta^{L_1 \times L_2} = \delta_{\text{geo}}^{L_1 \times L_2} + \tilde{\delta}^{L_1 \times L_2} \tag{4.13}$$

that does square to zero. Repeat this construction using $L_1 \times L_2$ and L_3 to construct $L_1 \times L_2 \times L_3$ and so on until arriving at $L_1 \times \dots \times L_d$ equipped with a differential $\delta^{L_1 \times \dots \times L_d}$ satisfying $\left(\delta^{L_1 \times \dots \times L_d}\right)^2 = 0$.

To find $\tilde{\delta}$ at each step, note that the possible maps it could contain are restricted by the equivariant degrees of the thimbles in $L_1 \times \dots \times L_d$. As a result, there are finite numbers of possibilities for $\tilde{\delta}$. One can make an ansatz for $\tilde{\delta}$ by adding all possible maps with undetermined coefficients and then using the condition $\delta^2 = 0$ to fix the coefficients. We only add maps where there is not already a map in δ_{geo}.

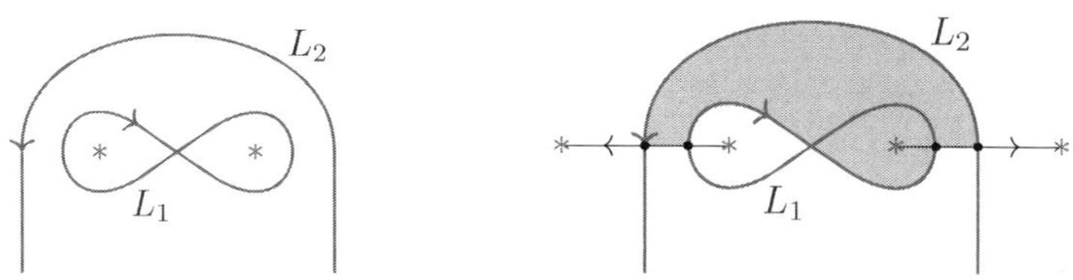

(A) Simple two-dimensional brane

(B) Disk from intersection with I_{13}

FIGURE 12. A simple example of a two-dimensional brane $L_1 \times L_2$ and the holomorphic disk that contributes to the Floer differential on intersection of $L_1 \times L_2$ with I_{13}

EXAMPLE 4.7. Consider the two-dimensional brane $L_1 \times L_2$ shown in Figure 12a. The resolutions of the one-dimensional branes $L_1 \cong E_2$ and L_2 are easily found to be

(4.14)

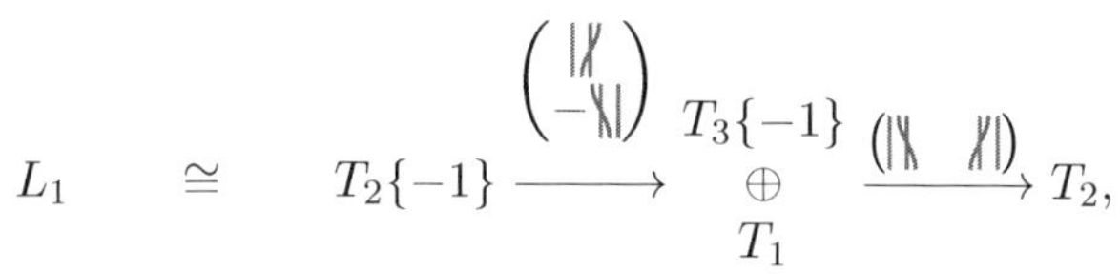

and

(4.15)

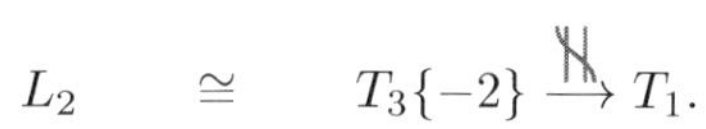

Taking the product of L_1 with L_2, we find
(4.16)

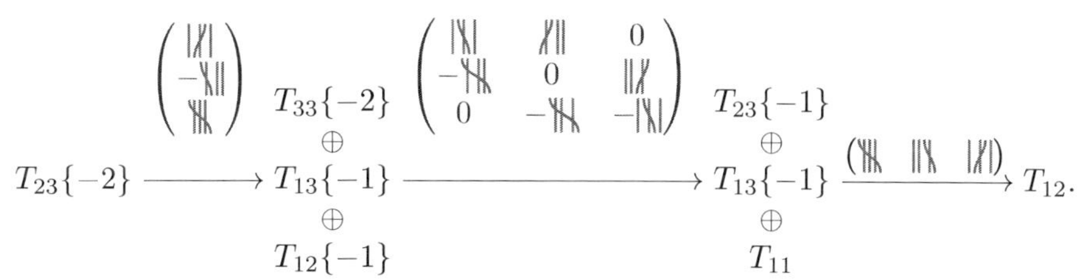

Notice that the equivariant degrees are different from the naive expectation because the crossing of two thimbles in the strand algebra has nonzero equivariant degree. The equivariant degrees of the thimbles are adjusted to stay consistent with the equivariant degrees of the maps.

One can verify that the differential on the product complex squares to zero only in $\mathscr{D}_{Y_0}$ but not in $\mathscr{D}_Y$, so we must add a correction to the differential. There are only two possible corrections allowed by degrees. With undetermined coefficients

x_i for the correction terms, the differential is
(4.17)

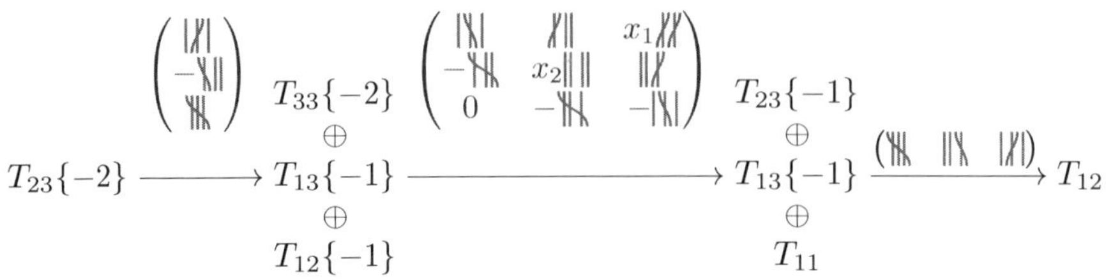

Solving $\delta^2 = 0$ in $\mathscr{D}_Y$ fixes $x_1 = 0$ and $x_2 = -u\hbar$, giving us a resolution for $L_1 \times L_2$.

As a check, we apply $Hom^{*,*}_{\mathscr{D}_Y}(-, I_{13})$ to the resolution and compare it to the geometric result. We find

$$Hom^{*,*}_{\mathscr{D}_Y}(L_1 \times L_2, I_{13}) = H^*\left(\mathbb{Z} \xrightarrow{u\hbar} \mathbb{Z}\right) = 0. \tag{4.18}$$

The corresponding geometric brane configuration is shown in Figure 12b. There are two intersection points connected by a holomorphic disk, shaded gray, so the resulting Floer complex matches the result we obtained algebraically.

REMARK 4.8. One may have noticed that the resolution of $L_1 \times L_2$ may have just as easily been computed by finding a chain map $f : E_2 \times T_3\{-1\} \to T_1 \times E_2$ and taking a cone over f. In most examples however, the method we used here is more tractable, so we felt there was value in explaining how it works for a simple example.

For more complicated examples, it is easiest to work iteratively in $\hbar$ to solve $\delta^2 = 0$. Expand δ as

$$\delta = \sum_{k=0}^{\infty} \hbar^k \delta_k \tag{4.19}$$

so that

$$\delta^2 = \delta_0^2 + \hbar(\delta_0\delta_1 + \delta_1\delta_0) + \hbar^2(\delta_1^2 + \delta_0\delta_2 + \delta_2\delta_0) + \cdots = 0 \tag{4.20}$$

is satisfied at each order in $\hbar$. Note that $\delta^2_{\text{geo}}|_{\hbar=0} = 0$, so we can identify $\delta_0 = \delta_{\text{geo}}$. Then $\tilde{\delta}$ contains terms of at least order $\hbar$ and we identify

$$\tilde{\delta} = \sum_{k=1}^{\infty} \hbar^k \delta_k. \tag{4.21}$$

Also notice that at order k, the equation $\delta^2 = 0$ is linear in δ_k once $\delta_0, \ldots, \delta_{k-1}$ are fixed, so finding the solution very tractable on a computer.

Addressing the issue of uniqueness of this construction requires considerable care. We claim that this construction is unique and refer the reader to [**4**] for the proof.

4.3. Proof of invariance. We give a brief sketch of a proof of Theorem 3.1 here. For a more detailed version, see [**4**], or for a proof based on homological mirror symmetry, see [**2**].

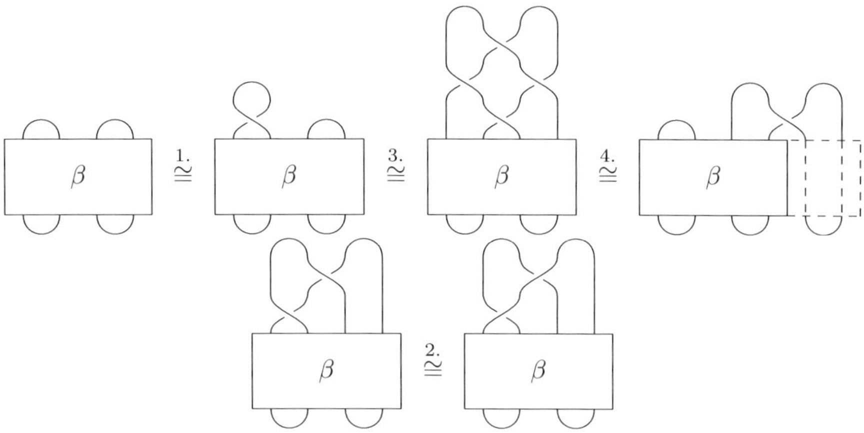

FIGURE 13. The four relations that must be satisfied for a quantity to be a link invariant

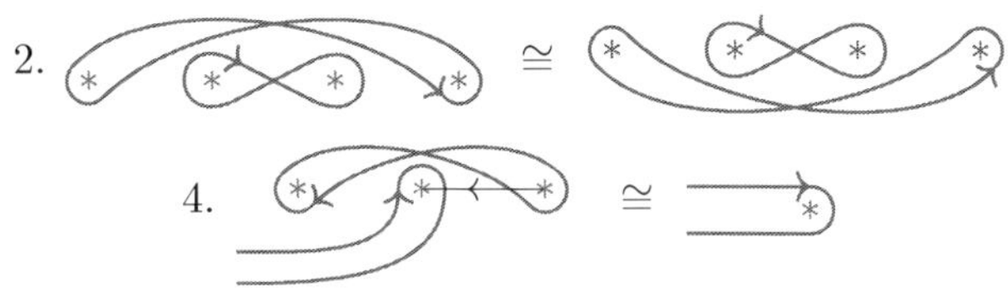

FIGURE 14. Invariance relations 2 and 4 shown in terms of branes

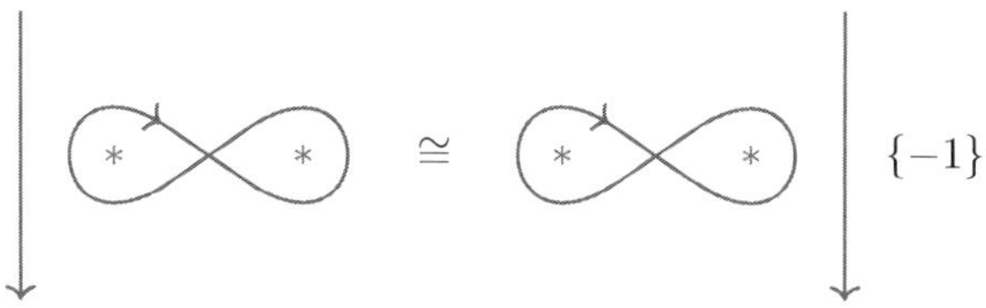

FIGURE 15. Two equivalent branes in $\mathscr{D}_Y$

SKETCH OF PROOF OF THEOREM 3.1. We will show that $Hom^{*,*}_{\mathscr{D}_Y}(\mathscr{B}E_{\mathcal{U}}, I_{\mathcal{U}})$ satisfy the four relations shown in Figure 13, which is sufficient to prove invariance [**6**], [**7**]. The fact that relations 1 and 3 are satisfied is immediate. One only has to be careful to keep track of the overall degree shifts.

Relations 2 and 4 require more work. These relations are shown in terms of A-branes in Figure 14. Both of these relations follow from the relation shown in Figure 15. Hence, it is sufficient to prove this relation. One can do so by constructing resolutions of the two branes using the method described above and then explicitly showing that the resolutions are homotopy equivalent. □

References

[1] M. Aganagic, *Knot categorification from mirror symmetry, part I: Coherent sheaves*, 2020. arXiv:2004.14518.

[2] M. Aganagic, *Knot categorification from mirror symmetry, part II: Lagrangians*, 2021. arXiv:2105.06039.

[3] M. Aganagic, I. Danilenko, Y. Li, V. Shende and P. Zhou, *Homological mirror symmetry for monopole moduli spaces*, to appear.

[4] M. Aganagic, E. LePage and M. Rapcak, *Homological knot invariants from Floer theory*, 2023. arXiv:2305.13480.

[5] Denis Auroux, *A beginner's introduction to Fukaya categories*, Contact and symplectic topology, Bolyai Soc. Math. Stud., vol. 26, János Bolyai Math. Soc., Budapest, 2014, pp. 85–136, DOI 10.1007/978-3-319-02036-5_3. MR3220941

[6] Stephen Bigelow, *A homological definition of the Jones polynomial*, Invariants of knots and 3-manifolds (Kyoto, 2001), Geom. Topol. Monogr., vol. 4, Geom. Topol. Publ., Coventry, 2002, pp. 29–41, DOI 10.2140/gtm.2002.4.29. MR2002601

[7] Joan S. Birman, *On the stable equivalence of plat representations of knots and links*, Canadian J. Math. **28** (1976), no. 2, 264–290, DOI 10.4153/CJM-1976-030-1. MR402715

[8] K. Hori, A. Iqbal and C. Vafa, *D-branes and mirror symmetry*, 2000. arXiv:hep-th/0005247

[9] Vaughan F. R. Jones, *A polynomial invariant for knots via von Neumann algebras*, Bull. Amer. Math. Soc. (N.S.) **12** (1985), no. 1, 103–111, DOI 10.1090/S0273-0979-1985-15304-2. MR766964

[10] Mikhail Khovanov, *A categorification of the Jones polynomial*, Duke Math. J. **101** (2000), no. 3, 359–426, DOI 10.1215/S0012-7094-00-10131-7. MR1740682

[11] Mikhail Khovanov and Lev Rozansky, *Matrix factorizations and link homology*, Fund. Math. **199** (2008), no. 1, 1–91, DOI 10.4064/fm199-1-1. MR2391017

[12] Peter Ozsváth and Zoltán Szabó, *An overview of knot Floer homology*, Modern geometry: a celebration of the work of Simon Donaldson, Proc. Sympos. Pure Math., vol. 99, Amer. Math. Soc., Providence, RI, 2018, pp. 213–249, DOI 10.1090/pspum/099/08. MR3838884

[13] Ben Webster, *Knot invariants and higher representation theory*, Mem. Amer. Math. Soc. **250** (2017), no. 1191, v+141, DOI 10.1090/memo/1191. MR3709726

[14] Edward Witten, *Quantum field theory and the Jones polynomial*, Comm. Math. Phys. **121** (1989), no. 3, 351–399. MR990772

Department of Physics, University of California, Berkeley, California, 94720
Email address: elepage@berkeley.edu

Proceedings of Symposia in Pure Mathematics
Volume **107**, 2024
https://doi.org/10.1090/pspum/107/01950

Analytic continuation and supersymmetry

Nikita Nekrasov

Abstract. Real-valued parameters of quantum field theory, such as Planck constant $\hbar$, the coupling constants $g^{-2}, \theta, \ldots$, temperature β^{-1} and spacetime metric $\mathbf{g}_{\mu\nu}$, chemical potentials or background gauge fields $\mathbf{A}^a_\mu$, can be made complex. In perturbative string theory, the worldsheet metric is sometime Lorentzian, somewhere Euclidean, and likely complex in between. Less often, integral, or quantized, parameters, such as the number n of flavors or species (as in replica method), the number of colors N (as in matrix models, master field), spacetime dimension D (dimensional regularization), or angular momentum l (Regge poles) are continued to complex values. We observe that many, if not all of these analytic continuations can be realized with conventional supersymmetric quantum field (or M-) theory. The supersymmetry is softly broken near a defect supporting the theory being analytically continued in some of its parameters.

Somewhat abridged talk at Strings-Math'2022 in Warsaw.

1. Introduction

It is interesting to study analytic continuations. It is especially interesting to study analytic continuations of quantum field theory. Usually the real-valued parameters, such as Planck constant $\hbar$, the coupling constants $g^{-2}, \theta, \ldots$, temperature β^{-1} and spacetime metric $\mathbf{g}_{\mu\nu}$, chemical potentials or background gauge fields $\mathbf{A}^a_\mu$, can be made complex. For example, to a quantum mechanical system with discrete spectrum Hamiltonian $\hat{H}$ we associate the thermal partition function

$$\mathcal{Z}(\beta) = \mathrm{Tr}\, e^{-\beta \hat{H}} = \oint_C dz e^{-\beta z}\, \mathrm{Tr} G_{\hat{H}}(z) \tag{1}$$

where $G_{\hat{H}}(z)$ is the resolvent [**39**], and C is the contour circling around the spectrum $\sigma(\hat{H})$. Using path integral

$$\mathcal{Z}(\beta) = \int_{L\mathcal{P}} Dp(t)Dq(t)\, e^{\mathrm{i}\oint pdq - \oint H(p(t),q(t))d\beta}\ , \tag{2}$$

we represent the partition function as an integral over the space $L\mathcal{P}$ of loops on a phase space, with the measure defined using a complex-valued one form $d\beta = \beta'(t)dt$ on S^1. If we eliminate momentum by its equations of motion, e.g. for H which is

Research was partly supported by the National Science Foundation by award NSF PHY Award 2310279. Any opinions expressed are solely our own and do not represent the views of the National Science Foundation.

quadratic in p, the saddle point dominating (2), does not, typically, fit in $\mathcal{P}$, since

$$p_i^{\rm saddle}(t) = \frac{1}{\mathrm{i}\beta'(t)}\, g_{ij}\frac{dq^j}{dt}\,, \qquad g^{ij} = \frac{\partial^2 H}{\partial p_i \partial p_j}$$

is complex unless $\beta'(t) \in \mathrm{i}\mathbb{R}$. Once $p(t)$ can become complex, so can $q(t)$. One is led to the idea that path integral contour $L\mathcal{P}$ can be deformed into the complex domain $L\mathcal{P}_{\mathbb{C}}$, as a middle dimensional contour Σ, which splits, in the appropriate homology group, as a sum of Lefschetz thimble cycles associated with the critical points of the exponential in (2).

The classification of such cycles in one- and two-dimensional quantum field theories has recently been revived in e.g. [**18**, **19**, **25**].

String worldsheet metric is sometime Lorentzian, somewhere Euclidean, and likely complex in between [**42**]. Less often, integral, or quantized, parameters, such as the number n of flavors or species (replica method), the number of colors N (matrix models, master field), spacetime dimension D (dimensional regularization), or angular momentum l (Regge poles) are continued into the complex domain. For example, the character (twisted partition function) of a particle on S^2 (phase space)

$$\chi_j(\theta) = \frac{\sin\left((j+\frac{1}{2})\theta\right)}{\sin(\frac{\theta}{2})} \tag{3}$$

admits analytic continuation to complex θ, j, etc. It was I. Gelfand's dream to understand the dimensional regularization used in computations of Feynman diagrams in quantum field theory and the subsequent renormalization, through the representation theory of the non-compact Lie group $SL(2,\mathbb{C})$ (whose real form is the Lorentz group of the four dimensional Minkowski space which looks like our space-time).

Theoretical physics often uses a trick of analytic continuation. In studying the disordered systems, where a free energy of a statistical mechanical model

$$F(J) = -\frac{1}{\beta}\log Z(\beta, J) \tag{4}$$

is to be averaged, e.g.

$$\langle F(J)\rangle_J = \int DJ\, e^{-\|J\|^2}\, F(J) \tag{5}$$

over some manifold of parameters J, e.g. in the spin glass models, where

$$Z(\beta, J) = \sum_{\{\sigma_i = \pm 1\}} e^{-\beta \sum\limits_{\langle i,j\rangle} J_{ij}\sigma_i\sigma_j} \tag{6}$$

In replica trick, one computes the averages of n'th powers of Z, then takes the limit $n \to 0$ to extract the free energy. Another popular application of the replica trick is in the computation of von Neumann entropy $\mathrm{Tr}\rho\log\rho$ which is extracted in the $n \to 1$ limit of $\mathrm{Tr}\rho^n$. Thus, one studies the statistical mechanics of n copies of the original system. For integer positive n this makes sense, but what is the meaning of the n complex?

Curiously, we find several examples of systems for which these analytic continuations can be realized with conventional supersymmetric quantum field (or M-) theory. The supersymmetry is softly broken near a defect supporting the theory being analytically continued in some of its parameters. We are interested in analytic continuations of spins, dimensions, temperatures, fugacities etc.

2. Models with $O(N)$ symmetry

In our first model we study the conformally coupled scalar living on S^{N-1}. Conformally coupled scalar field living on an N-dimensional Riemanian manifold M^N is described by the action

$$S = \frac{1}{2}\int_{M^N} d^N\xi\sqrt{g}\left(g^{mn}\partial_m\varphi\partial_n\varphi + \frac{N-2}{2N-2}R(g)\varphi^2\right) \tag{7}$$

In this paper we are interested in the conformally flat manifolds of the form $S^{N-1}\times S^1$ (more general case $S^{N_1-1}\times S^{N_2-1}$ with $N_1, N_2 \geq 3$ is also connected to supersymmetric systems, but the story is more complicated). We endow it with the metric

$$ds^2_{M^N} = \beta^2 dt^2 + R^2 d\Omega^2_{S^{N-1}} \tag{8}$$

with $t \sim t+1$,

$$\frac{\beta R^{N-1}}{2}\int_{S^{N-1}\times S^1} dt d^{N-1}\Omega\left[\frac{1}{\beta^2}(\partial_t\varphi)^2 + \frac{1}{R^2}\left(\nabla\varphi\cdot\nabla\varphi + \left(\frac{N}{2}-1\right)^2\varphi^2\right)\right] \tag{9}$$

We can decompose $\varphi(t,x) = \sum_{l=0}^{\infty}\sum_{A=1}^{d_l(N)}\varphi_{A,l}(t)Y_{A,l}(x)$ into the spherical functions, where

$$-\Delta_{S_R^{N-1}}Y_{A,l} = \frac{l(l+N-2)}{R^2}Y_{A,l}\ . \tag{10}$$

We see (9) is a Lagrangian describing an infinite set $(\varphi_{A,l}(t))_{l=0}^{\infty}{}_{A=1}^{d_l(N)}$ of harmonic oscillators with frequencies $\frac{l+\frac{N}{2}-1}{R}$. The multiplicity $d_l(N)$ is computed from the generating function

$$\sum_{l=0}^{\infty} x^l d_l(N) = \frac{1+x}{(1-x)^{N-1}} \tag{11}$$

(this is $1-x^2$ times the character of the space of polynomials in N variables, with x being the fugacity for dilatation symmetry). Thus

$$d_l(N) = \binom{N+l-2}{l} + \binom{N+l-3}{l-1}\,, \tag{12}$$

$$d_l(2) = 2-\delta_{l,0}\,,\ d_l(3) = 2l+1\,,\ d_l(4) = (l+1)^2\,,\ldots$$

Note that $d_l(N)$ admits analytic continuation to complex values of N, as

$$D_r(N) = \frac{2r}{\Gamma(N-1)}\frac{\Gamma(r+N/2-1)}{\Gamma(r-N/2+2)}\,, \qquad r = N/2-1+l \tag{13}$$

from which the symmetry

$$D_r(N) = (-1)^N D_{-r}(N) \tag{14}$$

is readily seen. This is related to the symmetry $x \mapsto x^{-1}$ of (11) accompanied by the $(-1)^N x^{N-2}$ prefactor.

We can thus write the partition function of conformal scalar on $S_R^{N-1} \times S_\beta^1$:

$$Z_N(q) = \mathrm{Tr}_{\mathcal{H}_{S_R^{N-1}}} e^{-\beta \hat{H}} \tag{15}$$
$$= q^{c_0} \prod_{r \geq \frac{N}{2}-1}^{\infty} \frac{1}{(1-q^r)^{D_r(N)}}$$
$$= q^{c_0} \exp \sum_{n=1}^{\infty} \frac{1}{n} \frac{q^{-n/2}+q^{n/2}}{(q^{-n/2}-q^{n/2})^{N-1}}$$

where q is the conformal invariant:

$$q = e^{-\frac{\beta}{R}} \tag{16}$$

and c_0/R is the Casimir energy.[1] We can refine (15) by introducing the general $O(N)$ twisted boundary conditions:

$$\varphi(t+1, x) = \varphi(t, \mathbf{g} \cdot x) \tag{18}$$

where $\mathbf{g} \in O(N)$ is characterized by $[N/2]$ angles $\theta_1, \ldots, \theta_{[N/2]}$ and, for odd N, in addition, by $\xi = \pm 1$:

(19)

$$\begin{cases} N = 2m\,, & Z_N(q, \theta_1, \ldots, \theta_m) = \exp \sum\limits_{n=1}^{\infty} \frac{q^{-n}-q^n}{n} \prod\limits_{a=1}^{m} \frac{1}{q^n - 2\cos(n\theta_a) + q^{-n}} \\ N = 2m+1\,, & Z_N(q, \theta_1, \ldots, \theta_m, \xi) = \exp \sum\limits_{n=1}^{\infty} \frac{q^{-n/2}+q^{n/2}\xi^n}{n} \prod\limits_{a=1}^{m} \frac{1}{q^n - 2\cos(n\theta_a) + q^{-n}} \end{cases}$$

For three dimensional conformally coupled scalar this formalism gives:

$$Z(q, \theta, \xi) = \exp \sum_{n=1}^{\infty} \frac{1}{n} \frac{q^{-n/2} + q^{n/2}\xi^n}{q^n + q^{-n} - 2\cos(n\theta)} \tag{20}$$

One can get similar formulas for the conformally coupled fermions. These three-dimensional partition functions are closely related to MacMahon function

$$M(q) = \prod_{n=1}^{\infty} \frac{1}{(1-q^n)^n} = \exp \sum_{n=1}^{\infty} \frac{1}{n} \frac{q^n}{(1-q^n)^2} \tag{21}$$

which counts plane partitions [**34**], is featured in representation theory of $W_{1+\infty}$-algebra [**2**, **3**], and in the theory of topological strings [**6**, **35**]. Recall [**32**] that the supersymmetric partition function of the Ω-deformed maximally supersymmetric $U(1)$ Yang-Mills theory on noncommutative $\mathbb{R}^6$ (or $D5 - D(-1)$ system) is equal to

$$Z_6(q, \varepsilon_1, \varepsilon_2, \varepsilon_3) = M(q)^{-\frac{(\varepsilon_1+\varepsilon_2)(\varepsilon_1+\varepsilon_3)(\varepsilon_2+\varepsilon_3)}{\varepsilon_1\varepsilon_2\varepsilon_3}} \tag{22}$$

where q is the $D(-1)$-instanton fugacity, and $\varepsilon_1, \varepsilon_2, \varepsilon_3$ are the three parameters of the $U(3)$ Ω-deformation. The group $U(3) \subset O(6)$ rotates the space-time $\mathbb{R}^6$ and the transverse $\mathbb{R}^2$ via determinant $U(1) \approx O(2)$. Thus, the replica of three dimensional bosons and fermions can be realized physically as six dimensional gauge theory

[1] One can compute c_0 as a regularized sum

$$c_0 = \frac{1}{2} \sum_{l=0}^{\infty} d_l(N)\,(l + N/2 - 1) := -\frac{1}{2} \mathrm{Coeff}_{\tau^1} \left. \frac{e^{\tau/2} + e^{-\tau/2}}{(e^{\tau/2} - e^{-\tau/2})^{N-1}} \right|_{\tau=0} \tag{17}$$

See [**1**] for more systematic treatment

by playing with the Ω-deformation parameters. Once the Ω-deformation preserves more supersymmetry, the exponent in (22) becomes 1.

Similarly, two dimensional boson partition function ($N = 2$)

$$Z(q) = q^{-\frac{1}{12}} \prod_{n=1}^{\infty} \frac{1}{(1-q^n)^2} \tag{23}$$

is related to Euler function

$$\phi(q) = \prod_{n=1}^{\infty} (1-q^n) \tag{24}$$

The replica of the two dimensional chiral bosons is given by the Ω-deformed $\mathcal{N} = 2^*$ four dimensional $U(1)$ super-Yang-Mills theory:

$$Z_4(q, \varepsilon_1, \varepsilon_2, \varepsilon_3) = \left(\frac{1}{\phi(q)} \right)^{\frac{(\varepsilon_1+\varepsilon_3)(\varepsilon_1+\varepsilon_2)}{\varepsilon_1\varepsilon_2}} \tag{25}$$

where ε_3 is the mass of the adjoint hypermultiplet [**28**, **31**]. One can refine the "replica theories" by lifting them to the supersymmetric theories in one dimension higher, compactified on a circle. The ε-parameters become q-parameters. For example, the two dimensional example refines to

$$Z_5(q, q_1, q_2, q_3) = \exp \sum_{n=1}^{\infty} \frac{q^n}{n} \frac{(1-q_1^n q_3^n)(1-q_2^n q_3^n)}{(1-q_1^n)(1-q_2^n)(1-q^n)} \tag{26}$$

which has two interpretations: as Witten index of five dimensional maximally supersymmetric Yang-Mills theory compactified on a circle with the Lorentz and R-symmetry twists compatible with a fraction of supersymmetry, or as elliptic genus of a six dimensional $(2,0)$-superconformal tensor theory compactified on a torus [**32**]. The refinement of the replica of three dimensional theory brings about the (linearized) eleven dimensional supegravity on $\mathbb{R}^{10} \times S^1$, subject to an $SU(5)$ twist [**29**, **32**]. Recently this partition function was reproduced using a twisted version of supergravity [**36**].

Our second problem related to S^{N-1} is the classical two dimensional $O(N)$-model, i.e. the sigma model with S^{N-1} as a target. We found, in [**18**, **19**], that the complex critical points of the action

$$\int_{T^2} \sqrt{h} h^{ab} \nabla_a \vec{n} \cdot \nabla_b \vec{n} \tag{27}$$

with $\vec{n} \in \mathbb{C}^N$, $\vec{n} \cdot \vec{n} = \sum_{i=1}^N n_i^2 = 1$, and complex flat metric h_{ab} on T^2 are described in terms of an algebraic integrable system. The vector $\vec{n}$ with N components is recovered from the so-called Baker-Akhiezer $\psi(k)$ function, which is an analytic function on an algebraic curve (called the Fermi-curve). The components n_i are related to the evaluation of ψ at N special points, singled out by an additional structure. In this way all $O(N)$ models are unified, and in a sense N is complexified (replaced by an algebraic curve).

3. Interpolating between SU, SO, Sp

In Dyson's original work on ensembles of random matrices with various time-reversal properties [**13**] one finds a discrete series of measures on the space of N indistinguishable particles on a real line (or a circle), which is naturally generalized into a continuous family of β-ensembles. The points $\beta = 1, 2, 4$ correspond to the unitary, orthogonal, and symplectic groups. The interpolating model is no longer a matrix model, yet it can be viewed, rather artificially, as a double-trace deformation of a more conventional matrix model.

In a simplest setting, the partition function of the β-ensemble is given by the integral

$$Z_\beta = \int_{\mathbb{R}^N/S(N)} dx_1 \ldots dx_N \prod_{i<j} |x_i - x_j|^{2\beta} e^{-\sum_{i=1}^N V(x_i)} \tag{28}$$

with some single-particle potential V. The beauty of (28) is that these models share many features with the fractional quantum Hall systems, they are amenable to the analysis via powerful Ward identities, more recently they were connected [**8, 30, 38**] to instanton partition functions [**24**] and topological recursion [**14**]. The connection between the partition sums of supersymmetric gauge theory and random matrix models was pointed out in [**28**] for $\varepsilon_1 = -\varepsilon_2$ background, where it was suggested that the $\mathcal{N} = 1$ Dijkgraaf-Vafa theories [**11**] solved by matrix models could potentially be analyzed systematically starting with the $\mathcal{N} = 2$ instanton count. It is however puzzling that the matrix models, at least in the asymptotic large $\hat{N}$ expansion, admit the β-deformation, while the $\mathcal{N} = 1$ theory seems to be confined to the anti-self-dual Ω-backgrounds. In a parallel series of developments, [**12**] proposed to describe the *refined topological strings* using β-deformed matrix models.

All these examples above get embedded in a conventional supersymmetric gauge theory setup (and string/M-theory via [**29**]) via identification

$$\beta = -\frac{\varepsilon_1}{\varepsilon_2}, \text{or } -\frac{\varepsilon_2}{\varepsilon_1} \tag{29}$$

The simplest way to understand the emergence of matrix-model like structure is by looking at the instanton measure written in the plethystic form [**31**], and observe that the contribution

$$I_{\mathrm{vm}} = \mathsf{E}\left[-\frac{S_{12}S_{12}^*}{(1-q_1^{-1})(1-q_2^{-1})} \right] \tag{30}$$

of the vector multiplets can be written as:

$$I_{\mathrm{vm}} = \mathsf{E}\left[-\frac{(1-q_1)M_{12}M_{12}^*}{(1-q_2^{-1})} \right] \tag{31}$$

where we write

$$S_{12} = N_{12} - K_{12}(1-q_1)(1-q_2) = M_{12}(1-q_1) \tag{32}$$

(see [**31**] for explanations of the notations) where, importantly, M is the pure character:

$$M_{12} = \sum_{\alpha=1}^{n} \sum_{j=1}^{\infty} e^{a_\alpha} q_2^{j-1} q_1^{\lambda_j^{(\alpha)t}} \sim \sum_I e^{x_I} \tag{33}$$

where x_I will play the role of the eigenvalues of some *effective matrix* $\mathcal{X}$. In the limit where all ε's, a's are therefore x_I's are uniformly scaled to zero, the measure (31) approaches the $\beta = -\varepsilon_1/\varepsilon_2$-deformed matrix ensemble for $\mathcal{X}$:

$$\Delta_{\mathcal{X}}^{2\beta} := \prod_{I \neq J} (x_I - x_J)^{-\varepsilon_1/\varepsilon_2} \tag{34}$$

where we are being cavalier with the infinite-dimensional products. Similar considerations apply to the contributions of the bi-fundamental and fundamental hypermultiplets. Thus quiver gauge theories become multi-matrix models.

One application of this imprecise dictionary is the matrix interpretation of the folded and crossed instantons of [**31**] and followup papers. For example, the folded instanton setup fuses $\mathcal{N} = 2^*$ gauge theory on $\mathbb{R}^4 = \mathbb{C}^2$ with Ω-background with parameters $\varepsilon_1, \varepsilon_2$, and mass of the adjoint hypermultiplet ε_3 with the $\mathcal{N} = 2^*$ theory on $\mathbb{C}^2$ with Ω-background with parameters $\varepsilon_2, \varepsilon_3$ and adjoint mass ε_1. A typical instanton measure in this case looks like

$$\mathsf{E}\Bigg[-(1-q_3)\frac{S_{12}S_{12}^*}{(1-q_1^{-1})(1-q_2^{-1})} - (1-q_1)\frac{S_{23}S_{23}^*}{(1-q_2^{-1})(1-q_3^{-1})} + q_3\frac{1-q_1^{-1}q_2^{-1}q_3^{-1}}{1-q_2^{-1}}S_{12}S_{23}^*\Bigg] \tag{35}$$

Now, introducing the effective matrices $\mathcal{X}, \mathcal{Y}$ whose eigenvalues (x_I), (y_J) are defined through

$$\sum_I e^{x_I} = \frac{S_{12}}{1-q_1}, \quad \sum_J = \frac{S_{23}}{1-q_3} \tag{36}$$

we rewrite (35) as

$$\frac{\Delta_{\mathcal{X}}^{2\beta}\Delta_{\mathcal{Y}}^{2/\beta}}{\mathrm{Det}(ad(\mathcal{X})+\varepsilon_3)^{\beta}\,\mathrm{Det}(ad(\mathcal{Y})+\varepsilon_3)^{1/\beta}} \times \frac{\mathrm{Det}(\mathcal{X}\otimes 1 - 1\otimes\mathcal{Y}+\varepsilon_3)\mathrm{Det}(\mathcal{X}\otimes 1 - 1\otimes\mathcal{Y}-\varepsilon_3)}{\mathrm{Det}(\mathcal{X}\otimes 1 - 1\otimes\mathcal{Y})^2} \tag{37}$$

in other words, a β-deformed supermatrix 3-matrix model (cf. [**16**]) with the superpotential

$$\mathrm{Tr}\left(\mathcal{Z}[\mathcal{X},\mathcal{Y}] + \varepsilon_3\mathcal{Z}^2 + V(\mathcal{X}) + W(\mathcal{Y})\right) \tag{38}$$

4. Complex spins

Spin chains, originally introduced to model the magnetic phenomena in metals, later generalized to models of ice etc. typically involve finite-dimensional representations of the spin group, e.g. $SU(2)$. In recent years an interest to spin chains with infinite dimensional representations has been growing. One source of interest is the integrable structure of two-dimensional conformal field theory [**4**], quantum hydrodynamics [**22, 33**]. Another source of interest is the integrable structure discovered in the planar $\mathcal{N} = 4$ super-Yang-Mills in four dimensions, where a spin chain based on the non-compact supergroup $PSU(2,2|4)$ is detected [**5**]. Finally, a growing body of evidence is provided by the supersymmetric sector of the Ω-deformed $\mathcal{N} = 2$ theories in four dimensions (and their five and six dimensional lifts). Namely, a mass deformed superconformal quiver theory in four dimensions is expected to be related to an integrable spin chain based on the Yangian of the Kac-Moody algebra associated with the quiver. Instead of stating the general conjecture, let us give a specific example: a super-QCD $\mathcal{N} = 2$ theory with $SU(N)$ gauge group, with $2N$ hypermultiplets in fundamental representation, with Ω-deformation in two out of four dimensions, is related to the $\mathfrak{sl}_2$ spin chain, with the space of states being a subspace of the tensor product of N infinite-dimensional modules:

$$H = V_{s_1,a_1}(u_1) \otimes V_{s_2,a_2}(u_2) \otimes \ldots \otimes V_{s_N,a_N}(u_N) \tag{39}$$

Here $V_{s,a}$ denotes a representation of the algebra $\mathfrak{sl}_2$, which has a eigenbasis $|n\rangle$ labelled by $n \in \mathbb{Z}$, with

$$L_0|n\rangle = (n+a)|n\rangle\,,\ L_+|n\rangle = (n+a-s)|n+1\rangle\,,\ L_-|n\rangle = (n+a+s)|n-1\rangle \tag{40}$$

The parameter $s \in \mathbb{C}$ is the spin of the representation, related to the Casimir by $L_0^2 - \frac{1}{2}(L_+L_- + L_-L_+) = s(s+1)$. The parameter $a \in \mathbb{C}$ determines a mid-point in the spectrum of L_0. The familiar finite-dimensional representations of $SU(2)$ are found for half-integer s and $a-s$, $a+s$ both non-negative integers.

The gauge theory corresponds to A_1 quiver. The a_i's are the Coulomb moduli, i.e. the eigenvalues of the adjoint scalar in the vector multiplet, while the spins s_i's are related to the masses by $m_i^+ - m_i^- = 2s_i$, $i = 1, \ldots, N$. The sums $u_i = \frac{1}{2}\left(m_i^+ + m_i^-\right)$ are the inhomogeneities, the evaluation parameters of the Yangian modules based on the $\mathfrak{sl}_2$ modules. Although the permutations of the $2N$ masses is a symmetry of the four dimensional theory, the specific two dimensional subsector defined by Ω-deforming and adding a surface defect, depends on the ordering of masses, which, in turn, defines the spins etc. See [**15, 20, 21**] for more details.

Another view on the same surface defect in the fully Ω-deformed theory reveals the connection to Knizhnik-Zamolodchikov equation borrowed from the $\widehat{\mathfrak{sl}_N}$-representation theory [**15, 21, 26**]. Again, complex spins and infinite-dimensional modules feature prominently.

5. Future directions

It would be nice to turn on interactions in the two and three dimensional theories. The supersymmetric methods of a different kind (Wegner and Efetov) are used in the study of disordered systems for many decades [**43**]. It would be nice to find a relation between our higher dimensional theories with conventional supersymmetry and the supercoset sigma models used in the studies of integer Hall plateau transitions [**44**].

The supersymmetric interfaces [**10**] in one, two, and three dimensional supersymmetric gauge theories with eight supercharges realize finite dimensional spin chains. A T-dual construction realizes the same systems in a four dimensional version of Chern-Simons theory [**9**]. It would be useful to find a four dimensional analogue of the interfaces [**10**] providing a geometric realization of R-matrix suitable for infinite dimensional representations of quiver Yangians and quantum loop algebras.

The most interesting and mysterious question is the role of complex metrics in string theory. In addition to the arguments of [**42**] one would like to use the complex worldsheet metric to avoid unphysical poles in the bulk of $\mathcal{M}_g$ which pop up in time-dependent cosmological string backgrounds [**23**], as well as the spacetime considerations of [**17, 40, 41**]. We hope to return to this question in the future (whatever it means in the complex world).

Acknowledgments

Discussions with M. Dedushenko, A. Grekov, S. Jeong, N. Lee, A. Okounkov, N. Seiberg, O. Tsymbaliuk, are gratefully acknowledged. Many thanks to the organizers of String-Math'2002 in Warsaw, especially to Piotr Sulkowski, for providing a stimulating working environment in troubled times.

References

[1] John L. Cardy, *Operator content and modular properties of higher-dimensional conformal field theories*, Nuclear Phys. B **366** (1991), no. 3, 403–419, DOI 10.1016/0550-3213(91)90024-R. MR1138650

[2] H. Awata, M. Fukuma, Y. Matsuo, and S. Odake, *Determinant formulae of quasi-finite representation of $W_{1+\infty}$ algebra at lower levels*, Phys. Lett. B **332** (1994), no. 3-4, 336–344, DOI 10.1016/0370-2693(94)91262-9. MR1286716

[3] Hidetoshi Awata, Masafumi Fukuma, Yutaka Matsuo, and Satoru Odake, *Representation theory of the $W_{1+\infty}$ algebra*, Progr. Theoret. Phys. Suppl. **118** (1995), 343–373, DOI 10.1143/PTPS.118.343. Quantum field theory, integrable models and beyond (Kyoto, 1994). MR1356527

[4] Vladimir V. Bazhanov, Sergei L. Lukyanov, and Alexander B. Zamolodchikov, *Integrable structure of conformal field theory, quantum KdV theory and thermodynamic Bethe ansatz*, Comm. Math. Phys. **177** (1996), no. 2, 381–398. MR1384140

[5] Niklas Beisert, Changrim Ahn, Luis F. Alday, and et al., *Review of AdS/CFT integrability: an overview*, Lett. Math. Phys. **99** (2012), no. 1-3, 3–32, DOI 10.1007/s11005-011-0529-2. MR2886411

[6] M. Bershadsky, S. Cecotti, H. Ooguri, and C. Vafa, *Kodaira-Spencer theory of gravity and exact results for quantum string amplitudes*, Comm. Math. Phys. **165** (1994), no. 2, 311–427. MR1301851

[7] Meng Cheng and Nathan Seiberg, *Lieb-Schultz-Mattis, Luttinger, and 't Hooft—anomaly matching in lattice systems*, SciPost Phys. **15** (2023), no. 2, Paper No. 051, 73, DOI 10.21468/scipostphys.15.2.051. MR4629094

[8] L. O. Chekhov, B. Eynard, and O. Marchal, *Topological expansion of the β-ensemble model and quantum algebraic geometry in the sectorwise approach*, Theoret. and Math. Phys. **166** (2011), no. 2, 141–185, DOI 10.1007/s11232-011-0012-3. Russian version appears in Teoret. Mat. Fiz. **166** (2011), no. 2, 163–215. MR3165804

[9] Kevin Costello, *Integrable lattice models from four-dimensional field theories*, String-Math 2013, Proc. Sympos. Pure Math., vol. 88, Amer. Math. Soc., Providence, RI, 2014, pp. 3–23, DOI 10.1090/pspum/088/01483. MR3330282

[10] M. Dedushenko and N. Nekrasov, *Interfaces and Quantum Algebras, I: Stable Envelopes*, `arXiv:2109.10941 [hep-th]` MR4642568
Interfaces and Quantum Algebras, II: Cigar Partition Function, `arXiv:2306.16434 [hep-th]`.

[11] R. Dijkgraaf and C. Vafa, *A Perturbative window into nonperturbative physics*, `arXiv:hep-th/0208048 [hep-th]`, 2002.

[12] R. Dijkgraaf and C. Vafa, *Toda Theories, Matrix Models, Topological Strings, and $\mathcal{N} = 2$ Gauge Systems*, `arXiv:0909.2453 [hep-th]`.

[13] Freeman J. Dyson, *The threefold way. Algebraic structure of symmetry groups and ensembles in quantum mechanics*, J. Mathematical Phys. **3** (1962), 1199–1215, DOI 10.1063/1.1703863. MR177643

[14] B. Eynard and N. Orantin, *Algebraic methods in random matrices and enumerative geometry*, `arXiv:0811.3531 [math-ph]`, 2008.

[15] Saebyeok Jeong, Norton Lee, and Nikita Nekrasov, *Intersecting defects in gauge theory, quantum spin chains, and Knizhnik-Zamolodchikov equations*, J. High Energy Phys. **10** (2021), Paper No. 120, 82, DOI 10.1007/jhep10(2021)120. MR4339806

[16] Vladimir A. Kazakov, Ivan K. Kostov, and Nikita Nekrasov, *D-particles, matrix integrals and KP hierarchy*, Nuclear Phys. B **557** (1999), no. 3, 413–442, DOI 10.1016/S0550-3213(99)00393-4. MR1713471

[17] Maxim Kontsevich and Graeme Segal, *Wick rotation and the positivity of energy in quantum field theory*, Q. J. Math. **72** (2021), no. 1-2, 673–699, DOI 10.1093/qmath/haab027. MR4271398

[18] I. Krichever and N. Nekrasov, *Towards Lefschetz Thimbles in Sigma Models, I*, J. Exp. Theor. Phys. **132**, no.4, 734-751 (2021) doi:10.1134/S1063776121040129 `arXiv:2010.15575` [hep-th].

[19] Igor Krichever and Nikita Nekrasov, *Novikov-Veselov symmetries of the two-dimensional $O(N)$ sigma model*, SIGMA Symmetry Integrability Geom. Methods Appl. **18** (2022), Paper No. 006, 37, DOI 10.3842/SIGMA.2022.006. MR4368996

[20] Norton Lee and Nikita Nekrasov, *Quantum spin systems and supersymmetric gauge theories. Part I*, J. High Energy Phys. **3** (2021), Paper No. 093, 86, DOI 10.1007/jhep03(2021)093. MR4262761

[21] S. Jeong, N. Lee and N. Nekrasov, *Parallel surface defects, Hecke operators, and quantum Hitchin system*, `arXiv:2304.04656 [hep-th]`, 2023.

[22] A. V. Litvinov, *On spectrum of ILW hierarchy in conformal field theory*, J. High Energy Phys. **11** (2013), 155, front matter+13, DOI 10.1007/JHEP11(2013)155. MR3132164

[23] N. A. Nekrasov, *Milne universe, tachyons, and quantum group*, Surveys High Energ. Phys. **17**, 115-124 (2002) doi:10.1080/0142241021000054176 `arXiv:hep-th/0203112` [hep-th].

[24] Nikita A. Nekrasov, *Seiberg-Witten prepotential from instanton counting*, Adv. Theor. Math. Phys. **7** (2003), no. 5, 831–864. MR2045303

[25] Nikita Nekrasov, *Tying up instantons with anti-instantons*, Ludwig Faddeev memorial volume, World Sci. Publ., Hackensack, NJ, 2018, pp. 351–388. MR3823594

[26] Nikita Nekrasov and Alexander Tsymbaliuk, *Surface defects in gauge theory and KZ equation*, Lett. Math. Phys. **112** (2022), no. 2, Paper No. 28, 53, DOI 10.1007/s11005-022-01511-8. MR4400676

[27] Nikita A. Nekrasov and Samson L. Shatashvili, *Supersymmetric vacua and Bethe ansatz*, Nuclear Phys. B Proc. Suppl. **192/193** (2009), 91–112, DOI 10.1016/j.nuclphysbps.2009.07.047. MR2570974

[28] Nikita A. Nekrasov and Andrei Okounkov, *Seiberg-Witten theory and random partitions*, The unity of mathematics, Progr. Math., vol. 244, Birkhäuser Boston, Boston, MA, 2006, pp. 525–596, DOI 10.1007/0-8176-4467-9_15. MR2181816

[29] Nikita Nekrasov and Andrei Okounkov, *Membranes and sheaves*, Algebr. Geom. **3** (2016), no. 3, 320–369, DOI 10.14231/AG-2016-015. MR3504535

[30] Takahiro Nishinaka and Chaiho Rim, *β-deformed matrix model and Nekrasov partition function*, J. High Energy Phys. **2** (2012), 114, front matter+21, DOI 10.1007/JHEP02(2012)114. MR2996093

[31] N. Nekrasov, *BPS/CFT correspondence: non-perturbative Dyson-Schwinger equations and qq-characters*, J. High Energy Phys. **03**, 181 (2016) `https://doi.org/10.1007/JHEP03(2016)181` `arXiv:1512.05388 [hep-th]`.

[32] Nikita A. Nekrasov, *Instanton partition functions and M-theory*, Jpn. J. Math. **4** (2009), no. 1, 63–93, DOI 10.1007/s11537-009-0853-9. MR2491283

[33] A. Okounkov, *Deep waters of quantum cohomology*, `https://scgp.stonybrook.edu/video_portal/video.php?id=524`

[34] Andrei Okounkov and Nikolai Reshetikhin, *Correlation function of Schur process with application to local geometry of a random 3-dimensional Young diagram*, J. Amer. Math. Soc. **16** (2003), no. 3, 581–603, DOI 10.1090/S0894-0347-03-00425-9. MR1969205

[35] Andrei Okounkov, Nikolai Reshetikhin, and Cumrun Vafa, *Quantum Calabi-Yau and classical crystals*, The unity of mathematics, Progr. Math., vol. 244, Birkhäuser Boston, Boston, MA, 2006, pp. 597–618, DOI 10.1007/0-8176-4467-9_16. MR2181817

[36] Surya Raghavendran, Ingmar Saberi, and Brian R. Williams, *Twisted eleven-dimensional supergravity*, Comm. Math. Phys. **402** (2023), no. 2, 1103–1166, DOI 10.1007/s00220-023-04745-2. MR4627318

[37] R. J. Szabo and M. Tirelli, *Noncommutative Instantons in Diverse Dimensions*, `arXiv:2207.12862 [hep-th]`.

[38] Piotr Sułkowski, *Matrix models for β-ensembles from Nekrasov partition functions*, J. High Energy Phys. **4** (2010), 063, 29, DOI 10.1007/JHEP04(2010)063. MR2673084

[39] L. A. Takhtajan, *Etudes of the resolvent* (Russian, with Russian summary), Uspekhi Mat. Nauk **75** (2020), no. 1(451), 155–194, DOI 10.4213/rm9917; English transl., Russian Math. Surveys **75** (2020), no. 1, 147–186. MR4070021

[40] E. Witten, *A Note On Complex Spacetime Metrics*, arXiv:2111.06514 [hep-th].
[41] Jonathan J. Halliwell and James B. Hartle, *Integration contours for the no-boundary wave function of the universe*, Phys. Rev. D (3) **41** (1990), no. 6, 1815–1834, DOI 10.1103/PhysRevD.41.1815. MR1048878
[42] Edward Witten, *The Feynman $i\epsilon$ in string theory*, J. High Energy Phys. **4** (2015), 055, front matter+24, DOI 10.1007/JHEP04(2015)055. MR3351314
[43] M. Zirnbauer, *The supersymmetry method of random matrix theory*, preprint arXiv:math-ph/0404057, 2004
[44] Martin R. Zirnbauer, *The integer quantum Hall plateau transition is a current algebra after all*, Nuclear Phys. B **941** (2019), 458–506, DOI 10.1016/j.nuclphysb.2019.02.017. MR3921203

SIMONS CENTER FOR GEOMETRY AND PHYSICS, STONY BROOK UNIVERSITY, STONY BROOK, NEW YORK 11794-3636

Proceedings of Symposia in Pure Mathematics
Volume **107**, 2024
https://doi.org/10.1090/pspum/107/01951

6D heterotic little string theories and F-theory geometry: An introduction

Michele Del Zotto, Muyang Liu, and Paul-Konstantin Oehlmann

Abstract. We review some aspects of our recent work about the geometric engineering of Heterotic little string theories using F-theory. Building on the seminal works by Aspinwall and Morrison as well as Intrilligator and Blum, we solve some longstanding open questions thanks to recent progress in our understanding of 6D (1,0) theories and their generalized symmetries. On the geometry side, these systems correspond to non-compact elliptically fibered Calabi-Yau varieties that must admit the structure of an elliptic K3 fibration. From fiberwise F-theory/Heterotic duality, the K3 plays a central role - it determines the 6D flavor group, as well as different T-dual LSTs via inequivalent elliptic fibration structures. The geometries we obtain are some finer versions of Kulikov degenerations: the point where the K3 fiber degenerates is the locus where the LST arises. This structure serves on one hand to check our field theory predictions on LST dualities via the match of Coulomb branch dimension, flavor symmetries, and 2-group structure constants, and also on the other hand to deduce novel LST models and their networks of dualities, thus allowing the exploration of non-geometric Heterotic regimes.

1. Introduction and summary

Six dimensional supersymmetric theories can be divided into three classes: supergravity theories (SUGRAs), little string theories (LSTs), and superconformal field theories (SCFTs). Of these, the latter two are distinguished as they are decoupled from 6D gravity. There is typically a hierarchical chain of 6D decoupling limits: SUGRAs come with a corresponding 6D Planck scale M_{PL}, which controls the strength of gravitational interactions; LSTs similarly come with an intrinsic string tension M_{LS} (much smaller than M_{PL}), while SCFTs have no scales whatsoever. In 6D, lowering the energy, one typically flows down from a SUGRA to an LST and then eventually to an SCFT. The LST string tension is an energy scale that breaks conformal invariance, at the price of allowing features usually found in string theories. In the latter, the Planck scale can mix with the Kaluza-Klein scale upon circle

---

2020 *Mathematics Subject Classification.* Primary 81T30.

The work of the first and second author received funding from the European Research Council (ERC) under the European Union's Horizon 2020 research and innovation program (grant agreement No. 851931). The first author also acknowledges support from the Simons Foundation Grant # 888984 (Simons Collaboration on Global Categorical Symmetries). The third author has received funding from the NSF CAREER grant PHY-1848089 and startup funding from Northeastern University by Fabian Ruehle.

reduction which realizes T-duality (as found i.e. in the $E_8 \times E_8 \leftrightarrow Spin(32)/\mathbb{Z}_2$ Heterotic duality). A similar mechanism applies in LSTs, which can be T-dual upon circle reduction since the KK scale can mix with the LST string tension, exchanging winding and momentum modes. This signals that these systems are not conventional quantum field theories, as T-duality indicates that they have multiple stress-energy tensors. Decoupling the little strings, considering energies well below M_{LS}, one ends up with an SCFT in the infrared [**21**, **41**, **50**–**52**, **66**].

Due to the advent of F-theory, powerful geometric methods became available to substantially improve our understanding of 6D systems. The hierarchy we outlined above is reflected by F-theory geometry: for SUGRA the 6D Planck scale is set by the volume of the F-theory base, with the latter becoming infinitely large when the 6D graviton decouples. This decoupling limit is equivalent to focusing on a shrinking surface: the latter can either shrink to a curve (the volume of such a curve sets the scale M_{LS} in the LST case) or to a point (the SCFT case). The resulting geometries are so constrained that they lead to a classification [**13**, **43**].[1] This approach allows us to recast the physics of LSTs and SCFTs into the geometrical properties of non-compact elliptic Calabi-Yau threefolds X_3 via standard F-theory techniques [**6**, **53**, **61**–**63**, **69**] (as well as stimulating the development of some novel features of F-theory [**14**, **68**]).

In this review we summarize some of the results in our recent papers [**31**–**33**]. In particular, we are interested in a special class of 6D LSTs that govern the worldvolumes of NS5 branes in Heterotic string theories in the presence of an ALE singularity of type $\mathbb{C}^2/\Gamma$, where Γ is a discrete subgroup of $SU(2)$. For the $Spin(32)/\mathbb{Z}_2$ cases, these models have been completely determined by [**17**, **50**] — since the theory of NS5 branes in the $Spin(32)/\mathbb{Z}_2$ case is Lagrangian [**70**], the theories corresponding to the worldvolume ALE instantons are orbifolds of the latter, which depend on the embedding of Γ into $Spin(32)/\mathbb{Z}_2$ (corresponding to a choice of a flat connection at infinity for the Heterotic bundle). However, the Heterotic $E_8 \times E_8$ instantons do not have a conventional Lagrangian description [**40**], and therefore the corresponding ALE instantons are much trickier to determine [**6**]. Thanks to recent advances in our understanding of 6D SCFTs, one can now characterize them in terms of generalized 6D quivers by exploiting the fusion of specific 6D building blocks [**30**, **45**]. The 6D SCFT building blocks of interest for us are 6D conformal matter models [**30**] and orbi-instanton 6D SCFTs [**37**, **45**, **59**], which have been completely classified. This strategy gives a full characterization of the Heterotic $E_8 \times E_8$ ALE instanton models and their dependence on choices of $E_8 \times E_8$ flat connections at infinity in terms of generalized 6D quiver diagrams. Moreover, each quiver diagram also encodes an elliptic fibration structure of the corresponding F-theory geometry.

As a consistency check, one should still verify that for each such model there is a corresponding T-dual $Spin(32)/\mathbb{Z}_2$ theory. To constrain the matching duals, we exploit the fact that, unlike gravitational theories, LSTs can have global symmetries,[2] which must be matched across dualities. In particular, a stringent constraint about the T-duality of LSTs is given by the matching of their 2-group symmetries [**27**]. 2-group symmetries are examples of generalized global symmetries [**39**], where

[1] e.g., [**44**] or Section 2 of [**34**] for a review and further references to recent works on the subject.

[2] In gravity theories it is widely believed that all symmetries must either be broken or gauged, which also applies to their generalized versions (see [**8**] for a general argument and [**3**, **19**] for a sample of application to 6D).

a symmetry acting on point operators mixes with a symmetry acting upon line operators.[3] Specifically, LSTs always have a universal continuous 2-group symmetry that couples the 2-form background field corresponding to the little string charge with the 1-form background fields for the Poincaré symmetry, the R-symmetry and the other global symmetries of the model [**23, 24, 27**]. Schematically, the LST 2-group is presented as follows:

$$ {}^{2}G_{\widehat{\kappa}_{\mathscr{P}},\widehat{\kappa}_{\mathscr{R}},\widehat{\kappa}_{\mathscr{F}_A}} = \left(\mathscr{P}^{(0)} \times \mathscr{R}^{(0)} \times \mathscr{F}_A^{(0)}\right) \times_{\widehat{\kappa}_{\mathscr{P}},\widehat{\kappa}_{\mathscr{R}},\widehat{\kappa}_{\mathscr{F}_A}} U(1)^{(1)}_{LS}, \tag{1.1}$$

where $U(1)^{(1)}_{LS}$ is the 1-form symmetry measuring the little string charge (whose background field is the 2-form mentioned above), $\mathscr{P}$ is the spacetime Poincaré symmetry, $\mathscr{R}^{(0)} \simeq SU(2)_R$ is the R-symmetry of the (1,0) supersymmetry, and various other flavor 0-form symmetries are denoted $\mathscr{F}_A^{(0)}$. Each component of this 2-group is characterized by a 2-group structure constant, denoted $\widehat{\kappa}$ above, which measures the "mixing" between these symmetries [**23, 24, 27**]. In particular, $\widehat{\kappa}_{\mathscr{P}}$ can have only two values: 0 or 2. In the former case, Poincaré symmetry does not mix with $U(1)_{LS}$; in the latter it does. Remarkably, this coincides with the number of M9 branes in an M-theory dual formulation.[4] It is easy to check that for the Heterotic LSTs, $\widehat{\kappa}_{\mathscr{P}}$ always equals 2, as expected from the HW description. On the contrary, $\widehat{\kappa}_{\mathscr{R}}$ has an interesting structure with much more constrained matching conditions [**35**]. The 2-groups provide a powerful novel invariance to explore the network of T-dual theories and their relations, also for the case of Heterotic ALE instanton theories. This gives a strategy to conjecture T-dual pairs of little string ALE instantons in the $E_8 \times E_8$ and the $Spin(32)/\mathbb{Z}_2$ Heterotic string theories, where the geometry of F-theory gives us the most powerful tool to verify these predictions.

However, this does not apply to exceptional singularities, which do not admit a description in perturbative string theory. Instead one must exploit F-theory geometry. In fact, F-theory allows us to reproduce (independently) the results from perturbative brane setups and to treat the exceptional cases on the same footing, thus giving a nice unifying perspective. In F-theory, the type IIB axio-dilation is realized as the torus modulus of an auxiliary elliptic fiber over the physical base space, which completes the total space to an elliptic Calabi-Yau (CY) threefold [**69**]. Due to its elegant relationship to M-theory, the F-theory framework enables us to geometrically demonstrate T-dualities [**5, 6, 13**]: M-theory on an elliptic threefold X_3 is dual to F-theory on the same CY times a circle. The radius of the circle in F-theory is inversely proportional to the volume of the elliptic fiber. Sending the radius to infinity and getting back to 6D, the elliptic fiber shrinks to zero size. The M-theory moduli space is therefore larger than the F-theory moduli space. In particular, the circle reduction introduces Wilson lines for the 6D gauge fields that become Coulomb branch parameters of the 5D theory, which add to the scalars from the 6D tensor branch to give rise to a larger 5D Coulomb branch. These parameters are identified with the Kähler moduli of the underlying threefold that allow one to resolve the singularities in the elliptic fiber (which one cannot do in 6D). A well-known feature of 5D compactifications is that the Coulomb branch

[3]For an incomplete list of references about higher symmetries in 6D systems, see, e.g., [**2, 15, 25, 29, 49**].

[4]Recall that there is a dual description of the $E_8 \times E_8$ Heterotic string in M-theory, where M-theory is placed on $S^1/\mathbb{Z}_2$ and two M9 branes are placed at the fixed points. This is the so-called Hořava-Witten (HW) description of the Heterotic string [**46, 47**].

splits into chambers separated by walls where some BPS particles become massless [**65**]. In geometry, these give flop transitions that lead to different *birationally equivalent* CYs being part of the same moduli space [**61**]. Together these form the so-called extended Kähler cone. With M-theory, one can probe the whole space that is not accessible in F-theory.

Two 6D LSTs are T-dual if they give rise to the same 5D theory upon circle compactification. This is realized from the F-theory/M-theory duality as follows: the elliptic CY giving rise to the two LSTs in question have two inequivalent elliptic structures which, however, belong to the same M-theory moduli space. The resulting geometry then must have two inequivalent ways to be lifted up to F-theory, thus leading to a 5D theory that has two inequivalent 6D origins. This is the hallmark of T-duality. It can happen that a single CY admits two inequivalent elliptic fibrations or that the two elliptically fibered CYs are birationally equivalent [**5**]; in either case, the two belong to the same M-theory moduli space while being disconnected in the F-theory moduli space [**13**]. Moreover, it can happen that the given CY admits more than just two inequivalent elliptic structures, leading to multiple 6D LSTs, which become equivalent theories in 5D [**12, 32**].[5]

The above approach gives a geometrical way to establish T-dualities and allows for a systematic search for novel duals. In the series of works [**31**–**33**] we have focused on Heterotic LSTs and identified a rich geometric substructure. This structure allows for general statements about this class of LSTs as well as potential new directions in mathematics, which we summarize as follows:

(1) **Universal elliptic K3 structure:** Each Heterotic LST admits a geometric engineering approach where the non-compact elliptic threefold X_3 admits a nested elliptic K3 fibration structure.
(2) **Flavor from the K3:** The K3 contains (compact) fibral divisors that are non-compact when pulled back to the threefold. These encode the (geometric) flavor group of the LST. For a given K3 polarization, the intersection form of the flavor divisors is fixed by the frame lattice W. This establishes a correspondence between Heterotic instantons labeled by flat connections at infinity and degenerations of K3s with a given frame lattice, *à la* Friedman-Morgan-Witten [**38**].
(3) **LST dualities from K3:** For a fixed X_3, all inequivalent elliptic fibrations of the flavor K3 fiber pull back to different fibration structures in X_3. This allows us to deduce T-dual LSTs corresponding to X_3 from the flavor K3 via powerful lattice techniques.[6]
(4) **Exotic LSTs and T-polyality**: K3s with large Picard numbers generally admit multiple inequivalent elliptic fibrations. This generalizes the

[5]Notice that since these models are equivalent 5D theories, the resulting 5D flavor symmetries must be identical. This often requires turning on a flavor Wilson line on the T-duality circle, which breaks the 6D flavor symmetries down to a common subgroup. In particular, the flavor group ranks must match, which is built in geometry: We denote d_{CB} the 5D Coulomb branch dimension and f the rank of the flavor symmetry $\dim H_2(X_3) = d_{CB} + f$, while $\dim H_4(X_3) = r_{5D}$. The Betti numbers match for birationally equivalent 3-folds.

[6]The extended Kähler cone has the potential for a much richer landscape of interconnections: one can have examples where X_3 is birational to a different elliptic CY, which has an inequivalent nested K3 fibration leading to another family of T-dual Heterotic theories. Remarkably however, in all the examples we have considered so far, a single nested K3 fibration seems to suffice to determine all the (non twisted) T-duals. See [**12**] for examples of twisted T-dualities.

order two T-duality of the $E_8 \times E_8 \leftrightarrow Spin(32)/\mathbb{Z}_2$ Heterotic strings to a polyality involving a number of 6D LSTs, which equals the number of inequivalent elliptic structures on the given flavor K3 – the same caveat in footnote 6 applies here: the actual T-duality network of a given LST might be even larger if we allow for birational transformations of that type.

(5) **LSTs and generalized Kulikov-like degenerations:** Viewing the CY as a K3 fibration, the LST is localized at a locus where the K3 fiber degenerates in a way much similar to a Kulikov degeneration [**56**]. However, LSTs require a much finer classification than the Kulikov classification, which is indicated by the possibility of realizing all Kodaira types of fibers. We stress here that **the 2-group structure constants are invariants of these fibrations** and therefore give a different approach towards a classification scheme *à la* Kulikov, but with labels given by the K3 frame lattice W, the 5D Coulomb branch dimension of the LST, and the 2-group structure constant $\widehat{\kappa}_{\mathscr{R}}$.

(6) **Flavor bounds and attractive K3s:** The frame lattice is a sublattice of $\mathrm{II}_{2,18}$ and it is bounded to have rank 18. The bound is saturated by the so-called attractive elliptic K3s. The models obtained from F-theory building on fibrations by attractive K3s generalize the HW setup slightly. Also in HW one obtains global symmetries at most of rank 18.[7] For the models arising from attractive K3s, one can obtain some exotic non-abelian enhancement of the flavor symmetry. When G_F is maximal and non-abelian, the underlying K3s are extremal.

The structure of this review is as follows. In section 2 we give a short condensed review of the field theoretical results from generalized 6D quivers, as well as explaining explicitly how to compute 2-group structure constants. In section 3 we give a quick review of the F-theory geometries involved in these constructions. In section 4 we give an explicit example of a T-duality and its geometric realization. In section 5 we discuss the T-hexality arising from the different inequivalent elliptic structures on an extremal attractive K3.

2. LSTs, 2-groups and fusion

In this section we recall the basic ingredients of how to engineer Heterotic LSTs via non-compact elliptic threefolds. For a nice review see [**44**].

First we denote the flavor Lie algebra of a rank n_T 6D LST as

$$\mathfrak{f}^{(0)} = \prod_{a=1}^{n_f} \mathfrak{f}_a, \tag{2.1}$$

where $\mathfrak{f}_a$ are irreducible factors. The dynamical degrees of freedom are denoted by a generalized quiver which is fixed by two sets of data:

- An $(n_T + 1 + n_f) \times (n_T + 1 + n_f)$ symmetric matrix for the quiver:

$$\begin{pmatrix} \eta^{IJ} & \eta^{IA} \\ \eta^{AI} & 0 \end{pmatrix} \qquad \begin{array}{l} I, J = 1, ..., n_T + 1 \\ A = 1, ..., n_f \end{array} \tag{2.2}$$

[7] One has one E_8 factor from each M9 brane and an $SO(4) = SU(2) \times SU(2)$ from the directions transverse to the M5s.

The diagonals η^{II} are positive integers that run from 0 to 12 and denote the Dirac pairings of the BPS strings that source the self-dual 2-form fields. All other entries are vanishing or negative integers.

- An $(n_T + 1 + n_f)$-tuple of Lie gauge and flavor algebras:

$$\mathfrak{g} = (\mathfrak{g}_1, \ldots, \mathfrak{g}_{n_T+1}, \mathfrak{f}_1, \ldots, \mathfrak{f}_{n_f}), \tag{2.3}$$

which result in dynamical couplings of the I-th tensor with the $\mathfrak{g}_I$-th gauge fields, where η^{II} is also identified with their anomaly coefficients.

Putting everything together, the generalized quiver is given by assigning a node to each nonzero η^{II} decorated with the Lie algebra $\mathfrak{g}_I$, and, whenever $\eta^{IA} \neq 0$, also with a factor $\mathfrak{f}_A$ of the symmetry Lie algebra. Schematically, we write

$$\cdots \quad \overset{\mathfrak{g}_I}{\underset{[\mathfrak{f}_A]_{(\eta^{IA})}}{\eta^{II}}} \quad \cdots \tag{2.4}$$

The off-diagonal entries of the η^{IJ} block are non-positive integers encoding the adjacency matrix for the generalized quiver. If $\eta^{IJ} \neq 0$, then two nodes η^{II} and η^{JJ} are neighbors and the corresponding BPS strings can form bound-states[8]. In terms of gauge and flavor algebra factors, such a non-trivial adjacency often yields bi-fundamental matter.

For 6D SCFTs the matrix η^{IJ} must be positive definite, while for 6D LSTs it has to be positive semi-definite with a single unique vanishing eigenvalue. The LST is a linear combination of BPS bound states that precisely correspond to that vanishing eigenvalue. One might thus consider η^{IJ} as a generalized *affine* Cartan matrix [**13**] such that

$$\eta^{IJ}\ell_J = 0 \qquad \gcd(\ell_1, \ldots, \ell_{n_T+1}) = 1 \qquad \ell_I > 0, \tag{2.5}$$

where ℓ_J is the generalized Kac label and identified as the LST $U(1)^{(1)}_{LS}$ charges of the BPS strings.

Such quiver data is sufficient to fix the 6D theory and to obtain the 2-group structure constants given as

$$\boxed{\widehat{\kappa}_{\mathscr{F}} = -\textstyle\sum_{I=1}^{r+1} \ell_I \eta^{IA} \qquad \widehat{\kappa}_{\mathscr{R}} = \textstyle\sum_{I=1}^{r+1} \ell_I h^\vee_{\mathfrak{g}_I} \qquad \widehat{\kappa}_{\mathscr{P}} = -\textstyle\sum_{I=1}^{r+1} \ell_I (\eta^{II} - 2)} \tag{2.6}$$

These can be deduced through an analysis of the Bianchi identity of the non-dynamical LST 2-form background field [**23, 27**], building on [**64**]. In [**27**], it was first proposed and verified for many examples that these structure constants are invariant across T-dual LSTs. In particular, the matching of $\widehat{\kappa}_{\mathscr{P}}$ and $\widehat{\kappa}_{\mathscr{R}}$ yields strong novel conditions to match T-dual LSTs.

LSTs can be grouped into two classes reflected by the two possible values 0 and 2 of $\widehat{\kappa}_{\mathscr{P}}$. As mentioned in the introduction, for Heterotic LSTs $\widehat{\kappa}_{\mathscr{P}} = 2$.

As the name suggests, a good starting point to construct Heterotic LSTs is the 10D $Spin(32)/\mathbb{Z}_2$ Heterotic string. In particular, the worldvolume theories governing stacks of N instantonic NS5 branes are well explored [**70**] and have a Lagrangian description with an $\mathfrak{sp}_N$ gauge algebra and $N_f = 16$ hypermultiplets in the fundamental representation and an additional hypermultiplet in the antisymmetric

[8] In this work the off-diagonal η^{IJ} are either 0 or -1. More general cases are given in [**13, 14**].

representation. The corresponding generalized 6D quiver is

$$[\mathfrak{so}(32)] \overset{\mathfrak{sp}_N}{0} [\mathfrak{su}(2)], \tag{2.7}$$

which highlights the dual F-theory picture that we discuss momentarily. Thanks to this Lagrangian description, the worldvolume theories of ALE Heterotic $Spin(32)/\mathbb{Z}_2$ can be easily determined by orbifolding [**16, 17, 50**]. Since $\pi_1(S^3/\Gamma_\mathfrak{g}) \simeq \Gamma_\mathfrak{g}$, the resulting theories also depend on the choice of a flat connection at infinity, encoded in a choice of a mapping:

$$\boldsymbol{\lambda} : \Gamma_\mathfrak{g} \to Spin(32)/\mathbb{Z}_2. \tag{2.8}$$

The singularity $\mathfrak{g}$ and a choice of flat connections[9] specify the theory, which we denote by

$$\widetilde{\mathcal{K}}_N(\boldsymbol{\lambda}; \mathfrak{g}). \tag{2.9}$$

The unbroken flavor group $F(\boldsymbol{\lambda})$ of the 6D theory is determined by the commutant of $\boldsymbol{\lambda}(\Gamma_\mathfrak{g})$ in $Spin(32)/\mathbb{Z}_2$.

Similarly one might consider the instantons on $E_8 \times E_8$. Such theories are closely related to the 6D $(2,0)$ SCFTs and hence have no conventional Lagrangian formulation. The Hořava-Witten formulation [**47**] yields a characterization of such theories: one can compactify M-theory on a finite interval with two 10D spacetime filling M9 branes that carry an E_8 worldvolume theory. The LST arises from a stack of N M5 branes that fill out six dimensions, parallel to the two M9 branes [**40**].

The strings that stretch between an M9 and an M5 are given by E-strings and those between two M5s are M-strings which we write [**40**] as the generalized quiver:

$$[\mathfrak{e}_8]\ 1\ 2\ 2 \ldots 2\ 2\ 1\ [\mathfrak{e}_8]. \tag{2.10}$$

Along the residual four transverse directions to the M5 branes, an ALE singularity $\mathbb{C}^2/\Gamma_\mathfrak{g}$ is located. The presence of the singularity leads to a fractionalization [**28, 47**], which also depends further on the choice of a flat connection at infinity for the two E_8 bundles. The latter are encoded in two group morphisms:

$$\boldsymbol{\mu}_a \colon \pi_1(S^3/\Gamma_\mathfrak{g}) \simeq \Gamma_\mathfrak{g} \to E_8, \tag{2.11}$$

with $a = 1, 2$ labeling the two M9 branes. The zero form global symmetry of the resulting LSTs is determined by the commutant of $\boldsymbol{\mu}_a(\Gamma_\mathfrak{g})$ in E_8. As such, we expect to have

$$F_a^{(0)} \equiv \{g \in E_8 \,|\, gh = hg, \forall h \in \boldsymbol{\mu}_a(\Gamma_\mathfrak{g})\} \qquad a = 1, 2. \tag{2.12}$$

The cases with global symmetry $F_{1,2}^{(0)} \simeq E_8$ correspond to the choice of trivial flat connections for the end of the world gauge fields. They are dual to certain geometries in F-theory that had been discovered by Aspinwall and Morrison [**6**]. In general, we denote the associated theories by

$$\mathcal{K}_N(\boldsymbol{\mu}_1, \boldsymbol{\mu}_2; \mathfrak{g}). \tag{2.13}$$

The tensor branches of all these theories can easily be obtained by noting that each 1 and 2 curve is an SCFT in itself. The fractionalizations of those SCFTs have been determined in [**30**] by exploiting F-theory techniques.

---

[9] An important subtlety here is that one could allow for boundary conditions without "vector-structure" in $Spin(32)/\mathbb{Z}_2$ via a positive second Stiefel-Whitney class $\widetilde{w}_2 \neq 0$ [**10**]. In this work we will only focus on configurations with vector structure, leaving the other cases for future work.

The resulting theories are generalized quiver theories of the form

$$\mathcal{K}_N(\boldsymbol{\mu}_1, \boldsymbol{\mu}_2; \mathfrak{g}) = \mathcal{T}(\boldsymbol{\mu}_1, \mathfrak{g}) \overset{\mathfrak{g}}{\text{------}} \mathcal{T}_{N-2}(\mathfrak{g}, \mathfrak{g}) \overset{\mathfrak{g}}{\text{------}} \mathcal{T}(\boldsymbol{\mu}_2, \mathfrak{g}) \,, \tag{2.14}$$

where

- $\mathcal{T}(\boldsymbol{\mu_a}, \mathfrak{g})$ is the minimal 6D orbi-instanton theory associated to the M9-M5 brane system in the presence of a $\mathbb{C}^2/\Gamma_{\mathfrak{g}}$ singularity transverse to the M5, including a choice of flat connection at infinity $\boldsymbol{\mu_a} : \Gamma_{\mathfrak{g}} \to E_8$;
- $\mathcal{T}_{N-2}(\mathfrak{g}, \mathfrak{g})$ is the 6D conformal matter theory associated to $(N-2)$ M5 branes probing a $\mathbb{C}^2/\Gamma_{\mathfrak{g}}$ singularity;
- $\overset{\mathfrak{g}}{\text{------}}$ denotes the operation of (diagonal) fusion of the common factors $\mathfrak{g}$ of the global symmetry of the corresponding 6D SCFTs, schematically at the level of the corresponding generalized quivers:

$$\cdots \overset{\mathfrak{g}'}{n'} \,[\mathfrak{g}] \overset{\mathfrak{g}}{\text{------}} [\mathfrak{g}]\, \overset{\mathfrak{g}''}{n''} \cdots \longrightarrow \cdots \overset{\mathfrak{g}'}{n'} \overset{\mathfrak{g}}{n} \overset{\mathfrak{g}''}{n''} \cdots \tag{2.15}$$

The theories $\mathcal{T}(\boldsymbol{\mu_a}, \mathfrak{g})$ have been classified in the literature [**30**, **37**, **43**, **59**] by identifying them with the minimal orbi-instanton models. When $N \geq 3$, the structure in (2.14) completely determines the tensor branches of all possible fractional $E_8 \times E_8$ Heterotic instantons for all possible singularities and holonomoies μ_a, complementing the results available for this class of models in the literature [**5**, **6**, **50**]. In [**32**] we explain a strategy to match all $Spin(32)/\mathbb{Z}_2$ instantonic LSTs with holonomy $\boldsymbol{\lambda}$ in terms of the $E_8 \times E_8$ duals with holonomies $\boldsymbol{\mu}_a$ for all singularities $\mathfrak{g}$. We focus on cases with exceptional flavor symmetries and gauge algebras, which cannot be realized in perturbative strings. At the end, we give a complete characterization of the simply laced cases.

3. Heterotic LSTs as elliptic K3 degenerations

In this section we want to use the geometric engineering framework of F-theory to describe Heterotic LSTs via elliptic CY threefolds

$$\begin{array}{c} T^2 \to X_3 \\ \downarrow \pi \\ B_2 \end{array} \tag{3.1}$$

over non-compact bases B_2. As stated in the introduction, this requires a base B_2 to be shrinkable to a curve Σ_{LS}, which is typically not irreducible. For Heterotic LSTs the irreducible components of Σ_{LS} are always rational curves intersecting transversally. Then η^{IJ} is identified with the intersection pairing $\eta^{IJ} = -w^I \cdot w^J$ of the base B_2 using $w^I \in H_2(B_2, \mathbb{Z})$ to denote the irreducible components of Σ_{LS}. The gauge algebra factors $\mathfrak{g}_I$ correspond to Kodaira singularities of the elliptic fiber along the irreducible component w^I. The irreducible components of the (geometric) flavor symmetry algebra are encoded in the Kodaira types of the non-compact components of the discriminant locus of the elliptic fibration[10]. It is then easy to show that the 2-group structure constant $\widehat{\kappa}_{\mathscr{P}}$ is birationally equivalent over the base B_2 and in particular $\widehat{\kappa}_{\mathscr{P}} = 2$ leads to the base of the schematic form

$$B_2 = BL^{n_T}(\mathbb{P}^1 \times \mathbb{C}), \tag{3.2}$$

[10] The free part of the Mordell-Weil group encodes possible Abelian flavor symmetries [**57**].

where we stress that most blowups occur at non-generic points. The LST curve is given as the linear combination dictated by (2.5):

$$\Sigma_{LS} = \sum_{I=1}^{n_T+1} \ell^I w^I, \tag{3.3}$$

which is by construction a rational curve of null self-intersection $\Sigma_{LS}^2 = 0$. This directly implies that for Heterotic LSTs the threefold X_3 also admits the structure of an elliptic K3 fibration

$$\begin{array}{ccc} K3 \hookrightarrow X_3 & & T^2 \hookrightarrow K3 \\ \downarrow \pi, & \text{with} & \downarrow \pi\,. \\ \mathbb{C}_u & & \Sigma_{LS} \end{array} \tag{3.4}$$

This is directly clear for $n_T = 0$ where $\Sigma_{LS} = \mathbb{P}^1$ in (3.2). We proceed by explaining first the properties of the K3 fiber that encodes the flavor symmetry and then the compact divisors at the origin.

The flavor K3: We denote the $K3$ fiber by S. When embedding the K3 into the whole threefold, we must fix a polarization of S, which fixes the Néron-Severi lattice NS(S) as a sublattice of $\Lambda_{K3} = II_{3,19}$ given as

$$NS(S) := H^{1,1}(S) \cap H^2(K3, \mathbb{Z}). \tag{3.5}$$

K3 being elliptic requires us to fix more structures of NS(S), namely to write it as

$$NS(S) = U \oplus W, \tag{3.6}$$

where U is the hyperbolic lattice corresponding to the section and the base, and W is the frame lattice. The frame lattice is spanned by divisors that do not intersect the zero section. The non-compact divisors $D \subset X_3$ restrict to divisors in the K3 that span NS(S) for a fixed polarization[11]. In the following we want to consider different elliptic fibration structures of the elliptic threefold. In the physics of F-theory, these structures correspond to different 6D F-theory compactifications with the same M-theory dual, i.e., dual theories upon circle reductions. First we recall that the locus $u = 1$ defines the K3 surface S. Now we assume that S admits different inequivalent fibration structures $\widetilde{\pi}$. These fibrations pull back to the threefold X_3 at the special locus $u = 0$ as well, where compact divisors reside. Note that we might have to perform flop-transitions to obtain a minimal and equi-dimensional elliptic fibration everywhere.

Upon performing such transitions, we obtain a new base $\widehat{B}_2$ with potentially non-trivial fibers, from which a new 6D F-theory quiver can be deduced. Since there is only a single non-compact direction in X_3 and, by construction, the rank of the $NS(S)$ lattice is fixed and yields the only non-compact divisors in the threefold, it is suggestive that all elliptic fibration structures of X_3 must be inherited from those of S. However, at this point, we cannot exclude that there might be other elliptic threefold fibrations $\widetilde{\pi}$ that may have been inherited from the single K3 fiber S.

We wish to come back to this interesting question in future work. However, for all elliptic fibration structures that come from the same K3 fibre, we find that those must always be of Heterotic LST type and therefore must have $\widehat{\kappa}_{\mathscr{P}} = 2$. For

[11] Note that we assume that we can display the full physical flavor group as geometrically resolvable singularities, which is not possible in all cases[**11**].

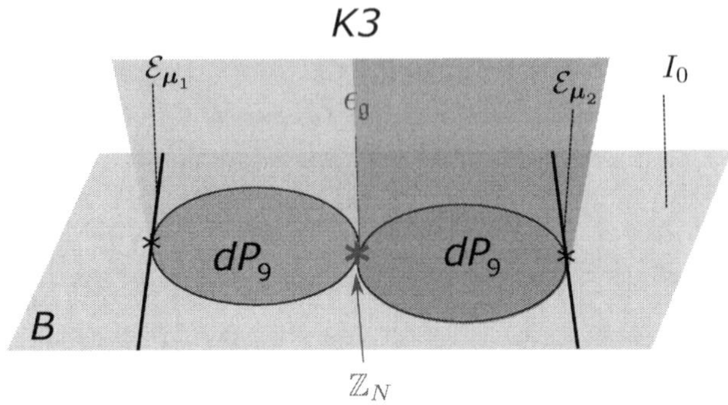

FIGURE 1. Heterotic LST from a stable degeneration of the K3 into two rational elliptic surfaces. At the intersection locus of the two -1 curves in the base, there is a $\mathbb{Z}_{N=n_T-1}$ singularity with another Kodaira type $\mathfrak{g}$ singularity on top. Flavor holonomies $\boldsymbol{\mu}_i$ for the two Hořava-Witten walls, dualize to $\mathcal{E}_{\boldsymbol{\mu}_a}$ Kodaira fibers of the rational elliptic surface.

this class of inequivalent elliptic fibration types, we have therefore geometrically proven the invariance of $\widehat{\kappa}_{\mathscr{P}}$ under T-duality[12]. Moreover, we have reduced the problem of finding inequivalent elliptic fibrations of the threefold X_3 to those of the compact K3 S with fixed polarization. Finding all elliptic fibrations for a polarized K3 is still highly non-trivial, but the self-dual property of the K3 lattice structure mitigates this problem substantially.

The degeneration locus: The compact divisors are located at the locus $u = 0$ of the base $\mathbb{C}$. Note that for $n_T = 1$ with no singularities, the base quiver takes the form

$$1\ 1. \tag{3.7}$$

This can be interpreted geometrically as follows: the restriction of the T^2 fibration along the -1 base curves gives a rational elliptic surface (the so-called del Pezzo 9 surface) into which the K3 fiber has degenerated into, see Figure 1. More specifically, note that the quiver given in (2.10) is nothing but a Kulikov degeneration of type II[13]. Note that we can shrink the -2 curves to yield a $\mathbb{C}^2/\mathbb{Z}_{n_T-1}$ singularity.

In a similar way, the two non-trivial ingredients discussed in Section 2 admit a geometric counterpart, which we now describe.

Firstly, decorating the quiver with gauge groups $\mathfrak{g}$ is achieved by taking a singularity $u = 0$ transversely, which yields the same type of singular elliptic fibers. Secondly, non-trivial flavor holonomies are obtained by rotating the flavor K3 to a smaller NS lattice.

Singularities of this type and their resolutions go beyond Kulikov degenerations. The structure, however, is similar enough to motivate a sufficient generalization of K3 degenerations that captures the full set of Heterotic LSTs mathematically. The full K3 degeneration picture of the threefold that corresponds to Heterotic LSTs is schematically depicted in Figure 2.

---

[12]In turn one may regard invariance of $\widehat{\kappa}_{\mathscr{P}}$ under T-duality as a prediction from physics, that all elliptic fibrations of X_3 must come from a K3 fibration structure.

[13]See [**56**] for a formal definition.

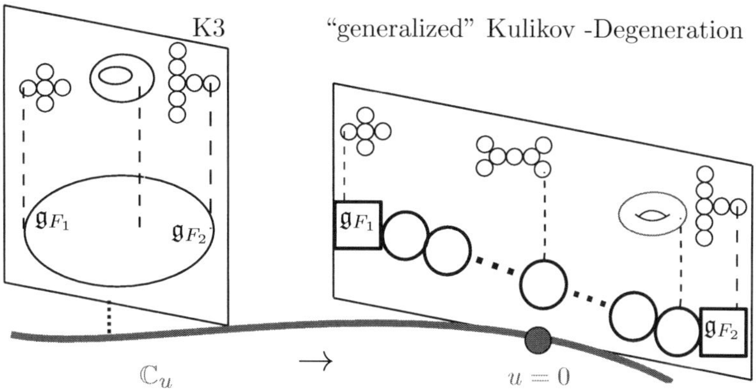

FIGURE 2. A little string theory as a "Kulikov-like" degeneration: Over a generic point in the base $\mathbb{C}_u$ there is an elliptic K3 fiber, whose frame lattice yields the flavor group. At the origin of $\mathbb{C}$, the K3 degenerates into compact surfaces with $\mathfrak{g}$ fibers corresponding to a fusion of $\mathcal{T}(\mathfrak{g}, \mathfrak{g}_{F_i})$ conformal matter.

The fact that Heterotic LST dualities descend from inequivalent elliptic fibration structures of the flavor K3, leads to a natural algorithm to find all duals for a Heterotic LST arising from a single chamber of the 5D Coulomb moduli space:

(1) Construct an arbitrary Heterotic LST and determine its flavor group $G_F = \prod_A \mathscr{F}_A$;
(2) Find the corresponding flavor K3 that realizes G_F as its frame lattice. Determine all inequivalent fibrations using lattice techniques and determine their frame lattices. This yields the collection of all possible dual flavor groups $\{G_F^{(e)}\}_{e \in E(S)}$ compatible with the given X_3, i.e., realized within the same chamber of the 5D moduli space. Here e is an index running over the possible elliptic structures $\mathbb{E}(S)$ of S;
(3) Starting from the SCFT classification results, construct a family of dual LSTs labeled by elements in $\mathbb{E}(s)$, obtained by fusing SCFTs with the constraints that *(i)* the dual flavor groups $G_F^{(e)}$ are compatible,[14] *(ii)* the 2-group structure constants match, and *(iii)* the 5D Coulomb branch dimension are the same.

We discuss an explicit example of this strategy at work to construct six dual LSTs in Section 5.

4. Example 1: Discrete Holonomy LSTs

In this section we exemplify the above considerations, using a modified version of Batyrev's construction to obtain the resolved non-compact threefold as a hypersurface in a non-compact ambient fourfold.

When doing so, we will start with a theory on the $E_8 \times E_8$ side, adding a singularity and flavor holonomies at infinity. Upon the toric resolution, we will

[14]This requires not only that the corresponding ranks are the same, but also that they have matching full subalgebras, and compatible $\widehat{\varkappa}_{\mathscr{F}_A}$ – sometimes this requires rescaling those by the index of embedding.

read off not only the T-dual on the $Spin(32)/\mathbb{Z}_2$ side, but also a third exotic dual Heterotic LST.

Starting with the E_8^2 we take $(N+1)$-many NS5 branes with an E_7 singularity. The $\mathbb{Z}_2$ center of E_7 allows us to choose a discrete holonomy $\mu_a = \mathbb{Z}_2$ which yields a breaking to the unbroken flavor group

$$G_F : E_8^2 \to (E_7 \times SU(2))^2/\mathbb{Z}_2. \tag{4.1}$$

Note that this choice yields a non-simply connected flavor group [**7, 36**]. Upon fractionalization, the full quiver then becomes

$$[\mathfrak{e}_7]\ 1\ \overset{\mathfrak{su}_2}{2}\ \overset{\mathfrak{so}_7}{3}\ \overset{\mathfrak{su}_2}{2}\ 1\ \ \underbrace{\overset{\overset{[\mathfrak{su}_2]}{1}}{\overset{\mathfrak{e}_7}{8}}\ 1\ \overset{\mathfrak{su}_2}{2}\ \overset{\mathfrak{so}_7}{3}\ \overset{\mathfrak{su}_2}{2}\ 1\ \overset{\mathfrak{e}_7}{8} \ldots \overset{\mathfrak{e}_7}{8}\ 1\ \overset{\mathfrak{su}_2}{2}\ \overset{\mathfrak{so}_7}{3}\ \overset{\mathfrak{su}_2}{2}\ 1\ \overset{\overset{[\mathfrak{su}_2]}{1}}{\overset{\mathfrak{e}_7}{8}}}_{\times N}\ \ 1\ \overset{\mathfrak{su}_2}{2}\ \overset{\mathfrak{so}_7}{3}\ \overset{\mathfrak{su}_2}{2}\ 1\ [\mathfrak{e}_7]. \tag{4.2}$$

The repeating chain $1\ \overset{\mathfrak{su}_2}{2}\ \overset{\mathfrak{so}_7}{3}\ \overset{\mathfrak{su}_2}{2}\ 1$ is the $\mathcal{T}(\mathfrak{e}_7, \mathfrak{e}_7)$ conformal matter required from the fractionalization of the M5 branes. This data is sufficient to compute the dimension of the Coulomb branch. Note that each of the N chains contributes 6 tensors and an $\mathfrak{e}_7 \times \mathfrak{su}_2^2 \times \mathfrak{so}_7$ gauge group, therefore contributing a total of 18 Coulomb branch parameters. From this we deduce the general expression

$$\dim(\text{CB}) = \mathbf{T} + \text{rk}(G) = 11 + 18N. \tag{4.3}$$

In order to compute the 2-group structure constants, we need the LS charge vector:

$$\vec{l}_{LS} = (1,1,1,2,3,\underbrace{\overset{1,}{1},4,3,2,3,4,1\ldots 1,4,3,2,3,4,\overset{1,}{1}}_{\times N},3,2,1,1,1), \tag{4.4}$$

which we depict in the same form as the quiver to simplify identification of LS charges with the corresponding quiver nodes.

In order to compute $\widehat{\kappa}_{\mathscr{R}}$ we also require the dual Coxeter numbers $h^\vee_{\mathfrak{g}}$ of the gauge algebra factors $\mathfrak{g}_I$ in the quiver, which are written in terms of a vector as

$$\vec{h}^\vee_{\mathfrak{g}_I} = (1,2,5,2,1,\underbrace{\overset{1}{18},1,2,5,2,1,18\ldots 18,1,2,5,2,1,\overset{1}{18}}_{\times N},1,2,5,2,1), \tag{4.5}$$

Note that we assign a non-trivial dual Coxeter number to trivial gauge groups as $h^\vee_{\mathfrak{su}_0} = 1$. Thus, from the repeating chain, we can then compute the 2-group contribution for each $\mathfrak{e}_7$ and conformal matter as

$$\widehat{\kappa}_{\mathscr{R}}(\overset{\mathfrak{e}_7}{8} \oplus \mathcal{T}(\mathfrak{e}_7, \mathfrak{e}_7)) = (1,4,3,2,3,4)^T \cdot (18,1,2,5,2,1) = 48. \tag{4.6}$$

Including N copies of the above contribution to $\widehat{\kappa}_{\mathscr{R}}$ we obtain for the full quiver

$$\widehat{\kappa}_{\mathscr{R}} = 2 + 48N. \tag{4.7}$$

The full data of the theory is given by the frame lattice W of the K3, which encodes the flavor group, the Coulomb branch dimension and 2-group structure constants given by

$$(W, \dim(\text{CB}), \widehat{\kappa}_{\mathscr{R}}, \widehat{\kappa}_{\mathscr{P}}) = \left(\frac{E_7^2 \times SU(2)^2}{\mathbb{Z}_2}, 11+18N, 2+48N, 2\right). \tag{4.8}$$

To deduce the T-duals, we explicitly resolve the above family of elliptic threefolds, starting with the compact flavor K3. The global structure of the flavor group is implemented by a $\mathbb{Z}_2$ Mordell-Weil (MW) group [**7**, **54**, **58**].

The K3 is specified by a 3D reflexive lattice polytope, with points given in Table 1. The elliptic K3 hypsersurface can easily be constructed using the Batyrev prescription (see [**9**] for more details) which generalizes to the non-compact case as well.

TABLE 1. Depiction of the toric rays of the flavor K3 polytope. The polytope admits three, 2D sub-polytopes which yield inequivalent elliptic fibrations with X, Y, Z, e_1 spanning the first. We have also highlighted the rays of the two $\mathfrak{e}_7$ tops.

fiber 1	
Z	(1,0,0)
X	(-1,2,0)
Y	(-1,-2,0)
e_1	(0,-1,0)

E_7 tops	
$\beta_{1,\pm}$	$(-1,2,\pm1)$
$\beta_{2,\pm}$	$(-1,2,\pm2)$
$\beta_{3,\pm}$	$(-1,2,\pm3)$
$\beta_{4,\pm}$	$(-1,2,\pm4)$
$\alpha_{3,\pm}$	$(-1,1,\pm3)$
$\alpha_{2,\pm}$	$(0,1,\pm2)$
$\alpha_{1,\pm}$	$(-1,1,\pm1)$
$\gamma_{2,\pm}$	$(-1,0,\pm2)$

The elliptic fibration of the K3 is evident from its 2D sub-polytope structure spanned by X, Z, Y, e_1. This sub-polytope highlights a property of the ambient variety, which itself admits a fibration by a 2D complex Fano variety. The torus fibration is thus inherited from the ambient space variety.[15] The residual rays resolve two E_7 singularities [**18**, **22**].

This is the first building block to construct a non-compact threefold. Consider now a 4D polytope Δ_4, in which the K3 is given as a 3D reflexive sub-polytope Δ_3 as

$$\Delta_4 = \left(\frac{\Delta_3 \mid 0}{\widehat{\Delta}_4}\right). \tag{4.9}$$

The compact surfaces are specified by the collection of vertices given by the collection $\widehat{\Delta}_4$, spelled out in detail in Table 2 of the Appendix and is sufficient to deduce the resolved quiver in (4.2).

By deducing two more 2D sub-polytopes in the 3D sub-polytope of the K3, we deduce the other toric fibrations, their flavor fibres and intersection structure of the compact divisors. The second 2D sub-polytope is spanned by the vertices $Z, \gamma_{2,+}$ and $\gamma_{2,-}$ which also admits a $\mathbb{Z}_2$ Mordell-Weil group.

After re-arranging all non-compact and compact divisors, we obtain the quiver

$$\begin{array}{c} \\ [\mathfrak{so}_8] \end{array} \begin{array}{c} \\ \overset{\mathfrak{sp}_{N-1}}{1} \end{array} \begin{array}{c} \\ \overset{\mathfrak{so}_{4N+4}}{4} \end{array} \begin{array}{c} \overset{\mathfrak{sp}_{2N-4}}{1} \\ \overset{\mathfrak{sp}_{3N-3}}{1} \end{array} \begin{array}{c} \\ \overset{\mathfrak{so}_{8N}}{4} \end{array} \begin{array}{c} \\ \overset{\mathfrak{sp}_{3N-1}}{1} \end{array} \begin{array}{c} \\ \overset{\mathfrak{so}_{4N+12}}{4} \end{array} \begin{array}{c} \\ \overset{\mathfrak{sp}_{N+5}}{1} \end{array} \begin{array}{c} \\ [\mathfrak{so}_{24}], \end{array} \tag{4.10}$$

[15]See [**1**, **48**] for more details on this construction for compact threefolds.

with LS charges

$$\begin{array}{c} 2, \\ \vec{l}_{LS} = (1,1,3,2,3,1,1). \end{array} \tag{4.11}$$

Note that this dual LST quiver admits the topology of an $\mathfrak{e}_7^{(1)}$, which is just the singularity that appears in the first fibration. The Coulomb branch dimension is simpler to determine here as the number of tensors is fixed to $T = 7$, which yields

$$\dim(\mathrm{CB}) = T + \mathrm{rk}(G) = 11 + 18N. \tag{4.12}$$

Using the dual Coxeter numbers we compute the 2-group structure constants:

$$h^\vee_{\mathfrak{so}_{2M}} = 2M - 2 \quad \text{and} \quad h^\vee_{\mathfrak{sp}_M} = M + 1. \tag{4.13}$$

In the end, the full data of the theory is

$$(W,\, \dim(\mathrm{CB}),\, \widehat{\kappa}_{\mathscr{R}},\, \widehat{\kappa}_{\mathscr{P}}) = \left(\frac{\mathrm{Spin}(24)\times\mathrm{Spin}(8)}{\mathbb{Z}_2},\, 11+18N,\, 2+48N,\, 2\right). \tag{4.14}$$

The third 2D sub-polytope is spanned by $e_1, \alpha_{2,+}$ and $\alpha_{2,-}$. This model admits also a $\mathbb{Z}_2$ group in addition to a free part, which yields a $\mathfrak{u}_1$ flavor contribution [**4, 57**]. Re-arranging all curves, one obtains the following quiver:

$$[\mathfrak{u}_{16}]\ \overset{\mathfrak{su}_{2N+10}}{2}\ \ \overset{\mathfrak{su}_{4N+4}}{2}\ \ \overset{\mathfrak{su}_{6N-2}}{2}\ \ \overset{\mathfrak{sp}_{4N-4}}{1}\ \ \overset{\mathfrak{so}_{4N+4}}{4}\ , \tag{4.15}$$

for which we have the LS charge vector

$$\vec{l}_{LS} = (1,2,3,4,1). \tag{4.16}$$

Note that the shape of this quiver does not resemble the shape of an affine ADE quiver but is closer to an $\mathfrak{e}_6^{(2)}$ (but with symmetric intersections.). Using $h^\vee_{\mathfrak{su}_M} = M$, the Coulomb branch dimension and 2-group structure constants can be readily computed. The full data of the theory is summarized as

$$(W,\, \dim(\mathrm{CB}),\, \widehat{\kappa}_{\mathscr{R}},\, \widehat{\kappa}_{\mathscr{P}}) = \left(\frac{SU(16)\times U(1)}{\mathbb{Z}_2},\, 11+18N, 2+48N,\, 2\right). \tag{4.17}$$

We have proven the explicit duality of all three LSTs. As expected, we observe that in all three models, there exist an explicit match of flavor rank, Coulomb branch dimension and 2-group structure constants summarized as

$$(\mathrm{rk}(\mathrm{G}_F),\, \dim(\mathrm{CB}),\, \widehat{\kappa}_{\mathscr{R}},\, \widehat{\kappa}_{\mathscr{P}}) = (16,\, 18N+11,\, 48N+2,\, 2). \tag{4.18}$$

5. Example 2: Extremal flavor rank and T-hexality

Using toric geometry in the section above, we often obtained more than two dual Heterotic LSTs. This might come as a surprise, as one might think of this duality as descending from the familiar 10D $E_8 \times E_8 \leftrightarrow Spin(32)/\mathbb{Z}_2$ duality. However, there might be even more fibrations that we could not access in the toric construction employed in the section before. Here we want to go beyond those restrictions and propose a T-hexality of LSTs that admit the maximal flavor rank 18. From the geometry, these large flavor groups descend from the frame lattice of the K3 that can

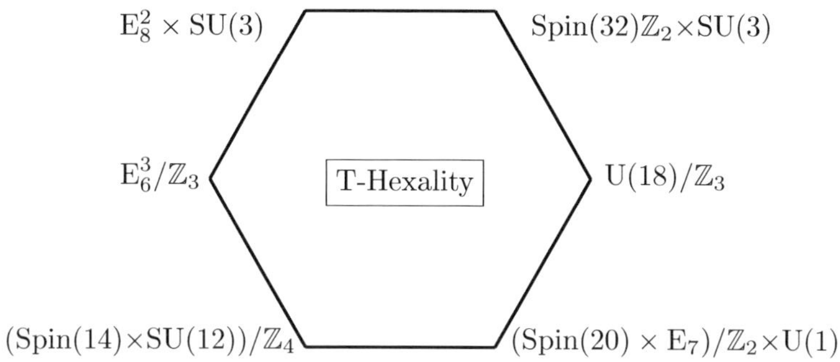

FIGURE 3. The six different flavor groups obtained from inequivalent elliptic fibration structures of the same attractive K3 [**20**]. Those map to non-geometric Heterotic string backgrounds and yield the universal flavor groups of six different 6D LSTs.

be as large as rank 18.[16] There are plenty of examples in the mathematics literature: elliptic K3s with a maximal non-Abelian group of rank 18 are called extremal and have been classified in [**60, 67**]. None of them have a toric construction, which makes a full resolution more complicated. The rigid structure of these K3s, on the other hand, makes a full classification of their elliptic fibrations in lattice theoretic terms feasible [**20**].

In the following we exploit this fact and make use of the geometric algorithm outlined in Section 4. We propose dual LSTs by using an attractive K3 where all six inequivalent fibrations are known, see Figure 3. For all six theories, the rank of flavor group, Coulomb branch dimension and 2-group structure constants

$$(\mathrm{rk}(G_F),\, \dim(\mathrm{CB}),\, \widehat{\kappa}_{\mathscr{R}},\, \widehat{\kappa}_{\mathscr{P}}) \;=\; (18, 21, 30, 2), \tag{5.1}$$

match, which we take as strong evidence for the existence of the duality. Below we list the quivers of the six LSTs we identified.

LST 1. $\mathbf{E_8^2} \times \mathbf{SU(3)}$ flavor: This configuration is very close to the HW setup, but admits an additional $SU(3)$ factor. This factor can be deduced by employing the 8D SUGRA theory, corresponding to the Heterotic string compactified on a self-dual $\mathfrak{su}_3$ torus. Using this 8D gauge group as a flavor group and conformal matter techniques [**28**], we deduce the following consistent LST quiver:

(5.2)

$$\begin{array}{c} [\mathfrak{su}_3] \\ 1 \\ [\mathfrak{e}_8]\; 1\; 2\; \overset{\mathfrak{sp}_1}{2}\; \overset{\mathfrak{g}_2}{3}\; 1\; \overset{\mathfrak{f}_4}{5}\; 1\; \overset{\mathfrak{g}_2}{3}\; \overset{\mathfrak{sp}_1}{2}\; 2\; 1\, [\mathfrak{e}_8] \end{array}, \qquad \begin{array}{c} 1, \\ \vec{l}_{LS} = (1,1,1,1,2,1,2,1,1,1,1,1). \end{array}$$

[16]This suggests that the actual duality of the LSTs originate from the eight dimensional SUGRA, where F-theory on K3 is dual to the Heterotic string on T^2 whereas the gauge symmetry becomes flavor . It would be interesting to reinterpret the 6d Heterotic LSTs as codimension 2 membranes for the 8d SUGRA - the Hagedorn density of states of LSTs might have some interesting implications for the physics of those systems.

LST 2. $\mathbf{SO(32)/\mathbb{Z}_2 \times SU(3)}$ flavor: When compared to the model before, this case reproduces the familiar 10D Heterotic duality upon which the additional $SU(3)$ self-dual torus is simply an observer. Using the fiber-base flip of the $\mathfrak{f}_4$ singularity yields a consistent quiver given as

$$[\mathfrak{so}_{32}]\ \overset{\mathfrak{sp}_8}{1}\ \overset{\mathfrak{so}_{16}}{4}\ 1\ 2\ \underset{[N_F=1]}{\overset{\mathfrak{su}_2}{2}}\ [\mathfrak{su}_3], \qquad \vec{l}_{LS} = (1,1,3,2,1). \tag{5.3}$$

Notably, the inserted conformal matter that starts from the $\mathfrak{so}_{16}$ curve looks like that of $\mathcal{T}(\mathfrak{e}_8, \mathfrak{su}_3)$. This on the other hand can also be deduced by noting that the $\mathbb{Z}_2$ quotient stops at the $\mathfrak{so}_{16}$ gauge group factor.

LST 3. $\mathbf{E_6^3/\mathbb{Z}_3}$ flavor: This case starts to deviate from the usual HW structure in more ways. The underlying singular K3 admits a very nice description in terms of toroidal orbifolds of type $(\mathbb{T}^2 \times \mathbb{T}^2)/\mathbb{Z}_3$ [**26, 42, 55**]. It is easy to see that the orbifold action fixes the complex structure of both tori to $\tau = e^{2\pi i/3}$. We identify the F-theory torus with one of those tori. As before, the rigid structure of the K3 only allows for singularities compatible with such a τ value. In addition there is a $\mathbb{Z}_3$ MW group, which also acts non-trivially on the possible fiber [**36**] singularities. This non-trivial global $\mathbb{Z}_3$ structure must also act non-trivially on all other gauge groups, which fixes the quiver almost uniquely:

$$\begin{array}{c} [\mathfrak{e}_6] \\ 1 \\ \overset{\mathfrak{su}_3}{3} \\ 1 \\ [\mathfrak{e}_6]\ 1\ \overset{\mathfrak{su}_3}{3}\ 1\ \overset{\mathfrak{e}_6}{6}\ 1\ \overset{\mathfrak{su}_3}{3}\ 1\ [\mathfrak{e}_6] \end{array}, \qquad \vec{l}_{LS} = \begin{array}{c} 1, \\ 1, \\ 2, \\ (1,\ 1,\ 2,\ 1,\ 2,\ 1,\ 1). \end{array} \tag{5.4}$$

This quiver also admits a manifest $\mathbb{S}_3$ permutation symmetry.

LST 4. $\mathbf{(SO(14) \times SU(12))/\mathbb{Z}_4)}$ flavor: Here the $\mathbb{Z}_4$ MW group restricts the admissible fibers again and, in particular, conformal matter theories to those classified in [**36**]. From those theories, a matching quiver can be constructed as

$$[\mathfrak{so}_{14}]\ \overset{\mathfrak{sp}_2}{1}\ \overset{\mathfrak{so}_{10}}{4}\ 1\ \overset{\mathfrak{su}_4}{2}\ \overset{\mathfrak{su}_8}{2}\ [\mathfrak{su}_{12}], \qquad \vec{l}_{LS} = (1,1,3,2,1). \tag{5.5}$$

LST 5. $\mathbf{U(18)/\mathbb{Z}_3}$ flavor: For this case, we again have restricted monodromies due to the $\mathbb{Z}_3$ MW group. That is, when adding an $\mathfrak{su}_3$ type of singularity, one obtains the chain

$$\overset{\mathfrak{su}_3}{3}\ 1\ \overset{\mathfrak{su}_6}{2}\ \overset{\mathfrak{su}_{12}}{2}\ [\mathfrak{u}_{18}], \qquad \vec{l}_{LS} = (1,3,2,1). \tag{5.6}$$

Note here that the $\mathfrak{u}_1$ flavor factor fits nicely into the full flavor group of the quiver diagram: The BPS string that wraps the 1 curve is an E-string for which we expect an E_8 flavor group. There we find an $\mathfrak{su}_3 \times \mathfrak{su}_6$ subgroup of the $\mathfrak{e}_8$ to be gauged by the neighbouring curves, which leaves a residual $\mathfrak{u}_1$ sub-group. This flavor factor is reflected in the rank one Mordell-Weil group.

LST 6. $\mathbf{(SO(20) \times E_7)/\mathbb{Z}_2 \times U(1)}$ flavor: The final configuration is similar as before, but with a $\mathbb{Z}_2$ MW group as well as an overall $\mathfrak{u}_1$ part [**36**] we obtain

$$[\mathfrak{so}_{20}]\ \overset{\mathfrak{sp}_4}{1}\ \overset{\mathfrak{so}_{12}}{4}\ 1\ \overset{\mathfrak{su}_2}{2}\ \overset{\mathfrak{so}_7}{3}\ \overset{\mathfrak{su}_2}{2}\ 1\ [\mathfrak{e}_7 \times \mathfrak{u}_1], \qquad \vec{l}_{LS} = (1,1,3,2,1,1,1). \tag{5.7}$$

There are two E-string curves: The rightmost has a maximal $\mathfrak{e}_7 \times \mathfrak{su}_2$ subgroup of $\mathfrak{e}_8$ gauge by the neighbouring curves. The middle one, however, admits only a

$\mathfrak{so}_{12} \times \mathfrak{su}_2$ subgroup to be gauged. Hence we have another residual $\mathfrak{u}_1$ flavor factor which is realized by the free part of the MW group in the geometry.

Appendix A. Toric rays of compact divisors in Example 1

In Table 2 we summarize the toric rays that correspond to the compact divisors $\widehat{\Delta}_4$ of the 4D ambient variety used in Example 1 of Section 4. Details of the analysis of their fibral structure under the three elliptic fibrations are given in [**33**].

TABLE 2. Summary of toric vertices used to resolve the 4D ambient variety used in the threefold of Example 1.

$\mathfrak{e}_7$ $i = 0 \ldots N$ gauge factors	
$\beta_{1,i}$	$(-1, 2, i, \pm 1)$
$\beta_{2,i}$	$(-1, 2, 2i, \pm 2)$
$\beta_{3,i}$	$(-1, 2, 3i, \pm 3)$
$\beta_{4,i}$	$(-1, 2, 4i, \pm 4)$
$\alpha_{3,i}$	$(-1, 1, 3i, \pm 3)$
$\alpha_{2,i}$	$(0, 1, 2i, \pm 2)$
$\alpha_{1,i}$	$(-1, 1, i, \pm 1)$
$\gamma_{2,i}$	$(-1, 0, i, \pm 2)$
$\mathcal{T}(\mathfrak{e}_7^i, \mathfrak{e}_7^{i+1})$ $i = 0 \ldots N-1$ Conformal matter	
$s_{\pm 1}$	$(-1, 2, 1+4i, 4)$
$s_{\pm 2}$	$(-1, 2, 1+3i, 3), (-1, 1, 1+3i, 3)$
$s_{\pm 3,i}$	$(-1, 2, 1+2i, 2), (-1, 0, 1+2i, 2), (0, 1, 1+2i, 2), (-1, 2, 2+2i, 4)$
$s_{\pm 4,j}$	$(-1, 2, 2+3i, 3), (-1, 1, 2+3i, 3)$
$s_{\pm 5}$	$(-1, 2, 3+4i, 4)$
0-th $\mathcal{T}(\mathfrak{e}_7, \mathfrak{e}_7)$ Conformal matter	
$s_{\pm 1}$	$(-1, 2, -3, 1)$
$s_{\pm 2}$	$(-1, 2, -2, 1), (-1, 1, -2, 1)$
$s_{\pm 3,i}$	$(-1, 2, -1, 1), (-1, 0, -1, 1), (0, 1, -1, 1), (-1, 2, -2, 2)$
$s_{\pm 4,j}$	$(-1, 2, -1, 2), (-1, 1, -1, 2)$
$s_{\pm 5}$	$(-1, 2, -1, 3)$
N-th $\mathcal{T}(\mathfrak{e}_7, \mathfrak{e}_7)$ Conformal matter	
$s_{\pm 1}$	$(-1, 2, 3+N, 1)$
$s_{\pm 2}$	$(-1, 2, 2+N, 1), (-1, 1, 2+N, 1)$
$s_{\pm 3,i}$	$(-1, 2, 1+N, 1), (-1, 0, 1+N, 1), (0, 1, 1+N, 1), (-1, 2, 2+N, 2)$
$s_{\pm 4,j}$	$(-1, 2, 1+2N, 2), (-1, 1, 1+2N, 2)$
$s_{\pm 5}$	$(-1, 2, 1+3N, 3)$
E-string Conformal Matter	
f_0	$(0, 0, 0, 1)$
f_M	$(0, 0, N, 1)$

Acknowledgments

We thank Andreas Braun, Jonathan Heckmann, Joseph Minahan, Thorsten Schimannek and Timo Weigand for discussions and Robert Moscrop for reading and comments.

References

[1] Lara B. Anderson, Xin Gao, James Gray, and Seung-Joo Lee, *Multiple fibrations in Calabi-Yau geometry and string dualities*, J. High Energy Phys. **10** (2016), 105, front matter+63, DOI 10.1007/JHEP10(2016)105. MR3578530

[2] Fabio Apruzzi, Lakshya Bhardwaj, Dewi S. W. Gould, and Sakura Schäfer-Nameki, *2-group symmetries and their classification in 6d*, SciPost Phys. **12** (2022), no. 3, Paper No. 098, 48, DOI 10.21468/scipostphys.12.3.098. MR4399578

[3] Fabio Apruzzi, Markus Dierigl, and Ling Lin, *The fate of discrete 1-form symmetries in 6d*, SciPost Phys. **12** (2022), no. 2, Paper No. 047, 47, DOI 10.21468/scipostphys.12.2.047. MR4378181

[4] Fabio Apruzzi, Marco Fazzi, Jonathan J. Heckman, Tom Rudelius, and Hao Y. Zhang, *General prescription for global $U(1)$'s in 6D SCFTs*, Phys. Rev. D **101** (2020), no. 8, 086023, 38, DOI 10.1103/physrevd.101.086023. MR4103192

[5] Paul S. Aspinwall, *Point-like instantons and the* $\mathrm{Spin}(32)/Z_2$ *heterotic string*, Nuclear Phys. B **496** (1997), no. 1-2, 149–176, DOI 10.1016/S0550-3213(97)00232-0. MR1457988

[6] Paul S. Aspinwall and David R. Morrison, *Point-like instantons in $K3$ orbifolds*, Nuclear Phys. B **503** (1997), no. 3, 533–564, DOI 10.1016/S0550-3213(97)00516-6. MR1482735

[7] Paul S. Aspinwall and David R. Morrison, *Nonsimply connected gauge groups and rational points on elliptic curves*, J. High Energy Phys. **07** (1998), 012.

[8] Tom Banks and Nathan Seiberg, *Symmetries and Strings in Field Theory and Gravity*, Phys. Rev. **D83** (2011), 084019.

[9] Victor V. Batyrev and Lev A. Borisov, *On Calabi-Yau complete intersections in toric varieties*, Higher-dimensional complex varieties (Trento, 1994), de Gruyter, Berlin, 1996, pp. 39–65. MR1463173

[10] Micha Berkooz, Robert G. Leigh, Joseph Polchinski, John H. Schwarz, Nathan Seiberg, and Edward Witten, *Anomalies, dualities, and topology of $D = 6$ $N = 1$ superstring vacua*, Nuclear Phys. B **475** (1996), no. 1-2, 115–148, DOI 10.1016/0550-3213(96)00339-2. MR1412098

[11] Marco Bertolini, Peter R. Merkx, and David R. Morrison, *On the global symmetries of 6D superconformal field theories*, J. High Energy Phys. **7** (2016), 005, front matter+46, DOI 10.1007/JHEP07(2016)005. MR3547281

[12] Lakshya Bhardwaj, *Discovering T-dualities of little string theories*, J. High Energy Phys. **2** (2024), Paper No. 46, DOI 10.1007/jhep02(2024)046. MR4706091

[13] Lakshya Bhardwaj, Michele Del Zotto, Jonathan J. Heckman, David R. Morrison, Tom Rudelius, and Cumrun Vafa, *F-theory and the classification of little strings*, Phys. Rev. D **93** (2016), no. 8, 086002, 37, DOI 10.1103/PhysRevD.93.086002. [Erratum: Phys.Rev.D 100, 029901 (2019) MR4017347]. MR3531134

[14] Lakshya Bhardwaj, David R. Morrison, Yuji Tachikawa, and Alessandro Tomasiello, *The frozen phase of F-theory*, J. High Energy Phys. **8** (2018), 138, front matter+47, DOI 10.1007/jhep08(2018)138. MR3861184

[15] Lakshya Bhardwaj and Sakura Schäfer-Nameki, *Higher-form symmetries of 6d and 5d theories*, J. High Energy Phys. **2** (2021), Paper No. 159, 45, DOI 10.1007/jhep02(2021)159. MR4259836

[16] Julie D. Blum and Kenneth Intriligator, *Consistency conditions for branes at orbifold singularities*, Nuclear Phys. B **506** (1997), no. 1-2, 223–235, DOI 10.1016/S0550-3213(97)00450-1. MR1491019

[17] Julie D. Blum and Kenneth Intriligator, *New phases of string theory and 6d RG fixed points via branes at orbifold singularities*, Nuclear Phys. B **506** (1997), no. 1-2, 199–222, DOI 10.1016/S0550-3213(97)00449-5. MR1491018

[18] Vincent Bouchard and Harald Skarke, *Affine Kac-Moody algebras, CHL strings and the classification of tops*, Adv. Theor. Math. Phys. **7** (2003), no. 2, 205–232. MR2015164

[19] Andreas P. Braun, *M-theory and orientifolds*, J. High Energy Phys. **9** (2020), 065, 35, DOI 10.1007/jhep09(2020)065. MR4203194

[20] Andreas P. Braun, Yusuke Kimura, and Taizan Watari, *On the Classification of Elliptic Fibrations modulo Isomorphism on K3 Surfaces with large Picard Number*, `arXiv:1312.4421`, 2013.

[21] Ilka Brunner and Andreas Karch, *Branes at orbifolds versus Hanany-Witten in six dimensions*, J. High Energy Phys. **3** (1998), Paper 3, 25, DOI 10.1088/1126-6708/1998/03/003. MR1619417

[22] Philip Candelas and Anamaría Font, *Duality between the webs of heterotic and type II vacua*, Nuclear Phys. B **511** (1998), no. 1-2, 295–325, DOI 10.1016/S0550-3213(96)00410-5. MR1606088

[23] Clay Córdova, Thomas T. Dumitrescu, and Kenneth Intriligator, *Exploring 2-group global symmetries*, J. High Energy Phys. **2** (2019), 184, front matter + 108, DOI 10.1007/jhep02(2019)184. MR3925280

[24] Clay Córdova, Thomas T. Dumitrescu, and Kenneth Intriligator, *2-group global symmetries and anomalies in six-dimensional quantum field theories*, J. High Energy Phys. **4** (2021), Paper No. 252, 45, DOI 10.1007/jhep04(2021)252. MR4276198

[25] Mirjam Cvetič, Jonathan J. Heckman, Max Hübner, and Ethan Torres, *0-form, 1-form, and 2-group symmetries via cutting and gluing of orbifolds*, Phys. Rev. D **106** (2022), no. 10, Paper No. 106003, 39, DOI 10.1103/physrevd.106.106003. MR4525437

[26] Keshav Dasgupta and Sunil Mukhi, *F-theory at constant coupling*, Phys. Lett. B **385** (1996), no. 1-4, 125–131, DOI 10.1016/0370-2693(96)00875-1. MR1410239

[27] Michele Del Zotto, Iñaki García Etxebarria, and Saghar S. Hosseini, *Higher form symmetries of Argyres-Douglas theories*, J. High Energy Phys. **10** (2020), 056, 22, DOI 10.1007/jhep10(2020)056. MR4204049

[28] Michele Del Zotto, Jonathan J. Heckman, David R. Morrison, and Daniel S. Park, *6D SCFTs and gravity*, J. High Energy Phys. **6** (2015), 158, front matter+44, DOI 10.1007/JHEP06(2015)158. MR3370177

[29] Michele Del Zotto, Jonathan J. Heckman, Daniel S. Park, and Tom Rudelius, *On the defect group of a 6D SCFT*, Lett. Math. Phys. **106** (2016), no. 6, 765–786, DOI 10.1007/s11005-016-0839-5. MR3500421

[30] Michele Del Zotto, Jonathan J. Heckman, Alessandro Tomasiello, and Cumrun Vafa, *6d conformal matter*, J. High Energy Phys. **2** (2015), 054, front matter+55, DOI 10.1007/JHEP02(2015)054. MR3321409

[31] article Michele Del Zotto, Muyang Liu, and Paul-Konstantin Oehlmann, *Back to heterotic string on ale spaces - part iii: Geometric engineering limits of the heterotic string*, To Appear.

[32] Michele Del Zotto, Muyang Liu, and Paul-Konstantin Oehlmann, *Back to heterotic strings on ALE spaces. Part I. Instantons, 2-groups and T-duality*, J. High Energy Phys. **1** (2023), Paper No. 176, 37, DOI 10.1007/jhep01(2023)176. MR4543652

[33] Michele Del Zotto, Muyang Liu, and Paul-Konstantin Oehlmann, *Back to heterotic strings on ALE spaces. Part II. Geometry of T-dual little strings*, J. High Energy Phys. **1** (2024), Paper No. 109, 95, DOI 10.1007/jhep01(2024)109. MR4694465

[34] Michele Del Zotto and Guglielmo Lockhart, *Universal features of BPS strings in six-dimensional SCFTs*, J. High Energy Phys. **8** (2018), 173, front matter+168, DOI 10.1007/jhep08(2018)173. MR3861149

[35] Michele Del Zotto and Kantaro Ohmori, *2-group symmetries of 6D little string theories and T-duality*, Ann. Henri Poincaré **22** (2021), no. 7, 2451–2474, DOI 10.1007/s00023-021-01018-3. MR4285948

[36] Markus Dierigl, Paul-Konstantin Oehlmann, and Fabian Ruehle, *Non-simply-connected symmetries in 6D SCFTs*, J. High Energy Phys. **10** (2020), 173, 62, DOI 10.1007/jhep10(2020)173. MR4203932

[37] Darrin D. Frey and Tom Rudelius, *6D SCFTs and the classification of homomorphisms* $\Gamma_{ADE} \to E_8$, Adv. Theor. Math. Phys. **24** (2020), no. 3, 709–756, DOI 10.4310/ATMP.2020.v24.n3.a4. MR4145058

[38] Robert Friedman, John Morgan, and Edward Witten, *Vector bundles and* F *theory*, Comm. Math. Phys. **187** (1997), no. 3, 679–743, DOI 10.1007/s002200050154. MR1468319

[39] Davide Gaiotto, Anton Kapustin, Nathan Seiberg, and Brian Willett, *Generalized global symmetries*, J. High Energy Phys. **2** (2015), 172, front matter+61, DOI 10.1007/JHEP02(2015)172. MR3321281

[40] Ori J. Ganor and Amihay Hanany, *Small* E_8 *instantons and tensionless non-critical strings*, Nuclear Phys. B **474** (1996), no. 1, 122–138, DOI 10.1016/0550-3213(96)00243-X. MR1404218

[41] Amihay Hanany and Alberto Zaffaroni, *Branes and six-dimensional supersymmetric theories*, Nuclear Phys. B **529** (1998), no. 1-2, 180–206, DOI 10.1016/S0550-3213(98)00355-1. MR1650334

[42] Hirotaka Hayashi, Patrick Jefferson, Hee-Cheol Kim, Kantaro Ohmori, and Cumrun Vafa, *SCFTs, holography, and topological strings*, Surveys in differential geometry 2018. Differential geometry, Calabi-Yau theory, and general relativity, Surv. Differ. Geom., vol. 23, Int. Press, Boston, MA, 2020, pp. 105–211. MR4451826

[43] Jonathan J. Heckman, David R. Morrison, Tom Rudelius, and Cumrun Vafa, *Atomic Classification of 6D SCFTs*, Fortsch. Phys. **63** (2015), 468–530.

[44] Jonathan J. Heckman and Tom Rudelius, *Top Down Approach to 6D SCFTs*, J. Phys. A **52** (2019), no. 9, 093001.

[45] Jonathan J. Heckman, Tom Rudelius, and Alessandro Tomasiello, *Fission, fusion, and 6D RG flows*, J. High Energy Phys. **2** (2019), 167, front matter + 38, DOI 10.1007/jhep02(2019)167. MR3925263

[46] Petr Hořava and Edward Witten, *Eleven-dimensional supergravity on a manifold with boundary*, Nuclear Phys. B **475** (1996), no. 1-2, 94–114, DOI 10.1016/0550-3213(96)00308-2. MR1412097

[47] Petr Hořava and Edward Witten, *Heterotic and type I string dynamics from eleven dimensions*, Nuclear Phys. B **460** (1996), no. 3, 506–524, DOI 10.1016/0550-3213(95)00621-4. MR1381609

[48] Yu-Chien Huang and Washington Taylor, *Fibration structure in toric hypersurface Calabi-Yau threefolds*, J. High Energy Phys. **3** (2020), 172, 14, DOI 10.1007/jhep03(2020)172. MR4089950

[49] Max Hübner, David R. Morrison, Sakura Schäfer-Nameki, and Yi-Nan Wang, *Generalized symmetries in F-theory and the topology of elliptic fibrations*, SciPost Phys. **13** (2022), no. 2, Paper No. 030, 46, DOI 10.21468/scipostphys.13.2.030. MR4481490

[50] Kenneth Intriligator, *New string theories in six dimensions via branes at orbifold singularities*, Adv. Theor. Math. Phys. **1** (1997), no. 2, 271–282, DOI 10.4310/ATMP.1997.v1.n2.a5. MR1605595

[51] Kenneth Intriligator, *Compactified little string theories and compact moduli spaces of vacua*, Phys. Rev. D (3) **61** (2000), no. 10, 106005, 6, DOI 10.1103/PhysRevD.61.106005. MR1790780

[52] Anton Kapustin, *Universality class of little string theories*, Phys. Rev. D (3) **63** (2001), no. 8, 086005, 11, DOI 10.1103/PhysRevD.63.086005. MR1831168

[53] Sheldon Katz, David R. Morrison, Sakura Schäfer-Nameki, and James Sully, *Tate's algorithm and F-theory*, J. High Energy Phys. **8** (2011), 094, 28, DOI 10.1007/JHEP08(2011)094. MR2876044

[54] Denis Klevers, Damian Kaloni Mayorga Pena, Paul-Konstantin Oehlmann, Hernan Piragua, and Jonas Reuter, *F-Theory on all Toric Hypersurface Fibrations and its Higgs Branches*, JHEP **01** (2015), 142.

[55] Finn Bjarne Kohl, Magdalena Larfors, and Paul-Konstantin Oehlmann, *F-theory on 6D symmetric toroidal orbifolds*, J. High Energy Phys. **5** (2022), Paper No. 64, 46, DOI 10.1007/jhep05(2022)064. MR4430228

[56] Vik. S. Kulikov, *Degenerations of K3 surfaces and Enriques surfaces* (Russian), Izv. Akad. Nauk SSSR Ser. Mat. **41** (1977), no. 5, 1008–1042, 1199. MR506296

[57] Seung-Joo Lee, Diego Regalado, and Timo Weigand, *6d SCFTs and* U(1) *flavour symmetries*, J. High Energy Phys. **11** (2018), 147, front matter+50, DOI 10.1007/jhep11(2018)147. MR3900435

[58] Christoph Mayrhofer, David R. Morrison, Oskar Till, and Timo Weigand, *Mordell-Weil torsion and the global structure of gauge groups in F-theory*, J. High Energy Phys. **10** (2014), 016, front matter+46, DOI 10.1007/JHEP10(2014)016. MR3270382

[59] Noppadol Mekareeya, Kantaro Ohmori, Yuji Tachikawa, and Gabi Zafrir, E_8 *instantons on type-A ALE spaces and supersymmetric field theories*, J. High Energy Phys. **9** (2017), 144, front matter+33, DOI 10.1007/jhep09(2017)144. MR3710470

[60] Rick Miranda and Ulf Persson, *On extremal rational elliptic surfaces*, Math. Z. **193** (1986), no. 4, 537–558, DOI 10.1007/BF01160474. MR867347

[61] David R. Morrison and Nathan Seiberg, *Extremal transitions and five-dimensional supersymmetric field theories*, Nuclear Phys. B **483** (1997), no. 1-2, 229–247, DOI 10.1016/S0550-3213(96)00592-5. MR1439440

[62] David R. Morrison and Cumrun Vafa, *Compactifications of F-theory on Calabi-Yau threefolds. I*, Nuclear Phys. B **473** (1996), no. 1-2, 74–92, DOI 10.1016/0550-3213(96)00242-8. MR1409284

[63] David R. Morrison and Cumrun Vafa, *Compactifications of F-theory on Calabi-Yau threefolds. II*, Nuclear Phys. B **476** (1996), no. 3, 437–469, DOI 10.1016/0550-3213(96)00369-0. MR1412112

[64] Kantaro Ohmori, Hiroyuki Shimizu, Yuji Tachikawa, and Kazuya Yonekura, *Anomaly polynomial of general 6d SCFTs*, PTEP **2014** (2014), no. 10, 103B07.

[65] Nathan Seiberg, *Five-dimensional SUSY field theories, non-trivial fixed points and string dynamics*, Phys. Lett. B **388** (1996), no. 4, 753–760, DOI 10.1016/S0370-2693(96)01215-4. MR1421732

[66] Nathan Seiberg, *New theories in six dimensions and matrix description of M-theory on T^5 and T^5/Z_2*, Phys. Lett. B **408** (1997), no. 1-4, 98–104, DOI 10.1016/S0370-2693(97)00805-8. MR1483580

[67] Ichiro Shimada, *On elliptic $K3$ surfaces*, Michigan Math. J. **47** (2000), no. 3, 423–446, DOI 10.1307/mmj/1030132587. MR1813537

[68] Yuji Tachikawa, *Frozen singularities in M and F theory*, J. High Energy Phys. **6** (2016), 128, front matter+8, DOI 10.1007/JHEP06(2016)128. MR3538766

[69] Cumrun Vafa, *Evidence for F-theory*, Nuclear Phys. B **469** (1996), no. 3, 403–415, DOI 10.1016/0550-3213(96)00172-1. MR1403744

[70] Edward Witten, *Small instantons in string theory*, Nuclear Phys. B **460** (1996), no. 3, 541–559, DOI 10.1016/0550-3213(95)00625-7. MR1381611

Department of Mathematics, Uppsala University, Uppsala, Sweden; and Department of Physics and Astronomy, Uppsala University, Uppsala, Sweden
Email address: michele.delzotto@math.uu.se

Department of Mathematics, Uppsala University, Uppsala, Sweden
Email address: muyang.liu@math.uu.se

Department of Mathematics, Uppsala University, Uppsala, Sweden
Current address: Department of Physics, Northeastern University, Boston, Massachusetts 02115
Email address: p.oehlmann@northeastern.edu

Proceedings of Symposia in Pure Mathematics
Volume **107**, 2024
https://doi.org/10.1090/pspum/107/01952

Vafa-Witten theory: Invariants, Floer homologies, Higgs bundles, a geometric Langlands correspondence, and categorification

Meng-Chwan Tan

ABSTRACT. We revisit Vafa-Witten theory in the more general setting whereby the underlying moduli space is not that of instantons, but of the full Vafa-Witten equations. We physically derive (i) a novel Vafa-Witten four-manifold invariant associated with this moduli space, (ii) their relation to Gromov-Witten invariants, (iii) a novel Vafa-Witten Floer homology assigned to three-manifold boundaries, (iv) a novel Vafa-Witten Atiyah-Floer correspondence, (v) a proof and generalization of a conjecture by Abouzaid-Manolescu about the hypercohomology of a perverse sheaf of vanishing cycles, (vi) a Langlands duality of these invariants, Floer homologies and hypercohomology, and (vii) a quantum geometric Langlands correspondence with purely imaginary parameter that specializes to the classical correspondence in the zero-coupling limit, where Higgs bundles feature in (ii), (iv), (vi) and (vii). We also explain how these invariants and homologies will be categorified in the process, and discuss their higher categorification. We thereby relate differential and enumerative geometry, topology and geometric representation theory in mathematics, via a maximally-supersymmetric topological quantum field theory with electric-magnetic duality in physics.

1. Vafa-Witten twist of $\mathcal{N} = 4$ gauge theory, and a Vafa-Witten invariant

In this section, we start by reviewing aspects of Vafa-Witten (VW) theory on a four-manifold M_4 with a real, simple and compact gauge group G necessary for this work, referring to [**VW94, LL97**].

1.1. Vafa-Witten Theory. VW theory is an $\mathcal{N} = 4$ topological quantum field theory (TQFT) on a four-manifold M_4 with a single scalar supercharge $\mathcal{Q}$ and complexified gauge coupling τ. The localization equations, also known as VW equations, are obtained by setting to zero $\{\mathcal{Q}, \text{fermion}\}$:

$$F^+_{\mu\nu} + \frac{1}{4}[B_{\mu\rho}, B_{\lambda\nu}]g^{\rho\lambda} = 0, \qquad \mathcal{D}_\nu B^{\nu\mu} = 0. \tag{1.1}$$

2020 *Mathematics Subject Classification.* Primary 57R56.

This proceeding is based on joint work with Z.-C. Ong in [**OT22**]. This proceeding was supported in part by the MOE AcRF Tier 1 grant R-144-000-470-114.

These constitute the BPS equations for the theory, with the zero modes of A_μ and $B_{\mu\nu}$ that satisfy (1.1) defining a moduli space which the path integral localizes on.[1] The bosonic part of the VW action involving only A and B will be of the form

$$\frac{1}{e^2}\int_{M_4} \mathrm{Tr}(|s|^2 + |k|^2) + \text{topological term}, \tag{1.2}$$

with

$$\begin{aligned} s_{\mu\nu} &= F^+_{\mu\nu} + \frac{1}{4}[B_{\mu\rho}, B^\rho_\nu], \\ k_\nu &= \mathcal{D}^\mu B_{\mu\nu}. \end{aligned} \tag{1.3}$$

We further note that the 2-form B need not vanish if the scalar curvature of *Kähler* M_4 and the gauge group G are not simultaneously non-negative and locally a product of $SU(2)$'s [**VW94**], and we will assume this to be the case here. The topological term in (1.2) is

$$-\frac{i\tau}{4\pi}\int_{M_4} \mathrm{Tr}\left(F\wedge F + dB\wedge\star DB + B\wedge d(\star DB)\right) \tag{1.4}$$

where we have taken the liberty to add a $\mathcal{Q}$-exact term after $F\wedge F$ that is null in the spectrum of VW theory given by the $\mathcal{Q}$-cohomology, for later convenience. Also, we have used the fact that B is self-dual, whence $\star B = B$, and here, $D = d + A$ where $\star DB$ is a one-form on M_4.

1.2. A Vafa-Witten invariant. VW theory is a balanced TQFT (same number of fermion pair zero modes), and the path integral localizes to a virtually zero-dimensional moduli space $\mathcal{M}_{\mathrm{VW}}$ of solutions to ((1.1)), whence the nonvanishing topological invariant is the partition function that can be interpreted as an integral of a virtual zero-form on virtually zero-dimensional $\mathcal{M}_{\mathrm{VW}}$:

$$\mathcal{Z}_{\mathrm{VW},M_4}(\tau, G) = \sum_k a_k q^{m_k}. \tag{1.5}$$

Here, $q = e^{2\pi i\tau}$, k denotes the kth sector of $\mathcal{M}_{\mathrm{VW}}$, the number a_k is given by

$$\boxed{a_k = \int_{\mathcal{M}^k_{\mathrm{VW}}} \Omega^0 \wedge e(T_{\mathcal{M}^k_{\mathrm{VW}}}), \quad \text{where} \quad \Omega^0(\mathcal{M}^k_{\mathrm{VW}}) = (1+B^4)^{dim_{\mathbb{C}}\mathcal{M}^k_{\mathrm{VW}}}} \tag{1.6}$$

B is a coordinate on $\mathcal{M}^k_{\mathrm{VW}}(A,B)$, e is the signed Euler class of the tangent bundle $T_{\mathcal{M}^k_{\mathrm{VW}}}$, and m_k is the corresponding VW number given by

$$\boxed{m_k = \frac{1}{8\pi^2}\int_{M_4} \mathrm{Tr}\left(F_{(k)}\wedge F_{(k)} + dB_{(k)}\wedge\star DB_{(k)} + B_{(k)}\wedge d(\star DB_{(k)})\right)} \tag{1.7}$$

Notice that $\mathcal{Z}_{\mathrm{VW},M_4}$ is a topological invariant of M_4 which is an algebraic count of VW solutions with corresponding weight given by $a_k q^{m_k}$ that we elaborated on above. This defines a novel τ-dependent Vafa-Witten invariant of M_4.[2]

---

[1] In the full VW equations, there is a bosonic scalar field C that is also a generator of gauge transformations. We thus set the zero mode of C to vanish since we wish to consider only irreducible connections, giving us (1.1) as the resulting BPS equations.

[2] A purely algebro-geometric definition of $\mathcal{Z}_{\mathrm{VW},M_4}$, in particular the a_k's, was first given by Tanaka-Thomas in [**TT17**], albeit for projective algebraic surfaces only. The novelty here is that we provide a purely differentio-geometric definition of the a_k's for a more general M_4.

When $B = 0$, a_k will become the Euler characteristic $\chi(\mathcal{M}^k_{\text{inst}})$, while m_k will become the instanton number. Then, $\mathcal{Z}_{\text{VW},M_4}$ will just become the usual partition function for instantons first derived in [**VW94**], as expected.

2. An $\mathcal{N} = (4,4)$ A-model, Higgs Bundles and Gromov-Witten theory

In this section, we will perform dimensional reduction of the 4d VW theory down to 2d. The four-manifold M_4 will be taken to be $M_4 = \Sigma \times C$, where Σ and C are both closed Riemann surfaces, and C is of genus $g \geq 2$.

2.1. Finiteness conditions, BPS equations in 2d and an $\mathcal{N} = (4,4)$ Sigma-model. We consider a block diagonal metric g for $M_4 = \Sigma \times C$,

$$g = \text{diag}\big(g_\Sigma, \epsilon g_C\big), \tag{2.1}$$

where ϵ is a small parameter to deform g_C. We shall use capital letters $A, B = x^1, x^2$ to denote coordinates on Σ, and small letters $a, b = x^3, x^4$ to denote coordinates on C. Taking the limit $\epsilon \to 0$ then gives us a 2d theory on Σ with $\mathcal{N} = (4,4)$ supersymmetry.

The topological term aside, terms in (1.2) with $\mu, \nu, \rho = A, B$ vanish as $\epsilon \to 0$, while those with $\mu, \nu, \rho = A, b$ survive. For $\mu, \nu, \rho = a, b$, each term must be set to zero individually as they are accompanied by a factor of ϵ^{-1}. Since the action (1.2) is a sum of squares of such terms, we will need to set them to zero. This constraint will give us the finiteness conditions.

Before we proceed further, we note the fact that $F^+_{\mu\nu} = \frac{1}{2}(F_{\mu\nu} + \frac{1}{2}\epsilon_{\mu\nu\rho\lambda}F^{\rho\lambda})$, and that $B_{\mu\nu}$ is an anti-symmetric and self-dual 2-form ($B_{\mu\nu} = \frac{1}{2}\epsilon_{\mu\nu\rho\lambda}B^{\rho\lambda}$) with 3 independent components which we can take to be B_{12}, B_{13} and B_{14}.

The first finiteness condition we obtain by using the self-duality property of $B_{\mu\nu}$ is

$$D_3 B_{12} = -D_4 B_{12} = 0. \tag{2.2}$$

The field B_{12} is a 0-form w.r.t rotations on both C and Σ, so (2.2) tells us that the 0-form B_{12} is covariantly constant on C, which means B_{12} generates infinitesimal gauge transformations while leaving A_C fixed. We can however set $B_{12} = 0$, since we require gauge connections to be irreducible to avoid complications on $\mathcal{M}^G_H(C)$.

Next, identifying B_{13} and B_{14} as the two components of a 1-form φ on C and using the self-duality properties of F^+ and $B_{\mu\nu}$, we obtain Hitchin's equations on C [**Hit87**] as the second finiteness condition, given by[3]

$$\begin{aligned} F_C - \varphi \wedge \varphi &= 0, \\ D\varphi = D^*\varphi &= 0, \end{aligned} \tag{2.3}$$

where

$$\varphi = B_{13}dx^3 + B_{14}dx^4 = \varphi_3 dx^3 + \varphi_4 dx^4. \tag{2.4}$$

The space of solutions of (A_C, φ) to (2.3) modulo gauge transformations then span Hitchin's moduli space $\mathcal{M}^G_H(C)$ for a connection A_C on a principal G-bundle P over the Riemann surface C, and a section $\varphi \in \Omega^1(C)$. The above equations leave the (x^1, x^2) dependence of A_C and φ arbitrary, and thus the fields (A_C, φ) define a map $\Phi : \Sigma \to \mathcal{M}^G_H(C)$. The target space $\mathcal{M}^G_H(C)$ is a hyper-Kähler manifold, whence the sigma-model on Σ has an $\mathcal{N} = (4,4)$ supersymmetry [**AGF81**].

---

[3] $D^*\varphi = \star D \star \varphi = D_\mu \varphi^\mu$, where $\star$ is the Hodge star operator.

To obtain the corresponding 2d BPS equations of the $\mathcal{N} = (4,4)$ sigma model on Σ, we perform dimensional reduction of (1.3) on C with $s = k = 0$. Noting the fact that only terms with mixed indices on $\Sigma \times C$ survive the reduction on C, together with the self-duality properties of $B_{\mu\nu}$, we obtain, from (1.3) and $s = k = 0$,

$$\begin{aligned} F^{+}_{Aa} &= 0, \\ \mathcal{D}_A B^{Aa} &= 0. \end{aligned} \tag{2.5}$$

Switching to complex coordinates, (2.5) can be written as $\partial_{\bar{z}} A_{\bar{w}} = \partial_{\bar{z}} \varphi_w = 0$.[4] With $A_{\bar{w}}$ and φ_w corresponding to bosonic scalars X^i and Y^i in the sigma-model, respectively, we get the 2d BPS equations as

$$\begin{aligned} \partial_{\bar{z}} X^i &= 0, \\ \partial_{\bar{z}} Y^i &= 0. \end{aligned} \tag{2.6}$$

After suitable rescalings, we can then rewrite (1.2) (with $idz \wedge d\bar{z} = |dz^2|$) as

$$\begin{aligned} S_{\text{2d}} = \frac{1}{e^2} \int_\Sigma |dz^2| g_{i\bar{j}} \Big(& \partial_z X^{\bar{i}} \partial_{\bar{z}} X^j + \partial_z X^i \partial_{\bar{z}} X^{\bar{j}} + \partial_z Y^{\bar{i}} \partial_{\bar{z}} Y^j + \partial_z Y^i \partial_{\bar{z}} Y^{\bar{j}} \Big) \\ & + \text{topological term}. \end{aligned} \tag{2.7}$$

Hence, the path integral of the 2d, $\mathcal{N} = (4,4)$ sigma model on Σ with action (2.7), localizes on the moduli space of holomorphic maps $\Phi(X^i, Y^i) : \Sigma \to \mathcal{M}^G_H(C)$:

$$\mathcal{M}_{\text{maps}} = \{\Phi(X^i, Y^i) : \Sigma \to \mathcal{M}^G_H(C) \mid \partial_{\bar{z}} X^i = \partial_{\bar{z}} Y^i = 0\}, \tag{2.8}$$

where we have a 2d $\mathcal{N} = (4,4)$ A-model on Σ with target $\mathcal{M}^G_H(C)$.

2.2. An A-model in complex structure I. The space of fields (A_C, φ) span an infinite-dimensional affine space $\mathcal{W}$. The cotangent vectors δA_C and $\delta\varphi$ to $\mathcal{M}^G_H(C)$ are solutions to the variations of equations (2.6). We can then introduce a basis $(\delta A_w, \delta\varphi_{\bar{w}})$ and $(\delta A_{\bar{w}}, \delta\varphi_w)$ in $\mathcal{W}$. From the BPS equations (2.6), which are $\partial_{\bar{z}} A_{\bar{w}} = 0$ and $\partial_{\bar{z}} \varphi_w = 0$, one can see that the complex structure relevant to the A-model is I, with linear holomorphic functions consisting of $A_{\bar{w}}$ and φ_w. In complex structure I, $\mathcal{M}^G_H(C)$ can be identified as the moduli space of stable Higgs G-bundles on C, $\mathcal{M}^G_{\text{Higgs}}(C)$. One can write the corresponding symplectic form as $\omega_I = \omega'_I - \delta\lambda_I$, where

$$\omega'_I = -\frac{1}{4\pi} \int_C \operatorname{Tr} \delta A_C \wedge \delta A_C \quad \text{and} \quad \lambda_I = \frac{1}{4\pi} \int_C \operatorname{Tr} \varphi \wedge \delta\varphi, \tag{2.9}$$

and ω_I is cohomologous to ω'_I. Comparing the 4d topological term in (1.4) to (2.9), we see that the topological term can be written as

$$i\tau \int_\Sigma \Phi^*(\omega_I). \tag{2.10}$$

The 2d action (2.7), including the topological term, is then

$$\begin{aligned} S_{\text{2d}} = \frac{1}{e^2} \int_\Sigma |dz^2| g_{i\bar{j}} \Big(& \partial_z X^{\bar{i}} \partial_{\bar{z}} X^j + \partial_z X^i \partial_{\bar{z}} X^{\bar{j}} + \partial_z Y^{\bar{i}} \partial_{\bar{z}} Y^j + \partial_z Y^i \partial_{\bar{z}} Y^{\bar{j}} \Big) \\ & + i\tau \int_\Sigma \Phi^*(\omega_I). \end{aligned} \tag{2.11}$$

[4]In complex coordinates, we have $z = x^1 + ix^2$ and $w = x^3 + ix^4$, where $A_{\bar{w}} = \frac{1}{2}(A_3 + iA_4)$ and $\varphi_w = \frac{1}{2}(B_{13} - iB_{14})$.

We thus have a 2d, $\mathcal{N} = (4,4)$ A-model on Σ with target $\mathcal{M}^G_{\text{Higgs}}(C)$, where the path integral localizes on

$$\mathcal{M}_{\text{maps}} = \{\Phi(X^i, Y^i) : \Sigma \to \mathcal{M}^G_{\text{Higgs}}(C) \mid \partial_{\bar{z}} X^i = \partial_{\bar{z}} Y^i = 0\}, \tag{2.12}$$

the moduli space of holomorphic maps $\Phi : \Sigma \to \mathcal{M}^G_{\text{Higgs}}(C)$.

2.3. Vafa-Witten invariants as Gromov-Witten invariants of Higgs bundles. The virtual dimension of $\mathcal{M}_{\text{maps}}$, like that of $\mathcal{M}_{\text{VW}}$, ought to also be zero. This is because the 2d A-model is obtained via a topological deformation that sets $C \to 0$ in the original 4d VW theory, whence the relevant index of kinetic operators counting the dimension of moduli space remains the same.

Like $\mathcal{Z}_{\text{VW},M_4}$ in 4d, $\mathcal{Z}^{\text{closed}}_{A,\Sigma}$ can be interpreted as an integral of a virtual zero-form on virtually zero-dimensional $\mathcal{M}_{\text{maps}}$, whence it can be evaluated as

$$\mathcal{Z}^{\text{closed}}_{A,\Sigma}(\tau, \mathcal{M}^G_{\text{Higgs}}(C)) = \sum_l \tilde{a}_l q^{\tilde{m}_l}. \tag{2.13}$$

Here, l denotes the lth sector of $\mathcal{M}_{\text{maps}}$ defined in (2.12) for *genus one* Σ, the rational number $\tilde{a}_l$ is given by

$$\boxed{\tilde{a}_l = \int_{\mathcal{M}^l_{\text{maps}}} e(\mathcal{V})} \tag{2.14}$$

where e is the signed Euler class of the vector bundle $\mathcal{V}$ with fiber $H^0(\Sigma, K \otimes \Phi^* T^* \mathcal{M}^l_{\text{maps}})$ and canonical bundle K on Σ, and $\tilde{m}_l$ is the corresponding worldsheet instanton number given by

$$\boxed{\tilde{m}_l = \frac{1}{2\pi} \int_\Sigma \Phi^*_l(\omega_I)} \tag{2.15}$$

Notice that $\mathcal{Z}^{\text{closed}}_{A,\Sigma}$ is an enumerative invariant which is an algebraic count of holomorphic maps with corresponding weight given by $\tilde{a}_l q^{\tilde{m}_l}$ that we elaborated on above. This coincides with the definition of the GW invariant, which then means that one can identify $\mathcal{Z}^{\text{closed}}_{A,\Sigma}$ as

$$\mathcal{Z}_{\text{GW},\Sigma}(\tau, \mathcal{M}^G_{\text{Higgs}}(C)) = \sum_l \tilde{a}_l q^{\tilde{m}_l} \tag{2.16}$$

where $\mathcal{Z}_{\text{GW},\Sigma}$ is a τ-dependent GW invariant of $\mathcal{M}^G_{\text{Higgs}}(C)$. From the topological invariance of the 4d theory, we have a 4d-2d correspondence of partition functions

$$\mathcal{Z}_{\text{VW},M_4}(\tau, G) = \mathcal{Z}_{\text{GW},\Sigma}(\tau, \mathcal{M}^G_{\text{Higgs}}(C)). \tag{2.17}$$

In other words, we have a correspondence between the VW invariant of $M_4 = \Sigma \times C$ and the GW invariant of $\mathcal{M}^G_{\text{Higgs}}(C))$.

In fact, recall that the numbers $\tilde{m}_l$ (in (2.16)) correspond to the numbers m_k (in (1.5)). Hence, (2.17) means that we have

$$\boxed{a_k = \tilde{a}_l} \tag{2.18}$$

where a_k and $\tilde{a}_l$ are given in (1.6) and (2.14), respectively. In other words, one can also determine the a_k's, the VW invariants of $T^2 \times C$, via the signed Euler class of a bundle $\mathcal{V}$ over $\mathcal{M}^l_{\text{maps}}$.[5]

[5]Computing the $\tilde{a}_l$'s and thus a_k's for $T^2 \times C$ explicitly is a purely mathematical endeavour that is beyond the scope of this physical mathematics proceeding which main objective is to furnish

3. A novel Floer homology from boundary Vafa-Witten theory

In this section, we will show how we can physically derive a novel Floer homology by considering boundary VW theory on $M_4 = M_3 \times \mathbb{R}^+$ to physically derive a VW Floer homology assigned to M_3.[6]

3.1. SQM interpretation of boundary Vafa-Witten theory. Let the manifold of the 4d theory in (1.2) be $M_4 = M_3 \times \mathbb{R}^+$, where the M_3 boundary is a closed three-manifold, and $\mathbb{R}^+$ is the 'time' coordinate. We also let spacetime indices take the values $\mu = 0, 1, 2, 3$, with $\mu = 0$ being the time direction, while $\mu = i, j, k = 1, 2, 3$ being the spatial directions.

Turning to the BPS equations (1.1) of boundary VW theory, we split the indices into space and time directions. Using $F^+_{\mu\nu} = \frac{1}{2}(F_{\mu\nu} + \frac{1}{2}\epsilon_{\mu\nu\rho\lambda}F^{\rho\lambda})$ and $B_{\mu\nu} = \frac{1}{2}\epsilon_{\mu\nu\rho\lambda}B^{\rho\lambda}$, we can reexpress the VW equations (1.1) as

$$\begin{aligned} \dot{A}^i + \frac{1}{2}\epsilon^{ijk}(F_{jk} - [B_j, B_k]) &= 0, \\ \dot{B}^i + \epsilon^{ijk}(\partial_j B_k + [A_j, B_k]) &= 0, \end{aligned} \tag{3.1}$$

where the temporal gauge $A^0 = 0$ is taken, $B^i = B^{0i}$, $\epsilon^{ijk} = \epsilon^{0ijk}$, and $A^i, B^i \in \Omega^1(M_3)$.[7]

Introducing a complexified connection $\mathcal{A} = A + iB \in \Omega^1(M_3)$, of a $G_{\mathbb{C}}$-bundle on M_3. We then find that (3.1) can be expressed as

$$\dot{\mathcal{A}}^i + \frac{1}{2}\epsilon^{ijk}\mathcal{F}_{jk} = 0, \tag{3.2}$$

where $\mathcal{F} \in \Omega^2(M_3)$ is the complexified field strength. Note that $\epsilon^{ijk}\mathcal{F}_{jk}$ is a gradient vector field of a complex Chern-Simons functional

$$V(\mathcal{A}) = CS(\mathcal{A}) = -\frac{1}{4\pi^2}\int_{M_3} \mathrm{Tr}\Big(\mathcal{A} \wedge d\mathcal{A} + \frac{2}{3}\mathcal{A} \wedge \mathcal{A} \wedge \mathcal{A}\Big), \tag{3.3}$$

where we have the gradient flow equation

$$\frac{d\mathcal{A}^i}{dt} + s g^{ij}_{\mathfrak{A}} \frac{\partial V(\mathcal{A})}{\partial \mathcal{A}^j} = 0. \tag{3.4}$$

The action for boundary VW theory in (1.2) can be rewritten as

$$\begin{aligned} S^{\text{bdry}}_{\text{VW}} = &\frac{1}{e^2}\int dt \int_{M_3} \mathrm{Tr}\Big(\dot{\mathcal{A}}^i + s g^{ij}_{\mathfrak{A}}\frac{\partial V(\mathcal{A})}{\partial \mathcal{A}^j}\Big)^2 \\ &- \frac{i\tau}{4\pi}\int_{M_3} \mathrm{Tr}\Big(A \wedge dA + \frac{2}{3}A \wedge A \wedge A + B \wedge \star DB\Big). \end{aligned} \tag{3.5}$$

their fundamental definitions via the expressions (2.14) and (1.6), respectively. The reader who seeks an explicit computation of these invariants may be happy to know that after our work appeared, this was done purely mathematically in [**N23**].

[6]To be precise, VW theory is still being defined on an M_4 with no boundary. However, to make contact with Floer theory, we will need to examine a hyper-slice of M_4, which we can topologically regard as $M_3 \times \mathbb{R}^- \cup_{M_3} M_3 \times \mathbb{R}^+$. As there is no time-evolution in our topological theory, it is sufficient to examine only $M_3 \times \mathbb{R}^+$, where M_3 can then be regarded as a boundary. This is consistent with the idea that categorification of topological invariants can be achieved via successive introductions of boundaries to M_4, which we will elaborate upon in §7.

[7]Using self-duality properties, we have $B^{0i} = B^i = \epsilon^{ijk}B_{jk}$.

We can interpret (3.5) as the action of an SQM model with target $\mathfrak{A}$, the space of complexified connections $\mathcal{A}$ on M_3 with metric $g_{\mathfrak{A}}^{ij}$, and a single nilpotent topological scalar supercharge $\mathcal{Q}$, where (3.4), which describes the VW equations, can be interpreted as a gradient flow equation between fixed (or time-invariant) critical points of $CS(\mathcal{A})$ on $\mathfrak{A}$. Assuming that the fixed critical points are isolated and nondegenerate in $\mathfrak{A}$,[8] the partition function of boundary VW theory will thus be an algebraic count of fixed critical points of $CS(\mathcal{A})$, i.e., fixed flat $G_{\mathbb{C}}$-connections on M_3, where there are VW flow lines between these fixed critical points described by the gradient flow equation.

The second term in (3.5) is a τ-dependent topological term that contributes to an overall factor in the path integral.

3.2. A novel Vafa-Witten Floer homology. For an M_4 with boundary $\partial M_4 = M_3$, one needs to specify boundary conditions on M_3 to compute the path integral by first defining a restriction of the fields to M_3, which we shall denote as Ψ_{M_3}, and then specifying boundary values for these restrictions. This is equivalent to inserting in the path integral an operator functional $F(\Psi_{M_3})$ that is nonvanishing in the $\mathcal{Q}$-cohomology. The partition function on M_4 can be expressed as

$$\mathcal{Z}_{\mathrm{VW},M_4}(\tau, G) = \langle 1 \rangle_{F(\Psi_{M_3})} = \sum_k \mathcal{F}^{G,\tau}_{\mathrm{VW}}(\Psi^k_{M_3}). \tag{3.6}$$

Here, the summation in 'k' is over all sectors of $\mathcal{M}_{\mathrm{VW}}$ labeled by the VW number m_k, and $\mathcal{F}^{G,\tau}_{\mathrm{VW}}(\Psi^k_{M_3})$ is the kth contribution to the partition function that depends on the expression of $F(\Psi_{M_3})$ in the bosonic fields on M_3 evaluated over the corresponding solutions of the VW equations restricted to M_3.

As an SQM model on $\mathfrak{A}$, the partition function can be expressed as

$$\mathcal{Z}_{\mathrm{VW},M_4}(\tau, G) = \sum_k \mathcal{F}^k_{\mathrm{VW\text{-}Floer}}(M_3, G, \tau), \tag{3.7}$$

where each $\mathcal{F}^k_{\mathrm{VW\text{-}Floer}}(M_3, G, \tau)$ can be identified with a class in what we shall henceforth call a Vafa-Witten Floer homology $\mathrm{HF}^{\mathrm{VW}}_{d_k}(M_3, G, \tau)$ assigned to M_3 of degree d_k, where d_k counts the number of outgoing VW flow lines from the corresponding fixed critical point of $CS(\mathcal{A})$.

In summary, from (3.6) and (3.7), we can write

$$\mathcal{Z}_{\mathrm{VW},M_4}(\tau, G) = \sum_k \mathcal{F}^{G,\tau}_{\mathrm{VW}}(\Psi^k_{M_3}) = \sum_k \mathrm{HF}^{\mathrm{VW}}_{d_k}(M_3, G, \tau) = \mathcal{Z}^{\mathrm{Floer}}_{\mathrm{VW},M_3}(\tau, G) \tag{3.8}$$

where 'k' sums from zero to the maximum number of fixed VW solutions on $M_3 \times \mathbb{R}^+$ that correspond to isolated and non-degenerate fixed critical points of $CS(\mathcal{A})$.[9]

The τ-dependence of $\mathcal{F}^{G,\tau}_{\mathrm{VW}}(\Psi^k_{M_3})$ and therefore $\mathrm{HF}^{\mathrm{VW}}_k(M_3, G, \tau)$ arises because in evaluating (3.6), there will be a factor of q^{s_k} for the kth term, where from the

[8]This is guaranteed (though not necessary) when all critical points are isolated and nondegenerate. This can be the case for an appropriate choice of G and M_3. For example, one could choose (1) G compact and M_3 of nonnegative Ricci curvature such as a three-sphere or its quotient, or (2) an M_3 with a finite $G_{\mathbb{C}}$ representation variety, and introduce physically-trivial $\mathcal{Q}$-exact terms to the action to perturb $V(\mathcal{A})$. We would like to thank A. Haydys for discussions on this.

[9]See footnote 8.

FIGURE 1. **Left:** M_3 as a connected sum of three-manifolds M_3' and M_3'' along a common Riemann surface C. **Right:** M_4 split along four-manifolds M_4' and M_4'' with corners.

action $S_{\text{VW}}^{\text{bdry}}$ in (3.5), the number

$$s_k = \frac{1}{8\pi^2}\int_{M_3} \text{Tr}\Big(A_{(k)} \wedge dA_{(k)} + \frac{2}{3}A_{(k)} \wedge A_{(k)} \wedge A_{(k)} + B_{(k)} \wedge \star DB_{(k)}\Big). \tag{3.9}$$

Here, the subscript '(k)' denotes that they are the kth fixed solution to the VW equations on $M_3 \times \mathbb{R}^+$ restricted to M_3.

4. A Vafa-Witten Atiyah-Floer correspondence

In this section, we continue with a Heegaard split of M_3 into M_3' and M_3'' along a Riemann surface C, as shown in Fig. 1 (left), allowing us to relate Vafa-Witten Floer homology obtained in the previous section to Lagrangian Floer homology, in what is a novel Vafa-Witten version of the Atiyah-Floer correspondence [**Ati88**] based on instantons. In doing so, we would be able to physically prove and generalize a conjecture by mathematicians Abouzaid-Manolescu about the hypercohomology of a perverse sheaf of vanishing cycles in the moduli space of irreducible flat $SL(2,\mathbb{C})$-connections on M_3. We can thus write $M_4 = (\mathbb{R}^+ \times I' \times C) \cup_C (\mathbb{R}^+ \times I'' \times C)$, where $M_3' = I' \times C$ and $M_3'' = I'' \times C$. This is illustrated in Fig. 1 (right), where taking $C \to 0$, we indeed have $\mathbb{R}^+ \times I'$ and $\mathbb{R}^+ \times I''$.

4.1. A Vafa-Witten version of the Atiyah-Floer correspondence. If $C \to 0$, we end up with an open A-model in complex structure I on $\mathbb{R}^+ \times I'$ and $\mathbb{R}^+ \times I''$, respectively, with target space $\mathcal{M}^G_{\text{Higgs}}(C)$. Because we have an A-model in complex structure I, the admissible branes are those of type $(A,*,*)$ Specifically, we need an $(A,*,*)$-brane in $\mathcal{M}^G_{\text{Higgs}}(C)$ that corresponding to a Higgs pair on C that can be extended to flat complex connections $\mathcal{A}$ on $M_3^{','' }$

Such an $(A,*,*)$-brane has indeed been obtained in [**K08**].[10] It is an (A,B,A)-brane $\alpha_{M_3^{','' }}$, that is simultaneously an A-brane in $\mathcal{M}^G_{\text{Higgs}}(C)$ and an A-brane in $\mathcal{M}^G_H(C)$ in complex structure K, i.e., $\mathcal{M}^{G_{\mathbb{C}}}_{\text{flat}}(C)$, the moduli space of flat $G_{\mathbb{C}}$-connections on C, where it corresponds to flat connections that can be extended to $M_3^{','' }$. It is middle-dimensional, and is therefore a Lagrangian brane. Let us henceforth denote this brane as L.

[10]The 4d theory considered in [**K08**] is not the VW but the GL theory of [**KW07**], albeit with parameter $t = 0$. However, both these 4d theories descend to the same 2d A-model with target $\mathcal{M}^G_{\text{Higgs}}(C)$ after dimensional reduction on C, and since our A-branes of interest are A-model objects within $\mathcal{M}^G_{\text{Higgs}}(C)$, the arguments used and examples stated in [**K08**] are applicable here.

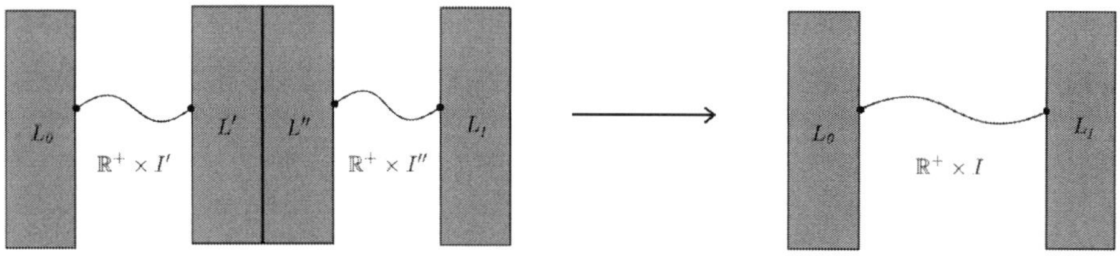

FIGURE 2. Identifying L' and L'' and gluing them together to form a single open string.

Now, with two split pieces $M_4^{','''}$, when $C \to 0$, we have two strings, each ending on pairs of Lagrangian branes (L_0, L') and (L'', L_1) (see Fig. 2.) We then glue the open worldsheets together along their common boundary $L^{','''}$, giving us a single A-model, with a single string extending from L_0 to L_1, which is equivalent to gluing $M_4^{','''}$ along $C \times \mathbb{R}^+$. (see Fig. 2 again.) As before, one can recast the A-model here as an SQM model, where $\mathbb{R}^+$ is 'time', and the target space is $\mathscr{P}(L_0, L_1)$, the space of smooth trajectories from L_0 to L_1 (arising from the interval I that connects them). The BPS equations for this A-model are (2.6), i.e., holomorphic maps from the worldsheet to the target space. They can be written as a gradient flow equation on the worldsheet

$$\frac{\partial Z^l}{\partial t} + i\frac{\partial Z^l}{\partial s} = 0, \tag{4.1}$$

where we have used real coordinates t and s (for $z = t + is$), and here, $Z^l = X^l + Y^l$. The fixed critical points of the underlying potential of the SQM model that contribute to the partition function are defined by $\dot{Z}^l = \partial Z^l / \partial s = 0$. Since '$s$' is the spatial coordinate of I, it would mean that the fixed critical points just correspond to fixed stationary trajectories in $\mathscr{P}(L_0, L_1)$, i.e., the intersection points of L_0 and L_1. Thus, the partition function of the A-model, which, from the SQM model perspective, is given by an algebraic count of the fixed critical points of its underlying potential, will be an algebraic count of the intersection points of L_0 and L_1, where there are flow lines between the intersection points that obey (4.1). These flow lines correspond to holomorphic Whitney disks.

From this description of the partition function, we have physically realized the Lagrangian Floer homology first defined in [**Flo88a**], where the intersection points of L_0 and L_1 actually generate the chains of the Lagrangian Floer complex, and the Floer differential, which counts the number of holomorphic Whitney disks, can be interpreted as the outgoing flow lines at each intersection point of L_0 and L_1 which number would be the degree of the corresponding chain in the complex.

Specifically, let $(L_0 \cap L_1)_i^{n_i}$ denote the ith point of the intersection $L_0 \cap L_1$ where there are n_i outgoing flow lines, whence we can identify

$$(L_0 \cap L_1)_i^{n_i} \in \mathrm{HF}_{n_i}^{\mathrm{Lagr}}\big(\mathcal{M}_{\mathrm{Higgs}}^G(C), L_0, L_1\big), \tag{4.2}$$

where $\mathrm{HF}_{n_i}^{\mathrm{Lagr}}\big(\mathcal{M}_{\mathrm{Higgs}}^G(C), L_0, L_1\big)$ is the Lagrangian Floer homology of (L_0, L_1) on $\mathcal{M}_{\mathrm{Higgs}}^G(C)$ of degree n_i. Then, the partition function of the A-model will be given by

$$\mathcal{Z}_{A,L}\big(\tau, \mathcal{M}_{\mathrm{Higgs}}^G(C)\big) = \sum_i \mathrm{HF}_{n_i}^{\mathrm{Lagr}}\big(\mathcal{M}_{\mathrm{Higgs}}^G(C), L_0, L_1, \tau\big), \tag{4.3}$$

A τ-dependency appears here because of a τ-dependent term in the A-model action.

Since the underlying boundary VW theory on $M_4 = M_3 \times \mathbb{R}^+$ is topological, we will have the following equivalence of partition functions:

$$\mathcal{Z}_{\text{VW},M_4}(\tau, G) = \mathcal{Z}_{A,L}\big(\tau, \mathcal{M}^G_{\text{Higgs}}(C)\big), \tag{4.4}$$

which, from (3.8) and (4.3), means that

$$\sum_k \text{HF}^{\text{VW}}_{d_k}(M_3, G, \tau) = \sum_i \text{HF}^{\text{Lagr}}_{n_i}\big(\mathcal{M}^G_{\text{Higgs}}(C), L_0, L_1, \tau\big). \tag{4.5}$$

The gradings 'd_k' and 'n_i' in (4.5) match. To understand this, recall that the VW flow lines between fixed critical points in $\mathfrak{A}$ are non-fixed solutions to the VW equations (1.1) on $M_3 \times \mathbb{R}^+$. Also, in §2.1, it was shown that the VW equations descend to the worldsheet instanton equations (2.6) defining holomorphic maps from the worldsheet to $\mathcal{M}^G_{\text{Higgs}}(C)$, the non-fixed solutions to which are the flow lines between fixed critical points in $\mathscr{P}(L_0, L_1)$. Thus, there is a one-to-one correspondence between the flow lines that define HF^{VW}_* and underlie the LHS of (4.5), and the flow lines that define $\text{HF}^{\text{Lagr}}_*$ and underlie the RHS of (4.5). Moreover, 'k' and 'i' obviously match, too.

We thus have a degree-by-degree isomorphism of the VW Floer homology and the Lagrangian Floer homology, whence we would have a Vafa-Witten Atiyah-Floer correspondence

$$\text{HF}^{\text{VW}}_*(M_3, G, \tau) \cong \text{HF}^{\text{Lagr}}_*\big(\mathcal{M}^G_{\text{Higgs}}(C), L_0, L_1, \tau\big). \tag{4.6}$$

4.2. A physical proof and generalization of a conjecture by Abouzaid-Manolescu about the hypercohomology of a perverse sheaf of vanishing cycles. A hypercohomology $\text{HP}^*(M_3)$ was constructed by Abouzaid-Manolescu in [**AM20**], where it was conjectured to be isomorphic to instanton Floer homology assigned to M_3 for the complex gauge group $SL(2, \mathbb{C})$. Its construction was via a Heegaard split of $M_3 = M_3' \cup_C M_3''$ along C of genus g, and the intersection of the two associated Lagrangians in the moduli space $X_{\text{irr}}(C)$ of irreducible flat $SL(2, \mathbb{C})$-connections on C (that represent solutions extendable to M_3' and M_3'', respectively), to which one can associate a perverse sheaf of vanishing cycles. $\text{HP}^*(M_3)$ is then the hypercohomology of this perverse sheaf of vanishing cycles in $X_{\text{irr}}(M_3)$, where it is an invariant of M_3 independent of the Heegaard split.

Based on the mathematical construction of $\text{HP}^*(M_3)$ described above, it would mean that a physical realization of (the dual of) $\text{HP}^*(M_3)$ ought to be via an open A-model with Lagrangian branes L_0 and L_1 in the target $X_{\text{irr}}(C)$, where the observables contributing to the partition function can be interpreted as classes in the Lagrangian Floer homology $\text{HF}^{\text{Lagr}}_*\big(X_{\text{irr}}(C), L_0, L_1, \tau\big)$.

First, note that there is an isomorphism between $\text{HF}^{\text{Lagr}}_*$ and the homology of Lagrangian submanifolds in $X_{\text{irr}}(C)$ [**Ped18**, Theorem 11], i.e.,

$$\text{HF}^{\text{Lagr}}_*\big(X_{\text{irr}}(C), L_0, L_1, \tau\big) \cong \text{H}_*(L, \mathbb{Z}_2)_{\otimes \mathbb{Z}_2}\Lambda, \tag{4.7}$$

where Λ is a scalar function over $\mathbb{Z}_2$, called the Novikov field, and L on the RHS can be taken as either L_0 or L_1. The homology cycles of the Lagrangian (i.e., middle-dimensional) submanifolds of $X_{\text{irr}}(C)$ have a maximum dimension of

$\frac{1}{2}\dim(X_{\text{irr}}(C))$, where $\frac{1}{2}\dim(X_{\text{irr}}(C)) = 2(3g-3)$.[11] Including the zero-cycle, the grading of $\text{H}_*(L,\mathbb{Z}_2)_{\otimes\mathbb{Z}_2}\Lambda$ and therefore $\text{HF}_*^{\text{Lagr}}\big(X_{\text{irr}}(C), L_0, L_1, \tau\big)$, goes as $0, 1, \ldots, 2(3g-3)$.

Second, note that in [**AM20**, Theorem 1.8], it was computed that HP^k is nonvanishing only if $-3g+3 \leq k \leq 3g-3$. In other words, the grading of HP^* goes as $-(3g-3), \ldots, 0, \ldots, (3g-3)$.

These two observations then mean that there is a one-to-one correspondence between the gradings of $\text{HP}^*(M_3)$ and $\text{HF}_*^{\text{Lagr}}$. Moreover, the generators of HP^* and $\text{HF}_*^{\text{Lagr}}$ both originate from the intersection points of L_0 and L_1 in $X_{\text{irr}}(C)$. Hence, we can identify HP^* with (the dual of) $\text{HF}_*^{\text{Lagr}}$, i.e.,

$$\text{HP}^*(M_3) \cong \text{HF}_*^{\text{Lagr}}\big(X_{\text{irr}}(C), L_0, L_1, \tau\big) \tag{4.8}$$

This agrees with [**BBD**$^+$**15**, Remark 6.15].

Notice from the Morse functional (3.3) and the gradient flow equation (3.4) that the definition of HF_*^{VW} coincides with the definition of the instanton Floer homology in [**Flo88b**], albeit for a *complex* gauge group $G_{\mathbb{C}}$. This means that we can also express the LHS of (4.6) as $\text{HF}_*^{\text{Inst}}\big(M_3, G_{\mathbb{C}}, \tau\big)$, the instanton Floer homology of $G_{\mathbb{C}}$ assigned to M_3. Also, recall that the Lagrangian branes L_0 and L_1 on the RHS of (4.6) are (A, B, A)-branes, i.e., they can also be interpreted as Lagrangian branes in $\mathcal{M}_H^G(C)$ in complex structure K, or equivalently, $\mathcal{M}_{\text{flat}}^{G_{\mathbb{C}}}(C)$, the moduli space of irreducible flat $G_{\mathbb{C}}$-connections on C.

These two points then mean that we can also write (4.6) as

$$\text{HF}_*^{\text{inst}}(M_3, G_{\mathbb{C}}, \tau) \cong \text{HF}_*^{\text{Lagr}}\big(\mathcal{M}_{\text{flat}}^{G_{\mathbb{C}}}(C), L_0, L_1, \tau\big) \tag{4.9}$$

In other words, the VW Atiyah-Floer correspondence in (4.6) can also be interpreted as an Atiyah-Floer correspondence for $G_{\mathbb{C}}$-instantons. It is now clear from (4.9) and (4.8), that for $G_{\mathbb{C}} = SL(2,\mathbb{C})$, we have

$$\text{HP}^*(M_3) \cong \text{HF}_*^{\text{inst}}(M_3, SL(2,\mathbb{C}), \tau) \tag{4.10}$$

for complex constant τ. This is exactly the conjecture by Abouzaid-Manolescu about $\text{HP}^*(M_3)$ in [**AM20**]! This agrees with their expectations in [**AM20**, sect. 9.2] that $\text{HP}^*(M_3)$ ought to be part of 3+1 dimensional TQFT based on the VW equations.

It was argued in [**AM20**, sect. 9.1] that the construction of $\text{HP}^*(M_3)$ can be generalized to $SL(N,\mathbb{C})$. Indeed, notice that (4.9) implies that there ought to be a $G_{\mathbb{C}}$ generalization of the Abouzaid-Manolescu conjecture in (4.10) to

$$\text{HP}^*(M_3, G_{\mathbb{C}}) \cong \text{HF}_*^{\text{inst}}(M_3, G_{\mathbb{C}}, \tau) \tag{4.11}$$

where the hypercohomology $\text{HP}^*(M_3, G_{\mathbb{C}})$ of the perverse sheaf of vanishing cycles in $\mathcal{M}_{\text{flat}}^{G_{\mathbb{C}}}(M_3)$ is such that

$$\text{HP}^*(M_3, G_{\mathbb{C}}) \cong \text{HF}_*^{\text{Lagr}}\big(\mathcal{M}_{\text{flat}}^{G_{\mathbb{C}}}(C), L_0, L_1, \tau\big) \tag{4.12}$$

which again agrees with [**BBD**$^+$**15**, Remark 6.15].

[11] It is a fact that $\dim(X_{\text{irr}}(C))$ is given by $4(N^2-1)(g-1)$ for $G_{\mathbb{C}} = SL(N,\mathbb{C})$, where g is the genus of C.

5. Langlands duality of Vafa-Witten invariants, Gromov-Witten invariants, Floer homologies and the Abouzaid-Manolescu hypercohomology

It is known that $\mathcal{N} = 4$ supersymmetric Yang-Mills theories has a $SL(2,\mathbb{Z})$ symmetry, with S- and T-duality, as mentioned in §1. In particular, the theory with complex coupling τ and gauge group G, is S-dual to a theory with complex coupling $-\frac{1}{n_{\mathfrak{g}}\tau}$ and Langlands dual gauge group ${}^LG$, i.e., we have, up to a possible phase factor of modular weights that is just a constant, a duality of VW partition functions

$$\mathcal{Z}_{\mathrm{VW},M_4}(\tau, G) \longleftrightarrow \mathcal{Z}_{\mathrm{VW},M_4}\Big(-\frac{1}{n_{\mathfrak{g}}\tau}, {}^LG\Big) \tag{5.1}$$

In other words, we have a Langlands duality of VW invariants of M_4, given by (5.1).

5.1. Langlands duality of Gromov-Witten invariants. Note that if $M_4 = \Sigma \times C$, from (5.1) and (2.17), 4d S-duality would mean that we have the 2d duality

$$\mathcal{Z}_{\mathrm{GW},\Sigma}\big(\tau, \mathcal{M}^G_{\mathrm{Higgs}}(C)\big) \longleftrightarrow \mathcal{Z}_{\mathrm{GW},\Sigma}\Big(-\frac{1}{n_{\mathfrak{g}}\tau}, \mathcal{M}^{{}^LG}_{\mathrm{Higgs}}(C)\Big) \tag{5.2}$$

where $\mathcal{M}^G_{\mathrm{Higgs}}$ and $\mathcal{M}^{{}^LG}_{\mathrm{Higgs}}$ are mirror manifolds. In other words, we have a Langlands duality of GW invariants that can be interpreted as a mirror symmetry of Higgs bundles, given by (5.2).

5.2. Langlands Duality of Vafa-Witten Floer Homology. If $M_4 = M_3 \times \mathbb{R}^+$, from (3.8) and (5.1), we have the duality

$$\mathcal{Z}^{\mathrm{Floer}}_{\mathrm{VW},M_3}(\tau, G) \longleftrightarrow \mathcal{Z}^{\mathrm{Floer}}_{\mathrm{VW},M_3}\Big(-\frac{1}{n_{\mathfrak{g}}\tau}, {}^LG\Big). \tag{5.3}$$

In turn, from (3.8), this means that we have the duality

$$\mathrm{HF}^{\mathrm{VW}}_*(M_3, G, \tau) \longleftrightarrow \mathrm{HF}^{\mathrm{VW}}_*(M_3, {}^LG, -1/n_{\mathfrak{g}}\tau) \tag{5.4}$$

In other words, we have a Langlands duality of VW Floer homologies assigned to M_3, given by (5.4).

5.3. Langlands duality of Lagrangian Floer homology. From (5.3) and (4.4), we have the duality

$$\mathcal{Z}_{A,L}\big(\tau, \mathcal{M}^G_{\mathrm{Higgs}}(C)\big) \longleftrightarrow \mathcal{Z}_{A,L}\Big(-\frac{1}{n_{\mathfrak{g}}\tau}, {}^LG\Big). \tag{5.5}$$

Then, from the RHS of the VW Atiyah-Floer correspondence in (4.6), which defines the state spectrum of $\mathcal{Z}_{A,L}$, we have the duality

$$\mathrm{HF}^{\mathrm{Lagr}}_*\big(\mathcal{M}^G_{\mathrm{Higgs}}(C), L_0, L_1, \tau\big) \longleftrightarrow \mathrm{HF}^{\mathrm{Lagr}}_*\big(\mathcal{M}^{{}^LG}_{\mathrm{Higgs}}(C), L_0, L_1, -1/n_{\mathfrak{g}}\tau\big) \tag{5.6}$$

In other words, we have a Langlands duality of Lagrangian Floer homologies of Higgs bundles, given by (5.6).

5.4. Langlands duality of the Abouzaid-Manolescu hypercohomology. From (4.11), the fact that its RHS can be identified with $\mathrm{HF}_*^{\mathrm{VW}}(M_3, G, \tau)$, and the relation (5.4), we have the duality

$$\mathrm{HP}^*(M_3, G_{\mathbb{C}}, \tau) \longleftrightarrow \mathrm{HP}^*(M_3, {}^L G_{\mathbb{C}}, -1/n_{\mathfrak{g}}\tau) \tag{5.7}$$

In other words, we have a Langlands duality of the Abouzaid-Manolescu hypercohomologies of a perverse sheaf of vanishing cycles in the moduli space of irreducible flat complex connections on M_3, given by (5.7).

6. A quantum and classical geometric Langlands correspondence

If we let $M_4 = I \times \mathbb{R}^+ \times C$ with $C \to 0$, S-duality gives a homological mirror symmetry of the category of A-branes. This implies a homological mirror symmetry of the τ-dependent category of A-branes:

$$\mathrm{Cat}_{A\text{-branes}}\big(\tau, \mathcal{M}^G_{\mathrm{Higgs}}(C)\big) \longleftrightarrow \mathrm{Cat}_{A\text{-branes}}\Big(-\frac{1}{n_{\mathfrak{g}}\tau}, \mathcal{M}^{^LG}_{\mathrm{Higgs}}(C)\Big) \tag{6.1}$$

where $\mathcal{M}^G_{\mathrm{Higgs}}$ and $\mathcal{M}^{^LG}_{\mathrm{Higgs}}$ are mirror manifolds.

For $\theta = 0$ ($\mathrm{Re}(\tau) = 0$), the category of τ-dependent A-branes can be identified with a category of twisted D-modules on $\mathrm{Bun}_{G_{\mathbb{C}}}(C)$ with parameter q. Thus, this mirror symmetry would mean that we have

$$\mathcal{D}^{\mathbf{c}}_{-h^\vee}\text{-mod}\big(q, \mathrm{Bun}_{G_{\mathbb{C}}}\big) \longleftrightarrow \mathcal{D}^{\mathbf{c}}_{-{}^Lh^\vee}\text{-mod}\Big(-\frac{1}{n_{\mathfrak{g}}q}, \mathrm{Bun}_{^LG_{\mathbb{C}}}\Big) \tag{6.2}$$

This is a quantum geometric Langlands correspondence for $G_{\mathbb{C}}$ with complex curve C and purely imaginary parameter q [**Fre07**, eqn. (6.4)].

On the other hand, in the zero-coupling, 'classical' limit of the 4d theory in G where $\mathrm{Im}(\tau) \to \infty$, we have $q \to \infty$. In this limit, the LHS of (6.2) can be identified with the category $\mathrm{Cat}_{\mathrm{coh}}\big(\mathcal{M}^{G_{\mathbb{C}}}_{\mathrm{flat}}(C)\big)$ of coherent sheaves on $\mathcal{M}^{G_{\mathbb{C}}}_{\mathrm{flat}}(C)$ [**Fre07**]. This 'classical' limit corresponds to the 'ultra-quantum' limit of the S-dual 4d theory in ${}^LG$, where ${}^Lq = -\frac{1}{n_{\mathfrak{g}}q} \to 0$. In this limit, the RHS of (6.2) can be identified with the category $\mathcal{D}^{\mathbf{c}}_{-{}^Lh^\vee}\text{-mod}\big(0, \mathrm{Bun}_{^LG_{\mathbb{C}}}\big)$ of critically-twisted D-modules on $\mathrm{Bun}_{^LG_{\mathbb{C}}}(C)$, giving us

$$\mathrm{Cat}_{\mathrm{coh}}\big(\mathcal{M}^{G_{\mathbb{C}}}_{\mathrm{flat}}(C)\big) \longleftrightarrow \mathcal{D}^{\mathbf{c}}_{-{}^Lh^\vee}\text{-mod}\Big(0, \mathrm{Bun}_{^LG_{\mathbb{C}}}\Big) \tag{6.3}$$

This is a classical geometric Langlands correspondence for $G_{\mathbb{C}}$ with complex curve C [**Fre07**, eqn. (6.4)].

7. Categorification and a novel web of mathematical relations

The mathematical procedure of categorification is realized in our physical framework, where the VW invariant is a number, the VW Floer homology is a vector (space), and the A-branes span a category of objects. Categorification can be physically understood as flattening a direction and then ending it on a boundary or boundaries. Explicitly in our case, the first step of categorification involves flattening a direction in M_4 and then ending it on an M_3 boundary, while the second step involves flattening a direction in M_3 and then ending it on two C boundaries. Therefore, one can also understand the procedure of categorifying as computing relative invariants[12] – computing the relative invariant of $\mathcal{Z}_{\text{VW}}$ give us HF_*^{VW}, and further computing the relative invariant of HF_*^{VW} gives us $\text{Cat}_{A\text{-branes}}$.

One could continue to further categorify the VW invariant of M_4 by flattening a direction along C and ending it on S^1 boundaries, i.e., let $C = I' \times S^1$. This should give us a 2-category, 2-Cat, consisting of objects, morphisms between these objects, and 2-morphisms between these morphisms. We thus have[13]

$$\begin{aligned}
\text{VW theory on } M_4 &\rightsquigarrow \text{number} \quad \mathcal{Z}_{\text{VW}}\\
\text{VW theory on } \mathbb{R}^+ \times M_3 &\rightsquigarrow \text{vector} \quad \text{HF}_*^{\text{VW}}\\
\text{VW theory on } \mathbb{R}^+ \times I \times C &\rightsquigarrow \text{1-category} \quad \text{Cat}_{A\text{-branes}}\\
\text{VW theory on } \mathbb{R}^+ \times I \times I' \times S^1 &\rightsquigarrow \text{2-category} \quad \text{2-Cat}\\
\text{VW theory on } \mathbb{R}^+ \times I \times I' \times [0,1] &\rightsquigarrow \text{3-category} \quad \text{3-Cat.}
\end{aligned} \tag{7.1}$$

As we go down the list, the categories get assigned to $M_3, C, \ldots$, and are determined by the category of boundaries of the effective 1d, 2d, ... theory on $\mathbb{R}^+, \mathbb{R}^+ \times I, \ldots$. Therefore, the 2-category will be determined by the category of 2d boundaries of the 3d theory on $\mathbb{R}^+ \times I \times I'$ given by VW theory compactified on S^1, that is assigned to S^1. These are surface defects that can be interpreted as objects; loop defects on the surface running around $I \times I'$ can be interpreted as morphisms between these objects; while opposing pairs of point defects on the loops can be interpreted as 2-morphisms between these morphisms. Also, note that the 3d TQFT in question is a 3d gauged A-model described in [**KSV10**, sect. 7],[14] and for abelian G and $\text{Re}(\tau) = 0$, the 2-category of surface defects have been explicitly determined in *loc. cit.* to be the 2-category $\text{2-Cat}_{\text{mod-cat}}(\text{FF-cat}(T^2))$ of module categories over the Fukaya-Floer category of T^2.[15]

4d S-duality also gives us a Langlands duality of the 2-category 2-Cat. According to [**KSV10**, sect. 7.4.1], 4d S-duality, which maps abelian G to its Langlands dual that is itself, will transform the symplectic area $\mathcal{A}$ of T^2 as

$$\mathcal{A} \to {}^L\mathcal{A} = \frac{4\pi^2}{\mathcal{A}}, \tag{7.2}$$

where ${}^L\mathcal{A}$ is the symplectic area of a torus ${}^LT^2$ that can be obtained from T^2 by inverting the radii of its two circles from $R \to \alpha'/R$ for some constant α'. In other words, ${}^LT^2$ is the T-dual torus to T^2, and $\text{FF-cat}(T^2)$, which is realized by a 2d open A-model with target T^2, will be invariant under T-duality of the target, i.e., $\text{FF-cat}(T^2) \cong \text{FF-cat}({}^LT^2)$. Thus, we have

$$\text{2-Cat}_{\text{mod-cat}}(\text{FF-cat}(T^2)) \longleftrightarrow \text{2-Cat}_{\text{mod-cat}}(\text{FF-cat}({}^LT^2)). \tag{7.3}$$

[12]A relative invariant is an invariant of an open manifold which was originally defined for a closed manifold.

[13]This perspective of categorifying topological invariants by successively introducing boundaries to the manifold was first pointed out in [**G09**].

[14]In [**KSV10**, sect. 7], the GL theory at $t = 0$ was considered, but it was shown in [**Set13**, sect. 5.2-5.3] that this theory compactified on S^1 is the same as VW theory compactified on S^1. Hence, their results are applicable to us.

[15]The 3d gauged A-model has a gauge and matter sector, where each sector can either have Dirichlet (D) or Neumann (N) boundary conditions. We have stated the result for the DD case, as this choice of boundary conditions allows us to describe the situation where line defects lie along the surface defects, which is the one relevant to us.

The last step to further categorify the VW invariant of M_4 is to flatten S^1, ending it on point boundaries, i.e., let $S^1 = [0,1]$. This should give us a 3-category, 3-Cat, consisting of objects, morphisms between these objects, 2-morphisms between these morphisms, and 3-morphisms between these 2-morphisms, giving us a 3-category of 3d boundary conditions of VW theory along $\mathbb{R}^+ \times I \times I'$ which is assigned to a point. These 3d boundary conditions can be realized by domain walls.

From the duality relations (5.1), (5.2), (5.4), (5.6), the correspondences (6.1), (6.2), (6.3), and the identifications (2.17), (3.6), (4.6), we will get Fig. 3 below.

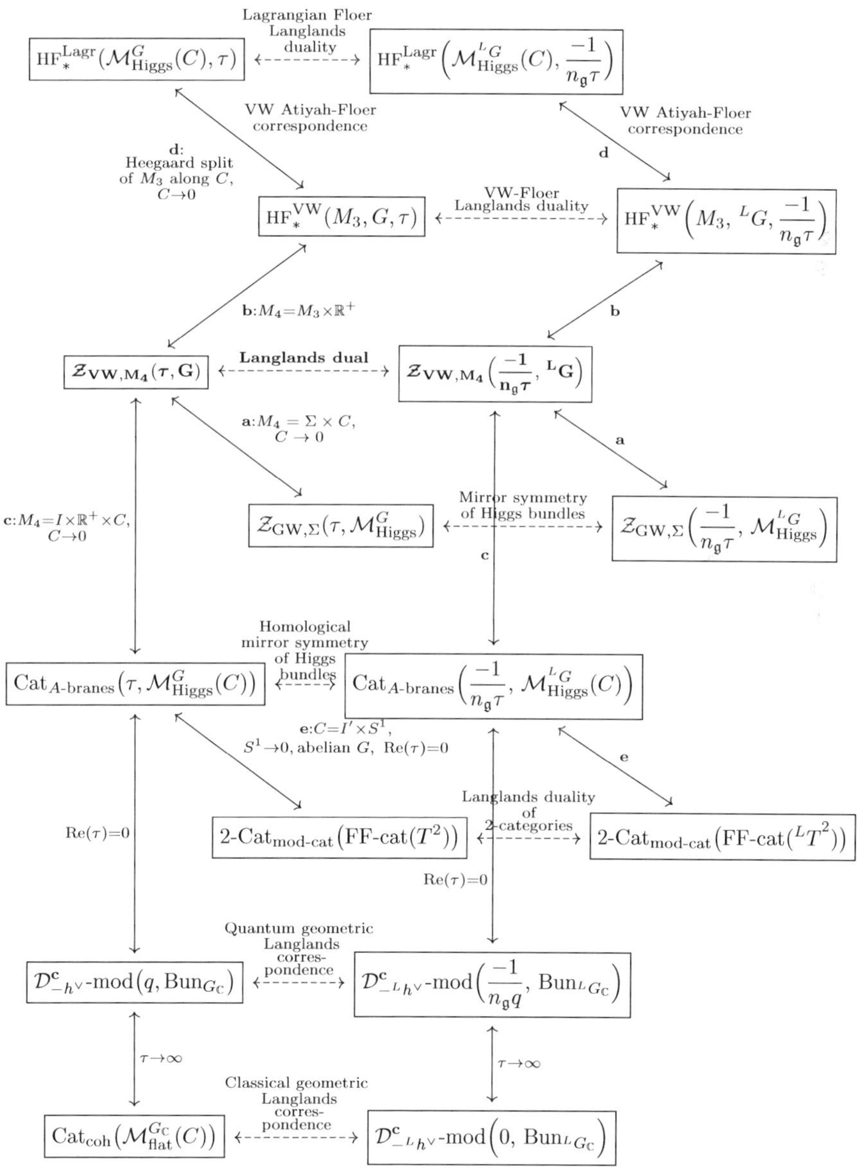

FIGURE 3. A novel web of mathematical relations stemming from Vafa-Witten theory that also involves higher categories.

Acknowledgment

I would like to thank the referee for questions which have led to further refinement of this proceeding.

References

[Ati88] Michael Atiyah, *New invariants of* 3- *and* 4-*dimensional manifolds*, The mathematical heritage of Hermann Weyl (Durham, NC, 1987), Proc. Sympos. Pure Math., vol. 48, Amer. Math. Soc., Providence, RI, 1988, pp. 285–299, DOI 10.1090/pspum/048/974342. MR974342

[AGF81] Luis Alvarez-Gaumé and Daniel Z. Freedman, *Geometrical structure and ultraviolet finiteness in the supersymmetric* σ*-model*, Comm. Math. Phys. **80** (1981), no. 3, 443–451. MR626710

[AM20] Mohammed Abouzaid and Ciprian Manolescu, *A sheaf-theoretic model for* $\mathrm{SL}(2,\mathbb{C})$ *Floer homology*, J. Eur. Math. Soc. (JEMS) **22** (2020), no. 11, 3641–3695, DOI 10.4171/jems/994. MR4167016

[BBD+15] C. Brav, V. Bussi, D. Dupont, D. Joyce, and B. Szendrői, *Symmetries and stabilization for sheaves of vanishing cycles*, J. Singul. **11** (2015), 85–151, DOI 10.5427/jsing.2015.11e. With an appendix by Jörg Schürmann. MR3353002

[Flo88a] Andreas Floer, *Morse theory for Lagrangian intersections*, J. Differential Geom. **28** (1988), no. 3, 513–547. MR965228

[Flo88b] Andreas Floer, *An instanton-invariant for 3-manifolds*, Comm. Math. Phys. **118** (1988), no. 2, 215–240. MR956166

[Fre07] Edward Frenkel, *Lectures on the Langlands program and conformal field theory*, Frontiers in number theory, physics, and geometry. II, Springer, Berlin, 2007, pp. 387–533, DOI 10.1007/978-3-540-30308-4_11. MR2290768

[G09] S. Gukov, *Surface operators and knot homologies*, In New Trends in Mathematical Physics: Selected contributions of the XVth International Congress on Mathematical Physics, 2009, (pp. 313-343).

[Hit87] N. J. Hitchin, *The self-duality equations on a Riemann surface*, Proc. London Math. Soc. (3) **55** (1987), no. 1, 59–126, DOI 10.1112/plms/s3-55.1.59. MR887284

[K08] A. Kapustin, *A note on quantum geometric Langlands duality, gauge theory, and quantization of the moduli space of flat connections*, preprint `arXiv:0811.3264`, 2008.

[KSV10] A. Kapustin, K. Setter, and K. Vyas, *Surface operators in four-dimensional topological gauge theory and Langlands duality*, preprint `arXiv:1002.0385`, 2010.

[KW07] Anton Kapustin and Edward Witten, *Electric-magnetic duality and the geometric Langlands program*, Commun. Number Theory Phys. **1** (2007), no. 1, 1–236, DOI 10.4310/CNTP.2007.v1.n1.a1. MR2306566

[LL97] J. M. F. Labastida and Carlos Lozano, *Mathai-Quillen formulation of twisted* $N = 4$ *supersymmetric gauge theories in four dimensions*, Nuclear Phys. B **502** (1997), no. 3, 741–790, DOI 10.1016/S0550-3213(97)00421-5. MR1477877

[N23] D. Nesterov, *Enumerative mirror symmetry for moduli spaces of Higgs bundles and S-duality*, `arXiv:2302.08379`, 2023.

[OT22] Z.-C. Ong, M.-C. Tan. *Vafa-Witten Theory: Invariants, Floer Homologies, Higgs Bundles, a Geometric Langlands Correspondence, and Categorification*, `arXiv:2203.17115`, 2022.

[Ped18] Andrés Pedroza, *A quick view of Lagrangian Floer homology*, Geometrical themes inspired by the N-body problem, Lecture Notes in Math., vol. 2204, Springer, Cham, 2018, pp. 91–125. MR3753117

[Set13] Kevin Setter, *Topological Quantum Field Theory and the Geometric Langlands Correspondence*, ProQuest LLC, Ann Arbor, MI, 2013. Thesis (Ph.D.)–California Institute of Technology. MR3121996

[TT17] Y. Tanaka and R.P. Thomas, *Vafa-Witten invariants for projective surfaces I: stable case*, `arXiv:1702.08487`, 2017.

[VW94] Cumrun Vafa and Edward Witten, *A strong coupling test of S-duality*, Nuclear Phys. B **431** (1994), no. 1-2, 3–77, DOI 10.1016/0550-3213(94)90097-3. MR1305096

Department of Physics, National University of Singapore, 2 Science Drive 3, S(117551), Singapore
Email address: mctan@nus.edu.sg

Proceedings of Symposia in Pure Mathematics
Volume **107**, 2024
https://doi.org/10.1090/pspum/107/01953

A holomorphic approach to fivebranes

Brian Williams

ABSTRACT. We present an approach to studying the algebra of local operators on a stack of finitely many fivebranes in M theory at the level of the holomorphic twist. Our approach is through the lens of twisted holography and utilizes a description of the minimal twist of eleven-dimensional supergravity.

These notes are an extension of a lecture given at String Math 2022 in Warsaw, Poland and mainly reports on recent work with Surya Raghavendran and Ingmar Saberi [**48**, **49**, **53**]. The main goal is to highlight an approach to understanding local operators of the ubiquitous six-dimensional superconformal field theory at the level of the holomorphic twist.

1. A model for twisted eleven-dimensional supergravity

Eleven-dimensional supergravity is the unique theory equipped with eleven-dimensional supersymmetry and is famously expected to describe the low energy limit of M-theory. In [**19**], Costello and Li gave a rigorous definition of what it means to *twist* supergravity, which is nicely compatible with the mathematically more familiar notion of twisting supersymmetric gauge theories [**12**, **28**, **56**]. In physical terms, twisted supergravity is supergravity in a background where a bosonic ghost for supersymmetry takes a nonzero value Q, where Q is a nilpotent supercharge. In principle, one can take the supercharge to be any square-zero covariantly constant spinor (so it satisfies the equations of motion of supergravity). A class of such supercharges lie in the supersymmetry algebra.

The eleven-dimensional supersymmetry algebra admits two types of twists characterized by the number of invariant directions that the supercharge determines:

- The $SU(5)$ twist. This twist leaves six real directions invariant and is stabilized by the subgroup $SU(5)$ of the eleven-dimensional Lorentz group.[1]
- The G_2 twist. This twist leaves nine real directions invariant and is preserved by the subgroup $SU(2) \times G_2$ of the Lorentz group.

In non-gravitational supersymmetric theories, like supersymmetric Yang–Mills theories, twists have the feature of admitting elegant descriptions in terms of moduli spaces internal to the worlds of complex or smooth geometry. The definition of Costello and Li can be applied to characterize the corresponding twists of eleven-dimensional supergravity, and one can attempt to put them on the same footing

[1]We will work in Euclidean signature throughout this note.

as in the case of gauge theories. Unfortunately, due to a lack of a mathematically rigorous description of the supersymmetric theory in terms of component fields, it is difficult to see this process through concretely. Nevertheless, using topological string theoretic arguments, Costello and Li were able to give conjectural descriptions of twists of a large class of ten-dimensional supergravity theories in this vein [**19**]. There is, however, an elegant proposed description of the supersymmetric theory using the language of pure spinors [**5**, **26**, **45**].

At the heart of the pure spinor formalism is the so-called "nilpotence variety" [**27**] which is simply the algebraic variety of square-zero supercharges of a given supersymmetry algebra $\mathfrak{g}_{susy}$. By definition, it is an affine subvariety

$$Y \stackrel{\text{def}}{=} \{Q \in \mathfrak{g}_{susy}^{odd} \mid Q^2 = 0\} \subset \mathfrak{g}_{susy}^{odd}. \tag{1.1}$$

The basic idea is that one can construct a supersymmetric field theory from the data of a sheaf on the nilpotence variety. For the structure sheaf, the basic fields in the pure spinor formalism are

$$\Phi \in C^\infty(\mathbf{R}^{d|k}) \otimes \mathcal{O}(Y). \tag{1.2}$$

Here $\mathbf{R}^{d|k}$ is the superspace with d bosonic directions, where d is the spacetime dimension, and k odd directions, where k a number determined by the amount of supersymmetry. The right action of $\mathfrak{g}_{susy}$ on superspace defines a natural cohomological vector field $\mathcal{D}$ on $\mathbf{R}^{d|k} \times Y$ and hence a differential on the space of fields. Using a twisted version of the pure spinor formalism, which amounts to deforming $\mathcal{D}$ in a way that depends on a fixed supercharge, Saberi and the author were able to completely characterize the free-limit of the twists of eleven-dimensional supergravity [**52**]. In the case of the $SU(5)$ twist the problem of characterizing the twisted theory reduced to describing the Koszul homology of the structure sheaf over the twisted nilpotence variety, which in this case is (the affine cone over) the Grassmannian $Gr(2,5)$.

In the next section we will give an overview of the $SU(5)$ twist of eleven-dimensional supergravity following [**48**].

1.1. The $SU(5)$ twist. In [**48**] the author together with Raghavendran and Saberi described a proposed mathematical model for the $SU(5)$-twist of eleven-dimensional supergravity. Many tests of this proposal were performed, including a consistency check that the dimensional reduction of the model agrees with proposals for twists of type IIA supergravity given in terms of topological strings [**19**].

The eleven-dimensional theory exists on any manifold which is locally of the form

$$\mathbf{R} \times Z$$

where Z is a Calabi–Yau fivefold. The model is topological along $\mathbf{R}$ in the sense that translations in this direction act in a homotopically trivial way on the theory. Likewise, locally on the Calabi–Yau fivefold, all anti-holomorphic translations act in a homotopically trivial way—in this sense, the theory is *holomorphic* along Z [**55**]. More generally, the eleven-dimensional theory can be constructed on any eleven-manifold equipped with a transversely holomorphic foliation which is equipped with an appropriate holomorphic volume form on the leaves of the foliation.

In [**48**, **49**] we describe how physical fields in supergravity are manifest in the twisted model. Here, we will jump straight to describing the twisted model. Fields generally live in spaces of the form

$$\Omega^\bullet(\mathbf{R}) \otimes \Omega^{0,\bullet}(Z, V) \tag{1.3}$$

where V is some holomorphic vector bundle on Z.[2] Since the theory is of topological/holomorphic nature, there is a linear part of the BRST differential of the form $\mathrm{d} + \overline{\partial}$ where d is the de Rham differential on $\mathbf{R}$ and $\overline{\partial}$ corresponds to the holomorphic structure on the bundle V.

Deformations of the background metric which survive the twist can be identified with Beltrami differentials, that is, sections of

$$\overline{\mathrm{T}}^*_Z \otimes \mathrm{T}_Z \cong \overline{\mathrm{T}}^*_Z \otimes \overline{\mathrm{T}}^*_Z.$$

More generally, the fields of this model include sections of the holomorphic tangent bundle with coefficients in Dolbeault forms on Z of arbitrary Dolbeault type and de Rham forms on $\mathbf{R}$, which we can write in superfield notation as

$$\mu = \mu^{\overline{I}}_j(t; z, \overline{z})\, \mathrm{d}\, \overline{z}_{\overline{I}} \otimes \partial_{z_j} + \mu^{\overline{I}}_{t,j}(t; z, \overline{z})\, \mathrm{d}\, t\, \mathrm{d}\, \overline{z}_{\overline{I}} \otimes \partial_{z_j}$$

where the sum over the multi-index $\overline{I}$ ranges over subsets of $\{1, \ldots, 5\}$. We view this as a section

$$\mu \in \Omega^\bullet(\mathbf{R}) \otimes \Omega^{0,\bullet}(Z, \Pi\, \mathrm{T}_Z). \tag{1.4}$$

In this notation, the component $\mu^{\overline{I}}_j$ is *odd* if $|\overline{I}| = 0, 2, 4$ is even and *even* if $|I| = 1, 3, 5$ is odd. The parity shift of the holomorphic tangent bundle is to ensure that deformations of the background metric which survive the twist are viewed as even fields. In particular we have odd fields $\mu_j(t; z, \overline{z})\partial_{z_j}$ which play the role of ghosts for infinitesimal changes of coordinates.

Part of the linear equations of motion in our model imply that the field μ satisfies a condition that it be *divergence-free* for the Calabi–Yau structure on the fivefold Z. This condition is imposed, cohomologically, by adding another field which is simply a differential form

$$\nu \in \Omega^\bullet(\mathbf{R}) \otimes \Omega^{0,\bullet}(Z) \tag{1.5}$$

and a linear part of the differential $\delta\nu = \partial_\Omega \mu$ encoding the divergence-free condition. Another piece of the linear equations of motion imply that it μ be constant along $\mathbf{R}$.

Ordinarily, for μ to describe a deformation of complex structure in the Z direction we would require that it satisfy the Maurer–Cartan equation. For twisted supergravity, however, we find a modification of this equation which involves the twisted analog of another familiar field—the higher form gauge field in supergravity.

In the BV formalism one will see all types of differential forms corresponding to a tower of ghosts for this higher gauge field together with antifields and antighosts. In local coordinates, the only components of this tower of differential forms which survives the twist are

$$\beta = \beta^{\overline{I}}\, \mathrm{d}\, \overline{z}_{\overline{I}} + \beta^{\overline{J}}_t\, \mathrm{d}\, t\, \mathrm{d}\, \overline{z}_{\overline{J}} \in \Omega^\bullet(\mathbf{R}) \otimes \Omega^{0,\bullet}(Z)$$
$$\gamma = \gamma^{i\overline{I}}\, \mathrm{d}\, z_i\, \mathrm{d}\, \overline{z}_{\overline{I}} + \gamma^{j\overline{J}}_t\, \mathrm{d}\, t\, \mathrm{d}\, z_j\, \mathrm{d}\, \overline{z}_{\overline{J}} \in \Omega^\bullet(\mathbf{R}) \otimes \Omega^{1,\bullet}(Z).$$

Fields $\beta^{\overline{I}}, \gamma^{i\overline{I}}_t$ with $|\overline{I}|$ =odd (resp. even) are *even* (resp. *odd*) and fields $\beta^{\overline{I}}_t, \gamma^{i\overline{I}}$ with $|\overline{I}|$ =odd (resp. even) are *odd* (resp. *even*).

[2]The tensor product $\otimes$ is the completed tensor product with respect to natural topologies present in the space of sections of a smooth vector bundle.

The most important equation of motion of the eleven-dimensional theory involves both the fields μ and γ. When $\partial_\Omega \mu = 0$ it takes the form

$$\overline{\partial}\mu + \frac{1}{2}[\mu, \mu] = \partial\gamma\partial\gamma. \tag{1.6}$$

Because of the term on the right hand side this equation is not exactly the usual Beltrami equation for deformations of complex structures. On the left hand side we are implicitly using an identification between the holomorphic tangent bundle T_Z and the bundle $\wedge^4 \mathrm{T}^*_Z$ granted by the holomorphic volume form on Z.

In **[48]** we find the above equation of motion together with its gauge symmetries by formulating the theory within the BV formalism. We briefly recount what this entails. In the BV formalism, the space of BV fields $\mathcal{E}$ is given as the sections of some graded vector bundle on spacetime. The grading is typically by the abelian group $\mathbf{Z}$, which is usually referred to as the cohomological grading. In cohomological degree zero of $\mathcal{E}$ sit the usual (sometimes called *physical*) fields. The ghosts comprise the cohomological degree -1 part of $\mathcal{E}$. In cohomological degree one sit the antifields, etc.. Another key piece of structure is a skew symmetric pairing ω_{BV} of cohomological degree -1 on $\mathcal{E}$ which is concretely a pairing between the fields and antifields (and ghosts and antighosts). The full BV action functional S_{BV} is a functional on the space of fields which encodes not only the original action functional on fields, but through its dependence on ghosts and antifields it also encodes the gauge symmetries (and gauge symmetries for gauge symmetries, etc.) of the theory.

For our eleven-dimensional model the BV action functional takes the following form

$$S_{BV} = \int_{\mathbf{R}\times Z} \left[\beta \wedge (\overline{\partial} + \mathrm{d})\nu + \gamma \wedge (\overline{\partial} + \mathrm{d})\mu + \beta \wedge \partial_\Omega \mu\right] + \frac{1}{2}\int_{\mathbf{R}\times Z} \mu^2 \vee \partial\gamma + \int_{\mathbf{R}\times X} \gamma\partial\gamma\partial\gamma + \cdots . \tag{1.7}$$

Here:

- the first line is the free part of the action. The operator $\overline{\partial}$ is the $\overline{\partial}$-operator on Z, the operator d is the de Rham operator on $\mathbf{R}$, and the operator ∂_Ω is the holomorphic divergence operator on Z corresponding to the holomorphic volume form.
- We implicitly use the holomorphic volume form in the above integrals.
- The term $\gamma\partial\gamma\partial\gamma$ is the holomorphic avatar of the famous Chern–Simons term of the higher form gauge field present in supergravity.
- The $\cdots$ indicate additional terms needed so that this functional satisfies the classical master equation (in this presentation there are, in fact, infinitely many terms). They take a form which is reminiscent of the genus zero BCOV action for the fivefold Z (we make a further comment on this below).

In **[48]** we have performed a number of consistency checks that this model indeed describes the $SU(5)$ twist of supergravity. Famously, M-theory on a circle is expected to be equivalent to type IIA string theory where the length of the circle is proportional to the string coupling constant. At the level of supergravity, the S^1-reduction of eleven-dimensional supergravity is ten-dimensional supergravity of type IIA. Here we utilize a description of the twist of type IIA strings using a version of Kodaira–Spencer theory developed in [**18, 19**].

From the point of view of the $SU(5)$ twist there are essentially two types of circles one can consider. The first is a 'holomorphic circle'. This is present when we assume our Calabi–Yau fivefold is of the form $Z = \mathbf{C}^\times \times Y$ with Y a Calabi–Yau fourfold. We assume that $\mathbf{C}^\times$ is equipped with the standard volume form $\mathrm{d}\, z$. The S^1 reduction along a circle in $\mathbf{C}^\times$ results in a twist of type IIA supergravity with global symmetry $SU(4)$ on the manifold

$$\mathbf{R} \times \mathbf{R} \times Y. \tag{1.8}$$

More generally, this twist can be defined when $\mathbf{R}^2$ is replaced by a smooth oriented two-dimensional manifold M^2. In [**19**], Costello and Li have given a conjectural description of this particular twist of type IIA (and many other twists) using methods of topological string theory. In this case, the conjecture of Costello and Li is that the twist is equivalent to a mixed topological string theory which is given by the A-model along M^2 and the B-model along Y. In perturbation theory, this admits a description in terms of Kodaira–Spencer theory along the fourfold Y. Via an explicit dimensional reduction of our model we find an exact match with this expectation.

PROPOSITION 1.1 ([**48**]). *The dimensional reduction of our eleven-dimensional model on*

$$\mathbf{R} \times \mathbf{C} \times Y \tag{1.9}$$

along $S^1 \subset \mathbf{C}$ is equivalent to the mixed topological A-model along $\mathbf{R}^2$ and B-model along Y as considered in [**19**]. *We heuristically write this model as*

$$\mathbf{R}^2_A \times Y_B. \tag{1.10}$$

The other type of reduction one can consider is that along a 'topological' circle. This corresponds to placing our eleven-dimensional model on

$$S^1 \times Z \tag{1.11}$$

and performing reduction along $S^1 \times Z \to Z$. What results is a theory with $SU(5)$ global symmetry. We conjecture that this is the $SU(5)$ twist of type IIA supergravity. There is no expected description of the $SU(5)$ twist in terms of topological strings like in the $SU(4)$ or $SU(2)$ cases.

1.2. The G_2 twist. We will give a brief description to of the G_2 twist of eleven-dimensional supergravity. Independent derivations of this twist have appeared in [**13, 25**]. The theory is defined on any manifold of the form

$$M \times Y \tag{1.12}$$

where M is a manifold with G_2 holonomy and Y is a manifold with $SU(2)$ holonomy. Unfortunately, the perturbative description we are about to give depends in no way on the G_2 structure and can be defined when M is any smooth seven-dimensional manifold. To see the full G_2 structure appearing one must take into account non-perturbative effects (much like in the target space description of the topological A-model).

The fundamental field is simply a differential form

$$\alpha \in \Omega^\bullet(M) \otimes \Omega^{0,\bullet}(Y)[1]. \tag{1.13}$$

The linear BRST differential is simply $\mathrm{d} + \bar{\partial}$ where d is the de Rham differential on M and $\bar{\partial}$ is the Dolbeault operator for Y.

Being a manifold with $SU(2)$ holonomy, the algebra of holomorphic fnuctions on Z is equipped with a holomorphic Poisson bracket $\{-,-\}$. Since this bracket involves only holomorphic differntial operators, it can be extended to a Poisson bracket on the dg algebra $\Omega^{0,\bullet}(Y)$. The theory utilizes this bracket in an essential way. Indeed, the full BV action is

$$\frac{1}{2}\int_{M\times Y} \Omega\, \alpha\, \mathrm{d}\, \alpha + \frac{1}{6}\int_{M\times Y} \Omega\, \alpha\{\alpha,\alpha\}. \tag{1.14}$$

where $\Omega \in \Omega^{2,0}(Y)$ is the holomorphic volume form. This description appeared in **[50]** and was independently derived from a further twist of the $SU(5)$ twist in **[48]**.

Like the $SU(5)$ twist, this theory is only $\mathbf{Z}/2$ graded. Nevertheless, one convenient way to think about this theory is as a higher dimensional Chern–Simons theory for a particular (dg) Lie algebra. Given any odd-dimensional smooth manifold M and a Lie algebra $\mathfrak{g}$ equipped with an invariant non-degenerate symmetric inner product one can consider a $\mathbf{Z}/2$ graded version of Chern–Simons theory whose BV action is given by the usual formula $\int_M CS(A)$. It is only in the case that $\dim M = 3$ that this theory admits a $\mathbf{Z}$-grading. The theory we are considering here can be interpreted as $\mathbf{Z}/2$ graded Chern–Simons theory on the seven-manifold M for the Lie algebra of holomorphic functions on Y equipped with the holomorphic Poisson bracket. Here, the invariant pairing is given by integrating against the holomorphic volume form Ω.

1.3. Exceptional ghosts. In any QFT, the BV action functional endows the shift of the fields $\mathcal{E}[-1]$ with the structure of an L_∞ algebra. The L_∞ structure maps can be read off from the BV action S_{BV} in the following way

$$S_{BV}(\Phi) = \sum_{k\geq 1} \omega_{BV}\left(\Phi, [\Phi,\ldots,\Phi]_k\right). \tag{1.15}$$

One way to think about $\mathcal{E}[-1]$ is as the dg Lie (really L_∞) algebra controlling infinitesimal gauge symmetries in perturbation theory near a fixed classical solution.

On flat space, the L_∞ algebras associated to twists of eleven-dimensional supergravity are explicit to describe. We denote these L_∞ algebras by $\mathcal{L}_{sugra}^{SU(5),G_2}$. Since we are working in a $\mathbf{Z}/2$ graded world, these L_∞ algebras are only $\mathbf{Z}/2$ graded.

With an eye towards local operators, we consider a slightly modified version of the L_∞ algebra which comes out of the BV formalism. Given a vector bundle $E \to M$, the bundle of ∞-jets $J^\infty E \to M$ is a ∞-dimensional pro vector bundle whose sections consist of ∞-jets of sections of E. If the fields of a theory are sections of E then $J^\infty_x E$ contains all possible jets of fields at $x \in M$. If $M = \mathbf{R}^d$ and E is translation invariant, then the bundle of ∞-jets can be identified with $E_0 \times \mathbf{C}[[x_i]]$ where $\{x_i\}$ is a chosen coordinate on $\mathbf{R}^d$. The process of taking ∞-jets is well-behaved for local Lie algebras: the ∞-jets of a translation invariant local Lie algebra on $\mathbf{R}^d$ at a point carries the natural structure of a Lie algebra. In what follows we consider the ∞-jets at $0 \in \mathbf{R}\times\mathbf{C}^5$ or $\mathbf{R}^7\times\mathbf{C}^2$ of the local L_∞ algebra $\mathcal{L}_{sugra}$ controlling twists of eleven-dimensional supergravity.

In the case of the $SU(5)$ twist, where we work on $\mathbf{R}\times\mathbf{C}^5$, this L_∞ algebra is closely related to the exceptional simple super Lie algebra $E(5|10)$ classified in **[36]**. This super Lie algebra admits the following formal geometric description.

The even part of $E(5|10)$ is the Lie algebra of divergence-free vector fields on the formal five-disk $\widehat{D}^5$. The odd part of $E(5|10)$ is the space of closed two-forms $\Omega^{2,cl}(\widehat{D}^5)$ on the formal five-disk. There is a natural action of divergence-free vector fields on closed two forms. Additionally there is a Lie bracket between two closed two-forms α, α' which produces a divergence-free vector field via the standard holomorphic volume form on the five-disk:

$$[\alpha, \alpha'] = \Omega_{\widehat{D}^5}^{-1} \vee (\alpha \wedge \alpha'). \tag{1.16}$$

We prove the following.

THEOREM 1.2 ([**48**]). *The ∞-jets of $\mathcal{L}_{sugra}^{SU(5)}$ at $0 \in \mathbf{R} \times \mathbf{C}^5$ is quasi-isomorphic to a one-dimensional central extension of the super Lie algebra $E(5|10)$:*

$$\mathbf{C} \to \widehat{E(5|10)} \to E(5|10). \tag{1.17}$$

The central extension is defined by the (totally even) three-linear cocycle

$$(\mu, \mu', \alpha) \mapsto \langle \mu \wedge \mu', \alpha \rangle_{z=w=0} \in \mathbf{C}. \tag{1.18}$$

Since this is a three-linear functional the model we use for the central extension is a super L_∞ algebra (with zero one-ary operation) rather than just a super Lie algebra.

As we mentioned above, the G_2 twist can be witnessed as a further twist of the $SU(5)$ model. In $E(5|10)$ this further twisting supercharge can be identified with the translation invariant Maurer–Cartan element

$$\mathrm{d}\, z_1 \,\mathrm{d}\, z_2 \in \Omega^{2,cl}(\widehat{D}^5) \subset E(5|10)^{odd}. \tag{1.19}$$

The variables z_1, z_2 correspond to the directions which remain holomorphic in the G_2-twist. At the level of ∞-jets of fields $\mathcal{L}_{sugra}^{G_2}$, the G_2 twist can be understood as the deformation of the central extension of $E(5|10)$ by the differential $\mathrm{d} = [\mathrm{d}\, z_1 \,\mathrm{d}\, z_2, -]$. The resulting dg Lie algebra is quasi-isomorphic to the Lie algebra of Hamiltonian functions on the formal two-disk $\widehat{D}^2$. It is useful to note that if we did not centrally extend $E(5|10)$ we would find the Lie algebra of Hamiltonian vector fields.

1.4. The S^{10} index. The relationship between M-theory and type IIA string theory suggests a deep connection between the index of eleven-dimensional supergravity and enumerative geometry [**39–42, 46**]. We remarked that the $SU(5)$ twist of the eleven-dimensional theory can be defined on any eleven-manifold which locally looks like the product of a Calabi–Yau fivefold times a smooth one-dimensional manifold. Such a property is ensured by a transverseley holomorphic foliation (THF) structure on the eleven-manifold. In particular we can define the theory on the manifold

$$(\mathbf{R} \times \mathbf{C}^5) - 0 \simeq \mathbf{R}_+ \times S^{10}. \tag{1.20}$$

Being an open submanifold of $\mathbf{R} \times \mathbf{C}^5$ this is equipped with a natural THF structure. To get a compact manifold we can quotient by the $\mathbf{Z}$-action generated by

$$(t, z_1, \ldots, z_5) \mapsto (t, q_1 z_1, \ldots, q_5 z_5) \tag{1.21}$$

where q_i are complex numbers satisfying $0 < |q_i| < 1$ and $\prod_i q_i = 1$. The last condition is to ensure that the holomorphic volume form on the leaves of the foliation is preserved by this action. The resulting quotient is a smooth eleven-dimensional manifold which is homeomorphic to $S^1 \times S^{10}$. One can think of this as a THF version of a Hopf manifold. As such, the partition function on this manifold has the interpretation of an index of local operators of the eleven dimensional model.

From this latter point of view, the index is identified with the $SU(5)$ character of local operators. The Cartan of $SU(5)$ is labeled by the fugacities $q_1, \ldots, q_5$ subject to the constraint that $\prod_{i=1}^5 q_i = 1$.

PROPOSITION 1.3. *The $S^1 \times S^{10}$ partition function of the eleven-dimensional theory is*

$$\prod_{i=1}^{5} \prod_{(m_i)\in \mathbf{Z}^5_{\geq 0}} \frac{1 - q_1^{m_1+1} \cdots q_i^{m_i} \cdots q_5^{m_5+1}}{1 - q_1^{m_1} \cdots q_i^{m_i+1} \cdots q_5^{m_5}}$$

where we recall the constraint that $\prod_i q_i = 1$.

This formula agrees with the index of eleven-dimensional supergravity computed in [**46**, Equation 6.11].

2. Twisted fivebranes

We next turn to extended objects in M-theory at the level of the holomorphic twist, with a particular focus on fivebranes. It is expected that the worldvolume theory of fivebranes in M-theory on flat space is equipped with $\mathcal{N} = (2,0)$ supersymmetry. Six-dimensional $\mathcal{N} = (2,0)$ supersymmetry admits two inequivalent classes of twists characterized by the number of directions which are left invariant:

- The holomorphic, or minimal, twist. This twist leaves three real directions invariant and is stabilized by the double cover of the group $U(3)$. The holomorphic twist of any $\mathcal{N} = (2,0)$ theory can be defined on any complex three-fold Z equipped with a square-root of its canonical bundle.
- The non-minimal twist. This twist leaves five real directions invariant and is stabilized by the group $Spin(4) \times U(1)$. The non-minimal twist of any $\mathcal{N} = (2,0)$ theory can be defined on a six-manifold of the form $M^4 \times \Sigma$ where M^4 is a smooth four-manifold and Σ is a Riemann surface.

Theories with $\mathcal{N} = (2,0)$ supersymmetry are extremely difficult to characterize by traditional methods partly due to the fact that they do not admit a Lagrangian descriptions like in supersymmetric Yang–Mills theories. This boils down to the fact that in $\mathcal{N} = (2,0)$ supersymmetry there is a 'chiral' [**57**] two-form whose curvature is self-dual. By string theory arguments, it is expected that such theories can be constructed in terms of the data of any simple Lie algebra of type ADE. The theory on a stack of $N > 1$ fivebranes in M-theory on flat space is anticipated to be the theory associated to the simple Lie algebra of type A_{N-1}. A concrete description of this theory in either mathematical or physical terms has remained elusive.

There is a further simplification of the $\mathcal{N} = (2,0)$ supersymmetric theory which has garnered more attention due to both its tractability and to its direct connection to famous 4d/2d correspondences such as Alday, Gaiotto, and Tachikawa (AGT) [**1**] and the mathematical work of Gronjowski and Nakajima. From one point of view, this simplification is obtained by placing the (non-minimal twist of the) theory in the so-called Ω-background [**43, 44**, ?]. Roughly this means one considers the theory on $\mathbf{R}^2_{\varepsilon_1} \times \mathbf{R}^2_{\varepsilon_2} \times \mathbf{C}$ and works equivariantly for the $S^1 \times S^1$-rotation acting on the $\mathbf{R}^2$-factors. The effect is that theory localizes to the fixed points of this action which is simply $0 \times 0 \times \mathbf{C}$. Here the ε_i are the same equivariant parameters that appear in Nekrasov's instanton partition function [**46**]. On the other hand, one can understand the simplification directly in the holomorphic twist as a certain deformation by a particular square-zero superconformal supercharge. This has a similar effect of localizing the theory to a complex codimension two subspace $\mathbf{C} \subset \mathbf{C}^3$.

From either point of view, the simplification results in a two-dimensional chiral conformal field theory. After this simplification the algebra of observables of the theory of type A_{N-1} is conjectured to be related to the W-algebra of type ADE. Costello has developed an approach to understanding this by investigating M-theory in the Ω-background [**13**]. In this situation, M-theory is an effective five-dimensional theory and fivebranes take the form of chiral defects. He finds that a certain planar limit of the chiral algebra on a stack of fivebranes in flat space is precisely the W_N algebra. Using ideas in string theory, the effective worldvolume theory of the fivebranes can be understood in terms of a particular free field realization of the W_N-algebra; it would be interesting to investigate a free field realization for the holomorphic twist of fivebranes as we consider here and in the next section.

2.1. Holomorphic twist of a single fivebrane. The abelian version of the $\mathcal{N} = (2,0)$ theory corresponds to the worldvolume theory on a single fivebrane. It admits a description in terms of the so-called six-dimensional tensor multiplet. In [**53**] we have formulated this theory within a degenerate version of the BV formalism and use this to give a mathematical description of the twists of the theory on a single fivebrane.

The holomorphic twist of the fivebrane theory is defined on any complex threefold X (which is not necessarily equipped with a Calabi–Yau structure). In the particular twist we will use we must assume, however, that X is equipped with a square-root of the canonical bundle $K_X^{1/2}$.

The theory is $\mathbf{Z} \times \mathbf{Z}/2$ graded where $\mathbf{Z}$ is the ghost number (or cohomological degree) and $\mathbf{Z}/2$ is parity. There are four fundamental fields of ghost number zero $(\alpha, \omega, \phi_1, \phi_2)$ which consist of even Dolbeault forms of type $(2,1)$ and $(3,0)$ on Z:

$$\alpha^{2,1} \in \Omega^{2,1}(X), \quad \alpha^{3,0} \in \Omega^{3,0}(X),$$

and a symplectic pair of odd $(0,1)$ forms twisted by the line bundle $K_X^{1/2}$:

$$\phi^{0,1} = (\phi_1^{0,1}, \phi_2^{0,1}) \in \Pi\Omega^{0,1}(X, K_X^{1/2}) \otimes \mathbf{C}^2.$$

The equations of motion read

$$\begin{aligned} \partial\alpha^{2,1} + \overline{\partial}\alpha^{3,0} &= 0 \\ \overline{\partial}\alpha^{2,1} = \overline{\partial}\phi_i^{0,1} &= 0. \end{aligned} \tag{2.1}$$

There are gauge symmetries for the fields $\alpha^{2,1}, \alpha^{3,0}$ determined by a ghost $\alpha^{2,0}$ which is a Dolbeault form of type $(2,0)$ which acts simply by

$$\begin{aligned} \alpha^{2,1} &\mapsto \alpha^{2,1} + \overline{\partial}\alpha^{2,0} \\ \alpha^{3,0} &\mapsto \alpha^{3,0} + \partial\alpha^{2,0}. \end{aligned} \tag{2.2}$$

There are also odd gauge symmetries for the odd field $\phi^{0,1}$ given by

$$\phi^{0,1} \mapsto \phi^{0,1} + \overline{\partial}\phi^0 \tag{2.3}$$

where ϕ^0 is an fermionic ghost living in $\Omega^0(X, K_X^{1/2}) \otimes \mathbf{C}^2$.

This theory notoriously does not admit a Lagrangian description. However, it can still be put in a degenerate form of the BV formalism where the space of fields above is equipped with an odd Poisson bivector [**53**]. In this setup the full space of BV fields of the theory is

$$\Omega^{\geq 2,\bullet}(X)[1] \oplus \Pi\Omega^{0,\bullet}(X, K_X^{1/2}) \otimes \mathbf{C}^2[1], \tag{2.4}$$

and the linear BRST differential is described by the following diagram

$$\begin{array}{ccc} \underline{-1} & & \underline{0} \\ & & \\ \Pi\Omega^{0,\bullet}(X, K_X^{1/2} \otimes \mathbf{C}^2) & & \\ & & \\ \Omega^{2,\bullet}(X) & \xrightarrow{\ \partial\ } & \Omega^{3,\bullet}(X), \end{array} \tag{2.5}$$

where the $\overline{\partial}$ operator acts implicitly on each summand. We continue to denote by ϕ and α fields which live in the first and second line above.

The free action functional takes the non-local form

$$\frac{1}{2}\int_X \alpha\overline{\partial}\partial^{-1}\alpha + \frac{1}{2}\int_X \phi\overline{\partial}\phi. \tag{2.6}$$

The non-locality present in the distributional operator ∂^{-1} traces back to the fact that within the BV formalism this theory is equipped with a degenerate shifted Poisson structure. On an affine patch in spacetime, however, we can formally make sense of this action in the following way. First, notice that all of the subtlety involves the field α (in fact, the field ϕ describes precisely the twist of the six-dimensional free hypermultiplet). Notice that α is required to be ∂-closed on-shell, this imposed cohomologically via the presence of the ∂-operator in the BV complex above. Thus, locally we can choose a primitive $\alpha = \partial\widetilde{\alpha}$ and recast the action as $\int \alpha\overline{\partial}\widetilde{\alpha}$. To turn this into a precise argument that the theory is well-posed we should take into account all possible choices of a primitive, how the primitives glue together, etc.. Thankfully, the BV formalism does this for us naturally; we comment more on this in the context of quantization below.

We remark that there is a slightly different presentation for the holomorphic twist. The twist of a theory depends on both the data of a square-zero supercharge and a *twisting* homomorphism which uses the R-symmetry to change how the fields transform. In the case of six-dimensional $\mathcal{N} = (2,0)$ supersymmetry the R-symmetry group is $Sp(4)$—the group of linear symplectic automorphisms of $\mathbf{C}^4$. The above description uses a rather simple twisting homomorphism which is

$$\phi\colon MU(3) \xrightarrow{\det^{1/2}} U(1) \xrightarrow{i\times\mathbb{1}} SU(2)\times SU(2) \hookrightarrow Sp(2). \tag{2.7}$$

Here the inclusion i is such that the holomorphic supercharge has weight $+1$. The last inclusion is the block diagonal corresponding to splitting $\mathbf{C}^4 = \mathbf{C}^2 \times \mathbf{C}^2$. The choice of a polarization of the second $\mathbf{C}^2$ determines a homomorphism $i'\colon U(1) \to SU(2)$. Replacing $i \times \mathbb{1}$ by $i \times i'$ in the above composition results in a different twisting homomorphism which actually descends to a homomorphism $\widetilde{\phi}\colon U(2) \to Sp(2)$. In other words, this variant twist exists on any complex threefold. Explicitly, this twist corresponds to replacing the bundle $K_X^{1/2} \otimes \mathbf{C}^2$ with the bundle $\mathcal{O}_X \oplus K_X$ in (2.5).

2.2. Coupling to supergravity and backreactions. Next we turn to coupling fivebranes in the holomorphic twist to our eleven-dimensional model described in the previous section. Recall that the near horizon geometry of backreacting fivebranes in supergravity on flat space is $AdS_7 \times S^4$. In the twisted context we see an analogous geometry which is natural from the point of view of complex geometry.

In the untwisted theory, the coupling of fivebranes located along $\mathbf{R}^6 \times 0 \subset \mathbf{R}^6 \times \mathbf{R}^5$ in eleven-dimensional supergravity is typically sourced by a term of the form

$$\int_{\mathbf{R}^6} \mathrm{d}^{-1}(\star \mathrm{d} C) + \cdots \tag{2.8}$$

where C is the three-form gauge field. Here, $\star$ is the Hodge star operator of eleven-dimensional spacetime. The $\cdots$ indicate additional terms involving degrees of freedom along the brane.

The naive starting place in the twisted theory is to start with our eleven-dimensional model on flat space $\mathbf{R} \times \mathbf{C}^5$ with some number of fivebranes along $0 \times \mathbf{C}^3$. More generally we can fix an arbitrary three-fold X and consider the eleven-dimensional model on $\mathbf{R} \times Z$ where

$$Z \overset{\text{def}}{=} \mathrm{Tot}\left(K_X^{1/2} \otimes \mathbf{C}^2 \to X\right). \tag{2.9}$$

The fivebranes wrap the zero section X of Z and are located at $0 \in \mathbf{R}$. The particular choice of rank two bundle is to ensure that the total space Z is a Calabi–Yau fivefold. At the level of the holomorphic twist we propose a first-order coupling of the following form

$$\int_X \partial^{-1}\mu = \int_{\mathbf{R}\times Z} \delta_X \wedge \partial^{-1}\mu. \tag{2.10}$$

Here we view μ as an element in $\Omega^\bullet(\mathbf{R}) \otimes \mathrm{PV}^{1,\bullet}(Z) \simeq \Omega^\bullet(\mathbf{R}) \otimes \Omega^{4,\bullet}(Z)$. Thus, formally we can view $\partial^{-1}\mu$ as an element of $\Omega^\bullet(\mathbf{R}) \otimes \Omega^{3,\bullet}(Z)$. When we integrate, only the component $\Omega^0(\mathbf{R}) \otimes \Omega^{3,3}(Z)$ will contribute.

The backreacted geometry will be given by a solution to the equations of motion upon deforming the eleven-dimensional action by the above coupling. Varying the potential $\partial_\Omega^{-1}\mu$, we obtain the following equation of motion involving the field γ:

$$\bar\partial\partial\gamma + \partial_\Omega\left(\frac{1}{1-\nu}\mu\right) \wedge \partial\gamma = \delta_{zero\ section}. \tag{2.11}$$

Notice that there is an extra derivative compared to the equation of motion arising from varying the field μ. This equation only depends on γ through its field strength $\partial\gamma$. There is another equation of motion involving $\partial\gamma \wedge \partial\gamma$.

In **[48]** we solve this equation using an explicit formula for the class representing the link of the fivebrane. Let

$$F_{M5} = \frac{1}{(2\pi i)^3} \frac{\overline{w}_1 \,\mathrm{d}\overline{w}_2 \wedge \mathrm{d}t - \overline{w}_2 \,\mathrm{d}\overline{w}_1 \wedge \mathrm{d}t + t\,\mathrm{d}\overline{w}_1 \wedge \mathrm{d}\overline{w}_2}{(\|w\|^2 + t^2)^{5/2}} \wedge \mathrm{d}w_1 \wedge \mathrm{d}w_2$$

where the w_i's are fiber coordinates of $Z \to X$ and t is the real coordinate. Then, $\partial\gamma = F_{M5}$ satisfies the equations of motion in the presence of a stack of N M5 branes sourced by the term $N\delta_{w_1=w_2=t=0}$:

$$\begin{aligned} \bar\partial(NF_{M5}) + \mathrm{d}_{\mathbb{R}}(NF_{M5}) &= N\delta_{w_1=w_2=t=0} \\ (NF_{M5}) \wedge (NF_{M5}) &= 0. \end{aligned}$$

Here, we set all components of the field μ equal to zero (as well as the fields ν, β). Notice that the first equation is equivalent to

$$\oint_{S^4} NF = N$$

for any four-sphere centered at 0 in the t, w_1, t_2 plane.

The complex form of the algebra of isometries for supergravity in the AdS_7 background is $\mathfrak{osp}(8|4)$. This agrees with the complex form of the 6d $\mathcal{N} = (2,0)$ superconformal algebra. The bosonic part of this algebra is isomorphic to $\mathfrak{so}(8) \oplus \mathfrak{sp}(4) \cong \mathfrak{so}(8) \oplus \mathfrak{so}(5)$. To provide evidence for the claim that this is the twisted analog of the AdS geometry we would like to know that the twist of this superconformal algebra is a symmetry of our deformed eleven-dimensional model in the presence of twisted fivebranes. At the level of the holomorphic twist, this super Lie algebra breaks down to $\mathfrak{osp}(6|1)$, so it is this Lie algebra that plays the role of the isometries in the twisted AdS background. In **[48]** we show that $\mathfrak{osp}(6|2)$ is a symmetry of the eleven-dimensional model in the presence of the source term $\partial\gamma = F_{M5}$ that we just introduced.

2.3. The higher "free boson". So far we have said very little about how to handle quantization. In our approach to QFT we study quantization at the level of observables, which we usually denote by $\mathcal{O}bs$. In a setting where perturbative QFT makes sense, the space of observables is defined over the ring $\mathbf{C}[[\hbar]]$. Semi-classically, the data on observables is an odd bracket which is commonly referred to as the BV bracket. This is the analog in deformation quantization of the Poisson bracket. Such a bracket endows the classical observables $\mathcal{O}bs/\hbar$, which is simply the commutative algebra of functions on the space of fields, with the structure of a so-called $\mathbb{P}_0$-algebra **[2, 10, 17, 54]**. In general, a $\mathbb{P}_n$-algebra is a commutative algebra equipped with a Poisson bracket of degree $1 - n$ so that $\mathbb{P}_1$-algebras coincide with ordinary Poisson algebras.

The quantum BV formalism sets up a way to study quantizations of $\mathbb{P}_0$ algebras and can be thought of as a shifted analog of ordinary deformation quantization. Given an ordinary Poisson (so $\mathbb{P}_1$) algebra $(A/\hbar, \{-,-\})$ a deformation quantization is a filtered associative algebra $(A, \star_\hbar)$ defined over $\mathbf{C}[[\hbar]]$ such that its associated graded modulo $\hbar$ returns the original Poisson structure. That is

$$\{a, b\} = \lim_{\hbar \to 0} (a \star_\hbar b - b \star_\hbar a). \tag{2.12}$$

It is expected that there is a higher analog of 'deformation quantization' where associative algebras are replaced by $\mathbb{E}_n$ algebras (the observables of an n-dimensional topological field theory) and Poisson algebras are replaced by $\mathbb{P}_n$ algebras.

So what should the 'quantization' of the $\mathbb{P}_0$ algebra of observables $\mathcal{O}bs/\hbar$ be? Let Q_0 denote the classical differential (this encodes the classical gauge symmetries and equations of motion of the theory) and let $\{-,-\}$ be the shifted Poisson bracket. Following the analogy, we should expect an $\mathbb{E}_0$ algebra $\mathcal{O}bs$, which is simply a (pointed) cochain complex; we denote the differential by Q. Modulo $\hbar$ the complex $\mathcal{O}bs$ should agree with the complex $\mathcal{O}bs/\hbar$, that is $Q = Q_0 + O(\hbar)$. Moreover, the relationship to the shifted Poisson bracket takes the following form. Let us assume we are in a situation where $Q = Q + \hbar\triangle$, where $\triangle$ is some operator also of cohomological degree $+1$. Then $\triangle$ is *not* a derivation for the product on $\mathcal{O}bs/\hbar$

but its failure is measured by the bracket

$$\triangle(ab) = \triangle(a)b \pm a\triangle(b) \pm \{a, b\}. \tag{2.13}$$

This is completely analogous to deformation quantization which says (to first order) that the failure for the quantum associative product to be commutative is precisely measured by the Poisson bracket. In this case, the failure for $\triangle$ to be a derivation (in which case Q would define a commutative dg algebra) is exactly $\{-,-\}$. For more details we refer to [**17**].

For ordinary gauge theories, like Yang–Mills theories, σ-models, and AKSZ theories, the $\mathbb{P}_0$-algebra of observables arises from a (-1)-shifted symplectic structure on the fields of the theory. In fact, this is true whenever one can present the moduli space of solutions as the (derived) critical locus of an action functional. For interacting theories finding a quantization is generally an intricate procedure where one must combine homological methods of the BV formalism together with renormalization, see [**11**]. However, for free theories there is a canonical BV quantization granted by the BV Laplacian: this is the order two differential operator acting on observables which on linear observables is simply the BV bracket

$$\Delta(ab) \stackrel{\text{def}}{=} \{a, b\}. \tag{2.14}$$

In field theoretic contexts the Poisson bracket generally involves a pairing with a distribution. So there is some care that must be taken in defining Δ rigorously. We refer to [**17**] for details about this.

We have already commented that the six-dimensional theory does not fit into the ordinary BV formalism stemming from the fact that the theory does not have a Lagrangian description. Nevertheless, the classical observables $\mathcal{O}\mathrm{bs}/\hbar$ are equipped with a $\mathbb{P}_0$ structure which is determined by a bivector

$$\Pi \stackrel{\text{def}}{=} \in \overline{\mathcal{E}}(X) \mathbin{\widehat{\otimes}} \overline{\mathcal{E}}(X)[1]. \tag{2.15}$$

In other words we can still make sense of the BV quantization via the formula (2.14). The hat on the tensor product means that we are considering a completion of the usual algebraic tensor product and the bars denote distributional sections. The [1] indicates that the bivector carries cohomological degree +1 (but note that the bivector is even with respect to parity). This bivector is degenerate in the sense that if we restrict Π to a fiber over $x \in X$ that it is degenerate. This bivector (or rather a slight variant of it) is an essential ingredient in Costello and Li's mathematical approach to the string field theory of the B-model in terms of Kodaira–Spencer gravity [**6, 18**]. In fact, when X is Calabi–Yau we have shown that our classical factorization algebra $\mathcal{O}\mathrm{bs}$ of the holomorphic twist of the abelian $\mathcal{N} = (2, 0)$ theory agrees with a certain free-limit of Kodaira–Spencer theory on X as a degenerate BV theory [**53**]. We expect this to be a shadow of the relationship between the six-dimensional superconformal theory and type IIB string theory on an ADE singularity.

This bivector can be described explicitly. First, notice that there is an obvious direct sum decomposition of (2.5): denote the top line by $\mathcal{E}_{hyper}$ and the bottom line by $\mathcal{E}_{tensor}$. The shifted Poisson tensor (2.15) then decomposes as

$$\Pi = \overline{\Pi}_{hyper} + \overline{\Pi}_{tensor} \in \overline{\mathcal{E}}_{hyper}(X) \mathbin{\widehat{\otimes}} \overline{\mathcal{E}}_{hyper}(X)[1] \oplus \overline{\mathcal{E}}_{tensor}(X) \mathbin{\widehat{\otimes}} \overline{\mathcal{E}}_{tensor}(X)[1]. \tag{2.16}$$

The bivector Π_{hyper} is essentially the δ-function along the diagonal in $X \times X$. If $\pi_{\mathbf{C}^2} = e_1 \wedge e_2$ denotes the standard Poisson bivector on $\mathbf{C}^2 = \text{span}\{e_1, e_2\}$ then

$$\Pi_{hyper} = \pi_{\mathbf{C}^2} \otimes \delta_{diag} \tag{2.17}$$

where δ_{diag} is the δ-function along the diagonal $X \hookrightarrow X \times X$. The bivector Π_{tensor} is a bit different as it involves a derivative of the δ-function

$$\Pi_{tensor} = (\partial \otimes \mathbb{1})\delta_{diag}. \tag{2.18}$$

This derivative is the entire source of the non-locality in our action functional describing the free theory (2.6).

With this explicit description in hand, the putative quantization of the holomorphic twist of the six-dimensional theory is of the form

$$\mathcal{O}\mathrm{bs}_{\mathfrak{gl}(1)}(X) = (\mathcal{O}(\mathcal{E})[[\hbar]], Q_0 + \hbar\triangle_{\Pi}), \tag{2.19}$$

where $(\mathcal{E}, Q_0)$ is the BV complex of fields in (2.5) and $\triangle_{\Pi}$ is the contraction with the shifted Poisson bivector (2.15). For the operator Δ_{Π} to be well-defined one should take observables which are non-distributional (by ellipticity of Q_0 this space of observables is quasi-isomorphic to the full space of observables).

So far the discussion of quantization was about global observables—observables supported on the entire manifold X. Taking into account the fact that the space of fields form a sheaf on spacetime gives rise to a rich algebraic structure on *local* observables—observables supported on arbitrary open sets in X. The precise mathematical notion of this structure is a factorization algebra [**16, 17**]. In fact, the approach of Costello and Gwilliam associates to any quantum field theory a factorization algebra on spacetime. This involves showing that BV quantization is compatible with the factorization algebra structure.

While we won't review the general theory of factorization algebras here, we point out a few examples all stemming from the structure obtained by looking at operations parameterized by very arbitrary small open sets. This is the connection between the general theory of observables and the theory of local operators. The local operators of any topological field theory gives rise to an algebra over the little n-disks operad. Local operators in a conformal field theory produce a vertex algebra. This can be seen directly using the theory of factorization algebras, see [**8, 16**]. For higher dimensional holomorphic field theories [**55**] the local operators have, in analogy with the topological case, the structure of a higher dimensional vertex algebra (or an algebra over an operad of holomorphic disks). It is absolutely necessary to work in the derived setting for such a theory to be nontrivial. The theory of higher dimensional vertex algebras has not yet been fully developed [**9, 24, 29, 33, 55**]. We leave the analysis of the higher, three-variable, vertex algebra structure present on the local operators of the fivebrane worldvolume theory to future work.

One advantage of the theory of factorization algebras (over the theory of simply local, point-like, operators) is that it makes sense to evaluate them on more general open sets than small neighborhoods of points. Relatedly, like sheaves, one can pushforward factorization algebras along morphisms—in the case that the morphism is proper this is akin to 'compactifying' the theory along the fibers. Let $\mathcal{O}\mathrm{bs}$ be the (quantum) factorization algebra of the holomorphic twist of the $\mathcal{N} = (2,0)$ theory. Suppose that Y is a smooth and projective complex surface and consider the theory on the three-fold

$$C \times Y \tag{2.20}$$

where C is a Riemann surface. For this geometry it is easiest to imagine the variant twist discussed at the end of §2.1 which exists on any three-fold. Compactification along $C \times Y \to C$ corresponds to pushing forward the factorization algebra $\mathcal{O}\mathrm{bs}$ to C. In more concrete terms, we are considering the OPE operators in the six-dimensional theory which wrap the surface Y and are point-like on C. When $C = \mathbf{C}$ this defines a (graded) vertex algebra $\mathbb{V}(Y)$. This vertex algebra depends on the cohomology of Y.

PROPOSITION 2.1 ([**53**]). *Let $\mathbb{V}_{boson}$ be the vertex algebra associated to the free chiral boson. There is an embedding of vertex algebras*

$$\mathbb{V}_{boson} \hookrightarrow \mathbb{V}(Y), \tag{2.21}$$

corresponding to the Kähler class in $H^{1,1}(Y)$.

There are (at least) two other senses in which the six-dimensional factorization algebra $\mathcal{O}\mathrm{bs}$ is a higher-dimensional enhancement of the free boson vertex algebra. Both are further localizations of the theory. One is more closely related to the non-minimal twist of the theory. The other follows an idea in [**4**] that there is a certain superconformal deformation of the supersymmetric theory which localizes the theory to a two-dimensional subspace of six-dimensional spacetime. It turns out that this deformation can be gleaned from the holomorphic twist, and we turn to that now.

2.4. Holomorphic superconformal symmetry and localization. The complex form of the six-dimensional superconformal algebra is $\mathfrak{osp}(8|4)$. The even part of this algebra contains the complexified conformal algebra $\mathfrak{so}(8, \mathbf{C})$ and the complexified R-symmetry algebra $\mathfrak{sp}(4, \mathbf{C})$ which acts by linear symplectomorphisms on $\mathbf{C}^4$. The six-dimensional $\mathcal{N} = (2, 0)$ supersymmetry algebra is a subalgebra of the superconformal algebra, hence we can view the holomorphic twisting supercharge Q^{hol} as acting by the commutator $[Q^{hol}, -]$ on $\mathfrak{osp}(8|4)$. By construction, the cohomology super Lie algebra with respect to this supercharge

$$H^\bullet(\mathfrak{osp}(8|4)) \simeq \mathfrak{osp}(6|2) \tag{2.22}$$

acts on the Q^{hol}-twist of any six-dimensional superconformal theory. The even part of this algebra is $\mathfrak{sl}(4, \mathbf{C}) \times \mathfrak{sl}(2, \mathbf{C})$. In the holomorphic twist it is accurate to think about $\mathfrak{sl}(4, \mathbf{C})$ as acting geometrically by constant, linear, and quadratic vector fields on $\mathbf{C}^3$; the $\mathfrak{sl}(2, \mathbf{C})$ factor acts non-geometrically.

For now we will work with the holomorphic theory defined on $\mathbf{C}^3$. There is an element $S_{\varepsilon_1,\varepsilon_2} \in \mathfrak{osp}(6|2)$ depending on two complex parameters $\varepsilon_1, \varepsilon_2$ which has the properties that it is square-zero and in the super Lie algebra $\mathfrak{osp}(6|2)$ one has

$$[S_{\varepsilon_1,\varepsilon_2}, Q^{nm}] = \varepsilon_1 z_1 \frac{\partial}{\partial z_1} + \varepsilon_2 z_2 \frac{\partial}{\partial z_2} + (\varepsilon_1 + \varepsilon_2) H \tag{2.23}$$

where $H \in \mathfrak{sl}(2, \mathbf{C})$. Here, Q^{nm} is the non-minimal supercharge which we will comment on more in the next section. What is important to notice from this formula is that once we further deform the holomorphic twist by $S_{\varepsilon_1,\varepsilon_2}$ the theory will localize in spacetime to the fixed points of the vector field $\varepsilon_1 z_1 \frac{\partial}{\partial z_1} + \varepsilon_2 z_2 \frac{\partial}{\partial z_2}$

(there is also a non-geometric localization occurring in the residual R-symmetry space corresponding to the last term in the expression above).

From this we conclude that for generic $\varepsilon_1, \varepsilon_2$ deforming a factorization algebra associated to the holomorphic twist of the $\mathcal{N} = (2,0)$ theory by $S_{\varepsilon_1,\varepsilon_2}$ results in a holomorphic factorization algebra that is completely localized to the plane $z_1 = z_2 = 0$ in $\mathbf{C}^3$. This can be viewed as a factorization algebra analog of the localization studied in [**4**]. The following will appear in upcoming work with Ingmar Saberi.

THEOREM 2.2. *The vertex algebra obtained by deforming the factorization algebra $\mathcal{O}bs_{\mathfrak{gl}(1)}$ associated to the free $\mathcal{N} = (2,0)$ theory on $\mathbf{C}^3$ by $S_{\varepsilon_1,\varepsilon_2}$ is equivalent to the free boson vertex algebra $\mathbb{V}_{boson}$.*

A future direction of research is to understand the general non-abelian six-dimensional $\mathcal{N} = (2,0)$ theory at the level of the holomorphic twist. In the next section we summarize partial work in this direction whereby we infer features of the (fully refined) character of local operators in the holomorphic twist by using methods of twisted holography. The general expectation is that the vertex algebra obtained by deforming by a generic $S_{\varepsilon_1,\varepsilon_2}$ above is equivalent to the principal W-algebra of type ADE. For type A_{N-1} this is a vertex algebra generated by fields of conformal weight $2, 3, \ldots, N$.

3. Holography and the superconformal index

This section is a summary of work with Raghavendran in [**49**] and on further ongoing work pertaining to this topic. The supersymmetric index counts local operators in the minimal (or holomorphic) twist [**51**]. From the factorization algebra point of view, local operators (operators with zero-dimensional support) are obtained by studying the value of the factorization algebra on arbitrarily small open sets. In the complex one-dimensional holomorphic setting it is this collection of observables that carries the structure of a vertex algebra. In higher dimensions we pointed out that this should carry the structure of a higher dimensional (derived) vertex algebra. We propose to access the holomorphic twist and its local operators using *twisted holography.*

3.1. Our approach via twisted holography. Introduced by Costello and Li in [**19**], the twisted holography proposal posits an avatar of the AdS/CFT correspondence that holds at the level of supersymmetric twists. Fundamentally, it conjectures a duality between the algebra of observables of a twisted gravitational theory and the algebra of observables of a twisted gauge theory describing the dynamics of a stack of branes coupled to the gravitational theory. There is an exciting body of work being developed around this program including tests of this proposal from both the gravitational and gauge theory sides.

It is expected that the six-dimensional $\mathcal{N} = (2,0)$ superconformal field theory of type A_{N-1} describes the low energy dynamics of a stack of fivebranes in M-theory on flat space. In the twisted situation, our ansatz for the low energy limit of M-theory is the minimal twist eleven-dimensional supergravity which we recalled in §1. In the presence of fivebranes we encounter a twisted version of the AdS_7/CFT_6 correspondence which will relate the large N limit of the minimally twisted six-dimensional $\mathcal{N} = (2,0)$ theory with minimally twisted eleven-dimensional supergravity on $AdS_7 \times S^4$.

The twisted holography proposal posits that holography can be understood as a concrete algebaic operation at the level of observables. We mentioned that the precise mathematical structure modeling observables of a quantum field theory is that of a factorization algebra. The predicted type of duality between the factorization algebras associated to a gravitational theory and to the worldvolume theory of a number of branes is a general version of *Koszul duality*. Ordinary Koszul duality for associative algebras (so quantum mechanical systems) associates to an (augmented) algebra A a dual algebra $A^!$ whose appropriate derived category of modules is the same as that of A. Following the work of [**19, 20**] (see also the review in [**47**]) there is a simple physical interpretation of Koszul duality. If A is the algebra of operators of some bulk quantum field theory (perturbatively we can even consider a theory of gravity) then $A^!$ is the algebra of operators on the universal topological line defect. Universal here means that algebra of operators on any other line defect which couples to the bulk system admits a unique map of algebras from $A^!$.

The general theory of Koszul duality for factorization algebras has not been developed. For holomorphic complex one-dimensional factorization algebras there is exciting work of Gui and Li [**32**] in this direction written in the language of chiral algebras. This sort of duality would allow one to make sense of universality statements as above for higher dimensional, possibly non-topological, defects in an arbitrary bulk quantum field theory. Nevertheless, we can make the following ansatz for the Koszul dual $\mathcal{F}^!$, which we refer to as the !-dual, of a factorization algebra $\mathcal{F}$ of observables of some bulk quantum field theory. It is the universal factorization algebra, along a specified defect, which couples to the bulk quantum field theory. While this heuristic definition sounds natural and is manifestly equipped with a universal property, it does not lend itself to an explicit description. However, for the particular kinds of factorization algebras of bulk quantum field theories an explicit description is furnished by a local version of Noether's theorem, see [**17**].

After taking into account the backreaction (see §2.2), the universal property of !-dual produces a map of factorization algebras

$$(\mathcal{O}\mathrm{bs}_{grav}|_X)^! \to \mathcal{O}\mathrm{bs}_{N-branes} \tag{3.1}$$

where the branes are supported on the submanifold X of the bulk gravitational theory. The proposal of twisted holography is that this becomes an equivalence in the large N limit.

One can relate this statement to traditional formulations of the AdS/CFT correspondence, where local operators of the CFT correspond to certain states of the gravitational theory on AdS. One can argue that $(\mathcal{O}\mathrm{bs}_{grav}|_Y)^!$ is a factorization algebra enhancement of the space of (twisted) multi-particle states of the gravitational theory on the backreacted geometry. Explicitly, this geometry refers to the manifold obtained by removing the brane together with the effect of the backreaction which deforms this geometry.

The above expectation can be tested in instances where both sides of the duality admit explicit descriptions. This has been carried out in many examples including:

- A stack of $D3$ branes in twisted Ω-deformed type IIB supergravity on flat space. The theory on the stack of $D3$ branes is dual to the closed string B-model on the deformed conifold [**15**]. This can be understood as a twisted Ω-deformed version of the physical AdS/CFT duality between 4d $\mathcal{N}=4$ super Yang-Mills and type IIB string theory on $AdS_5 \times S^5$. Here, the duality is formulated in terms of vertex algebras which are avatars of holomorphic factorization algebras on a Riemann surface.

- A program to understand celestial chiral algebras from the point of view of twisted holography has been undertaken in the works [**21**–**23**, **58**]. Here, the celestial chiral algebras are understood as complex one-dimensional holomorphic defects in holomorphic field theories on complex three-dimensional twistor space.
- A system of twisted $D1 - D5$ branes in a T^4-compactification of a twist of type IIB string theory. Upon compactifying along T^4 this can be understood as a variant of the above example where the deformed conifold is replaced by a certain superspace [**20**]. The duality is understood as a twisted version of the duality between type IIB supergravity on $AdS_3 \times S^3 \times T^4$ and the symmetric orbifold CFT $Sym^N(T^4)$.
- Membranes and fivebranes in twisted Ω-deformed M-theory on Taub-NUT space [**13**, **14**]. In the particular Ω-background, membranes are localized to a topological quantum mechanical system where the duality can be phrased in terms of associative algebras and ordinary Koszul duality. Moreover, the Ω-background localizes fivebranes to a complex plane and the observables of the localized theory are an affine W_N vertex algebra. The holomorphic factorization algebras we seek in the present paper may be thought of as enhancements of these vertex algebras to three-complex dimensions.

For branes in M-theory the Ω-background allows one to characterize subsectors of the relevant worldvolume theories in an explicit way [**13**, **14**]. For example, in [**13**], Costello uses string theory arguments to utilize a free field realization of the W_N-algebra to describe the theory on a stack of N fivebranes in the Ω-background. However, for the minimal twist, we are not so lucky. The dimensional reduction of the nonminimal twist will necessarily involve subtle, nonperturbative effects making an intrinsic characterization along the lines of the above more difficult.

Consequentially, rather than prove a version of the twisted holographic principle, our objective in [**49**] is to use the twisted holographic principal to surmise an explicit description of the factorization algebra associated to the worldvolume theory on a stack of minimally twisted fivebranes. Such an analysis reveals that twists of worldvolume theories of branes enjoy certain infinite dimensional symmetries, enhancing finite dimensional symmetries present in the untwisted theories.

Let's now turn to our situation. Let X be a complex threefold, which is where the twisted fivebrane theory lives. We consider twisted fivebranes wrapping a point in $\mathbf{R}$ and the zero section in the Calabi-Yau fivefold

$$Z = \mathrm{Tot}(K_X^{1/2} \oplus K_X^{1/2}). \tag{3.2}$$

The bulk factorization algebra on $\mathbf{R} \times Z$ associated to twisted eleven-dimensional supergravity is denoted $\mathcal{O}\mathrm{bs}_{grav}$. When we view this as a factorization algebra only along the threefold X we will denote it by $\mathcal{O}\mathrm{bs}_{grav}|_X$.

3.2. The large N limit. The theory on an infinite number of fivebranes in flat space is expected to be holographically dual to supergravity on $AdS_7 \times S^4$. It turns out that the holomorphic twist is enough to see the complete large N superconformal index which we compute using our explicit characterization of twisted eleven-dimensional supergravity.

In §2.2 we proposed a twisted version of this AdS geometry which corresponds to starting with fivebranes supported on $X = \mathbf{C}^3$. We recall how the twisted superconformal algebra $\mathfrak{osp}(6|2)$ is a symmetry of this geometry [48]. We only focus on the following bosonic part of this algebra

$$\mathfrak{sl}(3) \times \mathfrak{sl}(2) \times \mathfrak{gl}(1) \subset \mathfrak{osp}(6|2). \tag{3.3}$$

Choose coordinates on the eleven-manifold

$$\mathbf{R} \times Z \cong \mathbf{R}_t \times \mathbf{C}^2_w \times \mathbf{C}^3_z$$

with $z = (z_i), i = 1, 2, 3$ and $w = (w_a), a = 1, 2$. The stack of fivebranes wrap

$$w_1 = w_2 = t = 0. \tag{3.4}$$

Recall from §1.3 that part of the ghost system for our eleven-dimensional theory consists of divergence-free vector fields on $\mathbf{C}^5$ which are locally constant along $\mathbf{R}$. The subalgebra (3.3) of the twisted superconformal algebra $\mathfrak{osp}(6|2)$ embeds as ghosts in our eleven-dimensional model as follows.

- The subalgebra $\mathfrak{sl}(3) \subset \mathfrak{sl}(4)$ embeds as vector fields rotating the plane $\mathbf{C}^3_z$

$$\sum_{ij} A_{ij} z_i \frac{\partial}{\partial z_j} \in \mathrm{PV}^{1,0}(\mathbf{C}^5) \otimes \Omega^0(\mathbf{R}), \quad (A_{ij}) \in \mathfrak{sl}(3). \tag{3.5}$$

  By definition, these vector fields are automatically divergence-free. Notice that this vector field is divergence-free and restricts to the Euler vector field along $t = w_a = 0$.
- The subalgebra $\mathfrak{sl}(2)$ (R-symmetry) is mapped to the triple

$$w_1 \frac{\partial}{\partial w_2}, \quad w_2 \frac{\partial}{\partial w_1}, \quad \frac{1}{2}\left(w_1 \frac{\partial}{\partial w_1} - w_2 \frac{\partial}{\partial w_2}\right) \in \mathrm{PV}^{1,0}(\mathbf{C}^5) \otimes \Omega^0(\mathbf{R}). \tag{3.6}$$

- Scaling on the plane $\mathbf{C}^3$ embeds as the vector field

$$\Delta = \sum_{i=1}^3 z_i \frac{\partial}{\partial z_i} - \frac{3}{2} \sum_{a=1}^2 w_a \frac{\partial}{\partial w_a} \in \mathrm{PV}^{1,0}(\mathbf{C}^5) \otimes \Omega^0(\mathbf{R}). \tag{3.7}$$

This describes an embedding of the algebra

$$\mathfrak{sl}(3) \times \mathfrak{sl}(2) \times \mathfrak{gl}(1) \subset \mathfrak{sl}(4) \times \mathfrak{sl}(2) \tag{3.8}$$

into the ghosts of our model. The dimension of a Cartan subalgebra of $\mathfrak{sl}(3) \times \mathfrak{sl}(2) \times \mathfrak{gl}(1)$ is four and accordingly, the equivariant character we study has four fugacities. We choose these explicitly as follows:

- y_1, y_2, y_3 satisfying $y_1 y_2 y_3 = 1$ denote a generating set for the Cartan of $\mathfrak{sl}(3)$ which is spanned by divergence-free linear combinations of the vector fields

$$h_i = z_i \frac{\partial}{\partial z_i} \tag{3.9}$$

- q, y denotes a generating set for the Cartan of $\mathfrak{sl}(2) \times \mathfrak{gl}(1)$ corresponding to the vector fields

$$\Delta, \quad h = \frac{1}{2}\left(w_1 \frac{\partial}{\partial w_1} - w_2 \frac{\partial}{\partial w_2}\right) + \frac{1}{2}\Delta. \tag{3.10}$$

The space of single-particle twisted supergravity states is defined in terms of compactly supported distributions of the form

$$\alpha \mapsto \partial_{z_1}^{n_1}\partial_{z_2}^{n_2}\partial_{z_3}^{n_3}\alpha|_{z=0} \tag{3.11}$$

where α denotes a field. The full multiparticle states $\mathcal{H}_{sugra}$ is a free algebra on the single particle states and form a representation for $\mathfrak{osp}(6|2)$ [**48**]. In terms of Koszul duality, this is the space of local operators at $0 \in \mathbf{C}^3$ of the !-dual factorization algebra $(\mathcal{O}\mathrm{bs}_{grav}|_X)^!$. From the point of view of the bulk theory, this space is obtained through the KK reduction of twisted supergravity along the twisted version of AdS. So, following the twisted holographic principle (3.1) we interpret this as the superconformal index of the theory on an infinite number of fivebranes.

We enumerate supergravity states via the super trace of the operator $q^{\Delta}y_1^{h_1}y_2^{h_2}y_3^{h_3}y^h$ acting on $\mathcal{H}_{sugra}$:

$$f_{sugra}(y_i, y, q) = \mathrm{Tr}_{\mathcal{H}_{sugra}}(-1)^F q^{\Delta}y_1^{h_1}y_2^{h_2}y_3^{h_3}y^h. \tag{3.12}$$

PROPOSITION 3.1. *The single particle index of the space of twisted supergravity states $\mathcal{H}_{sugra}$ is given by the following expression*

$$f_{sugra}(y_i, y, q) = \frac{q^4(y_1+y_2+y_3) - q^2(y_1^{-1}+y_2^{-1}+y_3^{-1}) + (1-q^3)(yq+y^{-1}q^2)}{(1-y_1q)(1-y_2q)(1-y_3q)(1-yq)(1-y^{-1}q^2)}, \tag{3.13}$$

The full (multiparticle) index is defined to be the plethystic exponential

$$\mathrm{PExp}\left[f_{sugra}(y_i, y, q)\right]. \tag{3.14}$$

We note that this matches exactly with the result in [**38**, Eq. (3.23)] which computes the large N limit of the superconformal index in terms of the partition function of five-dimensional supersymmetric Yang–Mills theory on the five-sphere. Our formula also matches with [**7**, Eq. (3.24)].

The Schur limit of this index corresponds, with respect to our choice of Cartan generators, specializing this index to

$$\text{Schur limit}: \qquad y = 1, y_3 = 1. \tag{3.15}$$

Specializing in this way yields the famous MacMahon function

$$\mathrm{PExp}\left[f_{sugra}(y_1, y_1^{-1}, y_3 = 1, y = 1, q)\right] = \prod_{n\geq 1}\frac{1}{(1-q^n)^n}. \tag{3.16}$$

3.3. The fivebrane decomposition. To gain insight into the superconformal index at finite N we utilize a filtration of the factorization algebra of the !-dual of the gravitational observables

$$F^{\bullet}(\mathcal{O}\mathrm{bs}_{grav}|_X)^!. \tag{3.17}$$

This filtration is induced from the weight decomposition of the fields in the bulk gravitational theory defined by dilations in the directions transverse to the brane; so, along the fibers of the bundle Z over X.

We have seen in §1.3 that the exceptional super Lie algebra $E(5|10)$ plays an important role in the twist of supergravity. The algebra of local operators of twisted supergravity is the Lie algebra cohomology of a particular central extension of this exceptional super Lie algebra. In the case $X = \mathbf{C}^3$, the filtration above is closely related to a decomposition of the exceptional simple super Lie algebra $E(5|10)$

studied in **[37]**. Denote by $\{z_i\}$ the local coordinates along the three-fold $X = \mathbf{C}^3$ that the fivebrane wraps and $\{w_a\}$ for the transverse holomorphic coordinates to the zero section in the five-fold $\mathrm{Tot}(K_{\mathbf{C}^3}^{1/2} \otimes \mathbf{C}^2)$. Assign the following weights to the super Lie algebra $E(5|10)$.

- the coordinate z_i has weight zero.
- the coordinate w_a has weight $+1$.
- the parity of an element carries an additional weight of -1. Thus, for example, the odd element $[\mathrm{d}\, w_1 \,\mathrm{d}\, z_1] \in \Omega^{2,cl}(\widehat{D}^5)$ carries weight $+1-1=0$. (If we think about the odd part as the space of closed two-forms then equivalently this grading translates to the one-form symbol $\mathrm{d}(-)$ as carrying weight $-1/2$.)

The weight grading is concentrated in degrees ≥ -1. In particular, there is a decomposition of super vector spaces

$$E(5|10) = \widetilde{V}_{-1} \times \prod_{n \geq 0} V_n \tag{3.18}$$

with $\widetilde{V}_{-1}$ being the weight -1 subspace and V_n being the weight n subspace for $n \geq 0$. To extend this to a decomposition of the central extension of $E(5|10)$ considered in 1.3 we are forced to declare the central term has weight -1. This gives a grading of L_∞-algebras

$$\widehat{E(5|10)} = \prod_{n \geq -1} V_n \tag{3.19}$$

We refer to this as the *fivebrane decomposition* of $\widehat{E(5|10)}$. Here V_{-1} is a $\mathbf{C}$-extension of $\widetilde{V}_{-1}$ defined in the decomposition (3.18). Notice that for $n \geq 0$ the V_n's are the same as in the non centrally extended case.

We sketch how this decomposition translates at the level of factorization algebras and hence local operators. It turns out that, classically, the factorization algebra of twisted eleven-dimensional supergravity restricted to the brane is of the form

$$\mathcal{O}\mathrm{bs}_{grav}|_X = \mathrm{C}^\bullet(\mathcal{G}) \tag{3.20}$$

where $\mathcal{G}$ is a sheaf of L_∞ algebras on the worldvolume X of the fivebrane. In fact, $\mathcal{G}$ is the sheaf of sections of an explicit pro-vector bundle and carries a grading

$$\mathcal{G} = \prod_{k \geq -1} \mathcal{G}^{(k)}.$$

The !-dual of the factorization algebra is explicit to characterize classically. We have

$$(\mathcal{O}\mathrm{bs}_{grav}|_X)^! = \mathrm{C}_\bullet(\mathcal{G}_c). \tag{3.21}$$

To each open set $U \subset X$, this factorization assigns the cochain complex computing the Lie algebra homology of compactly supported sections $\mathcal{G}_c(U)$. The grading on $\mathcal{G}$ induces the filtration mentioned above on this factorization algebra

$$\mathrm{C}_\bullet(\mathcal{G}_c) = F^0\, \mathrm{C}_\bullet(\mathcal{G}_c) \supset F^1\, \mathrm{C}_\bullet(\mathcal{G}_c) \supset \cdots \tag{3.22}$$

At the level of associated gradeds

$$\mathrm{Gr}\, \mathrm{C}_\bullet(\mathcal{G}_c) = \mathrm{colim}_{N \geq 1}\, \mathrm{C}_\bullet(\mathcal{G}_{N,c}) \tag{3.23}$$

where $\mathcal{G}_N \stackrel{\text{def}}{=} \mathcal{G}/\mathcal{G}^{(\geq N-1)}$.

REMARK 3.2. As a vector space, the local operators of the factorization algebra $\mathrm{C}_\bullet(\mathcal{G}_c)$ is canonically isomorphic to the space of twisted supergravity states $\mathcal{H}_{sugra}$ that we introduced in the previous section. Manifestly, $\mathcal{H}_{sugra}$ breaks up into irreducible representations of the twisted superconformal algebra $\mathfrak{osp}(6|2)$. Representations for the full superconformal algebra have been studied extensively in the superconformal bootstrap program [**3**]. Through twisting, there should be a direct relationship between these twisted representations to the ones found via mass spectroscopy of untwisted supergravity [**30, 31**]. One feature of the representations appearing in our $\mathcal{H}_{sugra}$ is that they admit the structure of a representation for the infinite-dimensional exceptional Lie algebra $E(3|6)$ which will appear in the next section.

For $N \geq 1$, let $\mathcal{O}\mathrm{bs}_N$ be the factorization algebra of observables of the holomorphic twist of the worldvolume theory on a stack of N fivebranes wrapping a threefold Z. Our expectation is that at the level of cohomology the factorization algebra $\mathrm{C}\bullet(\mathcal{G}_{N,c})$ is a sub factorization algebra of $\mathcal{O}\mathrm{bs}_N$. In particular, for each open set $U \subset X$ we predict that

$$H_\bullet(\mathcal{G}_{N,c}(U)) \subset H^\bullet\left(\mathcal{O}\mathrm{bs}_N(U)\right) \tag{3.24}$$

where $H_\bullet$ is the Lie algebra homology. In particular, at the level of local operators we posit that the filtration of the states of twisted supergravity on twisted AdS space is related to the rank filtration on the local operators associated to the six-dimensional theory of type A_{N-1}.

REMARK 3.3. In the original version of the paper [**49**] we conjectured that the $H_\bullet(\mathcal{G}_{N,c}(U))$ is isomorphic to $\mathcal{O}\mathrm{bs}_N(U)$. We thank Jihwan Oh and Davide Gaiotto for discussions which led us to the expectation that this cannot be the case. The factorization algebra built from $\mathcal{G}_{N,c}$ only sees a perturbative contributions to local operators from the point of view of the gravitational theory. In ordinary (non-twisted) AdS_7 there are non-perturbative effects coming from black holes, for instance. These blackholes preserve the minimal amount of supersymmetry and have been recently studied in [**34**] So, while the Schur limit of the superconformal index is insensitive to such effects, the superconformal index and also the holomorphic twist of a theory on a stack of $N > 1$ twisted fivebranes should involve such phenomena.

3.4. A single fivebrane. The first term in the limit (3.23) is

$$\mathrm{C}_\bullet(\mathcal{G}_{1,c})$$

where $\mathcal{G}_1$ is the abelian local Lie algebra $\mathcal{G}_1 \cong \mathcal{G}^{(-1)}$—this is just the weight (-1) piece of the decomposition of $\mathcal{G}$. In §2 we recalled work with Saberi that explicitly characterizes the holomorphic twist of the worldvolume theory on a single fivebrane, that is, the six-dimensional superconformal theory associated to the abelian Lie algebra $\mathfrak{gl}(1)$. Here, we will denote the corresponding factorization algebra of classical observables on the three-fold Z by $\mathcal{O}\mathrm{bs}_1$.

PROPOSITION 3.4. *There is a quasi-isomorphism of factorization algebras commutative dg algebras on the threefold X*

$$\widetilde{\mathrm{C}}_\bullet(\mathcal{G}_{1,c}) \xrightarrow{\simeq} \mathcal{O}bs_1.$$

The left hand side is a modification of the factorization algebra $\mathrm{C}_\bullet(\mathcal{G}_{1,c})$ *after taking into account the backreaction. For more details see* [**49**].

From the explicit description of the $\mathfrak{gl}(1)$ theory given in §2 the character of local operators of $\widetilde{\mathrm{C}}_\bullet(\mathcal{G}_{1,c})$ is immediately computable. In terms of the parameters y_1, y_2, y_3, y, q the single particle character of local operators of the $\mathfrak{gl}(1)$ theory reads

$$f_1(y_1, y_2, y_3, y, q) = \frac{qy + q^2y^{-1} - q^2(y_1^{-1} + y_2^{-1} + y_3^{-1}) + q^3}{(1 - y_1q)(1 - y_2q)(1 - y_3q)}, \tag{3.25}$$

where we recall that $y_1y_2y_3 = 1$. The superconformal index is obtained by applying the plethystic exponential $\chi_1 = \mathrm{PExp}[f_1]$. The expression matches exactly with the index of the abelian six-dimensional superconformal theory. For example, compare with [**38**, Eq. (3.1)] or [**7**, Eq. (3.35)].

3.5. A conjectural description of operators on a stack of two fivebranes. We now turn to the factorization algebra $\mathrm{C}_\bullet(\mathcal{G}_{2,c})$. Since $\mathcal{G}$ is concentrated in weights ≥ -1 we see that $\widetilde{\mathcal{G}}_2$ is of the form

$$\mathcal{G}_2 = \mathcal{G}^{(0)} \ltimes \mathcal{G}^{(-1)}$$

We focus mostly on the factorization algebra $\mathrm{C}_\bullet(\mathcal{G}_c^{(0)})$.

What is special about this case is that $\mathcal{G}^{(0)}$ has the structure of a non-trivial local L_∞ algebra inherited from the L_∞ structure on $\mathcal{G}$. As a local Lie algebra it is equivalent to one of the form $\Omega^{0,\bullet}(Z, \mathcal{V}^{(0)})$. We will prove that it is equivalent to a local Lie algebra version of the locally compact infinite-dimensional exceptional super Lie algebra $E(3|6)$ classified by Kac [**36**]. By construction the twisted superconformal Lie algebra $\mathfrak{osp}(6|2)$ is a subalgebra of $E(3|6)$; so in this sense, the presence of this infinite-dimensional Lie algebra indicates an 'enhancement of symmetries' present in the holomorphic twist.

By local version, we mean a version of $E(3|6)$ which exists as a sheaf of super Lie algebras of the form $\Omega^{0,\bullet}(X, \mathcal{V}^{(0)})$ on any complex threefold X equipped with a square-root of its canonical bundle. The even part of this sheaf is simply a semi-direct product

$$\mathrm{Vect}^{hol}(X) \oplus \mathfrak{sl}(2) \otimes \mathcal{O}(X). \tag{3.26}$$

The odd part is

$$\Omega^{1,hol}(X, K_X^{-1/2}) \otimes \mathbf{C}^2 \tag{3.27}$$

The Lie bracket on this sheaf of super vector spaces is defined analogously to the bracket on $E(3|6)$, see [**36, 49**]. In particular, the bracket operations only involve holomorphic differential operators and hence extends to a local dg super Lie algebra structure on the Dolbeault resolution of the above holomorphic vector bundles. We denote this local dg super Lie algebra by $\mathcal{E}(3|6)$.

When $X = \mathbf{C}^3$ the ∞-jets at $0 \in \mathbf{C}^3$ of the local Lie algebra $\mathcal{E}(3|6)$ is equivalent to the exceptional super Lie algebra $E(3|6)$.

PROPOSITION 3.5. *The weight zero piece* $\mathcal{G}^{(0)} \subset \mathcal{G}$ *is equivalent to* $\mathcal{E}(3|6)$.

As a consequence of this we see that twisted supergravity states $\mathcal{H}_{sugra}$ on twisted AdS space form a representation for the exceptional algebra $E(3|6)$. We now state the formula for the character of local operators associated to the factorization algebra $\mathrm{C}_\bullet(\mathcal{E}(3|6))$.

PROPOSITION 3.6. *The character of local operators of the factorization algebra* $C_\bullet(\mathcal{G}_c^{(0)})$ *on* $\mathbf{C}^3$ *is given by the plethystic exponential of the following expression*
(3.28)

$$f_2(y_i, y, q) = \frac{q^4(y_1+y_2+y_3)+q^2(y^2+q+q^2y^{-2}) - q^3(y+qy^{-1})(y_1^{-1}+y_2^{-1}+y_3^{-1})}{(1-y_1q)(1-y_2q)(1-y_3q)}$$

where we recall that $y_1y_2y_3 = 1$.

We can now state our conjecture for local operators of the type A_1 theory at the level of superconformal indices.

CONJECTURE 3.7. *The superconformal index of the local operators in the six-dimensional superconformal theory of type* A_1 *which are dual to a gas of gravitons in* AdS_7 *is given by*

$$\widetilde{\chi}_2(y_i, y, q) = \mathrm{PExp}\left[\widetilde{f}_2(y_i, y, q)\right].$$

where $\widetilde{f}_2(y_i, y, q)$ *is as in* (3.28).

By 'dual to a gas of gravitons' we mean that under the holographic correspondence that this index is only sensitive to perturbative effects from the point of view of the bulk supergravity theory and potentially misses contributions from non-perturbative effects such as black holes.

We have some evidence for this conjecture based on superconformal indices which have been computed in the literature. If we expand our closed formula for $\widetilde{\chi}_2$ in powers of q we find an agreement with the expansions in [**38**] up to order q^3 (which is as far as this paper goes in the A_1 case). We also find agreement with the expansion in [**35**] (see equation (68)) except for a $\mathfrak{sl}(3)$-scalar term at order q^3.

The Schur limit $y = 1, y_3 = 1$ of $\widetilde{\chi}_2$ yields

$$\widetilde{\chi}_2(y_1, y_2, y_3 = 1, y = 1, q) = \prod_{n\geq 1} \frac{1}{(1-q^n)^2}$$

which is character of the Virasoro vacuum module.

In [**49**] we give a closed formula for the character of local operators associated to the factorization algebra $\mathcal{G}_{N,c}$ for any $N \geq 2$ as a function of four independent fugacities. Its Schur limit agrees with the vacuum character of the W_N algebra. We are led to a similar conjecture as above, stating that this local character should agree with the superconformal index of operators in the six-dimensional theory of type A_{N-1} which correspond to a gas of gravitons in the bulk under the holographic correspondence. In [**49**] we compare to formulas for the superconformal index found in [**35, 38**] in the type A_2 case as well.

Acknowledgment

We thank Ingmar Saberi and Surya Raghavendran for past and ongoing collaborations on many of the results that appear in this note.

References

[1] Luis F. Alday, Davide Gaiotto, and Yuji Tachikawa, *Liouville correlation functions from four-dimensional gauge theories*, Lett. Math. Phys. **91** (2010), no. 2, 167–197, DOI 10.1007/s11005-010-0369-5. MR2586871

[2] I. A. Batalin and G. A. Vilkovisky, *Gauge algebra and quantization*, Phys. Lett. B **102** (1981), no. 1, 27–31, DOI 10.1016/0370-2693(81)90205-7. MR616572

[3] Christopher Beem, Madalena Lemos, Leonardo Rastelli, and Balt C. van Rees, *The* $(2,0)$ *superconformal bootstrap*, Phys. Rev. D **93** (2016), no. 2, 025016, 37, DOI 10.1103/PhysRevD.93.025016. MR3481932

[4] Christopher Beem, Leonardo Rastelli, and Balt C. van Rees, $\mathcal{W}$ *symmetry in six dimensions*, J. High Energy Phys. **5** (2015), 017, front matter+37, DOI 10.1007/JHEP05(2015)017. MR3359380

[5] Nathan Berkovits, *Covariant quantization of the superparticle using pure spinors*, J. High Energy Phys. **9** (2001), Paper 16, 17, DOI 10.1088/1126-6708/2001/09/016. MR1867182

[6] M. Bershadsky, S. Cecotti, H. Ooguri, and C. Vafa, *Kodaira-Spencer theory of gravity and exact results for quantum string amplitudes*, Comm. Math. Phys. **165** (1994), no. 2, 311–427. MR1301851

[7] Jyotirmoy Bhattacharya, Sayantani Bhattacharyya, Shiraz Minwalla, and Suvrat Raju, *Indices for superconformal field theories in 3, 5 and 6 dimensions*, J. High Energy Phys. **2** (2008), 064, 42, DOI 10.1088/1126-6708/2008/02/064. MR2385981

[8] Daniel Bruegmann, *Vertex algebras and costello-gwilliam factorization algebras*, `arXiv:2012.12214` [math.QA], 2021.

[9] Kasia Budzik, Davide Gaiotto, Justin Kulp, Brian Williams, Jingxiang Wu, and Matthew Yu, *Semi-chiral operators in 4d* $N = 1$ *gauge theories*, `arXiv:2306.01039` [hep-th], 2023.

[10] Dylan Butson and Philsang Yoo, *Degenerate classical field theories and boundary theories*, `arXiv:1611.00311` [math-ph] 2016.

[11] Kevin Costello, *Renormalization and effective field theory*, Mathematical Surveys and Monographs, vol. 170, American Mathematical Society, Providence, RI, 2011, DOI 10.1090/surv/170. MR2778558

[12] Kevin Costello, *Notes on supersymmetric and holomorphic field theories in dimensions 2 and 4*, Pure Appl. Math. Q. **9** (2013), no. 1, 73–165, DOI 10.4310/PAMQ.2013.v9.n1.a3. MR3126501

[13] Kevin Costello, *M-theory in the Omega-background and 5-dimensional noncommutative gauge theory*, `arXiv:1610.04144` [hep-th] 2016.

[14] Kevin Costello, *Holography and Koszul duality: the example of the M2 brane*, `arXiv:1705.02500` [hep-th] 2017.

[15] Kevin Costello and Davide Gaiotto, *Twisted holography*, `arXiv:1812.09257` [hep-th] 2021.

[16] Kevin Costello and Owen Gwilliam, *Factorization algebras in quantum field theory. Vol. 1*, New Mathematical Monographs, vol. 31, Cambridge University Press, Cambridge, 2017, DOI 10.1017/9781316678626. MR3586504

[17] Kevin Costello and Owen Gwilliam, *Factorization algebras in quantum field theory. Vol. 2*, New Mathematical Monographs, vol. 41, Cambridge University Press, Cambridge, 2021, DOI 10.1017/9781316678664. MR4300181

[18] Kevin Costello and Si Li, *Quantization of open-closed BCOV theory, I*, `arXiv:1505.06703` [hep-th], 2015.

[19] Kevin Costello and Si Li, *Twisted supergravity and its quantization*, `arXiv:1606.00365` [hep-th], 2016.

[20] Kevin Costello and Natalie M. Paquette, *Twisted supergravity and Koszul duality: a case study in* AdS_3, Comm. Math. Phys. **384** (2021), no. 1, 279–339, DOI 10.1007/s00220-021-04065-3. MR4252878

[21] Kevin Costello and Natalie M. Paquette, *Celestial holography meets twisted holography: 4d amplitudes from chiral correlators*, J. High Energy Phys. **10** (2022), Paper No. 193, 67, DOI 10.1007/jhep10(2022)193. MR4531606

[22] Kevin Costello and Natalie M. Paquette, *Associativity of one-loop corrections to the celestial operator product expansion*, Phys. Rev. Lett. **129** (2022), no. 23, Paper No. 231604, 6, DOI 10.1103/physrevlett.129.231604. MR4522278

[23] Kevin Costello, Natalie M. Paquette, and Atul Sharma, *Top-down holography in an asymptotically flat spacetime*, Phys. Rev. Lett. **130** (2023), no. 6, Paper No. 061602, 7, DOI 10.1103/physrevlett.130.061602. MR4550721

[24] Kevin Costello and Claudia Scheimbauer, *Lectures on mathematical aspects of (twisted) supersymmetric gauge theories*, Mathematical aspects of quantum field theories, Math. Phys. Stud., Springer, Cham, 2015, pp. 57–87. MR3330239

[25] Richard Eager and Fabian Hahner, *Maximally twisted eleven-dimensional supergravity*, Comm. Math. Phys. **398** (2023), no. 1, 59–88, DOI 10.1007/s00220-022-04516-5. MR4544182

[26] Richard Eager, Fabian Hahner, Ingmar Saberi, and Brian R. Williams, *Perspectives on the pure spinor superfield formalism*, J. Geom. Phys. **180** (2022), Paper No. 104626, 47, DOI 10.1016/j.geomphys.2022.104626. MR4460345

[27] Richard Eager, Ingmar Saberi, and Johannes Walcher, *Nilpotence varieties*, Ann. Henri Poincaré **22** (2021), no. 4, 1319–1376, DOI 10.1007/s00023-020-01007-y. MR4229535

[28] Chris Elliott, Pavel Safronov, and Brian R. Williams, *A taxonomy of twists of supersymmetric Yang-Mills theory*, Selecta Math. (N.S.) **28** (2022), no. 4, Paper No. 73, 124, DOI 10.1007/s00029-022-00786-y. MR4468561

[29] Giovanni Faonte, Benjamin Hennion, and Mikhail Kapranov, *Higher Kac-Moody algebras and moduli spaces of G-bundles*, Adv. Math. **346** (2019), 389–466, DOI 10.1016/j.aim.2019.01.040. MR3910800

[30] Sudarshan Fernando, Murat Günaydin, and Seiji Takemae, *Supercoherent states of* $\mathrm{OSp}(8^*|2N)$*, conformal superfields and the* $\mathrm{AdS}_7/\mathrm{CFT}_6$ *duality*, Nuclear Phys. B **628** (2002), no. 1-2, 79–111, DOI 10.1016/S0550-3213(02)00076-7. MR1901224

[31] Sergio Ferrara and Emery Sokatchev, *Representations of superconformal algebras in the* $\mathrm{AdS}_{7/4}/CFT_{6/3}$ *correspondence*, J. Math. Phys. **42** (2001), no. 7, 3015–3026, DOI 10.1063/1.1374451. Strings, branes, and M-theory. MR1840329

[32] Zhengping Gui, Si Li, and Keyou Zeng, *Quadratic duality for chiral algebras*, arXiv:2212.11252 [math.QA], 2022.

[33] Owen Gwilliam and Brian R. Williams, *Higher Kac-Moody algebras and symmetries of holomorphic field theories*, Adv. Theor. Math. Phys. **25** (2021), no. 1, 129–239, DOI 10.4310/ATMP.2021.v25.n1.a4. MR4320072

[34] Seyed Morteza Hosseini, Kiril Hristov, and Alberto Zaffaroni, *A note on the entropy of rotating BPS* $\mathrm{AdS}_7 \times S^4$ *black holes*, J. High Energy Phys. **5** (2018), 121, front matter+19, DOI 10.1007/jhep05(2018)121. MR3815066

[35] Yosuke Imamura, *Analytic continuation for giant gravitons*, PTEP. Prog. Theor. Exp. Phys. **10** (2022), Paper No. 103B02, 24, DOI 10.1093/ptep/ptac127. MR4532727

[36] Victor G. Kac, *Classification of infinite-dimensional simple linearly compact Lie superalgebras*, Adv. Math. **139** (1998), no. 1, 1–55, DOI 10.1006/aima.1998.1756. MR1652530

[37] Victor G. Kac and Alexei Rudakov, *Representations of the exceptional Lie superalgebra* $E(3,6)$*. II. Four series of degenerate modules*, Comm. Math. Phys. **222** (2001), no. 3, 611–661, DOI 10.1007/PL00005581. MR1888090

[38] Hee-Cheol Kim, Seok Kim, Sung-Soo Kim, and Kimyeong Lee, *The general M5-brane superconformal index*, arXiv:1307.7660 [hep-th], 2013.

[39] D. Maulik, N. Nekrasov, A. Okounkov, and R. Pandharipande, *Gromov-Witten theory and Donaldson-Thomas theory. I*, Compos. Math. **142** (2006), no. 5, 1263–1285, DOI 10.1112/S0010437X06002302. MR2264664

[40] D. Maulik, N. Nekrasov, A. Okounkov, and R. Pandharipande, *Gromov-Witten theory and Donaldson-Thomas theory. II*, Compos. Math. **142** (2006), no. 5, 1286–1304, DOI 10.1112/S0010437X06002314. MR2264665

[41] Nikita Nekrasov, $\mathbf{Z}$*-theory: chasing* $\mathfrak{m}/f$ *theory* (English, with English and French summaries), C. R. Phys. **6** (2005), no. 2, 261–269, DOI 10.1016/j.crhy.2004.12.011. Strings 04. Part II. MR2121727

[42] Nikita Nekrasov and Andrei Okounkov, *Membranes and sheaves*, Algebr. Geom. **3** (2016), no. 3, 320–369, DOI 10.14231/AG-2016-015. MR3504535

[43] Nikita Nekrasov and Edward Witten, *The omega deformation, branes, integrability and Liouville theory*, J. High Energy Phys. **9** (2010), 092, i, 82, DOI 10.1007/JHEP09(2010)092. MR2776942

[44] Nikita A. Nekrasov, *Seiberg-Witten prepotential from instanton counting*, Adv. Theor. Math. Phys. **7** (2003), no. 5, 831–864. MR2045303

[45] Nikita A. Nekrasov, *Lectures on curved beta-gamma systems, pure spinors, and anomalies*, arXiv:hep-th/0511008, 2005.

[46] Nikita A. Nekrasov, *Instanton partition functions and M-theory*, Jpn. J. Math. **4** (2009), no. 1, 63–93, DOI 10.1007/s11537-009-0853-9. MR2491283

[47] Natalie M. Paquette and Brian R. Williams, *Koszul duality in quantum field theory*, Confluentes Math. **14** (2022), no. 2, 87–138. MR4561878

[48] Surya Raghavendran, Ingmar Saberi, and Brian R. Williams, *Twisted eleven-dimensional supergravity*, Comm. Math. Phys. **402** (2023), no. 2, 1103–1166, DOI 10.1007/s00220-023-04745-2. MR4627318

[49] Surya Raghavendran and Brian R. Williams, *A holographic approach to the six-dimensional superconformal index*, arXiv:2210.07910 [math-ph], 2022.

[50] Surya Raghavendran and Philsang Yoo, *Twisted S-duality*, arXiv:1910.13653 [math-ph], 2019.

[51] Ingmar Saberi and Brian R. Williams, *Twisted characters and holomorphic symmetries*, Lett. Math. Phys. **110** (2020), no. 10, 2779–2853, DOI 10.1007/s11005-020-01319-4. MR4146954

[52] Ingmar Saberi and Brian R. Williams, *Twisting pure spinor superfields, with applications to supergravity*, arXiv:2106.15639 [math-ph], 2021.

[53] Ingmar Saberi and Brian R. Williams, *Constraints in the BV formalism: six-dimensional supersymmetry and its twists*, Adv. Math. **412** (2023), Paper No. 108806, 87, DOI 10.1016/j.aim.2022.108806. MR4520424

[54] Pavel Safronov, *Lectures on shifted Poisson geometry* (English, with English and French summaries), Derived algebraic geometry, Panor. Synthèses, vol. 55, Soc. Math. France, Paris, [2021] ©2021, pp. 187–230. MR4509774

[55] Brian R. Williams, *Renormalization for holomorphic field theories*, Comm. Math. Phys. **374** (2020), no. 3, 1693–1742, DOI 10.1007/s00220-020-03693-5. MR4076086

[56] Edward Witten, *Topological quantum field theory*, Comm. Math. Phys. **117** (1988), no. 3, 353–386. MR953828

[57] Edward Witten, *Five-brane effective action in M-theory*, J. Geom. Phys. **22** (1997), no. 2, 103–133, DOI 10.1016/S0393-0440(97)80160-X. MR1451550

[58] Keyou Zeng, *Twisted Holography and Celestial Holography from Boundary Chiral Algebra*, arXiv:2302.06693 [hep-th] 2023.

Department of Mathematics and Statistics, Boston University, 665 Commonwealth Ave., Boston, Massachusetts 02215

Email address: bwill22@bu.edu

Proceedings of Symposia in Pure Mathematics
Volume **107**, 2024
https://doi.org/10.1090/pspum/107/01954

Algebras, regions, and observers

Edward Witten

ABSTRACT. In ordinary quantum field theory, one can define the algebra of observables in a given region in spacetime, but in the presence of gravity, it is expected that this notion ceases to be well-defined. A substitute that appears to make sense in the presence of gravity and that also is more operationally meaningful is to consider the algebra of observables along the timelike worldline of an observer. It is known that such an algebra can be defined in quantum field theory, and the timelike tube theorem of quantum field theory suggests that such an algebra is a good substitute for what in the absence of gravity is the algebra of a region. The static patch in de Sitter space is a concrete example in which it is useful to think in these terms and to explicitly incorporate an observer in the description.

1. Introduction

In ordinary quantum mechanics, we do not usually incorporate the observer as part of the system. That is fine for many purposes, but in the presence of gravity, one has to take into account the fact that the observer gravitates. In some contexts, this is unimportant, because the observer's gravity is negligible. For example, in an asymptotically flat spacetime, an observer more or less at rest in the asymptotic region is likely to have negligible gravitational influence on whatever is in the interior of the spacetime. On the other hand, in a closed universe, where the gravitational flux due to the observer has "nowhere to go," one might expect that it may be essential to include the observer in the description.

We will discuss an example of this – based on the paper [**1**]. But we begin with more general considerations. In ordinary quantum field theory, one can attach an algebra $\mathcal{A}_{\mathcal{U}}$ of observables to a rather general open set $\mathcal{U}$ in spacetime. In the presence of gravity, there are potentially two problems with this notion. First, when spacetime fluctuates, we may generically have difficulty saying just what we mean by the spacetime region $\mathcal{U}$. Second, it is not clear what is the logic of discussing an algebra $\mathcal{A}_{\mathcal{U}}$ unless there is someone who can make the observations corresponding to elements of that algebra. In the absence of gravity, we do not usually worry about this question; we just assume, in effect, that some observer external to the system has the relevant capability. But in the presence of gravity, anticipating that sometimes it will be necessary to include the observer in the description, we should take seriously the idea that it is only well-motivated to discuss an algebra of observables if this is the algebra of observables accessible to someone.

Research supported in part by NSF Grant PHY-2207584.

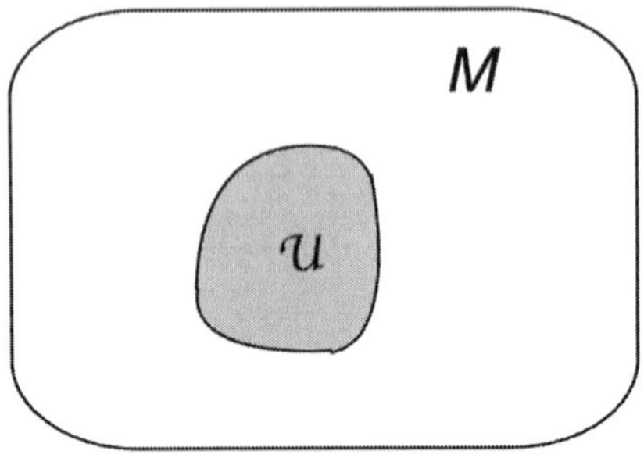

FIGURE 1. In the absence of gravity, one can consider any chosen open set $\mathcal{U}$ in a spacetime M, and define the algebra of observables in that region.

But what algebra is accessible to an observer? We will discuss this question in the context of ordinary quantum field theory, without gravity, but hoping to learn some lessons that are useful when gravity is included. We make use of two classic but not so well known results about ordinary quantum field theory. The first result [**2**], described in section 2, enables one to define an algebra of observables along the timelike worldline of an observer. The second result, described in section 3, is the "timelike tube theorem" [**3**–**6**], which among other things says that, in the absence of gravity, in a real analytic spacetime, the algebra of observables along the observer's worldline is the same as (roughly) the algebra of observables in the region causally accessible to the observer. The first result is more elementary, and more general (since it does not require a hypothesis of real analyticity), but in a real analytic spacetime, it can potentially be viewed as a corollary of the second.

The two results indicate that instead of discussing the algebra of observables in a spacetime region, we could discuss the algebra generated by the quantum fields along the observer's worldline. In a theory of gravity, the algebra of observables along the worldline would appear to be both better defined and operationally more meaningful than the algebra of a region.

In section 4, following [**1**], we consider the "static patch" in de Sitter space as an example in which it is necessary to explicitly include an observer in order to define a sensible algebra of observables. The algebra defined after taking the observer explicitly into account turns out to be a von Neumann algebra of Type II_1. This gives an abstract explanation of why "empty de Sitter space" is a state of maximum entropy, and in what sense the density matrix of empty de Sitter space is maximally mixed.

Presumably, in a full theory of the world, an observer cannot be added from outside but must emerge as part of the theory. In that context, what it means to include an observer in the description is that one considers a "code subspace" of states in which the observer is present, and one defines operators that are well-defined on the code subspace, though they would not be well-defined on all states of the theory. For considerations of the present article, however, however, it is not necessary to have such a full theory in hand.

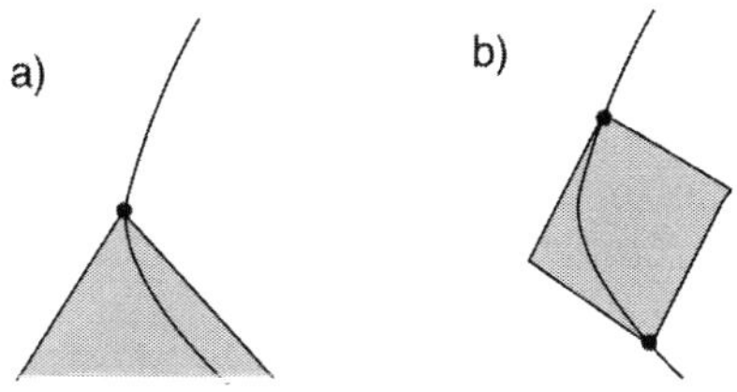

FIGURE 2. The worldline of an observer, with (a) the region of spacetime that is visible to the observer prior to a given time, or (b) the region that the observer can both see and influence in a given time interval. Time runs vertically and the boundaries of the chosen time intervals are marked by the black dots.

2. The algebra accessible to an observer

In ordinary quantum field theory in a spacetime M, we can arbitrarily specify any open set $\mathcal{U} \subset M$ and define an algebra $\mathcal{A}_{\mathcal{U}}$ of operators in $\mathcal{U}$ (fig. 1). In the presence of gravity, since spacetime fluctuates, it does not make sense to talk about the region $\mathcal{U}$ unless we have an invariant way to identify it. For example, in an asymptotically flat spacetime, in the presence of a black hole, we could talk about the region outside the black hole horizon. That is invariantly defined and presumably makes sense at least perturbatively even when spacetime fluctuates. We could introduce an observer who is more or less at rest near infinity and describe the region outside the horizon as the region visible to this observer. However, as already noted, in an asymptotically flat universe we do not expect that it is essential to incorporate the observer in the description.

If we assume the existence of an observer, we can invariantly identify various regions in spacetime. For example, if the observer carries a clock, then as in fig. 2, we can discuss the region that is visible to the observer prior to a given time, or, alternatively, the region that is causally accessible to the observer in a stated time interval (meaning that the observer can both see and influence this region during the interval in question).

But what can an observer actually measure? Here we will assume a very simple model in which the observer is described by a timelike worldline, and what the observer can measure are simply the quantum fields along this worldline.[1] This seems like a rather minimal model of what an observer is, and one could well worry that it is too crude. A realistic observer would presumably also carry measuring equipment, and a recording device, and would have access to operators that act on all that. But it turns out that in ordinary quantum field theory without gravity, the rather crude model in which an observer is just characterized by a worldline and the observables are the quantum fields along the worldline is sufficient, for many purposes.

This model raises two immediate questions:

(1) Can well-defined operators be defined by smearing a quantum field along a timelike worldline?

(2) Given a "yes" answer to the first question, what is the algebra generated by these operators?

[1]See [**8**] for a classic discussion in this framework, with a somewhat different motivation.

Let us elaborate a bit on the first question. We are accustomed in quantum field theory to talking about "local operators" $\phi(x)$, but a local operator is not really a Hilbert space operator, since acting on a Hilbert space state it takes us out of Hilbert space. In the case of the vacuum state Ω in Minkowski space, this is clear from the fact that $|\phi(x)|\Omega\rangle|^2 = \infty$ or equivalently

$$\langle\Omega|\phi(x)\phi(x)|\Omega\rangle = \infty, \tag{2.1}$$

due to a short distance singularity. Since the leading short distance singularity is universal, it is also true that $|\phi(x)|\Psi\rangle|^2 = \infty$ for any state Ψ in any spacetime M. The problem has nothing to do with the state Ψ, and purely reflects the fact the product $\phi(x) \cdot \phi(x)$ is not well-defined, since in fact a more general product $\phi(y) \cdot \phi(x)$ is singular for $y \to x$. This singularity is governed by the operator product expansion (OPE).

If we could measure $\phi(x)$, the answer would be one of its eigenvalues, but since $\phi(x)$ maps us out of Hilbert space, it does not have eigenvectors or eigenvalues, and we cannot measure it. What are actually measureable are suitable smeared versions of $\phi(x)$. Which ones? Suppose we are going to smear a real scalar field $\phi(x)$ over a set S to get a smeared "operator" $\phi_f = \int_S \mathrm{d}\mu f(x)\phi(x)$, where $f(x)$ is a complex-valued smooth function with support in S. If ϕ_f is actually going to make sense as an operator, the smearing has to be such that the $\phi(x') \cdot \phi(x)$ OPE singularity is integrable, when smeared in this fashion.

For example, spatial smearing will only succeed for an operator of rather low dimension. In $\mathsf{D} = d+1$ spacetime dimensions, with space coordinates $\vec{x}$ and a time coordinate t, spatial smearing at, say, $t = 0$, produces an expression $\phi_f = \int_{t=0} \mathrm{d}^d\vec{x} f(\vec{x})\phi(\vec{x},0)$, where $f(\vec{x})$ is a smooth function of the spatial coordinates $\vec{x}$ that we can assume to have compact support. Does ϕ_f make sense as an operator? In computing the product $\phi_{\overline{f}} \cdot \phi_f$, we run into the integral

$$\int_{t'=t=0} \mathrm{d}\vec{x}\mathrm{d}\vec{x}' \overline{f}(\vec{x}') f(\vec{x})\phi(\vec{x}',0)\phi(\vec{x},0). \tag{2.2}$$

For illustrative purposes, let us consider first the case of a conformal field theory, though with minor modifications, the following remarks apply much more widely. If ϕ is a conformal field of dimension Δ, then the leading singularity in the operator product $\phi(\vec{x}',0)\phi(\vec{x},0)$ for $\vec{x}' \to \vec{x}$ is proportional to $|\vec{x}' - \vec{x}|^{-2\Delta}$, where Δ is the dimension of the operator ϕ. The condition for this singularity to be integrable when inserted in (2.2) is $2\Delta < d$. If ϕ is a free scalar field, then $\Delta = (\mathsf{D}-2)/2 = (d-1)/2$, and the condition $2\Delta < d$ is satisfied.[2] But what about, say, QCD, in the real world with $d = 3$? QCD is asymptotically free, so short distance singularities have the behavior just discussed up to logarithms, which are inessential except in the borderline case $\Delta = d/2$. In QCD, the smallest value of Δ for any gauge-invariant operator is 3, corresponding to a quark bilinear such as $\overline{q}q$. So the condition $2\Delta < d = 3$ is never satisfied and QCD is an example of a theory in which no true operator can be produced by smearing of a "local operator" in space.

Smearing in Euclidean space is only slightly better. If we try to define a smeared operator $\phi_f = \int \mathrm{d}^{\mathsf{D}}x f(x)\phi(x)$ (where now the integration is over all D coordinates of Euclidean spacetime), we will run into the operator product singularity

[2]In dimension $\mathsf{D} = 2$, a free scalar has $\Delta = 0$ and the condition $\Delta < d/2$ is satisfied by any normal ordered polynomial $:\phi^n(x):$. This was important in early work on constructive field theory [**9**].

$\int \mathrm{d}^{\mathsf{D}}x \mathrm{d}^{\mathsf{D}}x' \overline{f}(x') f(x) \phi(x') \phi(x)$. This is integrable if and only if $2\Delta < \mathsf{D}$, a weaker condition but one that again cannot be satisfied in QCD, for example.

How then do we get true operators by smearing of "local operators"? The secret is smearing in real time. Though smearing in space is only effective in favorable cases, smearing in real time turns a "local operator" of any dimension into a true operator. This rather old result [**2**] was originally proved directly on the basis of the Wightman axioms of quantum field theory for the case that the timelike curve is a timelike geodesic in Minkowski space.

To understand the result, we begin again with the case of a conformal field theory. Suppose that at $\vec{x} = 0$, we smear a "local operator" $\phi(\vec{x}, t)$ by a compactly supported function $f(t)$ that depends only on t. Thus we define $\phi_f(0) = \int \mathrm{d}t\, f(t)\phi(0,t)$. In evaluating $|\phi_f|\Omega\rangle|^2$, we now run into the operator product $\int \mathrm{d}t' \mathrm{d}t \overline{f}(t') f(t) \phi(0,t')\phi(0,t)$. If ϕ has dimension Δ, the leading OPE singularity is $\phi(t')\phi(t) \sim (t' - t - \mathrm{i}\varepsilon)^{-2\Delta}$, so we have to consider the integral

$$\int \mathrm{d}t'\, \mathrm{d}t\, \overline{f}(t') f(t) \frac{1}{(t'-t-\mathrm{i}\varepsilon)^{2\Delta}}. \tag{2.3}$$

There are also subleading OPE singularities; they have the same form with different exponents and can be treated just as we are about to describe. The integral is obviously well-defined for $\varepsilon > 0$, and we want to show that no divergence appears in the limit $\varepsilon \to 0$. For this, we write[3]

$$\frac{1}{(t'-t-\mathrm{i}\varepsilon)^{2\Delta}} = C_n \frac{\partial^n}{\partial t^n} (t'-t-\mathrm{i}\varepsilon)^{n-2\Delta}, \tag{2.4}$$

for any integer $n > 0$, with a constant C_n. Inserting this in the integral (2.3) and integrating by parts n times, we replace the original integral with

$$(-1)^n C_n \int \mathrm{d}t'\, \mathrm{d}t\, \overline{f}(t') f^{[n]}(t)\, (t'-t-\mathrm{i}\varepsilon)^{n-2\Delta}. \tag{2.5}$$

For large enough n, this is manifestly convergent for $\varepsilon \to 0$.

In a general quantum field theory, consider the operator product expansion

$$\phi(0,t')\phi(0,t) \sim \sum_\alpha h_\alpha(t'-t-\mathrm{i}\varepsilon)\phi_\alpha(0,t). \tag{2.6}$$

The coefficient functions $h_\alpha(t'-t-\mathrm{i}\varepsilon)$ are holomorphic for $\mathrm{Im}(t'-t) < 0$. This follows from positivity of energy. Normally one can assume that the singularities of the h_α are bounded[4] by a power law, $|h_\alpha(t'-t)| < C_\alpha |t'-t|^{-n_\alpha}$, for some C_α, n_α. In the context of the Wightman axioms of quantum field theory, the correlation functions are usually assumed to be tempered distributions, which implies such a bound. Alternatively, if a theory is conformally invariant or asymptotically free in the ultraviolet, and the operators considered behave as fields of definite dimension in the ultraviolet (modulo logarithms in the asymptotically free case), this again implies such a bound. Given holomorphy in the lower half t' plane and a power law

[3]If $2\Delta \in \mathbb{Z}$, the following formula has to be slightly modified with a logarithmic factor on the right hand side. The derivation otherwise proceeds in the same way.

[4]An example of a local operator that would not satisfy this condition is $e^{\mathrm{i}\phi(x)}$, where ϕ is a scalar field in spacetime dimension $\mathsf{D} > 2$.

bound,[5] one can imitate the previous derivation to show the finiteness of

$$(2.7)\quad \int \mathrm{d}t'\mathrm{d}t\overline{f}(t')f(t)\phi(0,t')\phi(0,t) \sim \lim_{\varepsilon\to 0}\sum_\alpha \int \mathrm{d}t'\mathrm{d}t\overline{f}(t')f(t)h_\alpha(t'-t+\mathrm{i}\varepsilon)\phi_\alpha(t).$$

For this, one writes each $h_\alpha(t'-t-\mathrm{i}\varepsilon)$ as the nth derivative with respect to t', for some n, of a function $k_\alpha(t'-t-\mathrm{i}\varepsilon)$ that remains continuous (though not smooth) for $\varepsilon\to 0$, and then one integrates by parts as in the derivation of eqn. (2.5).

What happens if we consider an arbitrary timelike curve γ in Minkowski space, not necessarily a geodesic? Parametrize γ by the proper time τ and let $d(\tau',\tau)$ be the signed proper distance[6] in Minkowski space between points on γ labeled by τ' and by τ. The effect of replacing a timelike geodesic by an arbitrary timelike curve is to replace $\tau'-\tau-\mathrm{i}\varepsilon$ in the preceding formulas by $d(\tau',\tau)-\mathrm{i}\varepsilon$. Since $d(\tau',\tau)$ is a smooth function and $d(\tau',\tau)\sim\tau'-\tau+\mathcal{O}((\tau'-\tau)^2)$, this does not substantially affect the preceding analysis. The singularities at $\tau'\to\tau$ are of the same general form and are harmless.

What happens if we replace Minkowski space by a general spacetime M? Intuitively, one would not expect this to matter, since everything is determined by short distance behavior. In a curved spacetime, the operator product expansion becomes more complicated, with curvature dependent terms; see [**12**] for an axiomatic discussion. But the extra terms have similar singularities to what we have already considered, so one would expect the same result. For a proof, under reasonable assumptions about correlation functions in a curved spacetime, see [**13**]. See also further discussion in [**14**], [**6**].

3. The timelike tube theorem

3.1. The timelike envelope and the timelike tube theorem. In section 2, we learned that one can define operators by smearing a quantum field along the timelike worldline of an observer. Therefore we can consider the algebra generated by such operators (or more precisely by bounded functions of such operators). But what are the algebras that we make this way? In the context of quantum field theory without gravity, this question is answered by the "timelike tube theorem." This theorem was originally formulated for timelike geodesics in Minkowski space [**3, 4**]. It was generalized to free field theories in curved spacetime in [**5**]. For a version of the theorem suitable for non-free theories in curved spacetime, see [**6, 7**].

If $\mathcal{U}$ is an open set in spacetime, its "timelike envelope" $\mathcal{E}(\mathcal{U})$ consists of all points that can be reached by deforming timelike curves in $\mathcal{U}$ through a family of timelike curves, keeping the endpoints fixed (fig. 3). (Thus $\mathcal{E}(\mathcal{U})$ is contained in the intersection $J^+(\mathcal{U})\cap J^-(\mathcal{U})$ of the past and future of $\mathcal{U}$, but in general it is smaller. See section 3.3 for an example.) The timelike tube theorem asserts that

[5]The precise mathematical statement is that if $f(z)$ is a function holomorphic in the lower half plane, then a necessary and sufficient condition for the boundary values of $f(z)$ along the real axis to define a distribution is a bound $|f(z)| < k|\mathrm{Im}\, z|^{-c}$ for some constants k, c. For an elementary proof (and a bound on the distribution) see Proposition 4.2 in [**10**]. See also Theorem 1.1 in [**11**] for a d-dimensional generalization.

[6]The proper distance in Minkowski space between points along γ labeled by τ' and by τ is the proper time elapsed along a geodesic between the two points (not along the path γ). To get the signed proper distance $d(\tau',\tau)$, we multiply the proper distance by $+1$ if $\tau'>\tau$ and by -1 if $\tau>\tau'$.

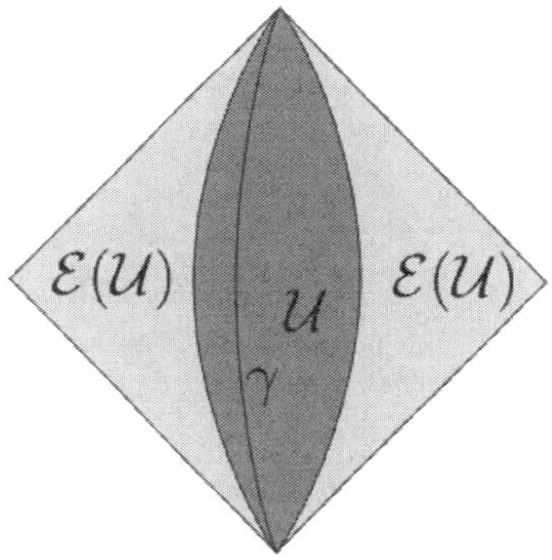

FIGURE 3. The timelike envelope $\mathcal{E}(\mathcal{U})$ of an open set $\mathcal{U}$ consists of all points that can be reached by deforming a timelike curve in $\mathcal{U}$ keeping its endpoints fixed. In general, one considers all possible timelike curves in $\mathcal{U}$, but in the case depicted, it suffices to consider segments of the particular timelike curve γ.

the algebra of operators[7] in $\mathcal{U}$ is the same as the algebra of operators in the possibly much larger region $\mathcal{E}(\mathcal{U})$. We will aim to give at least a hint of why this is true, after first explaining an implication, noted in [**5**].

Suppose that we are actually interested in a timelike curve γ, possibly one of finite extent with endpoints q, p or possibly an infinite or semi-infinite curve. The timelike envelope $\mathcal{E}(\gamma)$ consists of all points that can be reached by deforming γ through a family of timelike curves, keeping γ fixed near its ends. If γ has endpoints, they are omitted from $\mathcal{E}(\gamma)$ to ensure that $\mathcal{E}(\gamma)$ is an open set. We can thicken γ (minus its endpoints, if any) to an open set $\mathcal{U}$, in such a way that the timelike envelope $\mathcal{E}(\mathcal{U})$ does not depend on $\mathcal{U}$ – it is the same as the timelike envelope $\mathcal{E}(\gamma)$ of γ. See fig. 3. The timelike tube theorem says that the algebra $\mathcal{A}(\mathcal{U})$ of operators in region $\mathcal{U}$ does not really depend on $\mathcal{U}$ but only on γ. It coincides with $\mathcal{A}(\mathcal{E}(\gamma))$, the algebra of operators in $\mathcal{E}(\gamma)$. So we can define an algebra for every (possibly bounded) timelike curve γ. That in itself is not a surprise, since we arrived at the same conclusion in a more direct way[8] in section 2. But it is nice to see the consistency with what we learned from more elementary arguments.

Strictly speaking, the algebra associated to a curve γ in section 2 consists of the quantum fields smeared along γ, while the timelike tube theorem gives us an algebra $\mathcal{A}(\gamma)$ of operators that can be defined in an arbitrarily small neighborhood of γ. Presumably, these two algebras coincide (assuming that we include all possible local operators in the construction of section 2, including derivatives of operators). A proof might start with an axiomatic characterization of local operators at a point in terms of a limit of the operators in a small ball around that point. Were the two algebras to differ, one might argue that as the notion of an observer characterized by an infinitely thin worldline is an idealization, the physically more relevant algebra would be the one provided by the timelike tube theorem.

[7]For a more precise explanation of what algebra is intended here, see section 3.2.

[8]The timelike tube theorem depends on real analyticity of spacetime, as discussed shortly, and the reasoning in section 2 does not, so the previous analysis was more general.

The interpretation that we want to give is that the algebra $\mathcal{A}(\gamma)$ of operators supported on a curve γ is a good stand-in for the algebra $\mathcal{A}(\mathcal{E})$ of the spacetime region $\mathcal{E}$ associated to γ. The algebra $\mathcal{A}(\gamma)$ is more operationally meaningful than $\mathcal{A}(\mathcal{E})$, since it is more directly what an observer can measure. And it is also better defined in the presence of gravity than $\mathcal{A}(\mathcal{E})$ would appear to be. Once one takes into account fluctuations in spacetime, it is hard to see how one would directly define $\mathcal{A}(\mathcal{E})$, but as long as an observer is included in the description, the algebra of observables along the observer's worldline seems to be a meaningful notion, even when the observer's worldline fluctuates. So $\mathcal{A}(\gamma)$ seems like a good substitute for the algebras associated to open sets that one considers in the absence of gravity.

To try to gain some idea of why the timelike tube theorem is true, let us consider the classical limit. Suppose we have a reasonable relativistic wave equation like the Klein-Gordon equation $(\Box + m^2)\phi = 0$, where $\Box$ is the wave operator, or possibly a nonlinear modification of this equation. In fig. 4, we are given a solution in one region of a spacetime M, and we want to predict the solution in a larger region. We consider two cases. In fig. 4(a), the solution is given in a "spacelike pancake" $\mathcal{U}$, and we want to extend it over the domain of dependence $D(\mathcal{U})$ of $\mathcal{U}$. In fig. 4(b), the solution is given in a "timelike tube" $\mathcal{U}$ and we want to extend the solution over the corresponding "timelike envelope" $\mathcal{E}(\mathcal{U})$.

There is a basic asymmetry between the two cases.[9] The Holmgren uniqueness theorem of classical partial differential equations[10] asserts that the extension over the larger region is unique, if it exists, both in fig. 4(a) and in fig. 4(b). But existence is more special and only holds in fig. 4(a).

A simple counterexample (fig. 5) shows that an existence result cannot possibly hold in fig. 4(b). Let ℓ be a timelike curve that passes through the region $\mathcal{E}(\mathcal{U})$ but not through $\mathcal{U}$. Consider a solution of the equation $(\Box + m^2)\phi = 0$ with a delta function source along ℓ. If we consider Maxwell's equations rather than the Klein-Gordon equation, then ℓ can be the worldline of a point charge. The field ϕ blows up along ℓ (in dimension $\mathsf{D} > 2$; see footnote 9). Starting with the solution in the region $\mathcal{U}$, there is no way to extend it over $\mathcal{E}(\mathcal{U})$ as a solution, since if one tries to do this, one encounters the blowup along ℓ. The equation is not obeyed along ℓ, because of the delta function source.

The existence and uniqueness result in the case of fig. 4(a) is the basis for much of physics. It says that the solution can be predicted from initial data – physics is causal. But by contrast the uniqueness result without a guarantee of existence in fig. 4(b) is not usually useful at the classical level because generically the extension over $\mathcal{E}(\mathcal{U})$ of a solution on $\mathcal{U}$ does not exist and it is very hard to predict when it does.

---

[9] To be precise, this asymmetry holds in any spacetime dimension $\mathsf{D} > 2$. In $\mathsf{D} = 2$, spacetime has one dimension of space and one dimension of time, and there is a perfect symmetry between the two cases of fig. 4. In $\mathsf{D} > 2$, there are more space dimensions than time dimensions, and there is no such symmetry. The counterexample illustrated in fig. 5 fails in $\mathsf{D} = 2$ because in $\mathsf{D} = 2$, the solution with a delta function source along a worldline ℓ does not blow up along ℓ, but instead is discontinuous across ℓ. Starting with the solution on one side of ℓ, one can drop the discontinuity and smoothly continue the solution across ℓ. In $\mathsf{D} > 2$, a solution with such a delta function source blows up along ℓ and there is no way to remove the singularity without changing the solution in the original region $\mathcal{U}$.

[10] For an accessible exposition of this theorem, see chapter 5 of [**15**].

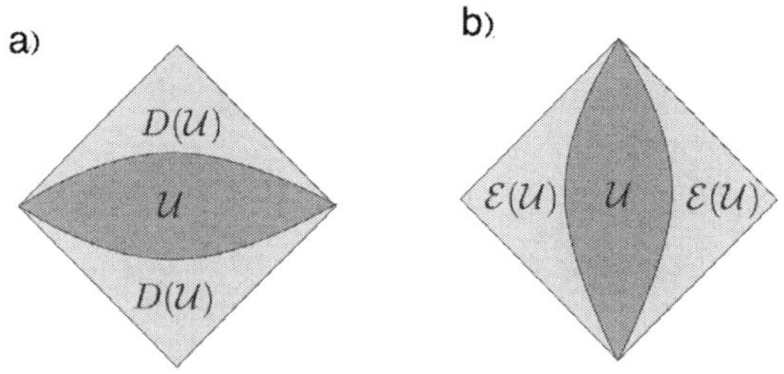

FIGURE 4. Two cases in which the solution of a hyperbolic wave equation is given in one open set (the dark shaded region) and one wishes to extend the solution over a larger open set (the more lightly shaded region). Time runs vertically and space runs horizontally. In (a), the solution is given in a "spacelike pancake" $\mathcal{U}$ and one wishes to extend it over the domain of dependence $D(\mathcal{U})$. In (b), the solution is given in a "timelike tube" $\mathcal{U}$, and one wishes to extend it over the "timelike envelope" $\mathcal{E}(\mathcal{U})$. In dimension 2, there is a perfect symmetry between these two cases, but in higher dimension, there is no such symmetry.

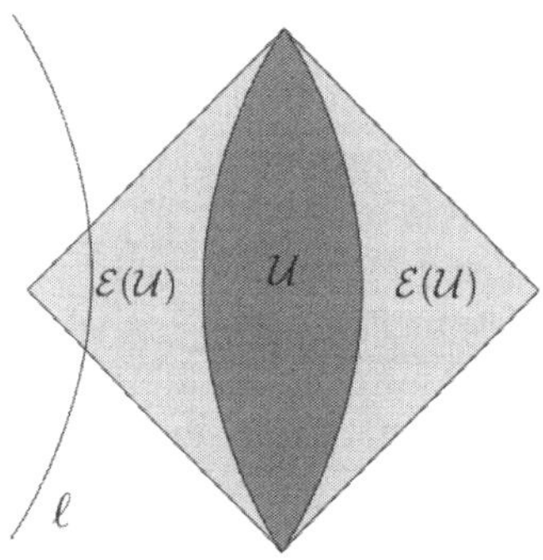

FIGURE 5. This picture illustrates the obstruction to existence in the setting of fig. 4(b). Starting with initial data in an open set $\mathcal{U}$, if one tries to extend a given solution in spatial directions over the timelike envelope $\mathcal{E}(\mathcal{U})$, it may develop singularities that will prevent the existence of the extended solution. In fact, this is the generic behavior. The singularity might arise on the worldline of a point charge that passes through $\mathcal{E}(\mathcal{U})$ but not through $\mathcal{U}$, as sketched here.

Suppose, however, that we are doing quantum field theory and for simplicity consider a free field ϕ with the action

$$I = \frac{1}{2}\int_M \mathrm{d}^{\mathsf{D}}x\sqrt{g}\left(-D_\mu\phi D^\mu\phi - m^2\phi^2\right). \tag{3.1}$$

In this case, we can view ϕ as an operator-valued solution of the Klein-Gordon equation $(\Box + m^2)\phi = 0$. If we are studying this quantum field theory on M, then the field $\phi(x)$ does exist throughout M and therefore existence of the extension from $\mathcal{U}$ to $\mathcal{E}(\mathcal{U})$ is not an issue.

But what does uniqueness mean? In some sense, uniqueness means "the field $\phi(x)$ for $x \in \mathcal{E}(\mathcal{U})$ is uniquely determined by $\phi(y)$ for $y \in \mathcal{U}$." As explained by

Borchers and Araki in the early 1960's, the quantum meaning of this statement is really "$\phi(x)$ for $x \in \mathcal{E}(\mathcal{U})$ is contained in the algebra generated by $\phi(y)$ for $y \in \mathcal{U}$" or equivalently the operator algebras of the two regions are the same:

$$\mathcal{A}(\mathcal{E}(\mathcal{U})) = \mathcal{A}(\mathcal{U}).$$

This is the timelike tube theorem.

Classically, we could just as well add higher order terms to the action and consider a field that satisfies a nonlinear partial differential equation. Holmgren uniqueness applies equally well to such an equation. Quantum mechanically, although a free quantum field can be viewed as an operator-valued solution of a classical field equation, that is not the case for a non-free quantum field, because of issues involving renormalization. Accordingly, proofs of the timelike tube theorem in non-free theories require additional ingredients, beyond what is needed in free theories. An approach that works for non-free theories in curved spacetime was presented in [**6**] (see [**7**] for an informal account).

In this discussion, we have skipped over two important points. First, we have not been precise about what is the algebra that is governed by the timelike tube theorem. See section 3.2. Second, we have omitted to explain a key assumption in the classical Holmgren uniqueness theorem and the quantum timelike tube theorem. These theorems hold in real analytic spacetimes. Real analyticity is not needed for existence and uniqueness in the setting of fig. 4(a), but it is needed for uniqueness in the setting of[11] fig. 4(b). Is this restriction important for the application of the timelike tube theorem to semiclassical gravity? In semiclassical gravity, one is usually studying the behavior of quantum fields in a background spacetime that in practice is normally real analytic. For instance, in section 4, we consider the static patch in de Sitter space as an example in which it is important to explicitly include an observer in order to define a sensible algebra of observables, and thus in which the timelike tube theorem provides important motivation. This is a typical example in which the starting point is a real analytic spacetime. After picking a semiclassical starting point, one then expands around it. In perturbation theory, the metric certainly fluctuates away from the real analytic starting point, but one expects that in perturbation theory, the gravitational field can be treated like any other quantum field in a real analytic background. Thus it appears that for typical applications of the timelike tube theorem to semiclassical gravity, the restriction of the theorem to real analytic spacetimes is not a problem.

3.2. The additive algebra. For a region $\mathcal{U}$ in spacetime, what precisely is the algebra of operators to which the timelike tube theorem applies? The smallest reasonable candidate is the algebra generated by local operators in $\mathcal{U}$ (or more precisely, by bounded functions of smeared local operators). If $\mathcal{U}$ is contractible, this is the natural algebra of observables in $\mathcal{U}$ in a generic quantum field theory. However, if $\mathcal{U}$ is not contractible, then in general, as we discuss shortly, there can be additional operators in $\mathcal{U}$, such as Wilson operators defined on noncontractible loops in $\mathcal{U}$, that cannot be constructed from local operators in $\mathcal{U}$.

The algebra generated by local operators has been called the additive algebra $\mathcal{A}_{\rm add}(\mathcal{U})$ [**17, 18**]. One can then reserve the name $\mathcal{A}(\mathcal{U})$ for the possibly larger algebra of all operators in $\mathcal{U}$. The timelike tube theorem is really a theorem about

[11] Actually, it suffices for the spacetime to be, in some coordinate system, real analytic in the time direction [**16**]. In free field theory, this is also true for the timelike tube theorem [**5**].

$\mathcal{A}_{\rm add}(\mathcal{U})$. Thus a more precise statement of the theorem than we have given so far is $\mathcal{A}_{\rm add}(\mathcal{U}) = \mathcal{A}_{\rm add}(\mathcal{E}(\mathcal{U}))$. This is clear from the proofs, which involve studying algebras of local operators. However, as we will discuss, while the distinction between $\mathcal{A}(\mathcal{U})$ and $\mathcal{A}_{\rm add}(\mathcal{U})$ exists in quantum field theory in general, there are reasons to believe that this distinction is absent in any theory that emerges at long distances from a full model of quantum gravity.

The motivation for the phrase "additive algebra" is as follows. Let $\mathcal{B}$ be a small open ball in $\mathcal{U}$. Then as $\mathcal{B}$ is contractible, there is no distinction between $\mathcal{A}(\mathcal{B})$ and $\mathcal{A}_{\rm add}(\mathcal{B})$. Now cover $\mathcal{U}$ by open balls $\mathcal{B}_\alpha$, with α ranging over some set S. The additive algebra of $\mathcal{U}$ is the same as the algebra generated by the $\mathcal{A}(\mathcal{B}_\alpha)$, $\alpha \in S$:

$$\mathcal{A}_{\rm add}(\mathcal{U}) = \vee_{\alpha\in S}\mathcal{A}(\mathcal{B}_\alpha). \tag{3.2}$$

In what situation will we have $\mathcal{A}(\mathcal{U}) \neq \mathcal{A}_{\rm add}(\mathcal{U})$, where $\mathcal{A}(\mathcal{U})$ is the algebra of all possible operators in $\mathcal{U}$, not necessarily built from local operators? For a typical example, consider a theory with a U(1) gauge field A, with curvature $F = {\rm d}A$. Let ℓ be a closed loop in $\mathcal{U}$, and consider the Wilson operator $W_\ell(A) = \exp({\rm i}\oint_\ell A)$. Is this operator contained in $\mathcal{A}_{\rm add}(\mathcal{U})$? The answer will be "yes" if ℓ is the boundary of an oriented two-manifold $D \subset \mathcal{U}$, for then $W_\ell(A) = \exp({\rm i}\int_D F)$ is a bounded function of a smeared local field in $\mathcal{U}$, namely $\int_D F$. But even if ℓ is not a boundary in $\mathcal{U}$, the answer can still be "yes," for a more subtle reason that was explained in [**19**]. Consider a theory that has a charge 1 field ϕ. Then for an open path $\gamma \subset \mathcal{U}$, say with endpoints q, p, one can consider the gauge-invariant operator $V_\gamma(A) = \overline{\phi}(p)\exp({\rm i}\int_\gamma A)\phi(q)$. In the case that γ is a very short path, with p very near q, V_γ can be expanded in terms of local operators at q and so is contained in the additive algebra of any open set containing γ. If ℓ is a closed loop, then $W_\ell(A)$ can be "cut" in the sense that if one omits from ℓ a number of open intervals of size ε, with the portion retained then being a disjoint union of closed intervals γ_i, $i = 1, \cdots, n$, then $W_\ell(A)$ appears as the most singular contribution in the operator product $\prod_{i=1}^n V_{\gamma_i}(A)$ for $\varepsilon \to 0$. Hence, in this situation $W_\ell(A)$ is contained in the additive algebra $\mathcal{A}_{\rm add}(\mathcal{U})$ for any region $\mathcal{U}$ containing ℓ.

However, if the theory has no field of unit charge, and ℓ is not a boundary in $\mathcal{U}$, then $W_\ell(A)$ is contained in $\mathcal{A}(\mathcal{U})$ but not in $\mathcal{A}_{\rm add}(\mathcal{U})$. This phenomenon is certainly not limited to U(1) gauge theory. In a theory with any gauge group G, if some representation R of G is "missing," in the sense that no local operator of the theory transforms in this representation, then for suitable regions $\mathcal{U}$ one will have $\mathcal{A}_{\rm add}(\mathcal{U}) \neq \mathcal{A}(\mathcal{U})$. A Wilson loop in the representation R can be present in $\mathcal{A}(\mathcal{U})$ but not in $\mathcal{A}_{\rm add}(\mathcal{U})$. A similar role can be played by a p-form gauge field B with $p > 1$. In a theory that has such a field, one can consider the operator $Q_S = \exp({\rm i}\int_S B)$, where S is a p-cycle in $\mathcal{U}$. If S is not a boundary in $\mathcal{U}$ and the theory does not have a string or membrane that couples to B (which would enable one to "cut" the operator Q_S in small pieces), then Q_S is contained in $\mathcal{A}(U)$ but not in $\mathcal{A}_{\rm add}(\mathcal{U})$. There is also an analog of this for magnetic charges (and their analogs for strings and membranes), with 't Hooft operators instead of Wilson operators.

We can now easily give an example that shows that it is necessary to specify that the timelike tube theorem applies to $\mathcal{A}_{\rm add}$, not to $\mathcal{A}$. Consider U(1) gauge theory without charged fields on a spacetime $\mathbb{R} \times S^1$ with product metric, where $\mathbb{R}$ parametrizes time and S^1 is parametrized by an angular variable ϕ. In this spacetime, there is a nontrivial Wilson operator $W_\ell = \exp({\rm i}\oint_\ell A)$, where ℓ is a closed loop that wraps around the S^1. Let p and q be two points at the same value of ϕ but different values of t, and let γ be the timelike geodesic that runs from q to p at fixed t. If $\mathcal{U}$ is a small neighborhood of γ, there is no nontrivial Wilson operator in $\mathcal{U}$, so $\mathcal{A}(\mathcal{U}) = \mathcal{A}_{\rm add}(\mathcal{U})$. But if p is separated in time from q by more than the circumference of the S^1, then the region $\mathcal{E}(\mathcal{U})$ includes a circle that wraps all the way around the S^1, so $\mathcal{A}(\mathcal{E}(\mathcal{U})) \neq \mathcal{A}_{\rm add}(\mathcal{E}(\mathcal{U}))$. Thus in this example, we have $\mathcal{A}_{\rm add}(\mathcal{U}) = \mathcal{A}_{\rm add}(\mathcal{E}(\mathcal{U}))$, as guaranteed by the timelike tube theorem, but $\mathcal{A}(\mathcal{U}) \neq \mathcal{A}(\mathcal{E}(\mathcal{U}))$.

Let us say that a quantum field theory which may have gauge fields or p-form gauge fields for $p > 1$ is "complete" if the electrically and magnetically charged particles, strings, and membranes coupling to those gauge fields are a maximal possible set consistent with all principles of low energy physics including Dirac quantization of electric and magnetic charge. Complete theories are precisely the ones with $\mathcal{A}_{\rm add}(\mathcal{U}) = \mathcal{A}(\mathcal{U})$ for all $\mathcal{U}$, so we might also call them "additive." It is believed that a quantum field theory that emerges at low energies from a full-fledged (ultraviolet complete) theory of quantum gravity is always complete in this sense. This was originally suggested based on experience with string theory [**20**]. Further arguments were given in [**21**], and the fullest known argument is based on holographic duality [**22**].

The claim that a theory that emerges from a full quantum gravity theory will be additive or complete has another interpretation. A theory with "missing" charges (or strings or membranes) has a p-form global symmetry in the language of [**23**] (where p depends on what is missing). So the statement that a full-fledged theory of quantum gravity is complete can be viewed as an extension to p-form global symmetry with $p > 1$ of the statement that there are no global symmetries in a full theory of quantum gravity. For discussion of this statement, see for example [**21, 24**].

In the spirit of the present article, one can motivate as follows the idea that an ultraviolet-complete theory of quantum gravity should have no missing charges. We assume that a complete gravity theory can be approximated, in an appropriate class of states, by an ordinary quantum field theory weakly coupled to gravity. Suppose that this ordinary quantum field theory is not complete, for example because it has a U(1) gauge field but no state of charge 1. Then if a loop ℓ is not a boundary in spacetime, the Wilson operator $W_\ell = \exp({\rm i}\oint_\ell A)$ is not measurable by any observer living in the spacetime and described by the theory. After all, given that ℓ is not a boundary, W_ℓ cannot be measured by measuring the curvature $F = {\rm d}A$. To measure W_ℓ when ℓ is not a boundary requires measuring the interference between different histories of a charge 1 particle that differ by the homology class of the worldline of the particle. But the assumption that the theory has no state of charge 1 means that such an experiment is not possible using the resources available in the theory.

What, then, do we mean in claiming that W_ℓ (or its hermitian part) is an "observable"? In ordinary quantum mechanics, we consider the observer to be external to the system, and we make no particular assumption about what resources the observer can use in probing the system. From that point of view, the observer might be able to introduce a massive charged particle to probe the system and measure W_ℓ. The observer, after all, is described by a more complete theory of the universe that might have the necessary charged particle. An ultraviolet-complete theory of quantum gravity, however, is expected to describe all that there is in the universe that it describes, with no way to add anything from outside. In the context of such a theory, the observer and any apparatus used by the observer must already be described by the theory. So in an ultraviolet complete theory of quantum gravity with a "missing" charge, we would be in the awkward situation that the theory would enable us to define an "observable" W_ℓ that would be well-defined in an appropriate, long distance limit, but that no one, in principle, could measure.

3.3. What else? In addition to the timelike tube theorem, which asserts that $\mathcal{A}_{\rm add}(\mathcal{U}) = \mathcal{A}_{\rm add}(\mathcal{E}(\mathcal{U}))$, one has causality, which, denoting the domain of dependence of a set $\mathcal{U}$ as $D(\mathcal{U})$, asserts that $\mathcal{A}(\mathcal{U}) = \mathcal{A}(D(\mathcal{U}))$. If the difference between $\mathcal{A}$ and $\mathcal{A}_{\rm add}$ is unimportant, these statements can be combined, relating $\mathcal{A}(\mathcal{U})$ to the algebra of a set that in general is larger than either $\mathcal{E}(\mathcal{U})$ or $D(\mathcal{U})$. This was noted by Araki [**4**] in one of the original papers on the timelike tube theorem. However, Araki actually conjectured a further generalization. In the context of quantum fields in Minkowski space, Araki wrote, " ...If ℓ is a snake-like line, connecting two mutually timelike points P_1 and P_2 but everywhere spacelike, and $\mathcal{B}$ is a tube around ℓ, of a small diameter, then it is rather likely that any solution of the [massless scalar] wave equation vanishing in $\mathcal{B}$ might always vanish in the double light cone spanned by P_1, and P_2. If such a conjecture turns out to be true, then we immediately have the corresponding theorem for $R(\mathcal{B})$." In our terminology, the proposal is that if Δ is the double cone or causal diamond with vertices P_1 and P_2, then $\mathcal{A}_{\rm add}(\mathcal{B}) = \mathcal{A}_{\rm add}(\mathcal{B}\cup\Delta)$. This follows from arguments in Araki's paper if it is true that any solution of the massless scalar wave equation vanishing in $\mathcal{B}$ also vanishes in Δ. This statement does not follow from Holmgren uniqueness in any obvious way and its validity remains unclear sixty years later.

In general, what extension of the timelike tube theorem is conceivable? For any set $\mathcal{U}$ in a spacetime M, one writes $\mathcal{U}'$ for the set of points in M that are spacelike separated from $\mathcal{U}$. Then $\mathcal{U}'' = (\mathcal{U}')'$ is called the causal completion of $\mathcal{U}$. For any open set $\mathcal{U}$, since $\mathcal{A}_{\rm add}(\mathcal{U})$ commutes with $\mathcal{A}_{\rm add}(\mathcal{U}')$, clearly any statement that $\mathcal{A}_{\rm add}(\mathcal{U}) = \mathcal{A}_{\rm add}(\mathcal{V})$ for some open set $\mathcal{V}$ implies that $\mathcal{A}_{\rm add}(\mathcal{V})$ commutes with $\mathcal{A}_{\rm add}(\mathcal{U}')$. If one is hoping to make a general statement that would apply to any quantum field theory, this really forces $\mathcal{V}\subset\mathcal{U}''$, since local operators outside of $\mathcal{U}'' = (\mathcal{U}')'$ typically do not commute with $\mathcal{A}_{\rm add}(\mathcal{U}')$. So the most optimistic statement that one might hope for would be $\mathcal{A}_{\rm add}(\mathcal{U}) = \mathcal{A}_{\rm add}(\mathcal{U}'')$.

For example, if $\mathcal{U}$ is a timelike curve in Minkowski space, then $\mathcal{E}(\mathcal{U}) = \mathcal{U}''$, and therefore the timelike tube theorem does indeed give $\mathcal{A}_{\rm add}(\mathcal{U}) = \mathcal{A}_{\rm add}(\mathcal{U}'')$. In general, $\mathcal{E}(\mathcal{U})\subset\mathcal{U}''$, but $\mathcal{U}''$ can be strictly bigger. Indeed, in Araki's example, $\mathcal{B}''$ is strictly bigger than $\mathcal{E}(\mathcal{B})$. His conjecture would be a special case of $\mathcal{A}_{\rm add}(\mathcal{U}) = \mathcal{A}_{\rm add}(\mathcal{U}'')$, in a situation in which that does not follow directly from the known statement of the timelike tube theorem.

A statement along the lines of $\mathcal{A}_{\rm add}(\mathcal{U}) = \mathcal{A}_{\rm add}(\mathcal{U}'')$ would be very attractive, as it would combine ordinary causality with the timelike tube theorem. Extending from a spacelike pancake to its domain of dependence (fig. 4(a)) and from a timelike tube to its timelike envelope (fig. 4(b)) can be regarded as two different cases of extending from $\mathcal{U}$ to $\mathcal{U}''$. However, there are some easy counterexamples to $\mathcal{A}_{\rm add}(\mathcal{U}) = \mathcal{A}_{\rm add}(\mathcal{U}'')$.

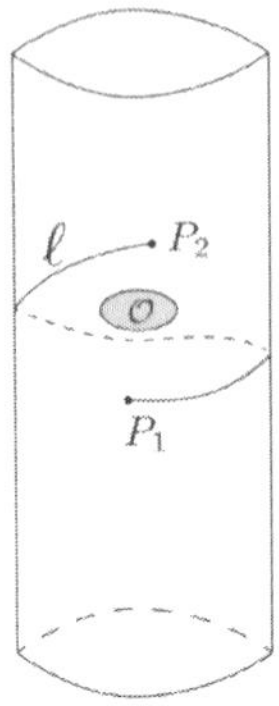

FIGURE 6. In the spacetime $\mathbb{R} \times S^1$, with $\mathbb{R}$ parametrizing the "time," ℓ is a spacelike curve that spirals around from a point P_1 to a point P_2 that is to its future. $\mathcal{O}$ is an open set in the "gap" between P_1 and P_2. If $\mathcal{B}$ (not drawn) is an open set that is a small "thickening" of ℓ, then a right-moving null geodesic through $\mathcal{O}$ does not intersect $\mathcal{B}$.

One type of counterexample[12] arises if $\mathcal{U}$ is not connected. In Minkowski space in any even dimension D, let $\mathcal{O}$ be a small neighborhood of the origin. Let $\mathcal{U}$ be any open set that contains points to the future of $\mathcal{O}$ and also points to the past of $\mathcal{O}$, but none that can be reached from $\mathcal{O}$ by a null geodesic. Then $\mathcal{O} \subset \mathcal{U}''$, so $\mathcal{A}_{\text{add}}(\mathcal{U}) = \mathcal{A}_{\text{add}}(\mathcal{U}'')$ would predict that $\mathcal{A}_{\text{add}}(\mathcal{O}) \subset \mathcal{A}_{\text{add}}(\mathcal{U})$. In the theory of a massless free scalar field ϕ, however, this statement is false. By Huygen's principle along with commutativity at spacelike separation, the commutator function $G(x,y) = [\phi(x), \phi(y)]$ in that theory is supported on the light cone (for D even), so it vanishes for $x \in \mathcal{O}$, $y \in \mathcal{U}$. Thus in this situation, $\mathcal{A}_{\text{add}}(\mathcal{O})$ commutes with $\mathcal{A}_{\text{add}}(\mathcal{U})$, rather than being contained in $\mathcal{A}_{\text{add}}(\mathcal{U})$.

In two dimensions, a small variant of this gives a counterexample with $\mathcal{U}$ connected. In fact, the setup is closely related to that suggested by Araki. Consider the spacetime $M = \mathbb{R} \times S^1$, with a product metric, where $\mathbb{R}$ parametrizes time and S^1 parametrizes space. Let the point P_2 be slightly to the future of P_1, and let ℓ be a spacelike curve that starts at P_1, spirals once around S^1 to the right, and ends at P_2. Let $\mathcal{B}$ be an open set that is a slight thickening of ℓ, and let Δ be the causal diamond with vertices P_1, P_2. Finally, let $\mathcal{O} \subset \Delta$ be an open set small enough that any right-moving null geodesic that intersects $\mathcal{O}$ does not intersect $\mathcal{B}$ (fig 6). Then $\mathcal{O} \subset \Delta \subset \mathcal{B}''$. In two-dimensional quantum field theory, it is possible to have a local field J with the property that $[J(x), J(y)]$ vanishes unless the points x and y are connected by a right-moving null geodesic. For example, J could be a right-moving conserved current (or a chiral component of the stress tensor) in a conformally invariant theory. Let f be a smearing function supported in $\mathcal{O}$, and consider the operator $J_f = \int \mathrm{d}^2x f(x) J(x) \in \mathcal{A}_{\text{add}}(\mathcal{O}) \subset \mathcal{A}_{\text{add}}(\mathcal{B}'')$. This operator commutes with $\mathcal{A}_{\text{add}}(\mathcal{B})$, rather than being contained in $\mathcal{A}_{\text{add}}(\mathcal{B})$, as one would expect based on $\mathcal{A}_{\text{add}}(\mathcal{B}) = \mathcal{A}_{\text{add}}(\mathcal{B}'')$.

[12]This example is relatively well known but the original reference is not clear. A version of the example of fig. 6 was discussed in [**5**].

In the context of the present article, we are not primarily interested in applying the timelike tube theorem or a hypothetical generalization to an arbitrary open set. We are primarily interested in observables along the timelike worldline γ of an observer. The timelike tube theorem tells us that it does not matter whether we consider a timelike curve γ or an open set that is a suitable slight thickening of it, so we will express the following in terms of the algebra associated to a curve. The example just discussed can be slightly modified to give a counterexample in that context. If P_2 is farther to the future of P_1 than was assumed so far, then the curve γ that wraps around the S^1 (in general any number of times) en route from P_1 to P_2 can be timelike, but almost null. Then the timelike envelope $\mathcal{E}(\gamma)$ is a small neighborhood of γ, but its causal completion $\mathcal{B}''$, which is the same as γ'' if $\mathcal{B}$ is suitably chosen, wraps all the way around the S^1. The statement $\mathcal{A}_{\rm add}(\mathcal{E}(\gamma)) = \mathcal{A}_{\rm add}(\gamma'')$ is false in general just as before.

In this example, γ'' is the same as the causal diamond[13] $\Delta(\gamma)$ with vertices P_1, P_2, so in particular $\mathcal{E}(\gamma) \subsetneq \Delta(\gamma)$. Thus, this example shows that in the statement of the timelike tube theorem, in general $\mathcal{E}(\gamma)$ cannot be replaced with[14] $\Delta(\gamma)$.

So some simple generalizations of the timelike tube theorem are false in a general quantum field theory, and others, such as Araki's conjecture, have unclear status. However, the following somewhat fanciful remarks come to mind. Let us go back to the example in which $\mathcal{U}$ is not connected and is, for example, the union of a small ball $\mathcal{B}_q$ around a point q and a small ball $\mathcal{B}_p$ around a point p to its future. (Similar remarks apply in the other examples.) In this case, for D even, massless free field theory provides a counterexample to $\mathcal{A}_{\rm add}(\mathcal{U}) = \mathcal{A}_{\rm add}(\mathcal{U}'')$, but the argument was limited to massless free field theory and would not generalize in an obvious way to any other theory. Consider then in a more generic theory an observer[15] who has the capability to manipulate the quantum fields in an arbitrary fashion, but only in $\mathcal{U} = \mathcal{B}_p \cup \mathcal{B}_q$. A strategy this observer can follow is to inject into spacetime a probe in the region $\mathcal{B}_q$, directed on a trajectory γ that will carry it to $\mathcal{B}_p$, and equipped and programmed to make some chosen measurement and record the result. Then the observer retrieves the probe at $\mathcal{B}_p$ and reads the answer. Does this construction along with the timelike tube theorem prove that $\mathcal{A}_{\rm add}(\mathcal{E}(\gamma)) \subset \mathcal{A}_{\rm add}(\mathcal{U})$? Though we assume no particular limit on the technological capabilities of the observer, we assume that whatever happens in a part of spacetime to which the observer does not have direct access is governed by the theory that is being probed. Thus in particular the probe that the observer injects into spacetime in $\mathcal{B}_q$ must be something that can be built in this theory. A complex probe cannot be built in free field theory, so there is no tension between existence of this protocol and our earlier observations about free field theory. However, in a sufficiently complex theory – possibly any theory that encompasses the Standard Model, for example – this protocol does indeed hint that $\mathcal{A}_{\rm add}(\mathcal{E}(\gamma)) \subset \mathcal{A}_{\rm add}(\mathcal{U})$. Actually, the observer can choose γ at will (assuming

[13]Here $\Delta(\gamma) = J^+(\gamma) \cap J^-(\gamma)$ is the intersection of the past and future of γ, or equivalently $J^-(P_2) \cap J^+(P_1)$.

[14]Similarly, $\mathcal{E}(\gamma) \subsetneq \Delta(\gamma)$ in a simply-connected spacetime $M = \mathbb{R} \times S^2$, with metric $\mathrm{d}s^2 = -\mathrm{d}t^2 + \mathrm{d}x^2 + \mathrm{d}y^2 + \epsilon \mathrm{d}z^2$, with $x^2 + y^2 + z^2 = 1$, $\epsilon \ll 1$. Embedding $\mathbb{R} \times S^1$ in $\mathbb{R} \times S^2$ at $z = 0$ and choosing γ as before, again $\Delta(\gamma)$ is much larger than $\mathcal{E}(\gamma)$ if P_2 is sufficiently far to the future of P_1. However, as in Araki's example, it is not clear in this case whether it is always true that $\mathcal{A}_{\rm add}(\gamma) = \mathcal{A}_{\rm add}(\Delta(\gamma))$. Free field theory does not provide an immediate counterexample.

[15]In this paragraph, unlike the rest of the article, we consider an observer probing the spacetime from outside, not an observer who propagates on a worldline in the spacetime.

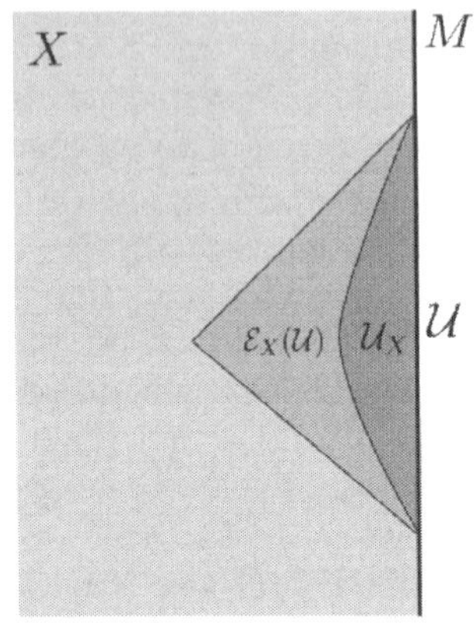

FIGURE 7. A region $\mathcal{U}$ in the boundary M of X, and a slight thickening $\mathcal{U}_X$ of $\mathcal{U}$ in the bulk spacetime X, chosen so that $\mathcal{E}_X(\mathcal{U}) = \mathcal{E}(\mathcal{U}_X)$.

that the probe can be equipped with a rocket engine and programmed to travel on a pre-chosen trajectory), and more generally the observer could inject into the system several probes with pre-chosen trajectories γ_i from $\mathcal{B}(q)$ to $\mathcal{B}(p)$, each designed and programmed to carry out a particular quantum operation, and then the observer can collect the probes in $\mathcal{B}(p)$ and process their (possibly quantum) output at will. All this suggests that in a theory that is sufficiently complex, possibly in the end $\mathcal{A}_{\text{add}}(\mathcal{U}) = \mathcal{A}_{\text{add}}(\mathcal{U}'')$ in general.

3.4. Causal wedge reconstruction. There is actually a manifestation of the timelike tube theorem that has been much discussed in the literature. This is causal wedge reconstruction, or HKLL reconstruction, in the context of the AdS/CFT correspondence [**25**–**33**].

In the AdS/CFT correspondence, one considers a conformal field theory on a globally hyperbolic spacetime M of dimension D. AdS/CFT duality says that an appropriate conformal field theory on M is equivalent to a gravitational theory formulated on a spacetime X that has M for its conformal boundary, and that is globally hyperbolic in the asymptotically AdS sense. In fact, in AdS/CFT duality, one has to consider all possible X's that have a given conformal boundary M, but for our purposes here, we can assume that a particular X is important. Causal wedge reconstruction applies in a semiclassical situation in which X can be viewed, in leading order, as a definite spacetime in which quantum fields are propagating.

Let S be a Cauchy hypersurface in M, B an open set in S, and $\mathcal{U}$ the domain of dependence of B in M. Then $\mathcal{U}$ is its own timelike envelope in M, so the timelike tube theorem, applied directly to the region $\mathcal{U}$ in the CFT on M, says nothing of interest. However, if M is the conformal boundary of X and $\mathcal{U} \subset M$, then it makes sense to consider the timelike envelope $\mathcal{E}_X(\mathcal{U})$ of $\mathcal{U}$ in X (fig. 7). In its simplest form, causal wedge reconstruction then asserts that $\mathcal{A}_{\text{add}}(\mathcal{U}) = \mathcal{A}_{\text{add}}(\mathcal{E}_X(\mathcal{U}))$. This can be viewed as a composite of two statements: (1) the basic AdS/CFT duality expressing local operators in M as limits of local operators in X, and (2) the timelike tube theorem applied to quantum fields in X.

To explain this, we can proceed as follows. First, it is possible to thicken $\mathcal{U}$ slightly to an open set $\mathcal{U}_X \subset X$ such that $\mathcal{E}_X(\mathcal{U}) = \mathcal{E}(\mathcal{U}_X)$. The timelike tube theorem says that $\mathcal{A}_{\rm add}(\mathcal{U}_X)$ does not depend on the precise choice of $\mathcal{U}_X$, and always equals $\mathcal{A}_{\rm add}(\mathcal{E}(\mathcal{U}_X))$. Since this is the case, we can take a limit in which $\mathcal{U}_X$ becomes arbitrarily "thin." The basic AdS/CFT relation between local operators on X and local operators on M can be interpreted as a statement

$$\lim_{\mathcal{U}_X \downarrow \mathcal{U}} \mathcal{A}_{\rm add}(\mathcal{U}_X) = \mathcal{A}_{\rm add}(\mathcal{U}), \tag{3.3}$$

where $\lim_{\mathcal{U}_X \downarrow \mathcal{U}}$ refers to a limit in which the bulk open set $\mathcal{U}_X$ collapses down to the boundary open set $\mathcal{U}$. On the left hand side, $\mathcal{A}_{\rm add}(\mathcal{U}_X)$ is the algebra generated by bulk local operators in the bulk region $\mathcal{U}_X$, and on the right hand side, $\mathcal{A}_{\rm add}(\mathcal{U})$ is the algebra generated by boundary local operators in the boundary region $\mathcal{U}$. Eqn. (3.3) is a way to express the fact that boundary local operators are the boundary limits of bulk local operators. Since the timelike tube theorem tells us that $\mathcal{A}_{\rm add}(\mathcal{U}_X) = \mathcal{A}_{\rm add}(\mathcal{E}(\mathcal{U}_X)) = \mathcal{A}_{\rm add}(\mathcal{E}_X(\mathcal{U}))$ regardless of the choice of $\mathcal{U}_X$, the limit in eqn. (3.3) is trivial and we learn that

$$\mathcal{A}_{\rm add}(\mathcal{E}_X(\mathcal{U})) = \mathcal{A}_{\rm add}(\mathcal{U}). \tag{3.4}$$

In a globally hyperbolic spacetime that satisfies the null energy condition, $\mathcal{E}_X(\mathcal{U})$ is the causal wedge of $\mathcal{U}$, and eqn. (3.4) is the usual statement of causal wedge reconstruction. However, this requires some explanation,[16] since the usual definition of the causal wedge is slightly different. The causal wedge of $\mathcal{U}$, which we will denote as $\mathcal{C}(\mathcal{U})$, is usually defined to consist of all points $r \in X$ that are contained in a causal curve $\gamma \subset X$ between two points $q, p \in \mathcal{U}$, say with p to the future of q. (Equivalently, $\mathcal{C}(\mathcal{U})$ is the intersection of the past and future of $\mathcal{U}$ in X.) Instead, $\mathcal{E}_X(\mathcal{U})$ contains all points $r \in X$ that are contained in a causal curve $\gamma \subset X$ between points $q, p \in \mathcal{U}$ such that γ can be deformed, through a family of causal curves with fixed endpoints, to a causal curve entirely in M (and therefore in $\mathcal{U}$). Thus to show that $\mathcal{E}_X(\mathcal{U}) = \mathcal{C}(\mathcal{U})$, we have to show that any causal curve $\gamma \subset X$ that starts and ends at points $q, p \in \mathcal{U}$ can be deformed, through a family of causal curves with those fixed endpoints, to a causal curve in M. To prove this, let $\mathcal{R}_{qp}$ be the space of causal curves in X with initial and final endpoints q, p. Pick a causal curve $\zeta_{qp} \subset \mathcal{U}$ from q to p, and let $\mathcal{T}_\zeta$ be the space of causal curves in X with initial endpoint q and any final endpoint $s \in \zeta_{qp}$. A causal curve from q to s can be converted to a causal curve from q to p by gluing on the segment ζ_{sp} of ζ_{qp}. This operation maps $\mathcal{T}_\zeta$ continuously to $\mathcal{R}_{qp}$; on the other hand, $\mathcal{R}_{qp} \subset \mathcal{T}_\zeta$. So a causal curve $\gamma \subset X$ with endpoints q, p can be deformed in $\mathcal{R}_{qp}$ to a curve entirely in M if and only if it can be so deformed in $\mathcal{T}_\zeta$. Hence, to prove that $\mathcal{E}_X(\mathcal{U}) = \mathcal{C}(\mathcal{U})$, it suffices to show that any causal curve $\gamma \subset X$ with endpoints q, p can be deformed in $\mathcal{T}_\zeta$ to a curve entirely in M. This question is topological in nature: is γ in the connected component of $\mathcal{T}_\zeta$ that contains curves that lie entirely in M? So the answer is invariant under an infinitesimal perturbation of X, and we

[16]The following proof is not needed in the rest of the article. Facts used in the proof are explained, for example, in [**34**–**36**]. On compactness of spaces of causal curves, see for instance sections 2 and 3.3 of [**36**]; on the fact that a causal curve that satisfies a promptness condition is a null geodesic without focal points, see sections 5.1 and 5.2 of that article; on the completeness of a null geodesic in an asymptotically AdS spacetime X whose ends are on the conformal boundary of X, see section 7.3; on the fact that a complete null geodesic along which the null energy condition is satisfied as a strict inequality must have focal points, see section 8.2.

can make such a perturbation to ensure that the null energy condition is satisfied in X as a strict inequality (not as an equality). We will see later exactly where in X the perturbation should be made. On $\mathcal{T}_\zeta$, one can define the following continuous function φ: if $\gamma_* \in \mathcal{T}_\gamma$ has endpoints $q, s \in \zeta_{qp}$, then $\varphi(\gamma_*)$ is the proper time elapsed from q to s along ζ_{qp}. An important detail is that we allow the case of a causal curve that consists of only one point. Thus in particular $\mathcal{T}_\zeta$ contains a point q_* that corresponds to a causal curve consisting only of the point q. The point q_* is the absolute minimum of φ, since $\varphi(q_*) = 0$, and at other points in $\mathcal{T}_\zeta$, $\varphi > 0$. The purpose of including the point q_* in the definition of $\mathcal{T}_\zeta$ is to ensure that $\mathcal{T}_\zeta$ is compact; this compactness follows from the fact that the initial endpoint of a curve $\gamma_* \in \mathcal{T}_\zeta$ is fixed and its final endpoint s ranges over the compact set ζ_{qp}, along with the fact that in a globally hyperbolic spacetime, the space of causal curves with specified endpoints is compact. Compactness of $\mathcal{T}_\zeta$ ensures that in every connected component of $\mathcal{T}_\zeta$, the nonnegative function φ has an absolute minimum. Any $\gamma \in \mathcal{T}_\zeta$ can be deformed in $\mathcal{T}_\zeta$ to the minimum of φ in its connected component. We will conclude the proof by showing that the point q_* is the unique local minimum of the function φ, implying that the local minimum of φ to which γ can be deformed in $\mathcal{T}_\zeta$ is actually the absolute minimum q_*. This in particular implies that γ can be deformed in $\mathcal{T}_\zeta$ to a causal curve entirely in M, as we wished to show. To show that φ has no other local minimum, we note that any point in $\mathcal{T}_\zeta$ other than q_* is a nontrivial causal curve $\widetilde{\gamma}$ with distinct endpoints in M. In general, if the initial point of a causal curve $\widetilde{\gamma}$ is specified (here q) and the final point of the curve is constrained by a condition of "promptness" (here $\widetilde{\gamma}$ is supposed to end along ζ_{qp}, and the condition for it to be a local minimum of φ means that it arrives on ζ_{qp} sooner than any nearby causal curve from q; this is the promptness condition), then $\widetilde{\gamma}$ must be a null geodesic without focal points. However, in an asymptotically AdS spacetime, any null geodesic between distinct points on the conformal boundary is complete in both directions. As remarked earlier, we can assume that the null energy condition in X is satisfied along $\widetilde{\gamma}$ as a strict inequality. A null geodesic that is complete in both directions and along which the null energy condition is satisfied as a strict inequality always has focal points. So the function φ can have no local minimum other than its absolute minimum q_*, completing the proof.

4. An algebra of observables for de Sitter space

Finally, we turn to a concrete example in which, in order to define a sensible algebra of observables, it is necessary to include an observer in the description [**1**]. De Sitter space in D dimensions or dS_D is the maximally symmetric solution of Einstein's equations with a positive cosmological constant. It can be described by the metric

$$\mathrm{d}s^2 = -\mathrm{d}t^2 + R^2 \cosh^2(t/R)\mathrm{d}\Omega^2, \tag{4.1}$$

where R is the radius of curvature and $\mathrm{d}\Omega^2$ is the metric of a round sphere $S^{\mathsf{D}-1}$ of unit radius. This sphere is compact, so dS_D is an example of a closed universe. At time t, the sphere has radius $R(t) = R\cosh(t/R)$, so it grows exponentially for $t \to +\infty$ or $t \to -\infty$. The exponential growth for $t \to +\infty$ is believed to be a good approximation to what is currently beginning to happen in the real world.

In the 1970's, Gibbons and Hawking [**37**] studied de Sitter space as a simple example of a spacetime with a cosmological horizon – in which an observer cannot

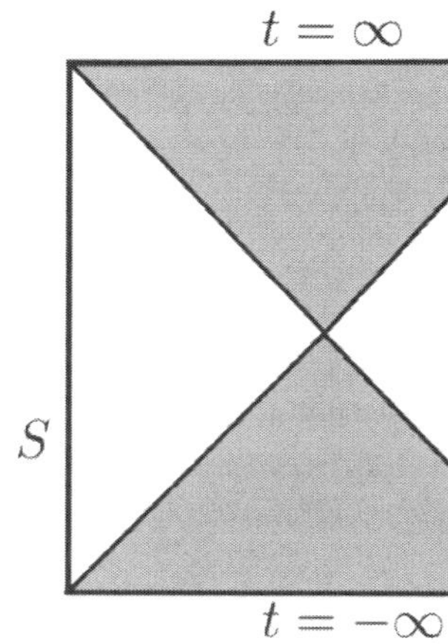

FIGURE 8. The Penrose diagram of de Sitter space. Past infinity is at the bottom; future infinity is at the top. The left and right boundaries represent timelike geodesics that can be viewed as the trajectories of observers who remains at rest on the south or north pole of $S^{\mathsf{D}-1}$, respectively. The diagonals represent the past and future horizons of those observers; the unshaded triangles on the left and right are the static patches that are causally accessible to these observers.

see the whole universe. They attached a temperature and entropy to the de Sitter horizon, as Bekenstein and Hawking had done not long before for the horizon of a black hole. The thermal interpretation is most obvious in Euclidean signature, where dS_{D} becomes simply a D-sphere S^{D}, with metric

$$ds_E^2 = d\tau^2 + R^2 \cos^2(\tau/R) d\Omega^2. \tag{4.2}$$

In ordinary quantum field theory in de Sitter space (and also in the presence of semiclassical gravity), there is a natural de Sitter state Ψ_{dS} such that correlation functions in this state can be obtained by analytic continuation from Euclidean signature. Let C be a great circle in S^{D} and let λ be a length parameter along C. C is an orbit of a $\mathrm{U}(1)$ symmetry of S^{D} (which is uniquely determined if we say that its fixed point set is a copy of $S^{\mathsf{D}-2}$). If we normalize the generator H of this $\mathrm{U}(1)$ to act as $\partial/\partial\lambda$ along C, then H obeys $\exp(-2\pi RH) = 1$. When continued to Lorentz signature, this leads to the striking statement that correlation functions in the state Ψ_{dS} have a thermal interpretation at the de Sitter temperature $T_{\mathrm{dS}} = 1/\beta_{\mathrm{dS}}$, where $\beta_{\mathrm{dS}} = 2\pi R$ [**37, 38**]. The thermal interpretation of de Sitter space has been extensively explored for nearly half a century. For a small sampling of the relevant literature, see [**39–53**].

In Lorentz signature, de Sitter space is conveniently understood via a Penrose diagram (fig. 8). The great circle C continues in Lorentz signature to a hyperbola that has two components, each of them a geodesic. Coordinates can be chosen so that these geodesics make up the left and right boundaries of the Penrose diagram. Given an observer traveling on a geodesic in dS_{D}, we can assume that the worldline of this observer is, say, the left boundary of the diagram. The observer then has past and future horizons which are the diagonals in the picture. The region causally accessible to the observer (the region the observer can see and also can influence) is bounded by these diagonals, along with the left boundary of the diagram. A similar region of the diagram is causally accessible to an observer on the right boundary. The operator H generates a symmetry of the Penrose diagram; with a suitable choice of sign, it maps the region accessible to the left observer forward in time and the region accessible to the right observer backwards in time. Near the bifurcate horizon where the two diagonals meet, H looks like the generator of a Lorentz boost.

If t measures the time along the left boundary, then on the left boundary of the figure, $H = \partial/\partial t$. So it is natural for an observer whose worldline is the left boundary to interpret H as a generator of time translations. With this interpretation, since H is a symmetry of the region causally accessible to the left observer, this region is time-independent and thus "static." This is why the region causally accessible to an observer in de Sitter space has been called a "static patch." However, this view of de Sitter space as being "static" is highly observer-dependent. In a global view, the form (4.1) of the de Sitter metric shows that $\mathrm{dS_D}$ is expanding exponentially both toward the future and toward the past, so globally one would definitely not call $\mathrm{dS_D}$ "static."

An ordinary quantum field theory in de Sitter space has a Hilbert space $\mathcal{H}_0$ of quantum states. In such a theory, we associate to the static patch (or any region) an algebra $\mathcal{A}_0$ of observables consisting of operators on $\mathcal{H}_0$ that act on the quantum fields in the region in question. The algebra of any local region, and in particular the algebra $\mathcal{A}_0$ of the static patch, is a possibly unfamiliar Type III von Neumann algebra. This is an algebra with an infinite amount of quantum entanglement built in, giving an abstract explanation of the fact that entanglement entropy is ultraviolet divergent in quantum field theory. Including weakly coupled gravitational fluctuations does not qualitatively change the picture. We simply include the weakly coupled gravitational field as one more field in the construction of $\mathcal{H}_0$ and $\mathcal{A}_0$. What does really change the picture is that in a closed universe, such as de Sitter space, the isometries have to be treated as constraints. In the case of the static patch, the important constraint is the Hamiltonian H. Imposing H as a constraint means that we should replace $\mathcal{A}_0$ by $\mathcal{A}_0^H$, its invariant subalgebra. But that does not work: the invariant subalgebra is trivial. Roughly, that is because anything that commutes with H can be averaged over all the thermal fluctuations and replaced by its thermal average, a c-number. A technical statement is that there are no nontrivial invariants in $\mathcal{A}_0$ because H generates the modular automorphism group of the state Ψ_{dS} for the algebra $\mathcal{A}_0$ [**39**], and therefore acts ergodically.

To get a reasonable algebra of observables, we include an observer in the analysis. Of course, as noted in the introduction, in principle an observer should really be described by the theory, not injected from outside. What it means to include an observer is that we consider a "code subspace" of states in which an observer is present in the static patch, and then we consider operators that can be defined in the low energy effective field theory in this code subspace, though they are not well-defined on the whole Hilbert space.

Should we be surprised that we need to include the observer in the analysis to get a sensible answer? As was also noted in the introduction, a gravitating system in a closed universe is the situation in which we are most likely to need to explicitly incorporate the observer in the analysis. That is exactly the situation here because de Sitter space is a simple model of a closed universe, that is, a universe with compact spatial sections.

Once we include an observer in the analysis, there is a rationale to study a particular static patch, namely the region that is causally accessible to that given observer. Moreover, the timelike tube theorem tells us that the algebra $\mathcal{A}_0$ of observables in the causally accessible region can be interpreted as the algebra generated by the quantum fields along the observer's worldline. Thus the algebra $\mathcal{A}_0$ of the static patch becomes, in a sense, operationally meaningful once the observer is included.

It turns out that to get a sensible answer, it suffices to consider a minimal model in which the observer is characterized just by a clock with Hamiltonian

$$H_{\mathrm{obs}} = q. \tag{4.3}$$

It is physically reasonable to assume that the observer's energy is bounded below. The precise lower bound will not be important and we will just take it[17] to be 0. With that choice, the effect of including the observer is to modify the Hilbert space by

$$\mathcal{H}_0 \to \mathcal{H}_1 = \mathcal{H}_0 \otimes L^2(\mathbb{R}_+), \tag{4.4}$$

where q is a multiplication operator on the positive half-line $\mathbb{R}_+$. The algebra is likewise extended from $\mathcal{A}_0$ to

$$\mathcal{A}_1 = \mathcal{A}_0 \otimes B(L^2(\mathbb{R}_+)). \tag{4.5}$$

The last factor is the (Type I) algebra of all bounded operators on $L^2(\mathbb{R}_+)$.

Finally the constraint becomes the total Hamiltonian of the quantum fields plus the observer:

$$H \to \widehat{H} = H + H_{\mathrm{obs}}. \tag{4.6}$$

The "correct" algebra of observables taking account of the presence of the observer is therefore

$$\mathcal{A} = \mathcal{A}_1^{\widehat{H}}, \tag{4.7}$$

that is, the $\widehat{H}$-invariant part of $\mathcal{A}_1$. To be more exact, this is the algebra of observables accessible to the observer in the limit $G_N \to 0$. In higher order in G_N, it will be necessary to incorporate direct couplings between the observer and the quantum fields; the constraint operator will not be a simple sum. Of course, the notion of the algebra of a spacetime region is presumably only well-defined in the limit $G_N \to 0$; the timelike tube theorem suggests that in going beyond that limit, we should reinterpret $\mathcal{A}$ as the algebra of observables along the observer's worldline. As explained presently, the important conclusions that we will draw about $\mathcal{A}$ are expected to be robust against perturbative corrections in G_N.

The answer (4.7) makes sense, unlike the previous one. The reason is that once an observer is present, we can "gravitationally dress" any operator to the observer's world-line. It is easiest to explain this if we momentarily ignore the lower bound $q \geq 0$. In fact, in the field of a black hole, there is a problem that is mathematically quite similar to what we are discussing here, but without the lower bound on q [**1**]. If we do ignore the lower bound on q, then the Hilbert space including the observer

[17] More realistically, the observer would have a mass m and minimum energy mc^2. If the observer is minimally coupled to gravity, then the observer worldline will be a geodesic. In the presence of an observer, the relevant static patch is the region causally accessible to the observer.

is $\mathcal{H}_1' = \mathcal{H}_0 \otimes L^2(\mathbb{R})$, and the algebra is $\mathcal{A}_1' = \mathcal{A}_0 \otimes B(L^2(\mathbb{R}))$. Here $B(L^2(\mathbb{R}))$ is generated by (bounded functions of) q and $p = -\mathrm{i}\frac{\mathrm{d}}{\mathrm{d}q}$. If Θ is the projection operator onto states with $q \geq 0$, then $\mathcal{H}_1 = \Theta\mathcal{H}_1'$, $\mathcal{A}_1 = \Theta\mathcal{A}_1'\Theta$. Similarly the $\widehat{H}$-invariant subalgebras $\mathcal{A} = \mathcal{A}_1^{\widehat{H}}$, $\mathcal{A}' = \mathcal{A}_1'^{\widehat{H}}$ are related by $\mathcal{A} = \Theta\mathcal{A}'\Theta$. So $\mathcal{A}$ is easily constructed once we understand $\mathcal{A}'$.

To construct $\mathcal{A}'$, we reason as follows. For any $\mathsf{a} \in \mathcal{A}_0$, the operator

$$\widehat{\mathsf{a}} = e^{\mathrm{i}pH}\mathsf{a}e^{-\mathrm{i}pH}$$

commutes with the constraint $\widehat{H} = H + q$. One more operator that commutes with the constraint is q itself (or equivalently $-H$, which equals q modulo the constraint). It follows from a classic result of Takesaki [**54**] that by coincidence was proved in the early days of black hole thermodynamics that (1) there are no more operators in $\mathcal{A}_1'$ that commute with the constraint, and (2) the algebra $\mathcal{A}'$ that is generated by the operators $\widehat{\mathsf{a}}$ and q is actually a von Neumann algebra of Type II_∞, with a trivial center (that is, its center consists only of c-numbers). In the case of the static patch in de Sitter space, we want to impose the constraint $q \geq 0$, so the appropriate algebra is not $\mathcal{A}'$ but $\mathcal{A} = \Theta\mathcal{A}\Theta$. This is an algebra of Type II_1, again with trivial center.

A basic introduction to the relevant facts about von Neumann algebras can be found in [**1**], among other places. For much more depth, see [**55**]. An important fact for our purposes is that a Type II algebra $\mathcal{A}$ (unlike one of Type III, which we would have in the absence of gravity) has a trace, that is a complex-valued linear function obeying $\mathrm{Tr}\,\mathsf{ab} = \mathrm{Tr}\,\mathsf{ba}$, for all $\mathsf{a}, \mathsf{b} \in \mathcal{A}$. Moreover, the trace in a Type II algebra is positive, in the sense that $\mathrm{Tr}\,\mathsf{aa}^\dagger > 0$ for all $\mathsf{a} \in \mathcal{A}$.

A von Neumann algebra with trivial center is called a factor. Thus the above analysis implies that in the limit $G_N = 0$, the algebra of observables in the static patch is a factor of Type II_1. A factor is the von Neumann algebra analog of a simple Lie group. A simple Lie group is rigid, in the sense that it has has no infinitesimal deformations. That is not true for a non-simple Lie group; for example, the symmetry group of $\mathbb{R}^n$ is non-simple, and can be deformed, by an arbitrarily small perturbation of the commutation relations in its Lie algebra, to the symmetry group $\mathrm{SO}(n+1)$ of S^n. A similar statement holds for von Neumann algebras: an algebra with a non-trivial center can potentially be deformed to an algebra of a different type, while making the center smaller,[18] but a factor is rigid and has no infinitesimal deformations. Though we have only analyzed the algebra of observables in the static patch in the limit $G_N \to 0$, because the answer that we obtained is a factor, perturbative corrections in G_N are not expected to modify the algebra up to isomorphism (they will modify the commutation relations among operators along the observer's worldline, but not the isomorphism class of the algebra that those operators generate). Nonperturbatively, matters are unclear and indeed it is quite unclear whether quantum de Sitter space makes sense nonperturbatively. If quantum de Sitter space does make sense nonperturbatively, then one expects to describe it by a finite-dimensional Hilbert space [**45, 46**], and the algebra will have to be of Type I.

[18] For example, the Type III algebra described by Leutheusser and Liu in the $G_N \to 0$ limit in the field of a black hole [**56, 57**] has a nontrivial center and is modified by perturbative corrections to a factor of Type II_∞ [**1, 58**].

A Type II algebra does not have pure states, but because it does have a trace, familiar ideas like density matrices and entropies make sense for a state of such an algebra [**59, 60**]. To a global state Ψ of de Sitter space plus the observer, reduced to the static patch, one can associate a density matrix $\rho \in \mathcal{A}$, characterized by

$$\langle \Psi | \mathsf{a} | \Psi \rangle = \operatorname{Tr} \mathsf{a} \rho, \ \forall \mathsf{a} \in \mathcal{A}. \tag{4.8}$$

Therefore, we can define a von Neumann entropy

$$S(\rho) = -\operatorname{Tr} \rho \log \rho. \tag{4.9}$$

There is no such definition in the absence of gravity, because without gravity, the observables in the static patch constitute the Type III algebra $\mathcal{A}_0$. The fact that gravity turns the Type III algebra into a Type II algebra gives an abstract explanation for why entropy is better defined in the presence of gravity than in ordinary quantum field theory. However, from a physical point of view, Type II entropy is a renormalized entropy from which an infinite constant has been subtracted.

As already remarked, if we put no constraint on the observer's energy, we get an algebra of Type II_∞; if we assume the observer's energy is bounded below, we get an algebra of Type II_1. In a Type II_1 algebra, the trace is defined for all elements of the algebra, while in an algebra of Type II_∞, the trace is finite only for a dense subset of elements of the algebra. In particular, for Type II_1, the identity element has a finite trace and we can normalize the trace so that the trace of the identity is 1, while in Type II_∞, the trace of the identity element is $+\infty$.

For our purposes, the important difference is that a Type II_1 algebra has a state of maximum possible entropy, which is the "maximally mixed" state with density matrix $\rho = 1$. This is consistent with $\operatorname{Tr} \rho = 1$, assuming that the trace has been appropriately normalized. Evaluating $S(\rho) = -\operatorname{Tr} \rho \log \rho$ for $\rho = 1$, we see that the state with $\rho = 1$ has entropy 0. It is not difficult to prove that all other states have negative entropy (for example, see [**1, 60**]). This tells us the meaning of the subtraction that is involved in defining the entropy of a state of a Type II_1 algebra so that $S(\rho) = 0$ for $\rho = 1$: the constant that is subtracted is the maximum possible entropy. In general, entropy of a state of a Type II_1 algebra is the entropy difference of that state relative to the entropy of the maximum entropy state. The maximum entropy state with $\rho = 1$ is the Type II_1 analog of a maximally mixed state in ordinary quantum mechanics, in which the density matrix is a multiple of the identity. By contrast, there is no upper bound on the entropy of a state of a Type II_∞ algebra.

It is felicitous that a Type II_1 algebra has a state of maximum entropy, because in fact de Sitter space is believed to have a state of maximum entropy – namely "empty de Sitter space," with all the entropy in the cosmological horizon [**40–42**]. So to get a reasonable model of de Sitter space, it is important to assume that the observer's energy is bounded below, which is a more reasonable assumption anyway, thereby ensuring that the algebra of observables is of Type II_1, not Type II_∞. One can explicitly construct a state $\Psi_{\max} \in \mathcal{H}_{\mathrm{dS}}$ that has maximum entropy (and thus density matrix $\rho = 1$) when reduced to the static patch. In fact, one can pick $\Psi_{\max} = \Psi_{\mathrm{dS}} \otimes e^{-\beta_{\mathrm{dS}} q/2} \sqrt{\beta_{\mathrm{dS}}}$. Thus empty de Sitter space, tensored with a state in which the observer's energy has a thermal distribution at the de Sitter temperature, has maximum entropy from the point of view of the Type II_1 algebra. In this sense the analysis based on the Type II_1 algebra agrees with the claim that empty de Sitter space has maximum entropy.

We can now compare with some further claims in the previous literature. First of all, since the maximum entropy state has $\rho = 1$, it has a "flat entanglement spectrum" (all eigenvalues of the density matrix are equal) and accordingly the Rényi entropies all vanish:

$$S_\alpha(\rho) = \frac{1}{1-\alpha} \log \operatorname{Tr} \rho^\alpha = 0. \tag{4.10}$$

This matches with a result that has been found using Euclidean path integrals to analyze the Rényi entropies of the static patch [**50**]. Given the assertion that de Sitter space has a state of maximum entropy, this result is what one should expect. In ordinary quantum mechanics, the maximum entropy state of a system is "maximally mixed," with a "flat entanglement spectrum" (the density matrix is a multiple of the identity and all its eigenvalues are equal) and its Rényi entropies are independent of α.

Now, suppose that the observer makes a measurement with two outcomes that correspond to the projection operators Π and $1 - \Pi$. The probabilities of the two outcomes are $\operatorname{Tr} \Pi$ and $\operatorname{Tr}(1-\Pi) = 1 - \operatorname{Tr} \Pi$. All values $0 \leq \operatorname{Tr} \Pi \leq 1$ are possible. If the outcome corresponding to Π is observed, then after this measurement, the density matrix is

$$\sigma = \frac{1}{\operatorname{Tr} \Pi} \Pi.$$

Since the two eigenvalues of σ are 0 and $1/\operatorname{Tr} \Pi$, one has $\sigma \log \sigma = \sigma \log(1/\operatorname{Tr} \Pi)$ so the entropy after the observation is

$$S(\sigma) = -\operatorname{Tr} \sigma \log \sigma = -\log(1/\operatorname{Tr} \Pi).$$

The entropy reduction from knowing the outcome is therefore $\Delta S = \log(1/\operatorname{Tr} \Pi)$, and this is related to the probability $p = \operatorname{Tr} \Pi$ of the given outcome by

$$p = e^{-\Delta S}.$$

However, the probability of a (low entropy) energy E fluctuation of the static patch is

$$p = e^{-\beta_{\rm dS} E},$$

according to the thermal interpretation of de Sitter space. Since also $p = e^{-\Delta S}$, we must have for consistency of the two descriptions

$$e^{-\beta_{\rm dS} E} = e^{-\Delta S}.$$

In other words, "'thermal" suppression of a fluctuation can be understood as purely entropic suppression. This is surprising, but it has been argued before on other grounds, notably by considering the case that the "fluctuation" is a small black hole at the center of the static patch [**49**].

Which part of this is unexpected? The formula $p = e^{-\Delta S}$ for the probability of an outcome is an inevitable consequence of having a maximum entropy state in which all states are equally probable. In other words, if all states are equally likely, then the probability of a given outcome is just proportional to the number of microstates that are compatible with that outcome. Here we use language appropriate for an ordinary quantum system with a finite-dimensional Hilbert space. For a system described by a Type II_1 algebra, the number of microstates compatible with any given outcome is infinite, and one has to express the argument in terms of traces, as we did earlier. In short, the surprise is not that $p = e^{-\Delta S}$, which one should expect for a maximum entropy state, but that after coupling to gravity and including the observer, the thermal state Ψ_{dS} can be promoted to a maximal entropy state $\Psi_{\max}$.

Let us see explicitly how this happens at the level of correlation functions.[19] First we recall some basic facts about time-dependent correlation functions in a thermal ensemble. The time dependence of an operator is defined in the usual way by $\mathsf{a}(t) = e^{\mathrm{i}Ht}\mathsf{a}e^{-\mathrm{i}Ht}$. A typical time-dependent two-point function is

$$F(t) = \langle \mathsf{a}(t)\mathsf{b}(0)\rangle_\beta = \frac{1}{Z}\mathrm{tr}\, e^{-\beta H}\mathsf{a}(t)\mathsf{b}(0) = \frac{1}{Z}\mathrm{tr}\, e^{-\beta H}e^{\mathrm{i}Ht}\mathsf{a}e^{-\mathrm{i}Ht}\mathsf{b}. \tag{4.11}$$

Here tr is the trace in the Hilbert space of a thermal system with Hamiltonian H, and Z is the partition function. It follows immediately from the definition that $F(t)$ is holomorphic in a strip $0 \geq \mathrm{Im}\, t \geq -\beta$ and moreover that the boundary value at $\mathrm{Im}\, t = \beta$ is the thermal correlator with the opposite ordering of the operators. In other words, let

$$G(t) = \langle \mathsf{b}(0)\mathsf{a}(t)\rangle_\beta = \frac{1}{Z}\mathrm{tr}\, e^{-\beta H}\mathsf{b}e^{\mathrm{i}Ht}\mathsf{a}e^{-\mathrm{i}Ht}. \tag{4.12}$$

This function is related to F by

$$G(t) = F(t - \mathrm{i}\beta). \tag{4.13}$$

The precise meaning of this statement is that there is a function holomorphic in the strip $0 \geq \mathrm{Im}\, t \geq -\beta$ whose boundary value on the upper boundary is F, while its boundary value on the lower boundary is G.

Let us express this in terms of Fourier transforms of the two functions. Suppose that

$$\begin{aligned} F(t) &= \int_{-\infty}^{\infty} \frac{\mathrm{d}\omega}{2\pi} e^{-\mathrm{i}\omega t} f(\omega) \\ G(t) &= \int_{-\infty}^{\infty} \frac{\mathrm{d}\omega}{2\pi} e^{-\mathrm{i}\omega t} g(\omega). \end{aligned} \tag{4.14}$$

Then eqn. (4.13) becomes

$$g(\omega) = e^{-\beta\omega} f(\omega). \tag{4.15}$$

These facts about an ordinary quantum system also hold in de Sitter space for time-dependent correlation functions in the state Ψ_{dS}, though more sophisticated proofs are required. To make precisely the same argument for thermal correlators in the static patch of de Sitter space that one can make for an ordinary thermal system, one should have a Hilbert space that describes the static patch, and tr should be the trace in this Hilbert space. Such a "one-sided" Hilbert space does not exist for quantum fields in de Sitter space, because the algebra of the static patch is of Type III, not Type I. Instead one only has a global Hilbert space describing all of de Sitter space. To deal with this situation requires more careful arguments using Tomita-Takesaki theory [**39**].

What happens after coupling to gravity? As described earlier, we introduce an observer with energy q, and canonical momentum $p = -\mathrm{i}/\mathrm{d}q$. Let $\Theta(x)$ be the function that is 1 for $x \geq 0$ and vanishes for $x < 0$, so that $\Theta = \Theta(q)$ is the projection operator onto states of $q \geq 0$. To get a model of de Sitter space with

---

[19] An error in a previous claim about this matter was pointed out by G. Penington.

weakly coupled gravity, we replace operators a, b of the original thermal system by gravitationally dressed versions

$$\widehat{\mathsf{a}} = \Theta e^{\mathrm{i}pH}\mathsf{a}e^{-\mathrm{i}pH}\Theta,\ \widehat{\mathsf{b}} = \Theta e^{\mathrm{i}pH}\mathsf{b}e^{-\mathrm{i}pH}\Theta. \tag{4.16}$$

If we do not include the projection operators Θ, the algebra would be of Type II_∞, with no maximum entropy state. Including the projection operators gives a Type II_1 algebra with a maximum entropy state. As claimed earlier, this state is

$$\Psi_{\mathrm{max}} = \Psi_{\mathrm{dS}} \otimes \Psi_\beta(q), \tag{4.17}$$

where Ψ_{dS} is the natural de Sitter invariant state in the absence of gravity, and

$$\Psi_\beta(q) = \begin{cases} e^{-\beta q/2}\sqrt{\beta} & q \geq 0 \\ 0 & q < 0 \end{cases} \tag{4.18}$$

is a state in which the observer's energy has a thermal distribution. Time dependence is introduced in the usual way, by, for example, $\widehat{\mathsf{a}}(t) = e^{\mathrm{i}Ht}\widehat{\mathsf{a}}e^{-\mathrm{i}Ht}$.

In the limit that gravity is weakly coupled, the de Sitter analogs of $F(t)$, $G(t)$ are supposed to be

$$\begin{aligned} \widehat{F}(t) &= \langle \Psi_{\mathrm{max}}|\widehat{\mathsf{a}}(t)\widehat{\mathsf{b}}|\Psi_{\mathrm{max}}\rangle = \langle \Psi_{\mathrm{max}}|\mathsf{a}e^{-\mathrm{i}Ht}e^{-\mathrm{i}pH}\Theta(q)e^{\mathrm{i}pH}\mathsf{b}|\Psi_{\mathrm{max}}\rangle \\ \widehat{G}(t) &= \langle \Psi_{\mathrm{max}}|\widehat{\mathsf{b}}\widehat{\mathsf{a}}(t)|\Psi_{\mathrm{max}}\rangle = \langle \Psi_{\mathrm{max}}|\mathsf{b}e^{-\mathrm{i}pH}\Theta(q)e^{\mathrm{i}pH}e^{\mathrm{i}Ht}\mathsf{a}|\Psi_{\mathrm{max}}\rangle. \end{aligned} \tag{4.19}$$

We have used the facts that $H\Psi_{\mathrm{max}} = 0$, and $\Theta\Psi_{\mathrm{max}} = \Psi_{\mathrm{max}}$. We observe now that

$$e^{-\mathrm{i}pH}\Theta(q)e^{\mathrm{i}pH} = \Theta(q-H) \tag{4.20}$$

so eqn. (4.19) simplifies to

$$\begin{aligned} \widehat{F}(t) &= \langle \Psi_{\mathrm{max}}|\mathsf{a}e^{-\mathrm{i}Ht}\Theta(q-H)\mathsf{b}|\Psi_{\mathrm{max}}\rangle \\ \widehat{G}(t) &= \langle \Psi_{\mathrm{max}}|\mathsf{b}\Theta(q-H)e^{\mathrm{i}Ht}\mathsf{a}|\Psi_{\mathrm{max}}\rangle. \end{aligned} \tag{4.21}$$

We expect $\widehat{F}(t) = \widehat{G}(t)$, since either of these functions is supposed to be $\mathrm{Tr}\,\widehat{\mathsf{a}}(t)\mathsf{b}$, where Tr is the trace of the Type II_1 algebra (as opposed to the trace in the Hilbert space of an underlying thermal system, which has been denoted tr – and which anyway does not really exist in the case of de Sitter space).

Now let us write a Fourier-transformed version of these last two equations. In the formula for $\widehat{F}(t)$, we see that a contribution that varies with t as $e^{-\mathrm{i}\omega t}$ comes from intermediate states with $H = \omega$, but in the formula for $\widehat{G}(t)$, such a contribution comes from states with $H = -\omega$. The upshot of this is that Fourier-transformed formulas can be written for $\widehat{F}$ and $\widehat{G}$ that are just analogous to those of eqn. (4.14), but with an extra factor involving the expectation value in the state Ψ_β of $\Theta(q \mp \omega)$:

$$\begin{aligned} \widehat{F}(t) &= \int_{-\infty}^{\infty} \frac{\mathrm{d}\omega}{2\pi} e^{-\mathrm{i}\omega t} f(\omega)\langle \Psi_\beta|\Theta(q-\omega)|\Psi_\beta\rangle \\ \widehat{G}(t) &= \int_{-\infty}^{\infty} \frac{\mathrm{d}\omega}{2\pi} e^{-\mathrm{i}\omega t} g(\omega)\langle \Psi_\beta|\Theta(q+\omega)|\Psi_\beta\rangle. \end{aligned} \tag{4.22}$$

Using the definition of Ψ_β, we can find explicit formulas for $u_\beta(\omega) = \langle\Psi_\beta|\Theta(q - \omega)|\Psi_\beta\rangle$ and $v_\beta(\omega) = \langle\Psi_\beta|\Theta(q+\omega)|\Psi_\beta\rangle$:

$$u_\beta(\omega) = \begin{cases} e^{-\beta\omega} & \omega \geq 0 \\ 1 & \omega < 0 \end{cases}$$

$$v_\beta(\omega) = \begin{cases} 1 & \omega \geq 0 \\ e^{\beta\omega} & \omega < 0. \end{cases} \tag{4.23}$$

Note that

$$u_\beta(\omega)v_\beta(\omega)^{-1} = e^{-\beta\omega}. \tag{4.24}$$

We have then

$$\widehat{F}(t) = \int_{-\infty}^{\infty} \frac{\mathrm{d}\omega}{2\pi} e^{-\mathrm{i}\omega t}\widehat{f}(\omega)$$

$$\widehat{G}(t) = \int_{-\infty}^{\infty} \frac{\mathrm{d}\omega}{2\pi} e^{-\mathrm{i}\omega t}\widehat{g}(\omega) \tag{4.25}$$

with

$$\widehat{f}(\omega) = u_\beta(\omega)f(\omega), \qquad \widehat{g}(\omega) = v_\beta(\omega)g(\omega). \tag{4.26}$$

So using eqns. (4.15) and (4.24), we have

$$\widehat{f}(\omega) = u_\beta(\omega)e^{\beta\omega}g(\omega) = u_\beta(\omega)v_\beta(\omega)^{-1}e^{\beta\omega}\widehat{g}(\omega) = \widehat{g}(\omega), \tag{4.27}$$

implying that $\widehat{F}(t) = \widehat{G}(t)$. This confirms that the coupling to gravity has converted the thermal expectation value of an operator into a trace, and moreover that this trace is the expectation value in the maximum entropy state $\Psi_{\max}$.

In sum, we have identified a concrete example of including an observer in order to get a sensible answer in a cosmological model with a closed universe. And, at least in the example of de Sitter space, we have understood that gravity makes the notion of entropy better defined than it is in ordinary quantum field theory. It is possible [**1**] to probe more deeply and show that entropy defined via the Type II_1 algebra agrees, up to an additive constant independent of the state, with the generalized gravitational entropy, as usually computed via quantum extremal surfaces. It is also possible to give an analogous treatment of a black hole [**1, 58, 61**], involving in this case an algebra of Type II_∞ and therefore no upper bound on the entropy.

Acknowledgments

The author thanks V. Hubeny, A. Jaffe, J. Kohn, H. Maxfield, R. Mazzeo, G. Penington, and A. Strohmaier for comments and advice.

References

[1] Venkatesa Chandrasekaran, Roberto Longo, Geoff Penington, and Edward Witten, *An algebra of observables for de Sitter space*, J. High Energy Phys. **2** (2023), Paper No. 82, 55, DOI 10.1007/jhep02(2023)082. MR4547343

[2] H.-J. Borchers, *Field operators as C^∞ functions in spacelike directions* (English, with Italian summary), Nuovo Cimento (10) **33** (1964), 1600–1613. MR192797

[3] H.-J. Borchers, *Über die Vollständigkeit lorentz-invarianter Felder in einer zeitartigen Röhre* (German, with English and Italian summaries), Nuovo Cimento (10) **19** (1961), 787–793. MR119866

[4] Huzihiro Araki, *A generalization of Borchers theorem*, Helv. Phys. Acta **36** (1963), 132–139. MR153340

[5] Alexander Strohmaier, *On the local structure of the Klein-Gordon field on curved spacetimes*, Lett. Math. Phys. **54** (2000), no. 4, 249–261, DOI 10.1023/A:1010927625112. MR1846957

[6] A. Strohmaier and E. Witten, *Analytic States in Quantum Field Theory on Curved Spacetimes*, `arXiv:2302.02709`.

[7] A. Strohmaier and E. WItten, *The Timelike Tube Theorem in Curved Spacetime*, `arXiv:2303.16380`.

[8] W. G. Unruh, *Notes on Black Hole Evaporation*, Phys. Rev. **D14** (1976) 870.

[9] Arthur Jaffe, *Wick polynomials at a fixed time*, J. Mathematical Phys. **7** (1966), 1250–1255, DOI 10.1063/1.1705027. MR220489

[10] Henning Bostelmann and Christopher J. Fewster, *Quantum inequalities from operator product expansions*, Comm. Math. Phys. **292** (2009), no. 3, 761–795, DOI 10.1007/s00220-009-0853-x. MR2551793

[11] Emil J. Straube, *Harmonic and analytic functions admitting a distribution boundary value*, Ann. Scuola Norm. Sup. Pisa Cl. Sci. (4) **11** (1984), no. 4, 559–591. MR808424

[12] Stefan Hollands and Robert M. Wald, *Axiomatic quantum field theory in curved spacetime*, Comm. Math. Phys. **293** (2010), no. 1, 85–125, DOI 10.1007/s00220-009-0880-7. MR2563799

[13] M. Keyl, *Quantum Fields on Timelike Curves*, `arXiv:math-ph/0012024`, 2000.

[14] C. J. Fewster, *Lectures on Quantum Energy Inequalities*, `arXiv:1208.5399`, 2012.

[15] Joel Smoller, *Shock waves and reaction-diffusion equations*, Grundlehren der Mathematischen Wissenschaften, vol. 258, Springer-Verlag, New York-Berlin, 1983. MR688146

[16] Daniel Tataru, *Unique continuation for operators with partially analytic coefficients*, J. Math. Pures Appl. (9) **78** (1999), no. 5, 505–521, DOI 10.1016/S0021-7824(99)00016-1. MR1697040

[17] Horacio Casini, Marina Huerta, Javier M. Magán, and Diego Pontello, *Entropic order parameters for the phases of QFT*, J. High Energy Phys. **4** (2021), Paper No. 277, 96, DOI 10.1007/jhep04(2021)277. MR4276173

[18] Horacio Casini and Javier M. Magán, *On completeness and generalized symmetries in quantum field theory*, Modern Phys. Lett. A **36** (2021), no. 36, Paper No. 2130025, 21, DOI 10.1142/S0217732321300251. MR4354464

[19] Daniel Harlow, *Wormholes, emergent gauge fields, and the weak gravity conjecture*, J. High Energy Phys. **1** (2016), 122, front matter+30, DOI 10.1007/JHEP01(2016)122. MR3471458

[20] Joe Polchinski, *Monopoles, duality, and string theory*, Internat. J. Modern Phys. A **19** (2004), no. February, suppl., 145–154, DOI 10.1142/S0217751X0401866X. MR2064571

[21] T. Banks and N. Seiberg, *Symmetries and Strings in Field Theory and Gravity*, Phys. Rev. **D83** (2011) 084019, `arXiv:1011.5120`.

[22] Daniel Harlow and Hirosi Ooguri, *Symmetries in quantum field theory and quantum gravity*, Comm. Math. Phys. **383** (2021), no. 3, 1669–1804, DOI 10.1007/s00220-021-04040-y. MR4244261

[23] Davide Gaiotto, Anton Kapustin, Nathan Seiberg, and Brian Willett, *Generalized global symmetries*, J. High Energy Phys. **2** (2015), 172, front matter+61, DOI 10.1007/JHEP02(2015)172. MR3321281

[24] E. Witten, *Symmetry and Emergence*, Nat. Phys. **14** (2018) 116-9, `arXiv:1710.01791`.

[25] T. Banks, M. R. Douglas, G. T. Horowitz, and E. Martinec, *AdS Dynamics From Conformal Field Theory*, `arXiv:hep-th/9808016`.

[26] Iosif Bena, *Construction of local fields in the bulk of* AdS_5 *and other spaces*, Phys. Rev. D (3) **62** (2000), no. 6, 066007, 6, DOI 10.1103/PhysRevD.62.066007. MR1791042

[27] Vijay Balasubramanian, Per Kraus, and Albion Lawrence, *Bulk versus boundary dynamics in anti-de Sitter spacetime*, Phys. Rev. D (3) **59** (1999), no. 4, 046003, 16, DOI 10.1103/PhysRevD.59.046003. MR1683985

[28] Alex Hamilton, Daniel Kabat, Gilad Lifschytz, and David A. Lowe, *Holographic representation of local bulk operators*, Phys. Rev. D (3) **74** (2006), no. 6, 066009, 15, DOI 10.1103/PhysRevD.74.066009. MR2276402

[29] Alex Hamilton, Daniel Kabat, Gilad Lifschytz, and David A. Lowe, *Local bulk operators in AdS/CFT correspondence: a holographic description of the black hole interior*, Phys. Rev. D **75** (2007), no. 10, 106001, 15, DOI 10.1103/PhysRevD.75.106001. MR2326827

[30] D. Kabat, G. Lifschytz, and D. A. Lowe, *Constructing Local Bulk Observables in Interacting AdS/CFT*, Phys. Rev. **D83** (2011) 106009, `arXiv:1102.2910`.

[31] I. Heemskerk, D. Marolf, J. Polchinski, and J. Sully, *Bulk and Transhorizon Measurements in AdS/CFT*, J. High Energy Phys. **10** (2012) 165, arXiv:1201.3664.
[32] I. A. Morrison, *Boundary-to-bulk Maps for AdS Causal Wedges and the Reeh-Schlieder Property in Holography*, J. High Energy Phys. **05** (2014) 053, arXiv:1403.3426.
[33] Veronika E. Hubeny, *Covariant residual entropy*, J. High Energy Phys. **9** (2014), 156, front matter+35, DOI 10.1007/JHEP09(2014)156. MR3267923
[34] Robert M. Wald, *General relativity*, University of Chicago Press, Chicago, IL, 1984, DOI 10.7208/chicago/9780226870373.001.0001. MR757180
[35] Sijie Gao and Robert M. Wald, *Theorems on gravitational time delay and related issues*, Classical Quantum Gravity **17** (2000), no. 24, 4999–5008, DOI 10.1088/0264-9381/17/24/305. MR1808809
[36] Edward Witten, *Light rays, singularities, and all that*, Rev. Modern Phys. **92** (2020), no. 4, 045004, 49, DOI 10.1103/revmodphys.92.045004. MR4209242
[37] G. W. Gibbons and S. W. Hawking, *Cosmological event horizons, thermodynamics, and particle creation*, Phys. Rev. D (3) **15** (1977), no. 10, 2738–2751, DOI 10.1103/PhysRevD.15.2738. MR459479
[38] Rodolfo Figari, Raphael Hoegh-Krohn, and Chiara R. Nappi, *Interacting relativistic boson fields in the de Sitter universe with two space-time dimensions*, Comm. Math. Phys. **44** (1975), no. 3, 265–278. MR384010
[39] Geoffrey Sewell, *Quantum fields on manifolds: PCT and gravitationally induced thermal states*, Ann. Physics **141** (1982), no. 2, 201–224, DOI 10.1016/0003-4916(82)90285-8. MR673980
[40] Kengo Maeda, Tatsuhiko Koike, Makoto Narita, and Akihiro Ishibashi, *Upper bound for entropy in asymptotically de Sitter space-time*, Phys. Rev. D (3) **57** (1998), no. 6, 3503–3508, DOI 10.1103/PhysRevD.57.3503. MR1679148
[41] Raphael Bousso, *Positive vacuum energy and the N-bound*, J. High Energy Phys. **11** (2000), Paper 38, 24, DOI 10.1088/1126-6708/2000/11/038. MR1806569
[42] Raphael Bousso, *Bekenstein bounds in de Sitter and flat space*, J. High Energy Phys. **4** (2001), Paper 35, 15, DOI 10.1088/1126-6708/2001/04/035. MR1834389
[43] T. Banks, *Cosmological breaking of supersymmetry?*, Strings 2000. Proceedings of the International Superstrings Conference (Ann Arbor, MI), Internat. J. Modern Phys. A **16** (2001), no. 5, 910–921, DOI 10.1142/S0217751X01003996. MR1827963
[44] T. Banks and W. Fischler, *M-Theory Observables For Cosmological Spacetimes*, arXiv:hep-th/0102077, 2001.
[45] T. Banks, *More Thoughts on the Quantum Theory of Stable de Sitter Space*, arXiv:hep-th/0503066, 2005.
[46] Tom Banks, Bartomeu Fiol, and Alexander Morisse, *Towards a quantum theory of de Sitter space*, J. High Energy Phys. **12** (2006), 004, 18, DOI 10.1088/1126-6708/2006/12/004. MR2276729
[47] T. Banks and W. Fischler, *The Holographic Model Of Cosmology*, arXiv:1806.01749, 2018.
[48] L. Susskind, *De Sitter Holography: Fluctuations, Anomalous Symmetry, and Wormholes*, Universe **7** (2021) 464, arXiv:2106.03964.
[49] L. Susskind, *Black Holes Hint Towards De Sitter-Matrix Theory*, arXiv:2109.01322, 2021.
[50] Xi Dong, Eva Silverstein, and Gonzalo Torroba, *De Sitter holography and entanglement entropy*, J. High Energy Phys. **7** (2018), 050, front matter+23, DOI 10.1007/jhep07(2018)050. MR3862003
[51] Patrick Draper and Szilard Farkas, *de Sitter black holes as constrained states in the Euclidean path integral*, Phys. Rev. D **105** (2022), no. 12, Paper No. 126022, 6, DOI 10.1103/physrevd.105.126022. MR4455314
[52] Tom Banks and Patrick Draper, *Comments on the entanglement spectrum of de Sitter space*, J. High Energy Phys. **1** (2023), Paper No. 135, 9, DOI 10.1007/jhep01(2023)135. MR4541052
[53] H. Lin and L. Susskind, *Infinite Temperature's Not So Hot*, arXiv:2206.01083, 2022.
[54] Masamichi Takesaki, *Duality for crossed products and the structure of von Neumann algebras of type III*, Acta Math. **131** (1973), 249–310, DOI 10.1007/BF02392041. MR438149
[55] J. Sorce, *Notes on the Type Classification of von Neumann Algebras*, arXiv:2302.01958, 2023.
[56] S. Leutheusser and H. Liu, *Causal Connectability Between Quantum Systems and the Black Hole Interior in Holographic Duality*, arXiv:2110.05497.

[57] S. Leutheusser and H. Liu, *Emergent Times in Holographic Duality*, arXiv:2112.12156.
[58] Edward Witten, *Gravity and the crossed product*, J. High Energy Phys. **10** (2022), Paper No. 8, 27, DOI 10.1007/jhep10(2022)008. MR4493837
[59] I. E. Segal, *A note on the concept of entropy*, J. Math. Mech. **9** (1960), 623–629, DOI 10.1512/iumj.1960.9.59036. MR121677
[60] R. Longo and E. Witten, *A Note On Continuous Entropy*, arXiv:2202.03357, 2022.
[61] Venkatesa Chandrasekaran, Geoff Penington, and Edward Witten, *Large N algebras and generalized entropy*, J. High Energy Phys. **4** (2023), Paper No. 9, 67, DOI 10.1007/jhep04(2023)009. MR4571907

School of Natural Sciences, Institute for Advanced Study, 1 Einstein Drive, Princeton, NJ 08540 USA

Proceedings of Symposia in Pure Mathematics
Volume **107**, 2024
https://doi.org/10.1090/pspum/107/01955

Invertible QFTs and differential Anderson duals

Mayuko Yamashita

Abstract. This is the proceeding of a talk given at Stringmath 2022. We introduce a Cheeger-Simons type model for the differential extension of Anderson dual to generalized homology theory with physical interpretations. This construction generalizes the construction of the differential Anderson dual to bordism homology theories, given in a previous work of Yonekura and the author [J. Gökova Geom. Topol. GGT 16 (2023), 1–64.].

1. Introduction

In this article, we introduce a Cheeger-Simons type model for the differential extension of Anderson dual to generalized homology theory with physical interpretations. This construction generalizes the construction of the differential Anderson dual to bordism homology theories, given in a previous work of Yonekura and the author [**YY23**].

The Anderson duals have been important in the recent study of quantum field theories (QFTs) in terms of algebraic topology. The Anderson duals to bordism homology theories are conjectured to classify invertible QFTs by Freed and Hopkins [**FH21**]. Based on this conjecture, Yonekura and the author gave a model $I\Omega^G_{\mathrm{dR}}$ for the Anderson dual to G-bordism homology theory with a physical interpretation [**YY23**]. Interestingly, it turns out that the natural object produced from QFTs is a *differential refinement* of it, and passing to topological group corresponds to taking deformation classes. The construction has a similarity to the Cheeger-Simons model of differential ordinary cohomology in terms of *differential characters* [**CS85**].

In this article, we explain this construction in a generalized form. We provide a Cheeger-Simons type models of the Anderson dual to generalized homology theories. More precisely, we construct a differential extension $\widehat{IE}^*_{\mathrm{dR}}$ of the Anderson dual IE^* to a spectrum E, provided that a differential *homology* $\widehat{E}_*$ is given.

Differential cohomology theories have been much studied, but it seems that the homology version has not been interested in so much. Elmrabty [**Elm16**] defines differential K-homology, and as far as the author know it is the only example existing explicitly in the literature. But certainly there are situations where we want to refine generalized homology in a differential way, for example in the case of bordism homology theories. Such situations also naturally arise in physics; for example D-branes fits into the framework of differential K-homology. Differential bordism theories naturally arise in (non-topological) QFT's, where physical quantities, such as partition functions, depend on differential structure on spacetime manifolds. Thus the author feels that it is valuable to have a general framework for differential refinements in the homology settings, as we do in Section 4.

Given a differential homology theory $\widehat{E}_*$, the corresponding differential Anderson dual cohomology $\widehat{IE}^*_{\mathrm{dR}}$ consists of pairs (ω, h), where $h\colon \widehat{E}_{n-1} \to \mathbb{R}/\mathbb{Z}$ is a group homomorphism which satisfies a compatibility with a closed differential form $\omega \in \Omega^n(X; E^\bullet(\mathrm{pt}) \otimes \mathbb{R})$. Applied to $E = MTG$, we recover the model $\widehat{I\Omega^G_{\mathrm{dR}}}$ in [**YY23**], where the homomorphism h is regarded as the complex phase of the *partition function* of an invertible QFT. Applied to $E = H\mathbb{Z}$, we recover the Cheeger-Simons differential character groups (note that we have the Anderson self-duality $IH\mathbb{Z} \simeq H\mathbb{Z}$).

This article is organized as follows. The first two sections are devoted to preliminaries. We give a brief introduction to differential cohomology theories in Section 2. We explain invertible QFTs and Anderson duals in Section 3. In Section 4, we introduce differential homology theories, and in Section 5 we construct the differential Anderson duals. In Section 6 we explain the physical interpretation of our model.

1.1. Notations and conventions. In this article we use differential forms and currents.

- $\Omega^*(X) = \Omega^*(X;\mathbb{R})$ denotes the space of differential forms on X with coefficients in $\mathbb{R}$. We denote by $\Omega_*(X)$ the space of compactly supported currents on X, defined as $\Omega_*(X) := \mathrm{Hom}_{\mathrm{conti}}(\Omega^*(X), \mathbb{R})$, the continuous dual to to $\Omega^*(X)$ in the sense of distributions.
- If $V_\bullet$ is a graded vector space over $\mathbb{R}$, we topologize it as the direct limit of all finite subspaces and define $\Omega^*(X; V_\bullet)$. We also set

$$\Omega_n(X; V_\bullet) := \oplus_{p+q=n} \mathrm{Hom}_{\mathrm{conti}}(\Omega^p(X), V_q).$$

- The de Rham chain complex is denoted as

$$\cdots \xrightarrow{\partial} \Omega_{n+1}(X; V_\bullet) \xrightarrow{\partial} \Omega_n(X; V_\bullet) \xrightarrow{\partial} \cdots .$$

- For a spectrum E, we set $V^E_\bullet = E_\bullet(\mathrm{pt}) \otimes \mathbb{R}$ and $V^\bullet_E := E^\bullet(\mathrm{pt}) \otimes \mathbb{R}$, so that $V^E_n = V^{-n}_E$. The homology Chern-Dold character for E_* is denoted by

$$\mathrm{ch}\colon E_*(-) \to H_*(-; V^E_\bullet).$$

2. Background: Differential cohomology

In this section we give a brief review of generalized differential cohomology theories. A *differential extension* of a generalized cohomology theory E^* is a refinement $\widehat{E}^*$ of the restriction of E^* to the category of smooth manifolds, which contains differential-geometric data.

2.1. Ordinary differential cohomology.

2.2. $\widehat{H}^2(X;\mathbb{Z})$. In order to illustrate the idea of differential cohomology, first we explain a model of the second ordinary differential cohomology. Recall that $H^2(X;\mathbb{Z})$ is identified as the group of isomorphism classes of complex line bundles over a space X. The differential group is defined by adding differential information on it.

DEFINITION 2.1. For a manifold X, we define $\widehat{H}^2_{\mathrm{geom}}(X;\mathbb{Z})$ to be the abelian group of isomorphism classes of hermitian line bundles with unitary connections over X.

The differential group is a refinement of the topological group, in the sense that there is a surjection (forgetful map)

$$I\colon \widehat{H}^2_{\mathrm{geom}}(X;\mathbb{Z}) \to H^2(X;\mathbb{Z}), \quad [L,\nabla] \mapsto [L]. \tag{2.2}$$

But $\widehat{H}^2_{\mathrm{geom}}$ has more information than H^2. We can extract the differential information by taking the curvature,

$$R\colon \widehat{H}^2_{\mathrm{geom}}(X;\mathbb{Z}) \to \Omega^2_{\mathrm{clo}}(X), \quad [L,\nabla] \mapsto F_\nabla/(2\pi\sqrt{-1}). \tag{2.3}$$

Connections on topologically trivial line bundles are identified with one form. This gives a map

$$a\colon \Omega^1(X) \to \widehat{H}^2_{\mathrm{geom}}(X;\mathbb{Z}), \quad \alpha \mapsto [\underline{\mathbb{C}}, d + 2\pi\sqrt{-1}\alpha]. \tag{2.4}$$

These maps are part of the *differential cohomology hexagon*,

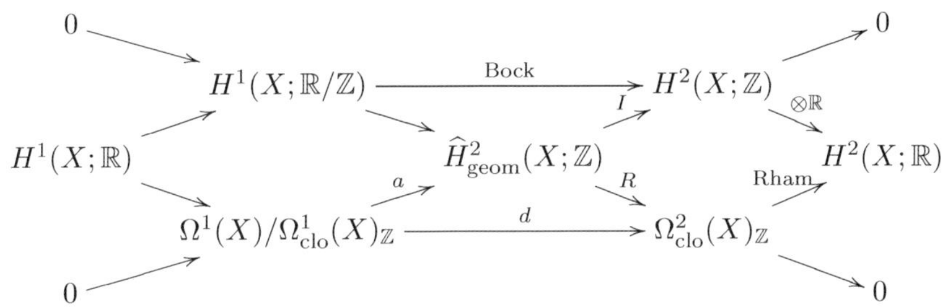

Here $\Omega^n_{\mathrm{clo}}(X)_{\mathbb{Z}}$ denotes the closed forms of integral periods. The above diagram is commutative, and the diagonal sequences are exact.

We would like to define $\widehat{H}^n(X;\mathbb{Z})$ for higher n, having analogous structure maps and hexagons.

2.2.1. *Cheeger-Simons differential characters.* Here we explain a model for a differential extension of $H\mathbb{Z}$ given by Cheeger and Simons, in terms of *differential characters* [**CS85**].

For a manifold X and a nonnegative integer n, the group of *differential characters* $\widehat{H}^n_{\mathrm{CS}}(X;\mathbb{Z})$ is the abelian group consisting of pairs (ω, k), where

- A closed differential form $\omega \in \Omega^n_{\mathrm{clo}}(X)$,
- A group homomorphism $k\colon Z_{\infty,n-1}(X;\mathbb{Z}) \to \mathbb{R}/\mathbb{Z}$,
- ω and k satisfy the following compatibility condition. For any

  $$c \in C_{\infty,n}(X;\mathbb{Z})$$

  we have

$$k(\partial c) \equiv \langle \omega, c\rangle_X \pmod{\mathbb{Z}}. \tag{2.5}$$

Here $C_{\infty,n}(X;\mathbb{Z})$ and $Z_{\infty,n-1}(X;\mathbb{Z})$ denote the smooth singular chains and cycles with integer coefficients.

We have homomorphisms

$$R_{\mathrm{CS}}\colon \widehat{H}^n_{\mathrm{CS}}(X;\mathbb{Z}) \to \Omega^n_{\mathrm{clo}}(X), \quad (\omega,k) \mapsto \omega$$
$$a_{\mathrm{CS}}\colon \Omega^{n-1}(X)/\mathrm{Im}(d) \to \widehat{H}^n_{\mathrm{CS}}(X;\mathbb{Z}), \quad \alpha \mapsto (d\alpha, \alpha).$$

and the quotient map gives

$$I_{\mathrm{CS}}\colon \widehat{H}^n_{\mathrm{CS}}(X;\mathbb{Z}) \to \widehat{H}^n_{\mathrm{CS}}(X;\mathbb{Z})/\mathrm{Im}(a_{\mathrm{CS}}) \simeq H^n(X;\mathbb{Z}).$$

We get the *differential cohomology hexagon*,

(2.6)

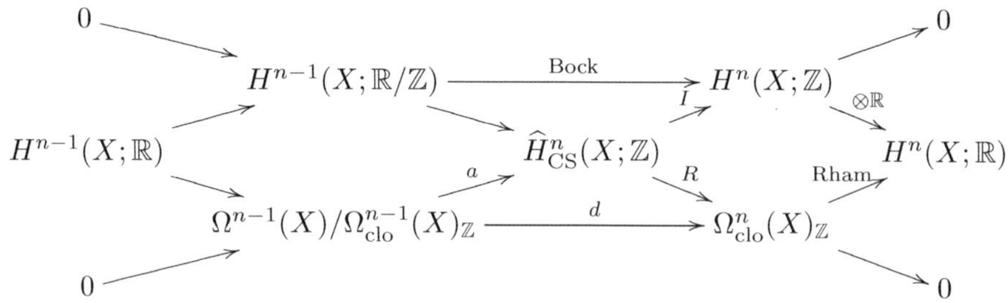

The above diagram is commutative. The diagonal sequences are exact.

We can construct an isomorphism between the two models $\widehat{H}^2_{\mathrm{geom}}$ and $\widehat{H}^2_{\mathrm{CS}}$ by taking the holonomy of connections,

$$\widehat{H}^2_{\mathrm{geom}}(X;\mathbb{Z}) \to \widehat{H}^2_{\mathrm{CS}}(X;\mathbb{Z}), \quad [L,\nabla] \mapsto (F_\nabla/(2\pi\sqrt{-1}), \mathrm{Hol}(L,\nabla)). \tag{2.7}$$

2.2.2. *Hopkins-Singer's differential cocycles.* Let X be a manifold. An n-th *differential cocycle* on X is an element

$$(c,h,\omega) \in Z^n_\infty(X;\mathbb{Z}) \times C^{n-1}_\infty(X;\mathbb{R}) \times \Omega^n_{\mathrm{clo}}(X)$$

such that

$$\omega - c_{\mathbb{R}} = \delta h. \tag{2.8}$$

Here C^*_∞ and Z^*_∞ denotes the groups of smooth singular cochains and cocycles. We introduce the equivalence relation $\sim$ on differential cocycles by setting

$$(c,h,\omega) \sim (c+\delta b, h - b_{\mathbb{R}} - \delta k, \omega)$$

for some $(b,k) \in C^{n-1}_\infty(X;\mathbb{Z}) \times C^{n-2}_\infty(X;\mathbb{R})$.

DEFINITION 2.9 ($\widehat{H}^*_{\mathrm{HS}}(X;\mathbb{Z})$ [**HS05**]). Set

$$\widehat{H}^n_{\mathrm{HS}}(X;\mathbb{Z}) := \{(c,h,\omega) : \text{differential } n\text{-cocycle on } X\}/\sim$$

The group $\widehat{H}^n_{\mathrm{HS}}(X;\mathbb{Z})$ also fits into the hexagon (2.6). In fact we have a natural isomorphism $\widehat{H}^n_{\mathrm{HS}} \simeq \widehat{H}^n_{\mathrm{CS}}$.

2.3. Differential K-theory. K-theory is the most studied generalized cohomology theory, other than $H\mathbb{Z}$, in the context of differential cohomology. There are various models for topological K-theories. Correspondingly, there are various models for differential K-theory. Here we briefly review the model constructed by Freed and Lott [**FL10**] in terms of vector bundles with connections. We will explain another model in Subsection 5.2.2 below.

Recall that the topological K-theory $K^0(X)$ can be defined in terms of stable isomorphism classes $[E]$ of complex vector bundles over X. The differential refinement is, roughly speaking, given by equipping vector bundles with connections. An element of $\widehat{K}^0_{\mathrm{FL}}(X)$ is represented by a triple (E, ∇^E, η), where (E,∇^E) is a hermitian vector bundle over X with a unitary connection and $\eta \in \Omega^{2\mathbb{Z}-1}(X)/\mathrm{im}(d)$.

We take equivalence classes with respect to a suitable equivalence relation including the relation

$$(E, \nabla^E_1, \eta) \sim (E, \nabla^E_0, \eta + \mathrm{CS}(\nabla^E_0, \nabla^E_1)), \tag{2.10}$$

where $\mathrm{CS}(\nabla^E_0, \nabla^E_1)$ is the Chern-Simons form measuring the difference of two connections. We have the structure homomorphisms

$$I\colon \widehat{K}^0_{\mathrm{FL}}(X) \to K^0(X), \quad [E, \nabla^E, \eta] \mapsto [E] \tag{2.11}$$

$$R\colon \widehat{K}^0_{\mathrm{FL}}(X) \to \Omega^{2\mathbb{Z}}_{\mathrm{clo}}(X), \quad [E, \nabla^E, \eta] \mapsto \mathrm{Ch}(\nabla^E) + d\eta \tag{2.12}$$

$$a\colon \Omega^{2\mathbb{Z}-1}(X)/\mathrm{im}(d) \to \widehat{K}^0_{\mathrm{FL}}(X), \quad \eta \mapsto [0, 0, \eta]. \tag{2.13}$$

We also have the similar hexagon as (2.6).

2.4. The axiomatic formulation of differential cohomology. We explain the axiomatic formulation of differential cohomology theory by Bunke and Schick [**BS12**]. We remark that there are more homotopy-theoretic formulations in terms of sheaves of spectra [**BNV16**].

Let E^* be a generalized cohomology theory. Let $N^\bullet$ be a graded vector space over $\mathbb{R}$ equipped with a transformation of cohomology theories

$$\mathrm{ch}\colon E^* \to H^*(-; N^\bullet). \tag{2.14}$$

The universal choice is $N^\bullet = E^\bullet_{\mathbb{R}}(\mathrm{pt}) =: V^\bullet_E$ with ch the Chern-Dold homomorphism ([**Rud98**, Chapter II, 7.13]) for E.

For a manifold X, set $\Omega^*(X; N^\bullet) := C^\infty(X; \wedge T^*M \otimes_{\mathbb{R}} N^\bullet)$ with the $\mathbb{Z}$-grading by the total degree. Let $d\colon \Omega^*(X; N^\bullet) \to \Omega^{*+1}(X; N^\bullet)$ be the de Rham differential. We have the natural transformation

$$\mathrm{Rham}\colon \Omega^*_{\mathrm{clo}}(X; N^\bullet) \to H^*(X; N^\bullet).$$

DEFINITION 2.15 (Differential extensions of a cohomology theory, [**BS12**, Definition 2.1]). A *differential extension* of the pair (E^*, ch) is a quadruple $(\widehat{E}, R, I, a)$, where

- $\widehat{E}$ is a contravariant functor $\widehat{E}\colon \mathrm{Mfd}^{\mathrm{op}} \to \mathrm{Ab}^{\mathbb{Z}}$.
- R, I and a are natural transformations

$$R\colon \widehat{E}^* \to \Omega^*_{\mathrm{clo}}(-; N^\bullet)$$
$$I\colon \widehat{E}^* \to E^*$$
$$a\colon \Omega^{*-1}(-; N^\bullet)/\mathrm{im}(d) \to \widehat{E}^*.$$

We require the following axioms.

- $R \circ a = d$.
- $\mathrm{ch} \circ I = \mathrm{Rham} \circ R$.
- For all manifolds X, the sequence

$$E^{*-1}(X) \xrightarrow{\mathrm{ch}} \Omega^{*-1}(M; N^\bullet)/\mathrm{im}(d) \xrightarrow{a} \widehat{E}(X) \xrightarrow{I} E^*(X) \to 0 \tag{2.16}$$

  is exact.

In the case $N^\bullet = V^\bullet_E$ and ch is the Chern-Dold homomorphism, we simply call it a *differential extension of* E^*.

Such a quadruple $(\widehat{E}, R, I, a)$ itself is also called a *generalized differential cohomology theory*. We usually abbreviate the notation and just write a generalized cohomology theory as $\widehat{E}^*$. The above axiom is equivalent to the half of the hexagon 2.6.

Hopkins and Singer gave a model of an differential extension for an arbitrary (E^*, ch) in terms of *differntial function spectra* ([**HS05**]). But we cannot guarantee that any differential cohomology theory is isomorphic to the Hopkins-Singer's model. For more about this uniqueness issue, we refer to [**BS10**].

3. Background : Invertible QFTs and Anderson duals

3.1. QFTs and invertibility. Basically, a quantum field theory (QFT) is a symmetric monoidal functor

$$\mathcal{T}\colon \mathrm{Bord}_n^{\mathcal{S}} \to \mathcal{C}. \tag{3.1}$$

Here n is the dimension of the theory, and $\mathcal{S}$ specifies a structure on manifolds, such as orientations, Riemannian metrics, spin structures and principal G-bundles with connections. The domain $\mathrm{Bord}_n^{\mathcal{S}}$ is the n-dimensional *Bordism category* of $\mathcal{S}$-manifolds. The mathematical formulation for it varies depending on the context. If we are talking about *non-extended topological QFTs (TQFTs)*, $\mathrm{Bord}_n^{\mathcal{S}}$ is an ordinary category whose objects are $(n-1)$-dimensional closed $\mathcal{S}$-manifolds and morphisms are given by bordisms by compact n-dimensional $\mathcal{S}$-manifolds ([**Ati88**], [**Seg04**]). In the case of *extended TQFTs*, $\mathrm{Bord}_n^{\mathcal{S}}$ is a higher category, up to (∞, n)-category in the fully extended case ([**Lur09**]). If we consider possibly *non-topological* QFTs, we need to impose some smoothness to the functor (3.1). For this we can use the smooth version of (higher) categories formulated by Grady and Pavlov [**GP20**].

According to the above variations, the formulation of the target $\mathcal{C}$ also depends on the context. Mathematically we can allow any symmetric monoidal ((∞, n)-, smooth, etc) category. However, the physically natural target category is the category of $\mathbb{Z}/2$-graded complex vector spaces $\mathrm{sVect}_{\mathbb{C}}$ in the non-extended case. In the extended case, it is natural to consider higher categories extending $\mathrm{sVect}_{\mathbb{C}}$, such as $\mathrm{sAlg}_{\mathbb{C}}$. We do not know the canonical target in general. Moreover, in order for a functor to be physically meaningful, we need other conditions such as reflection positivity.

A QFT $\mathcal{T}$ is called *invertible* if it factors as

$$\mathcal{T}\colon \mathrm{Bord}_n^{\mathcal{S}} \to \mathcal{C}^{\times} \subset \mathcal{C}, \tag{3.2}$$

where $\mathcal{C}^{\times}$ is the maximal Picard (∞-, smooth...) groupoid of $\mathcal{C}$. Here a *Picard groupoid* is a symmetric monoidal category whose all objects and morphisms are invertible under the monoidal product and the composition, respectively. This means that $\mathcal{T}$ assigns something invertible to all objects and morphisms of $\mathrm{Bord}_n^{\mathcal{S}}$. For example, if $\mathcal{C} = \mathrm{sVect}_{\mathbb{C}}$, we have $\mathcal{C}^{\times} = \mathrm{sLine}_{\mathbb{C}}$, the groupoid of one dimensional $\mathbb{Z}/2$-graded complex vector spaces and grading-preserving invertible linear maps.

Invertible QFTs are a special, but an important class of QFTs. For example they arise as *Symmetry Protected Topological phases (SPT phases)* in condensed matter physics. In that context, physical systems are described by Hamiltonians having uniquely gapped ground states, and the effective theories are considered as invertible theories. Invertible theories are also important in the study of *anomalies*. A $(n-1)$-dimensional anomalous theory is formulated as a boundary theory of a

n-dimensional invertible theory. Thus the classification of anomalies in $(n-1)$-dimension reduces to the classification of invertible theories in n-dimension.

3.2. Classification of invertible QFTs and Freed-Hopkins conjecture. Fully extended invertible QFTs are believed to be classified in terms of generalized cohomology. The argument uses the *stable homotopy hypothesis*, which gives equivalence between Picard ∞-groupoids with connective spectra. By this the functor (3.1) is transformed into a map of spectra. This means that $\mathcal{T}$ is regarded as an element of generalized cohomology represented by the mapping spectra.

For simplicity we focus on the case where the structure $\mathcal{S}$ is given by a tangential G-structure (for precise formulation see[**YY23**, Section 3]), where $G = \{G_d, s_d, \rho_d\}_{d\in\mathbb{Z}_{\geq 0}}$ is a sequence of compact Lie groups with homomorphisms $s_d\colon G_d \to G_{d+1}$ and $\rho_d\colon G_d \to \mathrm{O}(d;\mathbb{R})$ satisfying the compatibility with $\mathrm{O}(d;\mathbb{R}) \hookrightarrow \mathrm{O}(d+1,\mathbb{R})$.

By the work of Galatius, Madsen, Tillmann and Weiss [**GTMW09**], the spectrum corresponding to the groupoid completion of Bord_n^G is the (shift of) *Madsen-Tillmann spectrum.* Freed and Hopkins relates stability of the tangential structure and reflection positivity of QFTs [**FH21**, Section 7]. Correspondingly, the classification conjecture below uses the stable version of the Madsen-Tillmann spectrum MTG, which represents the stable tangential G-bordism homology theory Ω^G_*.

On the other hand, the spectrum corresponding to the target $\mathcal{C}^\times$ is unclear if we do not specify it. As explained above, we do not know the universal target of physically meaningful fully extended QFTs. Freed and Hopkins proposed an *ansatz* that the universal target for invertible *topological* QFTs corresponds to the spectrum $I\mathbb{C}^\times$, the *character dual* to the sphere (footnote 2). Based on this ansatz, they conjecture that

CONJECTURE 3.3 ([**FH21**, Conjecture 8.37]). *There is a* $1:1$ *correspondence*[1]

$$\left\{\begin{matrix}\textit{deformation classes of reflection positive invertible}\\ \textit{n-dimensional fully extended field theories with sym-}\\ \textit{metry type } G\end{matrix}\right\} \simeq (I\Omega^G)^{n+1}(\mathrm{pt}). \tag{3.4}$$

In fact they prove that *topological* ones are classified by the torsion part of $(I\Omega^G)^{n+1}(\mathrm{pt})$.

3.3. Anderson duals. In this subsection we collect basics on the Anderson duals for spectra. For more details, see for example [**HS05**, Appendix B] and [**FMS07**, Appendix B].

First note that the assignment $X \mapsto \mathrm{Hom}(\pi_*(X), \mathbb{R}/\mathbb{Z})$ for each spectrum X satisfies the Eilenberg-Steenrood axioms, so it is represented by an spectrum denoted by $I(\mathbb{R}/\mathbb{Z})$[2]. On the other hand we have $H^n(X;\mathbb{R}) \simeq \mathrm{Hom}(\pi_n(X);\mathbb{R})$. So we have a mod $\mathbb{Z}$ reduction map $H\mathbb{R} \to I(\mathbb{R}/\mathbb{Z})$.

The *Anderson dual to sphere* $I\mathbb{Z}$ is the spectrum defined as the homotopy fiber

$$I\mathbb{Z} \to H\mathbb{R} \to I(\mathbb{R}/\mathbb{Z}). \tag{3.5}$$

For a general spectrum E, we define its *Anderson dual* as the function spectrum,

$$IE := F(E, I\mathbb{Z}). \tag{3.6}$$

[1]Here *symmetry types* of QFT's in [**FH21**] are certain classes of G's in this paper which satisfy an additional set of conditions.

[2]If we replace $\mathbb{R}/\mathbb{Z}$ by $\mathbb{Q}/\mathbb{Z}$, we get the Brown-Comenetz dual to the sphere.

This implies that we have the exact sequence (Universal Coefficient Theorem)

$$0 \to \mathrm{Ext}(E_n(X), \mathbb{Z}) \to IE^{n+1}(X) \to \mathrm{Hom}(E_{n+1}(X), \mathbb{Z}) \to 0. \tag{3.7}$$

MTG represents the G-bordism homology theory Ω^G_*. The Anderson dual $I\Omega^G = F(MTG, I\mathbb{Z})$ fits into the exact sequence

$$0 \to \mathrm{Ext}(\Omega^G_n(X), \mathbb{Z}) \to (I\Omega^G)^{n+1}(X) \to \mathrm{Hom}(\Omega^G_{n+1}(X), \mathbb{Z}) \to 0. \tag{3.8}$$

We note that $\mathrm{Ext}(\Omega^G_n(X), \mathbb{Z})$ is the torsion part of $(I\Omega^G)^{n+1}(X)$ for reasonable spaces X. According to the Freed-Hopkins' result about the classification in *topological* case, this group classifies n-dimensional invertible *topological* QFTs on G-manifolds. Conjecture 3.3 says that we can use whole of $(I\Omega^G)^{n+1}(X)$ to classify possibly non-topological invertible QFTs.

In order to give a better understanding of this conjecture, Yonekura and the author gave a new model for $I\Omega^G$ with a physical interpretation ([**YY23**]). It turns out that the natural object produced from QFTs is a differential refinement of it, and passing to topological group corresponds to taking deformation classes. Also it has turned out that the analogous construction applies to give a "QFT-like" model of differential refinement of IE for general E, provided that a differential *homology* $\widehat{E}_*$ is given. In the rest of this article we explain this construction.

4. Differential homology theories

In this section we give a formulation of differential *homology* theories and explain examples.

4.1. The axiom. Here we give the axioms for differential extensions of generalized homology theories. It is just a straightforward homology-version of the axiom in [**BS10**]. We set $V^E_\bullet := \pi_\bullet(E) \otimes_{\mathbb{Z}} \mathbb{R}$. We have the homology Chern-Dold character homomorphism

$$\mathrm{ch}\colon E_* \to H_*(-; V^E_\bullet). \tag{4.1}$$

Here we focus on the differential extension in the coefficient $V^E_\bullet$ for our purpose, but the variant in coefficient can be formulated as in the cohomology case (Section 2.4).

Definition 4.2 (Differential extensions of homology theories). A *differential extension* of a generalized homology theory E_* is a quadruple $(\widehat{E}_*, R, I, a)$, where

- $\widehat{E}_*$ is a covariant functor $\widehat{E}_*\colon \mathrm{Mfd} \to \mathrm{Ab}^{\mathbb{Z}}$.
- R, I and a are natural transformations

$$R\colon \widehat{E}_* \to \Omega^{\mathrm{clo}}_*(-; V^E_\bullet)$$
$$I\colon \widehat{E}_* \to E_*$$
$$a\colon \Omega_{*+1}(-; V^E_\bullet)/\mathrm{im}(\partial) \to \widehat{E}_*.$$

- $R \circ a = \partial$.
- $\mathrm{ch} \circ I = \mathrm{Rham} \circ R$.
- For all manifolds X, the sequence

$$E_{*+1}(X) \xrightarrow{\mathrm{ch}} \Omega_{*+1}(M; V^E_\bullet)/\mathrm{im}(\partial) \xrightarrow{a} \widehat{E}_*(X) \xrightarrow{I} E_*(X) \to 0 \tag{4.3}$$

is exact.

4.2. Examples.

4.2.1. *Differential K-homology.* Elmrabty [**Elm16**] defines differential K-homology, by refining Baum-Douglas model of K-homology in terms of geometric cycles ([**BD82**]). This has a physical interpretation in terms of *D-branes.*

Recall that in the Baum-Douglas model, an element in $K_n(X)$ is represented by a *K-cycle*, a triple (M, E, f) consisting of closed Spin^c-manifold M with $\dim M \equiv n \pmod 2$, a complex vector bundle E over M and a continuous map $f\colon M \to X$. The K-homology group $K_n(X)$ is defined to be the group of equivalence classes of K-cycles under the equivalence relation generated by direct sum, bordism and "vector bundle modification".

Correspondingly, the differential K-homology group $\widehat{K}_n(X)$ is represented by a *differential K-cycle* $((M, \nabla^M), (E, \nabla^E), f, \psi)$, where in addition to the data of a K-cycle, M is equipped with a Spin^c-connection ∇^M, E is equipped with a hermitian metric and unitary connection ∇^E, f is required to be smooth, and we have additional data $\psi \in \Omega_{2\mathbb{Z}+n+1}(X)/\mathrm{im}(\partial)$. The group $\widehat{K}_n(X)$ is given by taking equivalence classes under the corresponding equivalence relation suitably modified with differential data. For example the bordism relation in this case is, if we have a $2\mathbb{Z}+n+1$-dimensional compact $((W, \nabla^W), (\mathcal{E}, \nabla^{\mathcal{E}}), f, 0)$ with boundary, we set

$$\left((\partial W, \nabla^W|_{\partial W}), (\mathcal{E}|_{\partial W}, \nabla^{\mathcal{E}}|_{\partial W}), f|_{\partial W}, \int_W \mathrm{Todd}(\nabla^E)\mathrm{Ch}(\nabla^{\mathcal{E}})f^*\right) \sim 0. \tag{4.4}$$

The axiom above is satisfied with the structure maps

$$R\left([(M, \nabla^M), (E, \nabla^E), f, \psi]\right) := \int_M \mathrm{Todd}(\nabla^M)\mathrm{Ch}(\nabla^E)f^* - \partial\psi, \tag{4.5}$$

$$I\left([(M, \nabla^M), (E, \nabla^E), f, \psi]\right) := [M, E, f], \tag{4.6}$$

$$a\left([\psi]\right) := [\varnothing, 0, (\varnothing \to X), \psi]. \tag{4.7}$$

where we identify $\Omega_n(X; V_\bullet^K) \simeq \Omega_{2\mathbb{Z}+n}(X)$.

4.2.2. *Differential ordinary homology.* We can construct a refinement $\widehat{H}^{\mathrm{HS}}_*(-;\mathbb{Z})$ of the ordinary integral homology given by a homology-version of differential cocycles [**HS05**]. Namely, a *differential n-cycles* over X is defined to be an element

$$(c, h, \omega) \in Z_n^\infty(X;\mathbb{Z}) \times C_{n+1}^\infty(X;\mathbb{R}) \times \Omega_n^{\mathrm{clo}}(X) \tag{4.8}$$

such that $\partial h = \omega - c$ as smooth currents[3]. We define a homomorphism

$$\partial\colon C_{n+1}^\infty(X;\mathbb{Z}) \times C_{n+2}^\infty(X;\mathbb{R}) \to Z_n^\infty(X;\mathbb{Z}) \times C_{n+1}^\infty(X;\mathbb{R}) \times \Omega_n^{\mathrm{clo}}(X) \tag{4.9}$$

by $\partial(b, k) := (\partial b, -\partial k - b, 0)$. We define

$$\widehat{H}_n^{\mathrm{HS}}(X;\mathbb{Z}) := \{\text{differential } n\text{-cocycles over } X\}/\mathrm{im}\partial. \tag{4.10}$$

It is straightforward to construct structure maps and check the axioms above.

4.2.3. *Differential G-bordism homology theory $\widehat{\Omega^G_*}$.* Here we construct a differential refinement of tangential G-bordism homology theories. We use the notations and conventions about differential G-structures introduced in [**YY23**, Section 3]. Let $G = \{G_d, s_d, \rho_d\}_{d \in \mathbb{Z}_{\geq 0}}$ be tangential structure groups. Then we get the *stable Madsen-Tillmann spectrum MTG*, which represents the *stable tangential G-bordism homology theory Ω^G_**. For details see [**Fre19**, Section 6.6] The topological group $\Omega_n^G(X)$ is the stable G-bordism group, the group consisting of bordism classes

[3]Note that we have $C_n^\infty \subset \Omega_n$.

$[M, g^{\mathrm{top}}, f]$, where (M, g^{top}) is a n-dimensional closed manifold with a stable tangential G-structure and $f\colon M \to X$ is a continuous map.

The differential refinement $\widehat{\Omega_n^G}(X)$ is constructed in terms of *differential stable tangential G-cycles*, as follows. A n-dimensional differential stable tangential G-cycle over X is a triple (M, g, f), where in this case g is a *differential* stable G-structure (i.e., equipped with G-connection) and $f\colon M \to X$ is required to be smooth (for details see [**YY23**, Section 3]).

We use the *Bordism Picard groupoid* $h\mathrm{Bord}_n^{G_\nabla}(X)$ defined in [**YY23**, Definition 3.8]. The objects are differential stable tangential G-cycles (M, g, f) of dimension n over X, and the morphisms are bordism class $[W, g_W, f_W]$ of bordisms of differential stable tangential G-cycles.

We need the Chern-Weil construction in this setting ([**YY23**, Subection 4.1.1]) We set[4]

$$N_G^\bullet := H^\bullet(MTG;\mathbb{R}) = \varprojlim_d H^\bullet(BG_d;\mathbb{R}_{G_d}) = \varprojlim_d (\mathrm{Sym}^{\bullet/2}\mathfrak{g}_d^* \otimes_{\mathbb{R}} \mathbb{R}_{G_d})^{G_d}. \tag{4.11}$$

In the case where G is oriented, i.e., the image of ρ_d lies in $\mathrm{SO}(d,\mathbb{R})$ for each d, the G_d-module $\mathbb{R}_{G_d}$ is trivial and $N_G^\bullet$ is the projective limit of invariant polynomials on $\mathfrak{g}_d$. In general cases, $N_G^\bullet$ can be regarded as the projective limit of polynomials on $\mathfrak{g}_d$ which change the sign by the action of G_d.

A differential stable tangential G-structure g on a manifold M defines a homomorphism ([**YY23**, Definition 4.4]),

$$\mathrm{cw}_g\colon \Omega^*(M;N_G^\bullet) \to \Omega^*(M;\mathrm{Ori}(M)), \tag{4.12}$$

where $\mathrm{Ori}(M)$ is the orientation line bundle of M.

An object (M, g, f) in $h\mathrm{Bord}_n^{G_\nabla}(X)$ gives a closed n-current

$$\mathrm{cw}(M, g, f) \in \Omega_n^{\mathrm{clo}}(X; V_\bullet^{MTG}) \subset \mathrm{Hom}_{\mathrm{conti}}(\Omega^n(X;N_G^\bullet),\mathbb{R}), \tag{4.13}$$

by, for $\omega \in \Omega^n(X;N_G^\bullet)$,

$$\mathrm{cw}(M, g, f)(\omega) := \int_M \mathrm{cw}_g(f^*\omega). \tag{4.14}$$

Similarly, a bordism $(W, g_W, f_W)\colon (M_-, g_-, f_-) \to (M_+, g_+, f_+)$ of differential stable tangential G-cycles of dimension n gives an $(n+1)$-current

$$\mathrm{cw}(W, g_W, f_W) \in \Omega_{n+1}(X; V_\bullet^{MTG}) \subset \mathrm{Hom}_{\mathrm{conti}}(\Omega^{n+1}(X;N_G^\bullet),\mathbb{R}), \tag{4.15}$$

by, for $\omega \in \Omega^{n+1}(X;N_G^\bullet)$,

$$\mathrm{cw}[W, g_W, f_W](\omega) := \int_W \mathrm{cw}_{g_W}({f_W}^*\omega), \tag{4.16}$$

If we have two such bordisms (W, g_W, f_W) and (W', g'_W, f'_W) which are bordant, the corresponding currents (4.15) differs by an image of ∂. Thus for a morphism $[W, g_W, f_W]$ in $h\mathrm{Bord}_n^{G_\nabla}(X)$ we get an element

$$\mathrm{cw}[W, g_W, f_W] \in \Omega_{n+1}(X; V_\bullet^{MTG})/\mathrm{im}(\partial). \tag{4.17}$$

[4] In the general notation introduced in (5.1) below, we have $N_G^\bullet = N_{MTG}^\bullet$. We abbreviate the notation.

DEFINITION 4.18 ($\widehat{\Omega^G_*}$). Let X be a manifold and n be an integer.

(1) We set

$$\widehat{\Omega^G_n}(X) := \{(M, g, f, \eta)\}/\sim, \tag{4.19}$$

where (M, g, f) is an object in $h\mathrm{Bord}_n^{G\triangledown}(X)$ and

$$\eta \in \Omega_{n+1}(X; V_\bullet^{MTG})/\mathrm{im}(\partial).$$

The relation $\sim$ is the equivalence relation generated by the relation

$$(M_-, g_-, f_-, \eta) \sim (M_+, g_+, f_+, \eta - \mathrm{cw}[W, g_W, f_W]) \tag{4.20}$$

for each morphism

$$[W, g_W, f_W] \colon (M_-, g_-, f_-) \to (M_+, g_+, f_+)$$

in $h\mathrm{Bord}_n^{G\triangledown}(X)$. We regard $\widehat{\Omega^G}$ as a functor $\mathrm{Mfd} \to \mathrm{Ab}^{\mathbb{Z}}$ in the obvious way.

(2) We define natural transformations R, I and a by

$$R \colon \widehat{\Omega^G_n}(X) \to \Omega_n^{\mathrm{clo}}(X; V_\bullet^{MTG}), \quad [M, g, f, \eta] \mapsto \mathrm{cw}(M, g, f) + \partial\eta,$$
$$I \colon \widehat{\Omega^G_n}(X) \to \Omega_n^G(X), \quad [M, g, f, \eta] \mapsto [M, g, f],$$
$$a \colon \Omega_{n+1}(X; V_\bullet^{MTG})/\mathrm{im}(\partial) \to \widehat{\Omega^G_n}(X), \quad \eta \mapsto [\varnothing, \eta].$$

We can easily check that the quadruple $(\widehat{\Omega^G_*}, R, I, a)$ satisfies the axiom in Definition 4.2.

5. The Anderson duals to differential homology

5.1. The construction. Let E be a spectrum and assume that we are given a differential extension $(\widehat{E}_*, R_{E_*}, I_{E_*}, a_{E_*})$ of E-homology. In this section we explain that it associates a model IE^*_{dR} of the Anderson dual cohomology theory IE and its differential extension of the pair $\left(IE^*, \mathrm{ch}'\right)$ (Definition 2.15), where ch' is defined below.

Set

$$N_E^\bullet := \mathrm{Hom}(\pi_{-\bullet}E, \mathbb{R}). \tag{5.1}$$

By the third arrow in (3.7), we have a canonical transformation of cohomology theories

$$\mathrm{ch}' \colon (IE)^*(X) \to H^*(X; N_E^\bullet) \simeq \mathrm{Hom}(E_*(X), \mathbb{R}), \tag{5.2}$$

For example if $E_n(\mathrm{pt})$ is finitely generated for all n, we have an isomorphism $V_{IE}^\bullet \simeq N_E^\bullet$ and (5.2) coincides with the Chern-Dold homomorphism.

We have the pairing

$$\langle -, - \rangle \colon \Omega_*(X; V_\bullet^E) \otimes \Omega^*(X; N_E^\bullet) \to \mathbb{R}.$$

Using this we regard $\Omega_*(X; V_\bullet^E) \subset \mathrm{Hom}_{\mathrm{conti}}(\Omega^*(X; N_E^\bullet), \mathbb{R})$.

Definition 5.3 ($(\widehat{IE}_{\mathrm{dR}})^*$ and IE^*_{dR} associated to $\widehat{E}_*$). Let E be a spectrum and assume that we are given a differential extension $(\widehat{E}_*, R_{E_*}, I_{E_*}, a_{E_*})$ of E-homology.

(1) For a manifold X and integer n, we set

$$(\widehat{IE}_{\mathrm{dR}})^n(X) := \{(\omega, h)\}, \tag{5.4}$$

where
- $\omega \in \Omega^n_{\mathrm{clo}}(X; N^\bullet_E)$.
- $h\colon \widehat{E}_{n-1}(X) \to \mathbb{R}/\mathbb{Z}$ is a group homomorphism.
- ω and h fits into the following commutative diagram.

$$\begin{array}{ccc} \Omega_n(X; V^E_\bullet)/\mathrm{im}(\partial) & \xrightarrow{(\mathrm{mod}\mathbb{Z})\circ\langle -,\omega\rangle} & \mathbb{R}/\mathbb{Z}\,. \\ & {\scriptstyle a_{E_*}}\searrow \qquad \nearrow{\scriptstyle h} & \\ & \widehat{E}_{n-1}(X) & \end{array} \tag{5.5}$$

(2) We define

$$a_{IE^*}\colon \Omega^{n-1}(X; N^\bullet_E)/\mathrm{im}(d) \to (\widehat{IE}_{\mathrm{dR}})^n(X), \quad \alpha \mapsto (0, (\mathrm{mod}\mathbb{Z}) \circ \langle -, \omega\rangle)\,.$$

We set

$$IE^n_{\mathrm{dR}}(X) := (\widehat{IE}_{\mathrm{dR}})^n(X)/\mathrm{im}(a_{IE^*}). \tag{5.6}$$

$(\widehat{IE}_{\mathrm{dR}})^*$ and IE^*_{dR} are regarded as functors $\mathrm{Mfd}^{\mathrm{op}} \to \mathrm{Ab}^{\mathbb{Z}}$ in the obvious way.

Definition 5.7 (Structure maps for $(\widehat{IE}_{\mathrm{dR}})^*$ and IE^*_{dR}). We define the following maps natural in X. The well-definedness is easily checked.

- We denote the quotient map by
$$I_{IE^*}\colon (\widehat{IE}_{\mathrm{dR}})^*(X) \to IE^*_{\mathrm{dR}}(X).$$
- We define
$$R_{IE^*}\colon (\widehat{IE}_{\mathrm{dR}})^*(X) \to \Omega^*_{\mathrm{clo}}(X; N^\bullet_E), \quad (\omega, h) \mapsto \omega.$$
- We define
$$\mathrm{ch}'\colon IE^*_{\mathrm{dR}}(X) \to H^n(X; N^\bullet_E)\,(\simeq \mathrm{Hom}(E_n(X), \mathbb{R}))\,, \quad I_{IE^*}((\omega, h)) \mapsto \mathrm{Rham}(\omega).$$
- We define
$$p\colon \mathrm{Hom}(E_{*-1}(X), \mathbb{R}/\mathbb{Z}) \to IE^*_{\mathrm{dR}}(X), \quad h \mapsto I_{IE^*}((0, h \circ I_{E_*})).$$

5.2. Examples. Here we list some examples. We apply the construction in the last subsection to each of the examples of differential homology theories in Subsection 4.2. We will see that they recover the existing Cheeger-Simons type models of differential cohomology theories.

5.2.1. *The differential ordinary cohomology in terms of differential characters by* [**CS85**]. The ordinary integral cohomology is Anderson self-dual, $H\mathbb{Z} \simeq IH\mathbb{Z}$. Thus the above construction produces a differential extension of $H\mathbb{Z}$. Indeed, it is easy to verify that the above construction applied to $\widehat{H}^{\mathrm{HS}}(-; \mathbb{Z})$ in Subsection 4.2.2 recovers the Cheeger-Simons model of differential ordinary cohomology in terms of differential characters.

5.2.2. *The differential K-theory in terms of "differential characters in K-theory" by* [**BM06**]. The K-theory is also Anderson self-dual, $K \simeq IK$. Thus the above construction applied to the geometric-cycle model of differential K-homology by [**Elm16**] mentioned in Subsection 4.2.2 produces a differential K-theory in terms of functions which assign an $\mathbb{R}/\mathbb{Z}$-value to each differential K-cycle. Indeed, this recovers the "differential characters in K-theory" by Benameur and Maghfoul [**BM06**].

A typical element of $\widehat{IK}^0_{\mathrm{dR}}(X)$ in this model comes from a hermitian vector bundle (F, ∇^F) with a unitary connection over X. Indeed, we get a pair $(h_{(E,\nabla^E)}, \mathrm{Ch}(\nabla^E)) \in \widehat{IK}^0_{\mathrm{dR}}(X)$, where $h_{(E,\nabla^E)} \colon \widehat{K}_{-1}(X) \to \mathbb{R}/\mathbb{Z}$ is defined by

(5.8)
$$h_{(E,\nabla^E)}[(M, \nabla^M), (E, \nabla^E), f, \psi] := \overline{\eta}(D_{E \otimes f^* F}) + \langle f_* \psi, \mathrm{Ch}(\nabla^F)\rangle \pmod{\mathbb{Z}}.$$

Here $D_{E\otimes f^*F}$ denotes the Spin^c Dirac operator on M twisted by $E \otimes f^*F$ with the connection $\nabla^E \otimes f^*\nabla^F$. $\overline{\eta}(D) = \frac{1}{2}(\eta(D) + \dim\ker D)$ is the reduced eta invariant. The well-definedness of the map (5.8) uses the Atiyah-Patodi-Singer's index theorem. Indeed, to be compatible with the bordism relation (4.4) of differential K-cycles, we need

$$\int_W \mathrm{Todd}(\nabla^W)\mathrm{Ch}(\nabla^{\mathcal{E}} \otimes f^*\nabla^F) \equiv \overline{\eta}(D_{(\mathcal{E}\otimes f^*F)|_{\partial W}}) \pmod{\mathbb{Z}}, \tag{5.9}$$

which is a consequence of the APS index theorem.

This element corresponds to the element $[E, \nabla^E, 0] \in \widehat{K}^0_{\mathrm{FL}}(X)$ in the Freed-Lott model (Subsection 2.3). An easy generalization of the above construction gives the isomorphism $\widehat{K}^0_{\mathrm{FL}}(X) \simeq \widehat{IK}^0_{\mathrm{dR}}(X)$.

5.2.3. *The differential Anderson dual to G-bordism theories by* [**YY23**]. Applying the construction for $\widehat{\Omega^G_*}$ in Subsection 4.2.3, we recover the following model given in [**YY23**]. Recall that we are using the abbreviation $N^\bullet_G := N^\bullet_{MTG}$ (4.11).

DEFINITION 5.10 ($(\widehat{I\Omega^G_{\mathrm{dR}}})^*$ and $(I\Omega^G_{\mathrm{dR}})^*$, [**YY23**]). Let n be a nonnegative integer.

(1) Define $(\widehat{I\Omega^G_{\mathrm{dR}}})^n(X)$ to be an abelian group consisting of pairs (ω, h), such that
   (a) ω is a closed n-form $\omega \in \Omega^n_{\mathrm{clo}}(X; N^\bullet_G)$.
   (b) h is a group homomorphism $h \colon \mathcal{C}^{G_\nabla}_{n-1}(X) \to \mathbb{R}/\mathbb{Z}$.
   (c) ω and h satisfy the following compatibility condition. Assume that we are given two objects (M_-, g_-, f_-) and (M_+, g_+, f_+) in $h\mathrm{Bord}^{G_\nabla}_{n-1}(X)$ and a morphism $[W, g_W, f_W]$ from the former to the latter. Then we have

$$h([M_+, g_+, f_+]) - h([M_-, g_-, f_-]) = \mathrm{cw}(\omega)([W, g_W, f_W]) \pmod{\mathbb{Z}}, \tag{5.11}$$

   where the right hand side is defined in (4.16).

   Abelian group structure on $(\widehat{I\Omega^G_{\mathrm{dR}}})^n(X)$ is defined in the obvious way.

(2) We define a homomorphsim of abelian groups,

$$\begin{aligned} a \colon \Omega^{n-1}(X; N^\bullet_G)/\mathrm{Im}(d) &\to (\widehat{I\Omega^G_{\mathrm{dR}}})^n(X) \\ \alpha &\mapsto (d\alpha, \mathrm{cw}(\alpha)). \end{aligned} \tag{5.12}$$

Here the homomorphism $\mathrm{cw}(\alpha)\colon \mathcal{C}^{G\nabla}_{n-1}(X) \to \mathbb{R}/\mathbb{Z}$ is defined by

$$\mathrm{cw}(\alpha)([M,g,f]) := \int_M \mathrm{cw}_g(f^*\alpha) \pmod{\mathbb{Z}}. \tag{5.13}$$

We set

$$(I\Omega^G_{\mathrm{dR}})^n(X) := (\widehat{I\Omega^G_{\mathrm{dR}}})^n(X)/\mathrm{Im}(a).$$

5.3. The proof of $IE \simeq IE_{\mathrm{dR}}$. The goal of the rest of the section is to prove that the functor IE^*_{dR} is actually the model of the Anderson dual cohomology IE to E. First we show that IE_{dR} fits into the exact sequence for the Anderson dual.

PROPOSITION 5.14. *For any manifold X and integer n, the following sequence is exact.*

$$\begin{aligned}\mathrm{Hom}(E_{n-1}(X),\mathbb{R}) \to \mathrm{Hom}(E_{n-1}(X),\mathbb{R}/\mathbb{Z}) &\xrightarrow{p} (IE_{\mathrm{dR}})^n(X)\\ &\xrightarrow{\mathrm{ch}'} \mathrm{Hom}(E_n(X),\mathbb{R}) \to \mathrm{Hom}(E_n(X),\mathbb{R}/\mathbb{Z}).\end{aligned} \tag{5.15}$$

PROOF. The proof is analogous to that for [**YY23**, Proposition 4.25]. □

THEOREM 5.16. *There is a natural isomorphism of the functors* $\mathrm{Mfd}^{\mathrm{op}} \to \mathrm{Ab}^{\mathbb{Z}}$,

$$F\colon IE_{\mathrm{dR}} \simeq IE,$$

which fits into the following commutative diagram.

$$\begin{array}{ccccccccc} 0 & \longrightarrow & \mathrm{Ext}(E_{n-1}(-),\mathbb{Z}) & \xrightarrow{p} & IE^n_{\mathrm{dR}} & \xrightarrow{\mathrm{ch}} & \mathrm{Hom}(E_n(-),\mathbb{Z}) & \longrightarrow & 0\\ & & \| & & \simeq\downarrow F & & \| & & \\ 0 & \longrightarrow & \mathrm{Ext}(E_{n-1}(-),\mathbb{Z}) & \longrightarrow & IE^n & \longrightarrow & \mathrm{Hom}(E_n(-),\mathbb{Z}) & \longrightarrow & 0\end{array} \tag{5.17}$$

Here the bottom row is the exact sequence in (3.7). *Moreover, the quintuple* $((\widehat{IE}_{\mathrm{dR}})^*, R, I, a)$ *in Definitions* 5.3 *is a differential extension of the pair* $((IE)^*, \mathrm{ch})$.

PROOF. The proof is essentially the same as the corresponding claim for the case $E = MTG$ given in [**YY23**, Theorem 4.51]. We use the model of IE^* in [**HS05**, Corollary B.17] in terms of functors of Picard groupoids.

Given an element $(\omega, h) \in (\widehat{IE}_{\mathrm{dR}})^n(X)$, we get the associated functor of Picard groupoids,

$$\widetilde{F}_{(\omega,h)}\colon \left(\Omega_n(X;V^E_\bullet)/\mathrm{im}(\partial) \xrightarrow{a_{E_*}} \widehat{E}_{n-1}(X)\right) \to (\mathbb{R} \xrightarrow{\mathrm{mod}\mathbb{Z}} \mathbb{R}/\mathbb{Z}) \tag{5.18}$$

by applying h on objects and ω on morphisms. This is well-defined thanks to the commutativity of the diagram (5.5). Moreover, given two elements (ω, h) and (ω', h'), and an element $\alpha \in \Omega^{n-1}(X;(V^E_\bullet)^\vee)/\mathrm{Im}(d)$ so that $(\omega',h') - (\omega,h) = a_{IE^*}(\alpha)$, we get the associated natural transformation,

$$\widetilde{F}_\alpha\colon \widetilde{F}_{(\omega,h)} \Rightarrow \widetilde{F}_{(\omega',h')}, \tag{5.19}$$

by $\widetilde{F}_\alpha := \langle R_{E_*}(-), \alpha\rangle$. Summarizing, we have defined the homomorphism which is functorial in X,

(5.20)

$$\widetilde{F}_X\colon IE^n_{\mathrm{dR}}(X) \to \pi_0\mathrm{Fun}_{\mathrm{Pic}}\left(\left(\Omega_n(X;V^E_\bullet)/\mathrm{im}(\partial) \xrightarrow{a_{E_*}} \widehat{E}_{n-1}(X)\right), (\mathbb{R} \to \mathbb{R}/\mathbb{Z})\right),$$

where $\pi_0\mathrm{Fun}_{\mathrm{Pic}}$ denotes the group of natural isomorphism classes of functors of Picard groupoids.

Now by [**HS05**, Corollary B.17] we have an isomorphism

$$IE^n(X) \simeq \pi_0\mathrm{Fun}_{\mathrm{Pic}}(\pi_{\leq 1}(L(X^+ \wedge E)_{1-n}), (\mathbb{R} \to \mathbb{R}/\mathbb{Z})). \tag{5.21}$$

Here we set L is the spectrification and $\pi_{\leq 1}$ is the fundamental Picard groupoid. We have $\pi_i(L(X^+ \wedge E)_{1-n}) = E_{n-1+i}(X)$, and the exact sequence in the bottom row of (5.17) becomes the canonical ones in this model (see the explanation in [**YY23**, Fact 2.6]).

As mentioned in [**HS05**, Example B.6], the Picard groupoids $\mathcal{C}$ coming from maps of Abelian groups are equivalent to $(\pi_1(\mathcal{C}) \xrightarrow{0} \pi_0(\mathcal{C}))$, so we have an equivalence

$$\left(\Omega_n(X; V_\bullet^E)/\mathrm{im}(\partial) \xrightarrow{a_{E_*}} \widehat{E}_{n-1}(X)\right) \simeq \left(\ker a_{E_*} \xrightarrow{0} \mathrm{coker}\ a_{E_*}\right), \tag{5.22}$$

and we have canonical isomorphisms

$$\ker a_{E_*} \simeq \mathrm{im}\left(E_n(X) \xrightarrow{\mathrm{ch}} H_n(X; V_\bullet^E)\right), \tag{5.23}$$

$$\mathrm{coker}\ a_{E_*} \simeq E_{n-1}(X), \tag{5.24}$$

by the exactness of (4.3).

Now we construct a (natural isomorphism class of) functor of Picard groupoids

$$\pi_{\leq 1}(L(X^+ \wedge E)_{1-n}) \to \left(\ker a_{E_*} \xrightarrow{0} \mathrm{coker}\ a_{E_*}\right) \tag{5.25}$$

as follows. Let $S\mathbb{R}/\mathbb{Z}$ denote the Moore spectrum for $\mathbb{R}/\mathbb{Z}$, so that we have $\pi_i(X^+ \wedge E \wedge S\mathbb{R}/\mathbb{Z}) = (E_{\mathbb{R}/\mathbb{Z}})_i(X)$. Let $(X^+ \wedge E \wedge S\mathbb{R}/\mathbb{Z})\langle n\rangle \to X^+ \wedge E \wedge S\mathbb{R}/\mathbb{Z}$ denote the n-connected cover. Composing it with the map $\Sigma^{-1}(X^+ \wedge E \wedge S\mathbb{R}/\mathbb{Z}) \to X^+ \wedge E$ coming from the extension of coefficient group $0 \to \mathbb{Z} \to \mathbb{R} \to \mathbb{R}/\mathbb{Z} \to 0$, we get a morphism of spectra,

$$\Sigma^{-1}\left((X^+ \wedge E \wedge S\mathbb{R}/\mathbb{Z})\langle n\rangle\right) \to X^+ \wedge E. \tag{5.26}$$

Let J be the homotopy cofiber of the map (5.26). We have $\pi_n(J) \simeq H_n(X; V_\bullet^E)$ and $\pi_{n-1}(J) \simeq E_{n-1}(X)$. Since $\pi_n(J)$ is torsion-free, the k-invariant for the Picard groupoid $\pi_{\leq 1}(LJ_{1-n})$ vanishes. Thus we have

$$\pi_{\leq 1}(LJ_{1-n}) \simeq \left(H_n(X; V_\bullet^E) \xrightarrow{0} E_{n-1}(X)\right). \tag{5.27}$$

Compose this equivalence with the functor of fundamental Picard groupoids associated to the map $X^+ \wedge E \to J$. Then the image is contained in the subgroupoid $\left(\ker a_{E_*} \xrightarrow{0} \mathrm{coker}\ a_{E_*}\right)$. So we get the desired functor (5.25). By construction, the functor (5.25) induces identity on π_0 and ch on π_1.

By composing (5.20) with the pre-composition of the functor (5.25) and equivalence (5.22), we get the desired functor F as

(5.28)

$$F_X \colon IE^n_{\mathrm{dR}}(X) \to \pi_0\mathrm{Fun}_{\mathrm{Pic}}\left(\pi_{\leq 1}(L(X^+ \wedge E)_{1-n}), (\mathbb{R} \to \mathbb{R}/\mathbb{Z})\right) \xrightarrow[\simeq]{(5.21)} IE^n(X).$$

The commutativity of (5.17) is obvious. Applying that diagram to any manifold X, we know that the top row is exact by Proposition 5.14, and bottom row is also exact. Thus by the five lemma we see that F is a natural isomorphism. The last statement is easily checked. This completes the proof of Theorem 5.16. □

6. The interpretation of $\widehat{IE}_{\mathrm{dR}}$ in terms of QFTs

In this section we explain the interpretation of our model $\widehat{IE}_{\mathrm{dR}}$ in terms of QFTs.

As explained in Subsection 3.2, the Anderson dual to the G-bordism theory is supposed to classify deformation classes of possibly non-topological invertible QFTs on G-manifolds. Indeed, the $\mathbb{R}/\mathbb{Z}$-valued function h of an element $(h, \omega) \in (\widehat{I\Omega^G_{\mathrm{dR}}})^{n+1}(X)$ can be regarded as the complex phase of the partition functions of an invertible QFTs for manifolds equipped with differential stable G-structures and smooth maps to X, under the identification $\mathbb{R}/\mathbb{Z} \simeq U(1)$. The forgetful map $I\colon (\widehat{I\Omega^G_{\mathrm{dR}}})^{n+1}(X) \to (I\Omega^G_{\mathrm{dR}})^{n+1}(X)$ is regarded as taking the deformation classes of such invertible QFTs.

This interpretation is based on the following empirical facts known for physically meaningful[5] invertible QFTs. Generally in a non-topological n-dimensional QFT $\mathcal{T}$, the partition function $Z_{\mathcal{T}}(M^n, \mathfrak{s}) \in \mathbb{C}$ for closed n-dimensional $\mathcal{S}$-manifold varies smoothly according to the smooth variation of the input $(M^n, \mathfrak{s})$. It is empirically known that, if $\mathcal{T}$ is *invertible*, the variation of the complex phase of $Z_{\mathcal{T}}$ can be measured by an integration of some characteristic form, i.e., there exists a $(d+1)$-dimensional characteristic polynomial $\omega_{\mathcal{T}}$ such that, for each bordism $(W^{d+1}, \mathfrak{s})\colon (M_-, \mathfrak{s}_-) \to (M_+, \mathfrak{s}_+)$ we have

$$\arg\left(\frac{Z_{\mathcal{T}}(M^d_+, \mathfrak{s}_+)}{Z_{\mathcal{T}}(M^d_-, \mathfrak{s}_-)}\right) = \int_W \mathrm{cw}(\omega_{\mathcal{T}})(W^{d+1}, \mathfrak{s}) \pmod{\mathbb{Z}}, \tag{6.1}$$

where the left hand side is well-defined since the partition function is nonzero in the invertible case. Note that a priori there is nothing to assign for $(d+1)$-dimensional bordism, since $\mathcal{T}$ is d-dimensional theory.

Moreover, it is also empirically known that the effect of smooth deformation of invertible theories can also be measured by local integrations. A smooth deformation $\mathcal{H}$ from $\mathcal{T}_0$ to $\mathcal{T}_1$ provides us a d-dimensional characteristic form $\alpha_{\mathcal{H}}$ so that

$$\arg\left(\frac{Z_{\mathcal{T}_1}(M^d, \mathfrak{s})}{Z_{\mathcal{T}_0}(M^d, \mathfrak{s})}\right) = \int_M \mathrm{cw}(\alpha_{\mathcal{H}})(M^d, \mathfrak{s}) \pmod{\mathbb{Z}}. \tag{6.2}$$

These empirical facts explain our interpretation of our model. Indeed, recall that $\Omega^*(X; N^\bullet_G)$ is the differential forms on X with coefficient in invariant polynomials on $\mathfrak{g}$, which can be regarded as characteristic polynomials for the structure $\mathcal{S}$ given by differential stable tangential G-structures and maps to X. The equation (6.1) corresponds to the compatibility condition (5.5) for $(\arg Z_{\mathcal{T}}, \omega_{\mathcal{T}})$, and the equation (6.2) says that the deformation corresponds to the addition by $a(\alpha_{\mathcal{H}})$ in (5.12).

Finally we comment on the possibility of interpreting the general construction of differential Anderson dual to differential homology $\widehat{E}_*$ also in terms of a kind of invertible QFTs. The idea is the following. Differential homology $\widehat{E}_*$ typically comes from a Picard groupoid $\mathcal{C}_\nabla$ with differential data. For example $h\mathrm{Bord}^{G_\nabla}_n(X)$, the groupoid of differential K-cycles (which has an interpretation in terms of *D-branes*),

[5] Note that the physically meaningfulness includes conditions such as reflection positivity and locality.

and the groupoid of differential ordinary cocycles. On morphisms of these categories, we can integrate differential forms. Differential Anderson dual $(\widehat{IE}_{\mathrm{dR}})^*(X)$ classifies functors

$$(\omega, h)\colon \mathcal{C}_\nabla(X) \to (\mathbb{R} \to \mathbb{R}/\mathbb{Z}) \tag{6.3}$$

which reflects the differential information. The map $h\colon \mathrm{Obj}(\mathcal{C}_\nabla) \to \mathbb{R}/\mathbb{Z}$ may be regarded as the phase of a partition function for some invertible QFTs with domain extending $\mathcal{C}_\nabla$.

References

[Ati88] Michael Atiyah, *Topological quantum field theories*, Inst. Hautes Études Sci. Publ. Math. **68** (1988), 175–186 (1989). MR1001453

[BD82] Paul Baum and Ronald G. Douglas, *K homology and index theory*, Operator algebras and applications, Part 1 (Kingston, Ont., 1980), Proc. Sympos. Pure Math., vol. 38, Amer. Math. Soc., Providence, RI, 1982, pp. 117–173. MR679698

[BM06] Moulay-Tahar Benameur and Mohamed Maghfoul, *Differential characters in K-theory*, Differential Geom. Appl. **24** (2006), no. 4, 417–432, DOI 10.1016/j.difgeo.2005.12.008. MR2231056

[BNV16] Ulrich Bunke, Thomas Nikolaus, and Michael Völkl, *Differential cohomology theories as sheaves of spectra*, J. Homotopy Relat. Struct. **11** (2016), no. 1, 1–66, DOI 10.1007/s40062-014-0092-5. MR3462099

[BS10] Ulrich Bunke and Thomas Schick, *Uniqueness of smooth extensions of generalized cohomology theories*, J. Topol. **3** (2010), no. 1, 110–156, DOI 10.1112/jtopol/jtq002. MR2608479

[BS12] Ulrich Bunke and Thomas Schick, *Differential K-theory: a survey*, Global differential geometry, Springer Proc. Math., vol. 17, Springer, Heidelberg, 2012, pp. 303–357, DOI 10.1007/978-3-642-22842-1_11. MR3289847

[CS85] Jeff Cheeger and James Simons, *Differential characters and geometric invariants*, Geometry and topology (College Park, Md., 1983/84), Lecture Notes in Math., vol. 1167, Springer, Berlin, 1985, pp. 50–80, DOI 10.1007/BFb0075216. MR827262

[Elm16] Adnane Elmrabty, *Differential K-homology and explicit isomorphisms between $\mathbb{R}/\mathbb{Z}$-K-homologies*, Bull. Belg. Math. Soc. Simon Stevin **23** (2016), no. 1, 1–19. MR3471975

[Fre19] Daniel S. Freed, *Lectures on field theory and topology*, CBMS Regional Conference Series in Mathematics, vol. 133, American Mathematical Society, Providence, RI, 2019. Published for the Conference Board of the Mathematical Sciences, DOI 10.1090/cbms/133. MR3969923

[FH21] Daniel S. Freed and Michael J. Hopkins, *Reflection positivity and invertible topological phases*, Geom. Topol. **25** (2021), no. 3, 1165–1330, DOI 10.2140/gt.2021.25.1165. MR4268163

[FL10] Daniel S. Freed and John Lott, *An index theorem in differential K-theory*, Geom. Topol. **14** (2010), no. 2, 903–966, DOI 10.2140/gt.2010.14.903. MR2602854

[FMS07] Daniel S. Freed, Gregory W. Moore, and Graeme Segal, *The uncertainty of fluxes*, Comm. Math. Phys. **271** (2007), no. 1, 247–274, DOI 10.1007/s00220-006-0181-3. MR2283960

[GTMW09] Søren Galatius, Ulrike Tillmann, Ib Madsen, and Michael Weiss, *The homotopy type of the cobordism category*, Acta Math. **202** (2009), no. 2, 195–239, DOI 10.1007/s11511-009-0036-9. MR2506750

[GP20] Daniel Grady and Dmitri Pavlov, *Extended field theories are local and have classifying spaces*, `arXiv:2011.01208` 2020.

[HS05] M. J. Hopkins and I. M. Singer, *Quadratic functions in geometry, topology, and M-theory*, J. Differential Geom. **70** (2005), no. 3, 329–452. MR2192936

[Lur09] Jacob Lurie, *On the classification of topological field theories*, Current developments in mathematics, 2008, Int. Press, Somerville, MA, 2009, pp. 129–280. MR2555928

[Rud98] Yuli B. Rudyak, *On Thom spectra, orientability, and cobordism*, Springer Monographs in Mathematics, Springer-Verlag, Berlin, 1998. With a foreword by Haynes Miller. MR1627486

[Seg04] Graeme Segal, *The definition of conformal field theory*, Topology, geometry and quantum field theory, London Math. Soc. Lecture Note Ser., vol. 308, Cambridge Univ. Press, Cambridge, 2004, pp. 421–577. MR2079383

[YY23] Mayuko Yamashita and Kazuya Yonekura, *Differential models for the Anderson dual to bordism theories and invertible QFT's, I*, J. Gökova Geom. Topol. GGT **16** (2023), 1–64. MR4641957

Department of Mathematics, Kyoto University, 606-8502, Kyoto, Japan
Email address: `yamashita.mayuko.2n@kyoto-u.ac.jp`

Selected Published Titles in This Series

107 **Ron Donagi, Adrian Langer, Piotr Sułkowski, and Katrin Wendland, Editors,** String-Math 2022, 2024

105 **A. Connes, C. Consani, B. I. Dundas, M. Khalkhali, and H. Moscovici, Editors,** Cyclic Cohomology at 40: Achievements and Future Prospects, 2023

104 **A. Kechris, N. Makarov, D. Ramakrishnan, and X. Zhu, Editors,** Nine Mathematical Challenges, 2021

103 **Sergey Novikov, Igor Krichever, Oleg Ogievetsky, and Senya Shlosman, Editors,** Integrability, Quantization, and Geometry, 2021

102 **David T. Gay and Weiwei Wu, Editors,** Breadth in Contemporary Topology, 2019

101 **Avraham Aizenbud, Dmitry Gourevitch, David Kazhdan, and Erez M. Lapid, Editors,** Representations of Reductive Groups, 2019

100 **Chiu-Chu Melissa Liu and Motohico Mulase, Editors,** Topological Recursion and its Influence in Analysis, Geometry, and Topology, 2018

99 **Vicente Muñoz, Ivan Smith, and Richard P. Thomas, Editors,** Modern Geometry, 2018

98 **Amir-Kian Kashani-Poor, Ruben Minasian, Nikita Nekrasov, and Boris Pioline, Editors,** String-Math 2016, 2018

97 **Tommaso de Fernex, Brendan Hassett, Mircea Mustaţă, Martin Olsson, Mihnea Popa, and Richard Thomas, Editors,** Algebraic Geometry: Salt Lake City 2015 (Parts 1 and 2), 2018

96 **Si Li, Bong H. Lian, Wei Song, and Shing-Tung Yau, Editors,** String-Math 2015, 2017

95 **Izzet Coskun, Tommaso de Fernex, and Angela Gibney, Editors,** Surveys on Recent Developments in Algebraic Geometry, 2017

94 **Mahir Bilen Can, Editor,** Algebraic Groups: Structure and Actions, 2017

93 **Vincent Bouchard, Charles Doran, Stefan Méndez-Diez, and Callum Quigley, Editors,** String-Math 2014, 2016

92 **Kailash C. Misra, Daniel K. Nakano, and Brian J. Parshall, Editors,** Lie Algebras, Lie Superalgebras, Vertex Algebras and Related Topics, 2016

91 **V. Sidoravicius and S. Smirnov, Editors,** Probability and Statistical Physics in St. Petersburg, 2016

90 **Ron Donagi, Sheldon Katz, Albrecht Klemm, and David R. Morrison, Editors,** String-Math 2012, 2015

89 **D. Dolgopyat, Y. Pesin, M. Pollicott, and L. Stoyanov, Editors,** Hyperbolic Dynamics, Fluctuations and Large Deviations, 2015

88 **Ron Donagi, Michael R. Douglas, Ljudmila Kamenova, and Martin Rocek, Editors,** String-Math 2013, 2014

87 **Helge Holden, Barry Simon, and Gerald Teschl, Editors,** Spectral Analysis, Differential Equations and Mathematical Physics: A Festschrift in Honor of Fritz Gesztesy's 60th Birthday, 2013

86 **Kailash C. Misra, Daniel K. Nakano, and Brian J. Parshall, Editors,** Recent Developments in Lie Algebras, Groups and Representation Theory, 2012

85 **Jonathan Block, Jacques Distler, Ron Donagi, and Eric Sharpe, Editors,** String-Math 2011, 2012

84 **Alex H. Barnett, Carolyn S. Gordon, Peter A. Perry, and Alejandro Uribe, Editors,** Spectral Geometry, 2012

83 **Hisham Sati and Urs Schreiber, Editors,** Mathematical Foundations of Quantum Field Theory and Perturbative String Theory, 2011

For a complete list of titles in this series, visit the
AMS Bookstore at **www.ams.org/bookstore/pspumseries/**.